Volume I

ENVIRONMENTAL IMPACT of SOIL COMPONENT INTERACTIONS

Natural and Anthropogenic Organics

Edited by

P.M. Huang

J. Berthelin

J.-M. Bollag

W.B. McGill

A.L. Page

CRC LEWIS PUBLISHERS

Boca Raton London Tokyo

Library of Congress Cataloging-in-Publication Data

Catalog record is available from the Library of Congress

Direct all inquiries to CRC Press, Inc., 2000 Corporate Blvd., N.W., Boca Raton, Florida 33431.

Lewis Publishers is an imprint of CRC Press

International Standard Book Number 0-87371-914-X
Printed in the United States of America 1 2 3 4 5 6 7 8 9 0
Printed on acid-free paper

PREFACE

Minerals, organic components, and microorganisms are integral parts of the pedosphere and associated environments. They profoundly affect the physical, chemical, and biological properties of soils and sediments. During the past few decades, scientific accomplishments in individual disciplines of the physics, chemistry, and biology of soils were impressive. However, information on interactions of soil minerals with organic components and microorganisms was fragmentary and scattered in the literature of soil and environmental sciences around the world. Yet these three groups of components are not separate "kingdoms", but rather a "united kingdom" constantly in close association with each other in the environment.

In view of the significance of the interactions of soil minerals with natural organic components and microorganisms, a special publication entitled *Interactions of Soil Minerals with Natural Organics and Microbes* was published in 1986 by the Soil Science Society of America. The second printing was published in 1989. This special publication provided an examination of these fundamental constituents and processes. It brought together new knowledge on how soil minerals affect the dynamics and transformations of natural organics and metabolic processes, the growth, adhesion, and ecology of microbes, and how natural organics and microbes affect the weathering transformations, properties, and reactivities of soil minerals.

To promote teaching and research on the interactions of soil minerals with organic components and microorganisms and the impact on the production of foodstuffs and fibers, the sustainability of the environment, and human health at the international level, the International Society of Soil Science (ISSS) established the Working Group MO entitled "Interactions of Soil Minerals with Organic Components and Microorganisms" in 1990. With the cosponsorship of the Canadian Society of Soil Science, the ISSS Working Group MO held the first workshop entitled *"Impact of Interactions of Inorganic, Organic and Microbiological Soil Components on Environmental Quality"* in Edmonton, Alberta, Canada, on August 11 to 15, 1992. The workshop focused on the impacts of interactions of soil minerals, organic components, and microorganisms on the dynamics and transformations of metals, other inorganics, xenobiotics, toxicity, microbial activities, pathogenesis, and related environmental quality.

The Organizing Committee received 86 papers from 20 countries located in North America, Central America, South America, western Europe, eastern Europe, Asia, Africa, and Oceania (Australia, Austria, Canada, Chile, China, Congo, Costa Rica, Egypt, France, Germany, India, Israel, Italy, Japan, Mexico, The Netherlands, New Zealand, Poland, U.K., and U.S.). Many scientists of international stature were invited to present key papers at the workshop. It was a major step forward in bringing together the fragmented literature on the environmental impact of the interactions of inorganic, organic, and microbial factors of soils. Such information served to identify gaps in our knowledge and as such provided future direction to stimulate scientific research in this important area of science. Thus, the workshop provided a forum for the exchange of information among soil and environmental scientists from around the world who have common interests and scientific pursuits, and for the subsequent technology transfer for the development of innovative management strategies for environmental sustainability. Furthermore, the workshop established avenues of communication on the subject matter among soil and environmental scientists in developing and developed countries to protect the global environment.

This book, composed of two volumes, is a compilation of 47 research papers and four position papers that were presented at the ISSS Workshop and have been accepted for publication after critical peer review. Volume I deals with the environmental impact of soil component interactions on general soil quality and transformations of natural and anthropogenic organic compounds. Volume II covers the environmental impact of soil component interactions on toxic metals, other inorganics, and microbial activity, pathogenesis, and biotechnology. This externally refereed book arising from the Workshop should be useful for teaching, research, and technology transfer on environmental protection at the international level. It is hoped that the book will serve as the seed for the germination, growth, and flourishing of this exciting area of scientific and educational pursuits for years to come.

We are grateful to the authors who have contributed chapters to this publication and to the external referees who have provided invaluable critical inputs to maintain the quality of this book. Gratitude is extended to the sponsors, Agriculture Canada, Alberta Oil Sands Technology and Research Authority, Canberra Packard Canada Ltd., Cominco Fertilizers, ESSO Chemical Canada, International Minerals & Chemical Corporation Canada Ltd., Natural Sciences and Engineering Research Council of Canada, Potash and Phosphate Institute of Canada, Summer Rain Ltd., and United Nations Environment Programme, who have provided the funding to make the workshop possible. Sincere appreciation is also extended to the Department of Soil Science, University of Saskatchewan, for the excellent support during the preparation of this publication.

P. M. Huang
J. Berthelin
J.-M. Bollag
W. B. McGill
A. L. Page

ABOUT THE EDITORS

P. M. Huang received his Ph.D. degree in soil science at the University of Wisconsin, Madison, in 1966. He is currently professor of soil science at the University of Saskatchewan, Saskatoon, Canada. His research work has very significantly advanced the frontiers of knowledge on the nature and surface reactivity of mineral colloids and organomineral complexes of soils and sediments and their role in the dynamics, transformations, and fate of plant nutrients, toxic metals, and xenobiotics in terrestrial and aquatic environments. His research findings, embodied in over 180 refereed scientific publications, including research papers, book chapters, and books, are fundamental to the development of sound strategies for managing land and water resources.

He has developed and taught courses in soil physical chemistry and mineralogy, soil analytical chemistry, and ecological toxicology. He has successfully trained and inspired M.Sc. and Ph.D. students, and postdoctoral fellows and received visiting scientists worldwide. He has served on numerous national and international scientific and academic committees. He also has served as a member of many editorial boards such as the *Soil Science Society of America Journal, Chemosphere,* and *Advances in Environmental Science.* He is an associate member of the Commission of Environmental Analytical Chemistry of the International Union of Pure and Applied Chemistry and Chairman of the Working Group MO "Interactions of Soil Minerals with Organic Components and Microorganisms" of the International Society of Soil Science. He is a fellow of the Canadian Society of Soil Science, the Soil Science Society of America, and the American Society of Agronomy, and is listed in *Who Is Who in the World.*

Jacques Berthelin is research director at the Centre National de la Recherche Scientifique (CNRS) at the Centre de Pédologie Biologique in Vandœuvre-les-Nancy, France.

He was born in 1940 and received a degree in biochemical engineering (qualified engineer) from the Ecole Nationale Supérieure de Brasserie, Malterie et Biochimie of Nancy, in 1964.

He prepared a thesis in the specialty of biochemistry, presented in 1967, and was also a midshipman and sublieutenant in the navy for 1½ years.

He joined the CNRS in 1968 in the Department of Earth Sciences at the Centre de Pédologie Biologique of the CNRS to conduct research in the field of soil biochemistry and microbiology, and has been a Doctor ès Science since 1976.

He is now the head of a group of Geomicrobiology and Soil Microbiology at the Centre of Pédologie Biologique. He has published more than 100 refereed scientific papers, chapters, and reviews, as author and coauthor, and edited a book titled *Diversity of Environmental Biogeochemistry.* He is a member of various national and international scientific and academic associations and editorial boards.

Jean-Marc Bollag is professor of soil biochemistry and co-director of the Center of Bioremediation and Detoxification, Environmental Resources Research Institute, at The Pennsylvania State University. He is the author or coauthor of more than 160 professional papers and serves on the editorial board of five international journals. He is a frequent lecturer at conferences and seminars throughout the world. A recipient of the Julius Baer Fellowship, the Gamma Sigma Delta Research Award, and the Badge of Merit from the Polish Ministry of Agriculture, Dr. Bollag is a fellow of the American Academy of Microbiology, the Soil Science Society of America, and the American Society of Agronomy. He is a member of the International Society of Soil Science, the International Humic Substances Society, and the American Society for Microbiology.

Dr. Bollag received a Ph.D. degree in plant physiology from the University of Basel, Switzerland, and conducted postdoctoral work at the Weitzmann Institute of Science, Rehovoth, Israel, and Cornell University, Ithaca, NY. He also was a visiting scientist in the Biochemistry Section of Agrochemicals at Ciba-Geigy, Basel, Switzerland. Most of his research is related to bioremediation problems: transformation of pesticides and other xenobiotics by microorganisms, incorporation of pollutants into soil organic matter as a detoxification method, and application of enzymes for pollution control.

W. B. McGill is a professor in the Department of Soil Science, University of Alberta, where he has been since 1971 after completing his Ph.D. degree at the University of Saskatchewan and M.Sc. and B.S.A. degrees at the University of Manitoba. He teaches Soil Microbiology and Biochemistry, Advanced Soil Biochemistry, and Waste Management and Utilization. In addition he has taught Soils and People and has contributed to short courses on environmental topics for industry. His research focuses on the fate of organic materials in soil and mechanisms controlling cycles of elements. He has used isotopic tracers to follow elements and simulation modeling to integrate diverse studies. This has led to knowledge of soil-plant systems on one hand, and questions about methods to monitor specific soil organisms on the other. He has supervised 22 graduate students to completion and has been an external examiner for seven dissertations from Canada and Australia; he has published over 65 refereed papers or reports and over 90 additional scientific contributions, and has given numerous invited lectures both nationally and internationally. He has international research/teaching experience in Australia, China, Thailand, and the U.S.

Dr. McGill was chair of the Department of Soil Science from 1979 to 1989, and associate dean, Faculty of Graduate Studies and Research, University of Alberta, from September 1991 through December 1993. He served as president of the Canadian Society of Soil Science during 1992–93. He was awarded the University of Manitoba Gold Medal for highest academic standing in Honors Agriculture upon graduation in 1967. In 1984–85 he received a Senior Research Fellowship from CSIRO, Division of Tropical Crops and Pastures, Brisbane.

Albert L. Page received his Ph.D. degree in Soil Science at the University of California, Davis, in 1960. He is currently professor of soil science in the Department of Soil and Environmental Sciences and chemist in the Agricultural Experiment Station, University of California, Riverside. His research has emphasized the chemistry of trace elements in soils and its influence on their bioavailability and mobility. He has served on numerous national and international committees in his area of specialization and presently serves as a member of the National Research Council-Water Science Technology Board committee on Planning and Remediation for Irrigation-Induced Water Quality Problems and chair of the committee on Use of Treated Municipal Wastewater Effluents and Sludge in the Production of Crops for Human Consumption. Dr. Page has been on the editorial boards of the *Journal of Environmental Quality, Soil Science Society of America Journal,* and *Science of the Total Environment;* presently, he serves on the Board of Directors of the Soil Science Society of America.

Dr. Page has received several professional awards. He is a fellow of the American Society of Agronomy and the Soil Science Society of America, the recipient of a Guggenheim Memorial Foundation Fellowship, and a Fulbright Fellowship. He has received a U.S. Department of Agriculture Distinguished Service Award, the American Society of Agronomy's Environmental Quality Research Award, the Association of Metropolitan Sewerage Agencies Environment Award, and a University of California, Riverside, Distinguished Teaching Award.

CONTRIBUTORS

D. W. Anderson
Department of Soil Science
University of Saskatchewan
Saskatoon, Saskatchewan, Canada

F. Andreux
Centre des Sciences de la Terre
Université de Bourgogne
6, boulevard Gabriel
Dijon, France

C. A. Bellin
E. I. DuPont
Agricultural Products Division
Stine-Haskell Research Center
Newark, Delaware

J.-M. Bollag
Laboratory of Soil Biochemistry
The Pennsylvania State University
University Park, Pennsylvania

S. A. Boyd
Department of Crop and Soil Sciences
Michigan State University
East Lansing, Michigan

J. A. Briceño-Salazar
CIA, Agronomy
University of Costa Rica
San José, Costa Rica

P. Ceccon
Dipartmento di Produzione Vegetale e
Tecnologie Agrarie dell'Universitá
Centro di studio sulla Chimica degli
Antiparassitari del C.N.R
Perugia, Italy

H. H. Cheng
Soil Science Department
University of Minnesota
St. Paul, Minnesota

C. Chenu
INRA
Station de Science du Sol
Versailles cedex, France

H. Claus
Institut für Wasser-, Boden-,
und Lufthygiene
des Bundesgesundheitsamtes
Aussentelle Langden
Langden, Germany

M. M. Cliath
USDA-ARS
University of California
Riverside, California

L. Deleón
CIA, Agronomy
University of Costa Rica
San José, Costa Rica

Z. K. Filip
Institut für Wasser-, Boden-,
und Lufthygiene
des Bundesgesundheitsamtes
Aussentelle Langden
Langden, Germany

P. Fusi
Dip. Scieza del Suolo e Nutrizone
della Pianta
Universitá di Firenze
Firenze, Italy

J. Gan
Soil Science Department
University of Minnesota
St. Paul, Minnesota

R. Giovanardi
Dipartmento di Produzione Vegetale e
Tecnologie Agrarie dell'Universitá
Centro di studio sulla Chimica degli
Antiparassitari del C.N.R
Perugia, Italy

E. G. Gregorich
Centre for Land and Biological
Resources Research
Agriculture Canada
Ottawa, Ontario, Canada

K. Haider
Institut für Pflanzenernährung
und Bodenkunde
der BFA Landwirtschaft (FAL)
Deisenhofen, Germany

L. K. Heng
Manaaki Whenua — Landcare Research
Lower Hutt, New Zealand

A. Herbillon
Centre de Pédologie Biologique
CNRS
Vandoeuvre-les-Nancy cedex, France

T. Higashi
Institute of Applied Biochemistry
University of Tsukuba
Tsukuba, Ibaraki, Japan

P. M. Huang
Department of Soil Science
University of Saskatchewan
Saskatoon, Canada

L. J. Jarvis
Soil Science Department
University of Minnesota
St. Paul, Minnesota

H. Kodama
Centre for Land and Biological Resources
Research
Agriculture Canada
Ottawa, Canada

W. C. Koskinen
USDA-ARS
Soil and Water Management Research Unit
Soil Science Department
University of Minnesota
St. Paul, Minnesota

L. S. Kotlyar
National Research Council of Canada
Institute for Environmental Chemistry
Ottawa, Ontario, Canada

J. S. K. Lee
Department of Soil Science
University of Saskatchewan
Saskatoon, Canada

L. S. Lee
Agronomy Department
Purdue University
West Lafayette, Indiana

G. Q. J. Lou
Department of Soil Science
University of Saskatchewan
Saskatoon, Canada

M. Mansour
Institut f Ökologische Chemie
GSF, Schulstrabe
Freising-Attaching, Germany

C. Marucchini
Dipartmento di Produzione Vegetale e
Tecnologie Agrarie dell'Universitá
Centro di studio sulla Chimica degli
Antiparassitari del C.N.R
Perugia, Italy

P. M. Mele
Rutherglen Research Institute
Department of Agriculture
Chiltern Valley Road
Rutherglen, Victoria, Australia

T. M. Miano
Istituto di Chimica Agraria
Universitá di Bari
Bari, Italy

C. M. Monreal
Centre for Land and Biological
Resources Research
Agriculture Canada
Ottawa, Ontario, Canada

C. Munier-Lamy
Centre de Pologie Biologique
CNRS, UPR 6831
Associée à l'Université de Nancy
Vandœuvre Cédex, France

I. P. Murarka
Electrical Power Research Institute
Palo Alto, California

C. Myers
Laboratory of Soil Biochemistry
The Pennsylvania State University
University Park, Pennsylvania

S. Nagatsuka
Laboratory of Soil Science
Institute of Applied Biochemistry
University of Tsukuba
Tsukuba City, Ibaraki Prefecture

J. M. Oades
Cooperative Research Centre for Soil
and Land Management
Department of Soil Science
Waite Agricultural Research Institute
University of Adelaide
Glen Osmond, South Australia

S. Pal
Laboratory of Soil Biochemistry
The Pennsylvania State University
University Park, Pennsylvania

R. L. Parfitt
Manaaki Whenua — Landcare Research
Lower Hutt, New Zealand

P. S. C. Rao
Soil and Water Science Department
University of Florida
Gainesville, Florida

G. G. Ristori
Centro di Studio per Suolo
C.N.R.
Firenze, Italy

M. Robert
INRA
Station de Science du Sol
Versailles cedex, France

R. Salas-Camacho
CIA, Agronomy
University of Costa Rica
San José, Costa Rica

I. Scheunert
Institut für Bodenökologie
Ingolstädter Landstrabe
Neuherberg, Germany

M. Schiavon
Ecole nationale Supérieure d'Agronomie
et des industires Alimentaires
Vandœuvre Cédex, France

M. Schnitzer
Land and Biological Resources
Research Center
Agriculture Canada
Ottawa, Ontario, Canada

R. Schutte
Research Department of Syncrude Ltd.
Edmonton, Alberta, Canada

N. Senesi
Istituto di Chimica Agraria
Universitá di Bari
Via Amendola, 165/A
70126 Bari
Italy

P. Sharma
Central Soil Salinity Research Institute
Karnal, India

R. C. Sims
Department of Civil and
Environmental Engineering
Utah Water Research Laboratory
Utah State University
Logan, Utah

W. J. Slattery
Rutherglen Research Institute
Department of Agriculture
Chiltern Valley Road
Rutherglen, Victoria, Australia

B. D. Sparks
National Research Council of Canada
Institute for Environmental Chemistry
Ottawa, Ontario, Canada

W. F. Spencer
USDA-ARS
University of California
Riverside, California

G. R. Steed
Rutherglen Research Institute
Department of Agriculture
Chiltern Valley Road
Rutherglen, Victoria, Australia

A. T. Stone
Department of Geography and
Environmental Engineering
G. W. C. Whiting School of Engineering
The Johns Hopkins University
Baltimore, Maryland

A. Swarp
Central Soil Salinity Research Institute
Karnal, India

M. Tani
Laboratory of Soil Science
Institute of Applied Biochemistry
University of Tsukuba
Tsukuba City
Ibaraki Prefecture, Japan

M. D. Taylor
Manaaki Whenua — Landcare Research
Lower Hutt, New Zealand

B. K. G. Theng
Manaaki Whenua — Landcare Research
Lower Hutt, New Zealand

A. Torrents
Department of Geography and
Environmental Engineering
G. W. C. Whiting School of Engineering
The Johns Hopkins University
Baltimore, Maryland

C. Vischetti
Dipartmento di Produzione Vegetale e
Tecnologie Agrarie dell'Universitá
Centro di studio sulla Chimica degli
Antiparassitari del C.N.R
Perugia, Italy

K. Wada
Headquarters
Kyushu University
Fukuoka, Japan

M. C. Wang
Department of Soil Science
National Chung Hsing University
Taichung, Taiwan

G. Whelan
Battelle
Pacific Northwest Laboratory
Richland, Washington

J. -M. Xu
Department of Soil and
Agricultural Chemistry
Zhejiang Agricultural University
Hangzhou, China

S. R. Yates
USDA-ARS
University of California
Riverside, California

K. -N. Yuan
Department of Soil and
Agricultural Chemistry
Zhejiang Agricultural University
Hangzhou, China

CONTENTS

PART I GENERAL SOIL QUALITY AS INFLUENCED BY INTERACTIONS OF SOIL MINERALS WITH ORGANICS AND MICROORGANISMS

PART II TRANSFORMATIONS OF NATURAL AND ANTHROPOGENIC ORGANIC COMPOUNDS AS AFFECTED BY SOIL MINERALS AND MICROORGANISMS

A. Natural Organics

Part I: General Soil Quality as Influenced by Interactions of Soil Minerals with Organics and Microorganisms

Chapter 1

Organic-Inorganic Interactions in Soils and Their Effects on Soil Quality

M. Schnitzer

CONTENTS

I. INTRODUCTION

In soils, organic and inorganic components are closely associated. Products resulting from these interactions are likely to affect the physical, chemical, and biological properties of soils and so exert significant effects on soil quality. The purpose of this chapter is to discuss some of these effects and to assess their possible significance.

Soil organic matter (SOM), as used in the context of this paper, refers to the sum total of all organic carbon-containing substances in soils. The organic matter (OM) of soils ranges from <0.1% in desert soils to close to 100% in organic soils. A typical agricultural soil may contain between 2 to 5% organic matter in the top 15 cm.

The following are estimates of the average composition of SOM in typical mineral Canadian soils: carbohydrates — 10%; N compounds (including proteins, peptides, amino acids, ammonia, purines, pyrimidines, and unidentified N substances) — 10%; alkanes, fatty acids, fats, waxes, resins, etc. — 10%; and humic substances (humic acid [HA], fulvic acid [FA], and humin) — 70%.[1] These percentages may vary depending on local conditions.

Thus, the bulk of SOM consists of humic substances. It is for this reason that the role of these materials in organic–inorganic interactions in soils is emphasized in this chapter.

0-87371-914-X/95/$0.00+$.50

Table 1 Analytical Characteristics of HA and FA[2]

	HA	FA
Element (g kg^{-1})		
C	564	509
H	55	33
N	41	7
S	11	3
O	329	448
Functional Groups (cmol kg^{-1})		
Total acidity	660	1240
COOH	450	910
Phenolic OH	210	330
Alcoholic OH	280	360
Quinonoid C=O	250	60
Ketonic C=O	190	250
OCH_3	30	10
E_4/E_6	4.3	7.1

II. ANALYTICAL CHARACTERISTICS OF HUMIC SUBSTANCES

The elemental composition and functional group content of an HA extracted from a Mollisol Ah horizon and a FA extracted from the Bh horizon of a Spodosol are shown in Table 1.[2] These two humic materials have been investigated in our laboratory over a period of many years, and hence are emphasized in this chapter.

A more detailed analysis of the data shows that (1) the HA contains more C but less O than does the FA, (2) the HA contains more H, N, and S than the FA, (3) the total acidity and COOH content of the FA are appreciably higher than those of the HA, (4) the FA is richer in phenolic OH, alcoholic OH, and ketonic C=O groups than the HA, but the latter contains more quinonoid C=O groups per unit weight, and (5) about 90% of the O in the HA can be accounted for in functional groups, but all of the O in the FA is similarly distributed. The E_4/E_6 ratio (ratio of absorbances at 465/665 nm) of the FA is considerably higher than that of the HA, which means that the FA has a lower particle or molecular weight than the HA. From data published in the literature, it appears that the analytical characteristics of humins are similar to those of HAs.[3,4]

A. ^{13}C-NMR SPECTROSCOPY OF HA AND FA

Solid-state ^{13}C-NMR spectra of a HA extracted from a Mollisol Ah horizon and of a FA extracted from a Spodosol Bh horizon are shown in Figure 1. Spectrum (a) exhibits several fairly distinct peaks in the aliphatic (0- to 105-ppm), aromatic (106- to 150-ppm), phenolic (near-155-ppm), and carboxyl (171- to 190-ppm) regions. The signals at 16.5, 21.1, 25.0, 27.1, and 31.3 ppm are most likely due to aliphatic carbons in alkyl chains. The peak at 16.5 ppm is characteristic of terminal methyl groups, and that at 31.3 ppm of $(CH_2)n$ in long alkyl chains. The resonance at 40.2 ppm could include contributions from both alkyl carbons and amino acid carbons. In the 50- to 105-ppm regions, aliphatic carbons substituted by O and N are usually observed. The broad signal at 52.6 ppm and the sharper one at 58.5 ppm may be due to C in OCH_3. Amino acid-C may also contribute in this region.[5] Carbohydrates in the HA would be expected to produce signals in the 60- to 65-, 70- to 80-, and 90- to 105-ppm regions, although other types of aliphatic C bonded to O could also do so. The aromatic region contains a relatively sharp maximum at 131.8 ppm, which can be assigned to ring carbons in

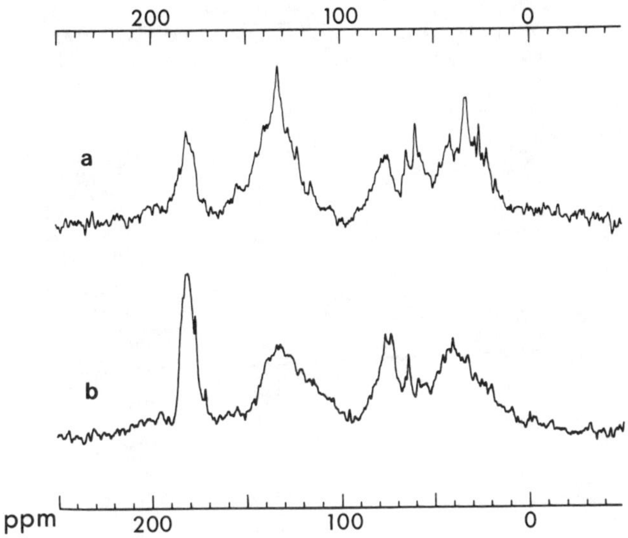

Figure 1 ^{13}C-NMR spectra of (a) Chernozem HA and (b) Podzol Bh FA.

which the ring is free of deshielding by strong electron donor substituents, such as O and N. Alkyl benzenes could yield such a resonance, as would unsubstituted aromatic C-H.[5] The small peak near 155 ppm in spectrum (a) indicates the presence of O- and N-substituted aromatic C (phenolic OH or aromatic NH_2 groups). The broad peak near 180 ppm is attributed to C in CO_2H groups. Amides and esters could contribute to this peak.

The ^{13}C-NMR spectrum of the FA (Figure 1b) consists of a number of aliphatic resonances in the 20- to 50-ppm region, followed by signals from OCH_3 groups, amino acids, amino sugars, and carbohydrates between 50 and 85 ppm. The presence of aromatic carbons is indicated by a broad signal with maximum intensity at 130 to 133 ppm. The preponderance of CO_2H groups is shown by signals between 171 and 182 ppm. In general, fewer sharp signals are observed in the ^{13}C-NMR spectrum of the FA than in that of the HA, possibly because of more H bonding in the FA.

The ^{13}C-NMR data are summarized in Table 2. The spectra are divided into the following regions: 0 to 40 ppm (C in straight-chain, branched, and cyclic aliphatics), 41 to 60 ppm (C in branched aliphatics, amino acids, and OCH_3 groups), 61 to 105 ppm (C in carbohydrates, and in aliphatics containing C bonded to OH, ether oxygens, or occurring in five- or six-membered rings bonded to O), 105 to 150 ppm (aromatic C), 151 to 170 ppm (phenolic C), and 171 to 190 ppm (C in CO_2H) groups.

An inspection of the data in Table 2 shows a similar C distribution in the two humic materials. The HA is slightly more aromatic than the FA, but the FA is considerably richer in CO_2H groups, which appears to be the main difference between the two substances. Another, possibly significant difference between the HA and FA is that the HA contains, per unit C, more paraffinic C (0 to 40 ppm) than the FA.

B. PYROLYSIS-FIELD IONIZATION MASS SPECTROMETRY (Py-FIMS) OF THE HA AND FA

Additional support for similarities in the main structural features of the HA and FA comes from the Py-FI mass spectra of these two materials.[6] The major components identified in these spectra are carbohydrates, phenols, monomeric and dimeric lignins, n-fatty acids,

Table 2 Distribution of C (%) in the Mollisol HA and Spodosol FA as Determined by ^{13}C-NMR

Chemical shift range (ppm)	% of C HA	FA
0–40	24.0	15.6
41–60	12.5	12.8
61–105	13.5	19.3
106–150	35.0	30.3
151–170	4.5	3.7
171–190	10.5	18.3
Aliphatic C (0–105 ppm)	50.0	47.7
Aromatic C (106–150 ppm)	39.5	34.0
Phenolic C (151–170 ppm)	4.5	3.7
Aromaticity[a]	44.1	41.6

[a] $\left(\dfrac{\text{aromatic C} + \text{phenolic C}}{\text{aromatic C} + \text{phenolic C} + \text{aliphatic C}}\right) \times 100$

n-alkanes, n-alkyl-benzenes, and heterocyclic N-compounds. The spectrum of the HA also shows the presence of small amounts of n-alkylmonoesters, while that of the FA indicates the presence of small amounts of sterols.

III. REACTIONS OF HUMIC SUBSTANCES WITH METALS AND MINERALS

One of the most striking characteristics of humic substances is their ability to interact with metal ions, metal oxides, metal hydroxides, other minerals, and organics to form water-soluble and water-insoluble associations of widely differing chemical and biological stabilities. The following types of reactions have been observed to occur in terrestrial and aquatic systems: (1) dissolution of minerals, (2) formation of water-soluble simple metal complexes, (3) formation of water-soluble mixed ligand complexes, (4) adsorption on and desorption from water-insoluble HAs and metal-humate complexes, (5) adsorption on external mineral surfaces, and (6) adsorption in clay interlayers.[3] The different metal-organic complexes exert physical, chemical, and biological effects on soil quality. The formation and characteristics of each of the complexes and their effects on soil quality are discussed in greater detail in the following paragraphs.

A. DISSOLUTION OF MINERALS BY HA AND FA AND BY LOW-MOLECULAR-WEIGHT ORGANIC ACIDS OF KNOWN CHEMICAL STRUCTURES

Several workers postulate that in soils, low-molecular-weight organic acids such as oxalic and salicylic acids, rather than humic substances, are the effective agents of mineral degradation.[7-9]

The capacity of many low-molecular-weight acids of known chemical structures to maintain the solubility of relatively insoluble inorganic compounds has been demon-

Table 3 Action of HAs on Minerals and Metals[17]

Sample	Element determined	Micrograms of metal extracted in 1 h[a]					
		H_2O-CO_2	HA-1	HA-2	HA-3	HA-4	HA-5
Galena	Pb	1	200	210	80	230	110
Sphalerite	Zn	<1	30	90	80	40	30
Bornite	Cu	<1	190	230	180	200	130
Chalcocite	Cu	10	3,800	5,100	1,300	5,600	2,700
Bismuthinite	Bi	<1	550	1,600	410	1,000	460
Stibnite	Sb	10	45	340	10	50	<5
Pararammelsbergite	Ni	90	9,800	9,600	5,400	9,200	6,000
Hematite	Fe	<1	470	90	410	<5	490
Pyrolusite	Mn	1	1,000	1,800	290	1,040	440
Calcite	Ca	50	10,500	9,600	6,400	11,100	730
Copper	Cu	<1	5,700	5,800	3,200	6,100	4,800
Lead	Pb	2	27,400	39,400	23,000	33,000	24,400

[a] Concentrations of all extractants: 0.1% w/v.

strated by several workers.[10-13] On the other hand, the possibility of humic substances interacting with and dissolving minerals was already considered 175 years ago by Sprengel,[14] and later by Thenard[15] and Julien,[16] who suggested that humic substances could attack silicates and that they played important roles in many geological processes. To resolve the two conflicting views, Baker[17] undertook a comprehensive investigation of the role of humic substances in the weathering of minerals by comparing the capacity of HAs to dissolve metals from minerals with that of a number of low-molecular-weight organic acids.

The data in Table 3 provide details on the action of five HAs from under different vegetation covers (eucalyptus, sphagnum, fern, and pine) on a number of minerals and metals. The results show that while all HAs are active in mineral degradation, there are considerable variations in their efficiencies, but these variations show no consistent pattern. For example, HA-4 exhibits a low activity with regard to Fe in hematite but a very high activity in dissolving Ca from calcite. The higher solubility of Mn from pyrolusite relative to Fe from hematite is not universal; it is exhibited only by HA-1, HA-2, and HA-4. The interaction of HAs with bismuthinite is greater than that with stibnite, although HA-2 dissolves an appreciable amount of Sb from stibnite.

In Table 4, the metal-dissolving capacity of a HA is compared with those of a number of organic acids. The HA is more efficient in dissolving Pb from galena and Fe from hematite, and equally efficient with salicylic acid for extracting Zn from sphalerite, Ni from parammelsbergite, and Ca from calcite, but less effective for dissolving the other elements than the organic acids. On the whole, the metal-dissolving capacity of HA appears to be of the same order as that of the organic acids.

Another point that needs to be considered in the context of this discussion is the relative concentrations of humic substances and organic acids in soils.

As mentioned previously, humic substances constitute approximately 70% of the weight of SOM. By contrast, concentrations of low-molecular-weight organic acids in soils are difficult to assess. Their concentrations depend largely on the extent of microbial activity, but a high microbial activity will enhance their formation as well as their degradation. Whitehead[18] detected a number of aromatic acids in aqueous extracts of several soils. These acids included *p*-hydroxybenzoic, *p*-coumaric, vanillic, and ferulic acids. He estimated that the total concentration of these acids accounted for 0.007 to 0.008% of the total weight of SOM. Even if we multiply the concentrations of these acids

Table 4 Action of Various Organic Compounds on Minerals and Metals[17]

		Micrograms of metal extracted in 1 h[a]				
Sample	Element determined	HA-1	Salicylic acid	Oxalic acid	Pyro-gallol	Alanine
Galena	Pb	200	130	95	35	<5
Sphalerite	Zn	30	30	20	8	20
Bornite	Cu	190	260	650	55	15
Chalcocite	Cu	3,800	4,450	9,750	920	1,530
Bismuthinite	Bi	550	180	4,820	1,640	55
Stibnite	Sb	45	<5	580	<5	<5
Pararammelsbergite	Ni	9,800	10,500	7,620	2,380	1,730
Hematite	Fe	470	<3	80	10	20
Pyrolusite	Mn	1,000	4,200	15,500	5,150	520
Calcite	Ca	10,500	11,900	980	2,040	1,400
Copper	Cu	5,700	5,500	2,620	1,190	700
Lead	Pb	27,400	41,800	660	1,470	240

[a] Concentration of all extractants: 0.1% w/v.

by ten, the concentrations of total humic substances exceed those of organic low-molecular-weight acids by at least a factor of 10^3. As the data in Table 4 show, however, the low-molecular-weight organic acids exhibit metal-dissolving capacities that are of the same order as that of HA. From the latter information and because of their low concentrations in soils, it becomes apparent that in bulk soils, the role of low-molecular-weight organic acids, compared to that of HA and FA, in the weathering of minerals is limited. It is possible that low-molecular-weight organic acids are more important in the rhizosphere in this regard.

Table 5 presents data on reactions of HA-1 with silicate minerals. While these silicates are more resistant to attack by the organic acids than the other minerals, the amounts of metals dissolved by HA-1, compared to water, are noteworthy. High extractions of Na, K, Mg, Ca, and Fe compared to Si and Al suggest that these metals are removed from the silicate framework.

Baker[17] concludes that HAs are powerful solvents of a number of minerals and are very active in the weathering cycle. Their maximum effectiveness would be under a cool-temperate, high-rainfall climate.

B. FORMATION OF WATER-SOLUBLE SIMPLE METAL COMPLEXES

Reactions in water near pH 7 between di- and trivalent metal ions and HA or FA are likely to proceed by either one or more, or simultaneously all four, of the reaction mechanisms shown in Figure 2, taking the divalent metal ion M^{2+} as an example.[19] According to Equation 1, one CO_2H group reacts with one metal ion to form an organic salt or monodentate complex. Equation 2 describes a reaction in which one CO_2H and one adjacent OH group react simultaneously with the metal ion to form a bidentate complex or chelate. According to Equation 3, two adjacent CO_2H groups interact simultaneously with the metal ion to form a bidentate chelate. Equation 4 shows a situation in which the metal ion M^{n+} is linked to the FA, in addition to electrostatic bonding, through a water molecule in its primary hydration shell via a hydrogen bond to a C=O group. This type of interaction is of special importance when the cation has a high solvation energy and so retains its primary hydration shell.

Equations 2 and 3 describe the formation of strong inner-sphere complexes, whereas Equation 4 refers to the formation of a weaker outer-sphere complex.

Table 5 Reaction of H_2O and HA-1 with Silicates[17]

Mineral	Element determined	Micrograms of metal extracted in 5 d: H_2O/atm CO_2	0.1% (w/v) HA-1
Feldspar	Na	340	1,260
(perthite)	K	190	640
(Na, K) Al Si_3 O_8	Al	<10	220
	Si	<5	<5
Biotite	K	60	100
K (Mg, Fe)$_3$	Mg	20	90
(Al Si_3 O_{10}) $(OH)_2$	Fe	<5	240
	Al	<10	300
	Si	<5	5
Enstatite	Mg	80	620
(Mg, Fe^{2+})SiI_3	Fe	<5	220
	Si	<5	<5
Actinolite	Mg	25	80
Ca (Mg, Fe^{2+})$_5$	Fe	<5	960
(Si_8 O_{22})(OH, F)$_2$	Ca	750	2,900
	Si	<5	40
Epidote	Ca	1,800	10,700
Ca Fe^{2+} $Al_2O{\cdot}OH$	Fe	10	1,300
(Si_2 O_7) (SiO_4)	Al	40	320
	Si	<5	5

The stability constants of water-soluble metal-HA and metal-FA complexes have been determined by a number of workers. The subject has been reviewed by Stevenson.[20] A number of serious problems have been encountered in the analysis and interpretation of data. Probably the most serious obstacle to progress in this area is our lack of adequate knowledge of the chemical structure and dissociation behavior of HAs and FAs.

C. MIXED LIGAND COMPLEXES

The formation of metal-FA-phosphate complexes was first described by Lévesque and Schnitzer.[21] It is likely that in soils, an appreciable portion of the total P exists in the form of such complexes, but it is difficult to demonstrate this because of the low P content of soils. The formation and stability of mixed ligand complexes of the type Cu^{2+}-FA-secondary ligand (Y) has been studied by Manning and Ramamoorthy.[22] The secondary ligands Y investigated were citrate, tartrate, salicylate, phosphate, NTA (nitrilotriacetate), aspartate, and glycine. In neutral and weakly acidic solutions, mixed complexes predominate over simple complexes. The values of equilibrium constants for mixed complexes with citrate, phosphate, and NTA are particularly high compared to simple complexes. If phosphate functions in the same way as other oxyanions, the relatively high concentration of HCO_3^- and SO_4^{2-} in some soil solutions should lead to the formation of mixed Cu^{2+}-FA-HCO_3 and Cu^{2+}-FA-HSO_4 complexes. There is increasing evidence that mixed complexes are important in the transport of trace metal ions in the soil and in the uptake of cations and anions by plant roots. Furthermore, the formation of mixed complexes will prevent the precipitation of metal ions by hydrolysis at elevated pH values and will also interfere with the precipitation of insoluble metal phosphates, sulfates, or carbonates. More information on the occurrence and properties of mixed ligand complexes in soils is needed.

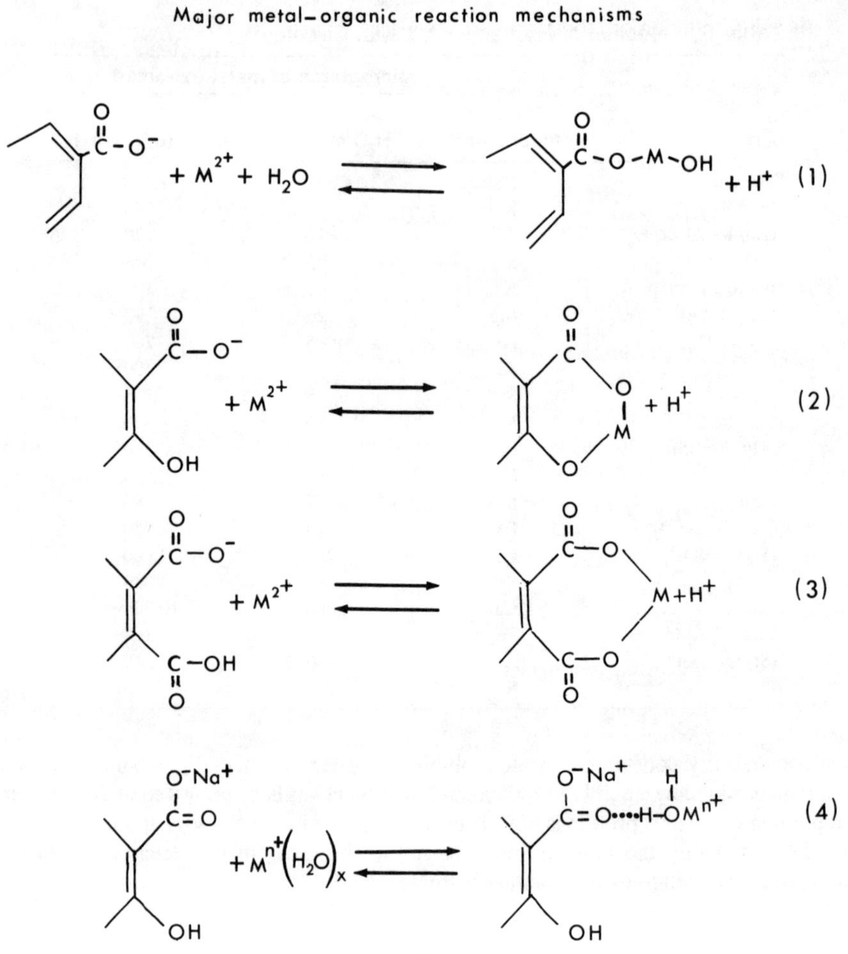

Figure 2 Major metal-HA and metal-FA reaction mechanisms.[19]

D. SORPTION AND DESORPTION

One of the major characteristics of HAs is their ability to sorb inorganic and organic substances. I shall confine this discussion to the sorption of metal ions. Kerndorff and Schnitzer[23] examined the interaction of HA with a solution containing equimolar concentrations of 11 different metal ions. They report that the efficiency of sorption of metal ions on HA increases with rises in pH and HA concentrations and a decrease in metal concentration.

At pH 2.4, the order of sorption is

$$\text{Hg} > \text{Fe} > \text{Pb} > \text{Cu} = \text{Al} > \text{Ni} > \text{Cr} = \text{Zn} = \text{Cd} = \text{Co} = \text{Mn}$$

At pH 3.7, the order is

$$\text{Hg} > \text{Fe} > \text{Al} > \text{Pb} > \text{Cu} > \text{Cr} > \text{Cd} = \text{Zn} = \text{Ni} = \text{Co} = \text{Mn}$$

At pH 4.7, the order is

$$Hg = Fe = Pb = Cu = Al = Cr > Cd > Ni = Zn > Co > Mn$$

At pH 5.8, the order is

$$Hg = Fe = Pb = Al = Cr = Cu > Cd > Zn > Ni > Co > Mn$$

Hg(II) and Fe(III) are always sorbed most readily by HA, while Co and Mn are sorbed least readily. The different metal ions appear to compete for active sites (COOH and phenolic OH groups) on the HA. The sorption mechanism is complex. Not only do the 11 metal ions plus H^+ ions (a total of 12 ions) interact with the HA, they also interact with each other. The mechanisms involved are likely to include ion exchange, coprecipitation, and the formation of inner-sphere and outer-sphere complexes. The affinities of the 12 ions for sorption on HA do not correlate with their atomic weights, atomic numbers, or crystal and hydrated ionic radii.

The metal ions sorbed by HA can subsequently be desorbed through complexation by water-soluble FA. Similarly, FA will desorb metal ions sorbed on inorganic soil constituents such as clays and hydrous oxides. Thus, FA can effectively change the sorption, desorption, and precipitation characteristics of metals.

E. ADSORPTION ON EXTERNAL MINERAL SURFACES

The extent of adsorption of humic materials on mineral surfaces depends on the physical and chemical characteristics of the surface, the pH of the system, and its water content. One can visualize the formation of a wide range of mineral-humic associations, involving chemical bonding with widely differing strengths. Another mechanism of considerable importance is hydrogen bonding, the occurrence of which is clearly indicated in the IR spectra of HA- and FA-mineral complexes.[4] These reactions are likely to involve H and O of CO_2H and OH groups in HA and FA and O and H on mineral surface and edges. It is probable, as mentioned above, that cations with high solvation energies on mineral surfaces react via water bridges with HA and FA functional groups. Van der Waals forces may also contribute to the adsorption of humic substances on mineral surfaces. Kodama and Schnitzer[24] report high adsorption of FA on sepiolite surfaces. Sepiolite has a channel-like structure formed by the joining of edges of long, slender talc structures. In untreated sepiolite, the channels are occupied by bound and/or zeolitic water, which apparently can be displaced by undissociated FA.

F. ADSORPTION IN CLAY INTERLAYERS

The evidence currently available shows that the interlayer adsorption of FA by expanding clay minerals is pH dependent, being greatest at low pH and no longer occurring at pH >5.0 (Figure 3).[25] Adsorbed FA cannot be displaced from clay interlayers by leaching with 1 *M* NaCl; an inflection occurs in the adsorption-pH curve near the pH corresponding to the pK of the FA.[25] On the basis of these criteria, the adsorption could be classified as a "ligand-exchange" reaction.[26] This type of complex appears to be similar to an inner-sphere complex. In this type of reaction, the anion is thought to penetrate the coordination shell of the dominant cation in the clay and displace water coordinated to the dominant cation in the clay interlayer.

The ease with which water can be displaced will depend on the affinity for water of the dominant cation with which the clay is saturated and also on the degree of dissociation

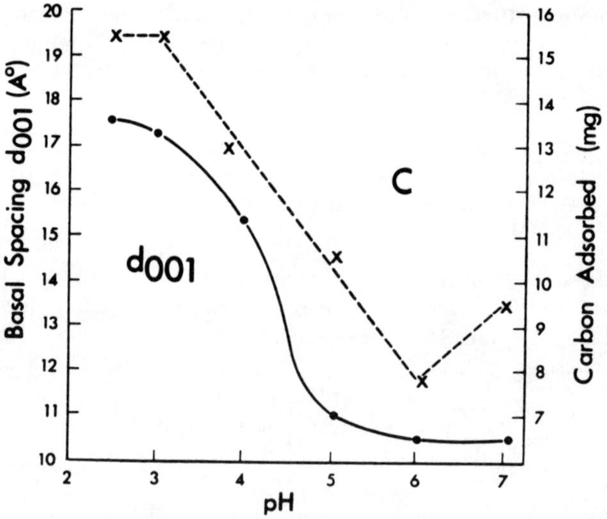

Figure 3 Effect of pH on d001 and adsorption of FA (% CX2).[25]

of the FA. Since the latter is very low at low pH, interlayer adsorption of FA is greatest at low pH levels, as Schnitzer and Kodama have observed.[25] Concurrently, the FA can dissolve a proportion of the dominant cation in the clay by forming a soluble complex and replacing the removed cation by H^+. If this process continues over long periods, the FA will eventually degrade the clay structure.

IV. THE SPECIAL ROLE OF FA

FA is soluble in water at any pH and is therefore the most abundant humic material in acid soils and in aquatic systems. It also has a great capacity for complexing mono-, di-, tri-, and tetravalent metal ions and to interact with silica.[3] In this manner, FA may increase the concentrations of dissolved minerals in aqueous solutions far in excess of their normal solubilities. This may lead to the synthesis of new minerals by permitting the complexed and dissolved metals and silica to form new combinations.[19] Also, active surfaces of clays or hydrous oxides may catalyze either the degradation or synthesis of humic substances.[27,28]

Dilute FA solution is much more efficient for dissolving Fe, Al, and Mg from chlorites than is distilled water at the same pH under identical conditions[29] (Table 6). Table 7 presents similar data for three micas.[30] It is noteworthy that 0.25 g l^{-1} of FA is relatively more effective for degrading soil minerals than are 1.0 and 2.0 g l^{-1} of FA (Figure 4), indicating that very dilute FA solutions have considerable efficiency for dissolving minerals. One point of special interest is illustrated by the data in Tables 6 and 7: minerals rich in Fe are more susceptible to attack by FA than are those free of Fe or containing little Fe. Thus, the complexing of Fe by humic materials appears to have an adverse effect on the structural stability of Fe-rich minerals, so that along with Fe, other major constituent elements, such as Mg, Al, K, and Si, are also brought into solution. Senesi et al.[31] combined ESR spectroscopy with chemical treatments to obtain information on the oxidation states and site symmetries of iron bound by HA and FA. At least two, possibly three, different binding sites for iron occur in humic materials. (1) Fe^{3+} is strongly bound and protected by tetrahedral and/or octahedral coordination. This form of iron exhibits considerable resistance to complexation by known seques-

Table 6 Dissolution of Metals (Fe, Al, Mg) from Chlorites by Aqueous FA Solution (2.0 g l^{-1}) After 360 h of Shaking at pH 2.5[29]

	Percent of samples dissolved by	
	H_2O	FA (2.0 g l^{-1})
Leuchtenbergite (Fe poor)	2	4
Thuringite (Fe rich)	6	26

tering agents and chemical reduction. (2) Fe^{3+} is adsorbed on the external surfaces of humic materials, weakly bound octahedrally, and easily complexed and reduced. These observations were made on laboratory-prepared Fe-HA and Fe-FA complexes and on a Fe-rich layer taken from a soil also rich in organic matter. These findings were confirmed more recently by Kodama et al.,[32] who studied laboratory-prepared 57 Fe(III)-FA complexes by Mössbauer spectroscopy and chemical methods. They found that the Fe(III) in these complexes occurred partly in the form of organic complexes and partly as inorganic species.

SOM, one of five factors of soil formation, is much involved in soil development and contributes to the differentiation of horizons.[33] In Spodosols, FA constitutes up to 75% of SOM and is very active in Podzolization.[3] In the surface horizons of these soils, FA forms water-soluble multidentate complexes with Fe, Al, and other di- and trivalent metals. These complexes move down the profiles and are coagulated lower in the profiles in B horizons through additional reactions with the same metals that block most of the oxygen-containing functional groups (COOH and OH) in the FA. This process has been referred to as "cheluviation" by Swindale and Jackson.[34] This term refers to the dissolution of metals from minerals by chelation, followed by eluviation of the metal-organic complexes.

Some of the reactions and processes active in Podzolization may also play important roles in the formation of other soils such as Gray-Wooded, Gray-Brown Podzolic, and Red and Yellow Podzolic soils.[35]

Table 7 Dissolution of Micas by Aqueous FA Solution (2.0 g l^{-1}) (pH 2.5)[30]

Mineral	Element dissolved	Percent dissolved after 710 h
Biotite	Fe	11.5
(Fe rich)	Al	14.5
	Mg	17.0
	K	18.5
	Si	14.0
Phlogopite	Al	8.0
(Fe poor)	Mg	8.6
	K	8.0
	Si	9.0
Muscovite	Al	2.3
(Fe poor)	K	1.8
	Si	0.6

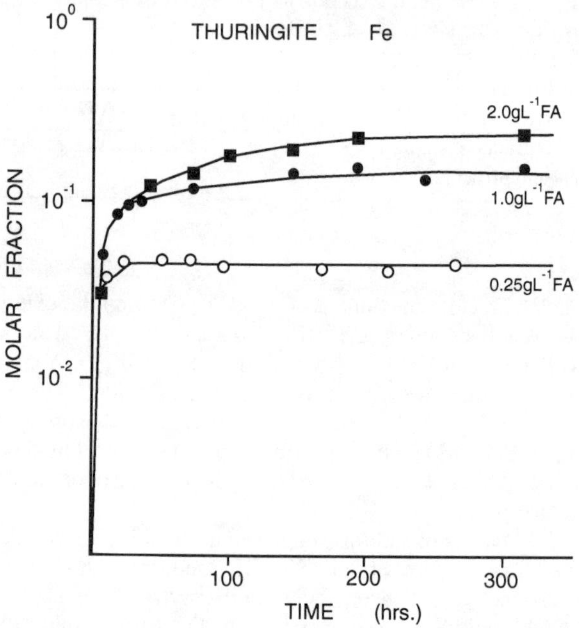

Figure 4 The release of Fe from thuringite by 2.0, 1.0, and 0.25 g l^{-1} of FA.[19]

V. SOIL ORGANIC MATTER AND SOIL STRUCTURE

In the surface layers of many agricultural soils, SOM and SOM bonded to clay particles through complexing with aluminum or iron appear to play major roles in stabilizing soil aggregates.[36] While it has been suggested that polysaccharides are important to aggregate stability, it has not yet been possible to clearly determine whether one SOM component was more important than any other.

According to Tisdall and Oades,[37] organic binding agents can be classified as: (a) transient, mainly polysaccharides; (b) temporary, roots and fungal hyphae; and (c) persistent, aromatic components complexed with polyvalent metal cations. We now discuss (c) in more detail.

Persistent binding agents are degraded, partly aromatic, humic materials complexed with amorphous iron, aluminum, and aluminosilicates.[37] These binding agents probably include complexes of clay-polyvalent metal organic matter, CPOM and $(CPOM)_x$, both of which are <250 μm in diameter.[37,38] SOM is believed to be the center of the aggregate, with particles of fine clay sorbed onto it rather than SOM sorbed onto the clay surface.[37] Thus, metal-organic complexing appears to play an important role in the binding of soil particles.

A. EFFECTS OF WAXES ON SOIL AGGREGATION

According to Schreiner and Shorey,[39] waxy materials are responsible for the hydrophobic properties of soils, and their presence positively affects soil aggregation and aggregate stability but exerts negative effects on water retention. Further support for the role of waxes in soil aggregation comes from the work of Giovannini et al.,[40] who extracted hydrophobic fractions from five soil samples with benzene and concluded that both hydrophobic components and metal-organic interactions are involved in soil aggregation.

Analyses by pyrolysis-field ionization (Py-FI) and pyrolysis-field desorption (Py-FD) mass spectrometry of n-hexane, chloroform, and supercritical n-pentane extracts of fine clay fraction separated from Aquoll, Udic Boroll, and Typic Boroll soils indicate the presence of n-alkanes, n-fatty acids, n-alcohols, n-diols, sterols, n-alkyl monoesters, n-alkyl diesters, and n-alkyl triesters (Table 8).[41] As the data in Table 8 show, similar materials are extracted from fine clay fractions from three different soils. The major components in all extracts are n-alkanes, n-fatty acids, the C_{16} and C_{24} diols, and n-alkyl monoesters. Chemical structures of the major types of compounds identified are shown in Figure 5.

Especially noteworthy are (1) the wide range of n-alkanes from m/z 156 (C_{11}) to m/z 1318 (C_{94}), (2) the relatively narrow range of n-fatty acids from m/z 214 (C_{13}) to m/z 508 (C_{34}), and (3) the relatively high molecular weight of the C_{99} n-alkyl triester, which is 1478. The C-even to C-odd atom ratio of the n-alkanes in the three clay extracts is 0.96, which is close to that of hydrocarbons of bacteria, phyto- and zooplankton, which suggests a microbial or plant origin for the n-alkanes.[42] The C-even to C-odd atom ratio of the n-fatty acids is 1.37; the n-fatty acids range from C_{13} to C_{34}. According to Kvenvolden,[42] even-C-numbered n-fatty acids in the C_4 and C_{26} range predominate in most living organisms, whereas C_{26} and C_{38} even-C-numbered acids are found principally in waxes of insects and plants. These observations suggest at least two different sources for the n-fatty acids. All of the components identified (Table 8) have been reported to occur in waxes.[43,44] These observations suggest that the compounds identified herein are components of natural waxes which are prominent in clay-associated SOM. These waxes originate most likely from coniferous and deciduous vegetation and, protected by close association with clay minerals, survive over long periods of time after carbohydrates, proteins, phenols, and lignins have decomposed. The net result is a selective enrichment in waxes. Because of their hydrophobic properties, these materials resist biodegradation and chemical decomposition and have therefore long residence times in soils. Jambu et al.[45] suggested that these waxy materials exert beneficial effects on the structural stability by forming continuous films over the surfaces of soil aggregates. To what extent metals are involved in these interactions remains to be investigated.

VI. EFFECTS OF METAL- AND CLAY-OM (HUMIC) COMPLEXES ON SOIL QUALITY

Finally, we consider to what extent the presence of metal- and clay-OM (humic) complexes affects soil quality. For this purpose, we first classify these effects into physical, chemical, and biological effects and define each effect in the following manner:

1. Physical effects concern soil aggregation, erosion, drainage, tilth, aeration, water-holding capacity, bulk density, permeability, mechanical properties, etc.
2. Chemical effects relate to exchange capacity, buffering capacity, supply and availability of N, P, S, and micronutrients, adsorption of pesticides and other agricultural chemicals, etc.
3. Biological effects are involved with activities in the soil of bacteria, fungi, actinomycetes, earthworms, roots, root hairs, soil biomass, etc.[46]

From what has been presented in this chapter and from information available in the literature, it is quite clear that the formation of metal- and clay-OM (humic) complexes exerts all of the three principal effects on soil quality described above. More is known at this time about the physical and chemical than about the biological effects, but more information is needed about all three effects.

Table 8 Major Components Identified in Hexane, Chloroform, and n-Pentane Extracts of Fine Clay Fractions Separated from the Bainsville, Melfort, and Regina Soils[41]

Component	Bainsville clay	Melfort clay	Regina clay
n-Alkanes	C_{15}–C_{85}	C_{17}–C_{94}	C_{11}–C_{89}
n-Fatty acids	C_{14}–C_{32}	C_{14}–C_{34}	C_{13}–C_{28}
n-Alcohols	C_{22}, C_{32}–C_{34}	C_{22}	—
n-Diols	C_{16}, C_{24}	C_{16}, C_{24}	C_{16}, C_{24}
Sterols	C_{27}, C_{29}	C_{27}, C_{29}	C_{27}, C_{29}
n-Alkyl monoester	C_{38}–C_{67}	C_{37}–C_{62}	C_{39}–C_{62}
n-Alkyl diester	C_{62}–C_{90}	C_{56}–C_{91}	C_{66}–C_{86}
n-Alkyl triester	C_{83}	C_{86}, C_{89}, C_{92}	C_{99}

VII. CONCLUSIONS

1. The chemical structures of HAs and FAs are similar, but FAs have lower molecular weights and contain, per unit weight, more oxygen and CO_2H groups.
2. Low-molecular-weight organic acids are as active in mineral weathering as HAs and FAs, but their low concentrations in bulk soils limit their role in weathering. They may be more important in the rhizosphere.
3. HAs and FAs are powerful solvents for a number of minerals and are very active in the weathering cycle. Especially effective in this regard is FA, which is a water-soluble and efficient metal complexer.
4. The ability of HAs and FAs to form water-soluble and water-insoluble metal-organic complexes is directly related to their high contents of CO_2H and phenolic OH groups.
5. In surface layers of many agricultural soils, humic substances bonded to clay particles through complexation with Al or Fe play a major role in the formation of stable soil aggregates. Other possible organic-binding agents are polysaccharides.
6. Until recently, the role of waxes and other hydrophobic organics in binding soil particles has been largely overlooked. Analyses of these materials by pyrolysis-soft ionization mass spectrometry show that their major components are alkanes, fatty acids, alcohols, and n-alkyl mono-, di-, and triesters, all constituents of natural waxes. These hydrophobic materials resist biodegradation and chemical decomposition and therefore have long residence times in soils. The waxy materials appear to exert beneficial effects on the structural stability of soil aggregates.
7. From the data presented in this chapter, it is apparent that the formation of metal- and clay-OM (humic) complexes is a fundamental reaction in soil chemistry that has profound physical, chemical, and biological effects on soil quality.

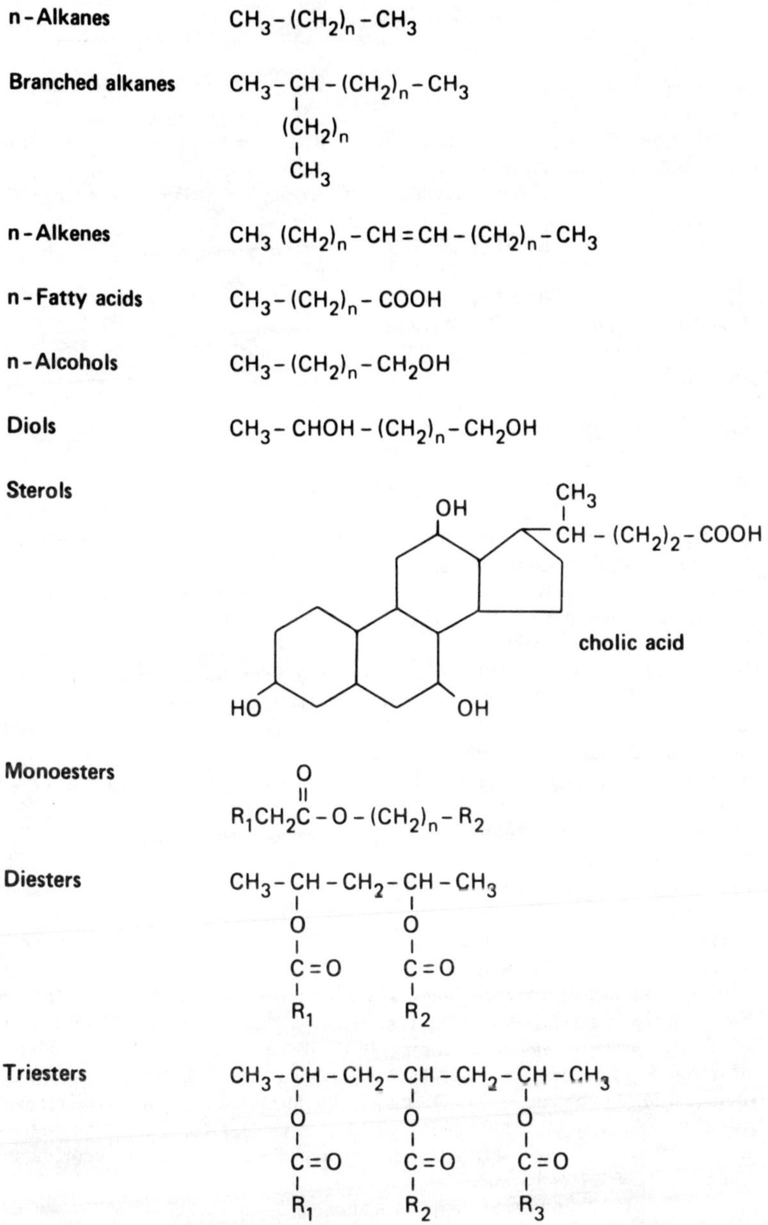

Figure 5 Major types of compounds identified in Py-FI mass spectra of organic extracts of fine clay fractions separated from several soils.

REFERENCES

1. **Schnitzer, M.,** Soil organic matter — the next 75 years, *Soil Sci.,* 151, 41, 1991.
2. **Schnitzer, M. and Kodama, H.,** Reactions of minerals with soil humic substances, in *Minerals and their Roles in the Soil Environment,* Dixon, J. B. and Weed, S. B. Eds., Soil Science Society of America, Madison, WI, 1977, 741.
3. **Schnitzer, M.,** Humic substances: chemistry and reactions, in *Soil Organic Matter,* Schnitzer, M., and Khan, S. U., Eds., Elsevier, Amsterdam, 1978, chap. 1.
4. **Schnitzer, M. and Khan, S. U.,** *Humic Substances in the Environment,* Marcel Dekker, New York, 1972.
5. **Breitmaier, E. and Voelter, W.,** *^{13}C NMR Spectroscopy,* Verlag Chemie, Weinheim, 1978, 247.
6. **Schnitzer, M. and Schulten, H.-R.,** The analysis of soil organic matter by pyrolysis-field ionization mass spectrometry, *Soil Sci. Soc. Am. J.,* 56, 1811, 1992.
7. **Harrar, N. J.,** Solvent effects of certain organic acids upon oxides of iron, *Econ. Geol.,* 24, 50, 1929.
8. **Murray, A. N. and Love, W. W.,** Action of organic acids on limestone, *Bull. Am. Assoc. Petrol. Geol.,* 13, 1467, 1929.
9. **Fetzer, W. G.,** Humic acids and true organic acids as solvents of minerals, *Econ. Geol.,* 26, 421, 1946.
10. **Schatz, A., Cheronis, N. D., Schatz, V., and Trecolawney, G. S.,** Chelation (sequestration) as a biological weathering factor in pedogenesis, *Penn. Acad. Sci.,* 28, 44, 1954.
11. **Schalscha, E. B., Appelt, H., and Schatz, A.,** Chelation as a weathering mechanism. I. Effect of complexing agents on the solubilization of iron from minerals and granodiorite, *Geochim. Cosmochim. Acta,* 31, 587, 1967.
12. **Huang, W. H. and Keller, W. D.,** Dissolution of rock-forming silicate minerals in organic acids: simulated first-stage weathering of fresh mineral surfaces, *Am. Mineral.,* 55, 2076, 1970.
13. **Huang, W. H. and Keller, W. D.,** Dissolution of clay minerals in dilute organic acids at room temperature, *Am. Mineral.,* 56, 1082, 1971.
14. **Sprengel, C.,** Über Pflanzen-humus, Humussäure und Humussäure Salze, *Kastner's Arch. Ges. Naturlehre,* 8, 145, 1826.
15. **Thenard, P.,** Observations sur le mémoire de M. Friedel, *Compt. Rend.,* 70, 1412, 1870.
16. **Julien, A. A.,** On the geological action of humus acids, *Proc. Am. Assoc. Adv. Sci.,* 28, 311, 1879.
17. **Baker, W. E.,** The role of humic acids from Tasmanian Podzol soils in mineral degradation and metal mobilization, *Geochim. Cosmochim. Acta,* 37, 269, 1973.
18. **Whitehead, D. C.,** Identification of p-hydroxybenzoic, vanillic, p-coumaric and ferulic acids in soils, *Nature (London),* 202, 417, 1963.
19. **Schnitzer, M.,** Binding of humic substances by soil colloids, in *Interactions of Soil Minerals with Natural Organics and Microbes,* Huang, P. M. and Schnitzer, M., Eds., SSSA Spec. Publ. No. 17, Soil Science Society of America, Madison, WI, 1986, chap. 4.
20. **Stevenson, F. J.,** *Humus Chemistry,* John Wiley & Sons, New York, 1982.
21. **Lévesque, M. and Schnitzer, M.,** Organo-metallic interactions in soils. VI. Preparation and properties of fulvic acid-metal-phosphates, *Soil Sci.,* 103, 183, 1966.
22. **Manning, P. G. and Ramamoorthy, S.,** Equilibrium studies of metal ion complexes of interest to natural waters. VII, *J. Inorg. Nucl. Chem.,* 35, 1577, 1973.
23. **Kerndorff, H. and Schnitzer, M.,** Sorption of metals on humic acid, *Geochim. Cosmochim. Acta,* 44, 1701, 1980.
24. **Kodama, H. and Schnitzer, M.,** Adsorption of fulvic acid by non-expanding clay minerals, in 10th Int. Congr. Soil Science, (Moscow), Trans. II, 1974, 51.
25. **Schnitzer, M. and Kodama, H.,** Montmorillonite: effect of pH on its adsorption of a soil humic compound, *Science,* 153, 70, 1966.
26. **Greenland, D. J.,** Interactions between humic and fulvic acids and clays, *Soil Sci.,* 111, 34, 1971.
27. **Wang, M. C. and Huang, P. M.,** Significance of Mn(IV) oxide in the abiotic ring cleavage of pyrogallol in natural environments, *Sci. Total Environ.,* 113, 147, 1992.
28. **Huang, P. M.,** Role of soil minerals in transformations of natural organics and xenobiotics in soil, in *Soil Biochemistry, Vol. 6,* Bollag, J.-M. and Stotzky, G., Eds., Marcel Dekker, New York, 1990, 29.

29. **Kodama, H. and Schnitzer, M.,** Dissolution of chlorite minerals by fulvic acid, *Can. J. Soil Sci.,* 53, 240, 1973.
30. **Schnitzer, M. and Kodama, H.,** The dissolution of micas by fulvic acid, *Geoderma,* 15, 381, 1976.
31. **Senesi, N., Griffith, S. M., Schnitzer, M., and Towsend, M. G.,** Binding of the Fe^{3+} by humic materials, *Geochim. Cosmochim. Acta,* 41, 969, 1977.
32. **Kodama, H., Schnitzer, M., and Murad, E.,** An investigation of iron (III)-fulvic acid complexes by Mössbauer spectroscopy and chemical methods, *Soil Sci. Soc. Am. J.,* 52, 994, 1988.
33. **Simonson, R. W.,** Outline of a generalized theory of soil genesis, *Soil Sci. Soc. Am. Proc.,* 23, 152, 1959.
34. **Swindale, L. D. and Jackson, M. L.,** Genetic processes in some residual Podzolized soils of New Zealand, *Trans. Int. Congr. Soil Sci.,* A, 233, 1956.
35. **Stobbe, P. C. and Wright, J. R.,** Modern concepts of the genesis of Podzols, *Soil Sci. Soc. Am. Proc.,* 23, 161, 1959.
36. **Hamblin, A. and Greenland, D. J.,** Effect of organic constituents and complexed metal ions on aggregate stability of some East Anglican soils, *J. Soil Sci.,* 28, 410, 1977.
37. **Tisdall, J. M. and Oades, J. M.,** Organic matter and water-stable aggregates in soils, *J. Soil Sci.,* 33, 141, 1982.
38. **Edwards, A. P. and Bremner, J. M.,** Microaggregates in soil, *J. Soil Sci.,* 18, 64, 1967.
39. **Schreiner, C. and Shorey, E. C.,** Some acid constituents of soil humus, *J. Am. Chem. Soc.,* 32, 1674, 1910.
40. **Giovannini, G., Lucchesi, S., and Corvelli, S.,** Water-repellent substances and aggregate stability in hydrophobic soil, *Soil Sci.,* 135, 110, 1983.
41. **Schulten, H.-R. and Schnitzer, M.,** Aliphatics in soil organic matter in fine-clay fractions, *Soil Sci. Soc. Am. J.,* 54, 98, 1990.
42. **Kvenvolden, K. A.,** Molecular distribution of normal fatty acids and paraffins in some lower cretaceous sediments, *Nature (London),* 20, 573, 1966.
43. **Hawthorne, S. B. and Miller, J. D.,** Analysis of commercial waxes using capillary supercritical fluid chromatography-mass spectrometry, *J. Chromatogr.,* 388, 397, 1987.
44. **Schulten, H.-R., Murray, K. E., and Simmleit, N.,** Natural waxes investigated by soft ionization mass spectrometry, *Z. Naturforsch. Teil C,* 42, 178, 1987.
45. **Jambu, P. E., Fustec, E., and Jacquesy, R.,** Les lipides des sols: nature, origine, évolution, propriétés, *Sci. Sol,* 4, 229, 1978.
46. **Schnitzer, M.,** Soil organic matter and soil quality, in *Soil Quality in the Canadian Context,* Mathur, S. P. and Wang, C., Eds., Technical Bulletin 1991-1E, Agriculture Canada, Ottawa, Ontario, 1991, 33.

Chapter 2

Sorption Phenomena Between Inorganic and Organic Compounds in Soils: Impacts on Transformation Processes

K. Haider

CONTENTS

I. INTRODUCTION

Soils are complex microbial habitats because they are dominated by a solid phase surrounded by aqueous and gaseous phases that fluctuate markedly with time and space. The solid phase is a three-dimensional system composed of minerals of various particle sizes intermingled with plant, animal, and microbial residues in various stages of decay and a living and metabolizing microbiota. Among the most fascinating phenomena in soil as a microbial habitat are the associations between organic materials and inorganic soil particles and amorphous metal oxides. These associations markedly influence microbial activities in decomposing plant, animal, or microbial residues and metabolites, as well as the transformation and degradation of anthropogenic chemicals that come into contact with soils.

Soil organic matter is intimately associated in organomineral complexes and particularly with the clay fraction. Greenland[1] tabulated results from various workers which showed that 50 to 100% of the soil organic matter was present as organomineral complexes. Turchenek and Oades[2] reported that in samples of red-brown earth or black chernozem soils, 37 to 48% of the organic carbon and 43 to 57% of the organic nitrogen occurred in the clay fractions. Anderson et al.,[3] using ultrasonic dispersion and size fractionation techniques, indicated that 96% of the organic C and 93% of the organic N and similar amounts of the carbohydrate fraction were associated with mineral soil particles ranging from sand to the fine clay fraction (Table 1).

0-87371-914-X/95/$0.00+$.50

Table 1 Distribution of Organic Carbon, Nitrogen, Carbohydrates, and C/N Ratios Among Particle Size Fractions in a Black Chernozemic Soil[3]

Fraction	Whole soil (2 mm)	Sand (0.05–2 mm) (37% w/w)	Coarse silt (5–50 μm) (35.3% w/w)	Fine silt (2–5 μm) (4.1% w/w)	Coarse clay (0.2–2 μm) (14.3% w/w)	Fine clay (0.2 μm) (8.3% w/w)	% Recovery
		Percentage Distribution Among Particle-Size Fractions					
Organic C (%)	3.30	3.9	25.0	16.4	37.6	17.1	96.3
Organic N (%)	0.34	2.3	21.2	14.7	39.1	22.6	93.2
Carbohydrate C (%)	0.32	2.6	19.3	14.1	36.6	27.5	95.5
C/N	9.8	17.0	11.9	11.2	9.7	7.6	—

Expressed on the basis of whole soil, only 3 to 4% of the organic matter C was associated with the sand fraction, substantially greater amounts with the coarse and fine silt fraction, and the greatest amounts with the coarse and fine clay fraction. Organic N and the carbohydrates followed a trend similar to organic C, but the C/N ratios decreased by a factor of two with a decrease in particle size, indicating that the material combined with the fine clay was richest in N. This fraction, therefore, provides a store of humus that acts as a source of nitrogen and appears to be of considerable importance in the medium-term cycling of nutrients. This view has been supported by Tiessen and Stewart,[4] who demonstrated that organic matter associated with the fine clay (<0.2 μm) of soil was more rapidly depleted during 60 years of continuous cultivation than organic materials associated with fine silt (5 to 2 μm) and coarse clay.

II. TURNOVER RATES AND STRUCTURE OF SOIL ORGANIC MATTER — EFFECTS OF SOIL MANAGEMENT

A. GENERAL VIEWS

The ^{14}C-dating technique provides evidence that many modern soils contain a pool of very stable soil organic carbon with a turnover time of up to 1000 years or more and a much younger pool not older than several decades.[5-8] Cultivation usually results in lower organic matter levels and involves changes in the association of soil organic matter with clays and in soil structure.[9] This decrease, caused by accelerated decomposition of the younger pool, results in an increased mean age of the remaining stable soil humus.[5] Using the ^{14}C-dating technique, Hsieh[10] estimated the pool sizes of the active and passive organic carbon in the surface soil layer of differently managed subplots of the Morrow plots at Illinois and the Sanborn fields at Missouri (Table 2).

At the Morrow plots, the passive pool comprised 36% of the total organic C under continuous blue grass, 85% under continuous corn, and from 55 to 65% under corn-oat-clover crop rotations. Similarly, at the Sanborn fields, a greater stable pool was observed under continuous wheat with no fertilizer than with fertilizer addition. Furthermore, Balesdent et al.,[11] using a ^{13}C biomarker technique, demonstrated that the surface soil of the Sanborn fields contained a pool of stable soil humus that remained virtually unchanged after decades of cultivation.

B. INFLUENCE OF MANAGEMENT ON SOIL ORGANIC MATTER STRUCTURE

Microbial biomass and its products are combined with the young and easily decomposable humus and determine the flow of carbon and nutrients.[12] Doran and Smith[13] and Doran[14] demonstrated that differences in soil N mineralization, after 25 years of various continuous management practices, were more closely related to the amounts of young organic matter than to the total N content in the soil. Management conditions influence mostly these easily decomposable fractions and, as is becoming more and more apparent, the amount and quality of the microbial biomass.[15] As a result, the humus composition and the availability of humus becomes altered.[16,17]

The effect of different management practices on levels of soil organic matter, humus structure, and microbial biomass can best be studied in croplands under long-term continuous management, where C and N contents, microbial biomass and activities, and turnover and humification rates of plant and organic residues are most likely to approach equilibrium.[18] Recently, we studied the influence of various long-term continuous managements at several sites installed by the Bavarian Provincial Agricultural Institute.[16,17] These sites include plots located at Puch, near Fürstenfeldbruck in southern Bavaria, which were established by Springer in 1949.[19] Here, several subplots are under extreme management, including bare fallow, continuous monocultures of potato or wheat, with or

Table 2 Relative Pool Sizes of Active and Stable Organic C, Calculated from ^{14}C Specific Activity in Surface Soil Samples (0–20 cm) from the Morrow Plots (University of Illinois) and the Sanborn Fields (University of Missouri)[10]

Crop rotation	Active pool Amount (g kg^{-1})	Active pool % of soil C	Stable pool Amount (g kg^{-1})	Stable pool % of soil C
	Morrow Plots: 1988			
Continuous corn, no fertilizer	1.9	15	11	85
Corn-oat-clover, no fertilizer	6.0	35	11	65
Corn-oat-clover, fertilizer	9.0	45	11	55
Continuous bluegrass	19.6	64	11	36
	Sanborn Fields: 1989			
Continuous wheat, fertilizer	4.6	39	7.3	61
Continuous wheat, no fertilizer	0.9	11	7.3	89

without additions of organic manure, and continuous grassland; other subplots are under crop rotations with diverse organic and inorganic fertilization. Table 3 shows that as a result of the extreme differences in management, soil C and N contents, as well as microbial biomass, differ considerably. Another site, located at Neuhof near Donauwörth and more related to current agricultural practices, was initiated in 1977[20,21] and has since been under continuous management. It is characterized by graded intensities of agricultural production, ranging from plots with intensive inorganic fertilization, comparable to modern industrialized agriculture without livestock, to more extensively inorganic fertilized and managed plots that are oriented to the demands of "integrated" agriculture, to plots with mixed organic/inorganic fertilization, and to plots with only organic fertilization according to the requirements of "organic" farming without pesticide application and appropriate crop rotations. In the intensively fertilized and integrated managed plots, crop residues are added to the soil with green manure from occasional intercroppings, whereas straw from the manure-fertilized plots is regularly removed and only stubble and the leaves from tuber crops remain in the field. Therefore, annual carbon inputs into the differently managed plots are similar and average about 3 mg/ha/year. This is also indicated by rather similar C and N contents in the differently managed Neuhof plots, with a slightly decreasing tendency with increasing intensity of management (Table 3).

From 1987 to 1990, soil cores (20-cm depth) were taken randomly from each plot in spring before fertilization and in fall after harvest. They were analyzed as bulk samples at field moist conditions for their contents of microbial biomass, the enzymatic activities of the biomass, and the turnover rates of plant residues and nitrogen metabolism.[22,23] The data indicated that the respective management conditions influence significantly the quantitative and qualitative activities of the microbial biomass. They also showed that the biomass in intensively managed plots, particularly in monocultures, used the carbon from the plant residues less efficiently and that the soil had a smaller capacity to supply nitrogen to crops.[22]

Impacts of management on the structure of soil humus of the Puch, Neuhof, and other conventional and organically managed plots were investigated by ^{13}C-NMR spectrometry

Table 3 Management Types of the Plots from Puch and Neuhof[16]

Abbreviation	Management	C_{org} (g kg^{-1})	N_{org} (g kg^{-1})	Biomass (mg C^a kg^{-1}) Spring 87	Spring 88
	Puch: Luvisol Derived from Loess Loam				
1 BF	Bare fallow	7.7	0.9	110	70
2 PM(–)	Potato monocultures, N-P-K	9.6	1.2	220	200
3 PM(+)	Potato monocultures, N-P-K, stable manure	10.4	1.3	220	220
4 CR_1	Barley, N-P-K	11.9	1.5	300	290
5 CR_2	Winter wheat, N-P-K, green manure	12.5	1.6	320	310
6 CR_3	Potato N-P-K + stable manure	13.3	1.5	350	340
7 GL	Continuous grassland	17.6	2.0	970	1130
	Soil C, 1956	13.5[16]			
	Neuhof: Gleysol Derived from Loess Loam				
8 CR/N_3 intense	Crop rotation 182 kg ha^{-1} N,P-K straw + crop residues	11.2	1.5	330	330
9 CR/N_2 integrated	Crop rotation 124 kg ha^{-1} N,P-K straw + green manure	11.9	1.5	370	330
10 CR/N_1 org. min.	Crop rotation 12 t ha^{-1} manure + 37 kg ha^{-1} N, P-K	12.8	1.4	600	420
11 CR/N_0 organic	Crop rotation 18 t ha^{-1} manure + P-K	12.9	1.5	680	510
	Soil C, 1979	12.5[20]			

and by pyrolysis-field ionization mass spectrometry.[16,17] Although the ^{13}C-NMR spectra from each of the differently managed plots looked rather similar, integration of the relative carbon contents in distinct spectral areas of both soil and the extracted humic acids revealed apparent influences of management.[16,17]

The humic acids from Puch samples with rather drastic differences in management conditions, and with sometimes very low input of organic C residues, showed a positive correlation of the relative aromatic carbon contents (range between 110 and 160 ppm) with the respective C content of the samples.[16] The long-term continuous monocultures and bare fallow managements probably have led to a considerable destruction in the aromatic structures of the humic matrix. The less-graded management system at Neuhof with nearly equal soil C contents showed a slight increase of the relative aromatic C contents with increasing management intensity[17] (Table 4). There was an inverse, but not as straight, correlation of the management intensities with the relative C contents in the aliphatic region (0 to 45 ppm), including that of the CO/CN region (60 to 80 ppm). The aromaticity calculated according to the procedure of Schnitzer and Preston[24] therefore

Table 4 Integration of the CP-MAS ^{13}C-NMR of Humic Substances from Puch and Neuhof Soils[16,17]

Sample[a]	Soil C_{org} (g kg^{-1})	Aliphatic region[b] (0–110 ppm)	Benzene and phenolic region[b] (110–140–160 ppm)	Aromaticity[c]
1 BF	7.5	73.4	14.4	16.4
2 PM(–)	9.6	71.2	16.4	18.7
3 PM(+)	10.4	70.7	18.3	20.6
4 CR_1	11.9	70.3	17.8	20.2
5 CR_2	12.5	70.5	20.5	22.5
6 CR_3	13.3	69.7	20.1	22.4
7 GL	17.6	68.8	21.1	23.4
8 CR/N_3	11.2	63.6	22.6	26.2
9 CR/N_2	11.8	64.8	21.7	25.1
10 CR/N_1	11.8	66.1	20.3	23.5
11 CR/N_0	12.7	67.0	20.8	23.7

[a] See Table 3 for explanation of abbreviations.
[b] Percent of total area.
[c] Aromaticity expresses aromatic C as a percent of aliphatic + aromatic C.

decreased with intensity of management for the samples from Puch, whereas it increased for those of the Neuhof system.

The decrease in aromaticity of the soil samples and extracted humic substances with decreasing soil C contents in the system at Puch indicates that exploitative soil management leads even to a degradation of the polyphenolic structure of humic compounds. This agrees with ^{13}C-NMR spectroscopic observations by Skjemstad et al.[25] on cultivation effects on several soils in Australia, which indicated a relative decrease in the phenolic region and a simultaneous increase in the aliphatic region of intensively managed soils. Furthermore, Schulten and Schnitzer[26] showed that humic acids extracted from clays were of a more aliphatic nature than those extracted from larger particle-size fractions or from the whole soil. They[26] proposed that because of their hydrophobic properties, these compounds resist biodegradation and have long residence times in soil. Furthermore, pyrolysis mass spectra of the soil samples and their humic compounds[16] revealed that mass signals related to phenols, carbohydrates, or amino acid compounds were less pronounced in the intensively than in the extensively managed plots, even if the C contents in the respective plots did not differ significantly in C contents but differed in the intensity of the long-term management. Pyrolysis mass spectrometry also revealed differences in N-containing fragments from soils under different management. The intensities of fragments from amino acids or proteins were generally positively correlated with increasing N contents of the soil samples. In contrast, however, fragments from heterocyclic N compounds were negatively correlated with soil N contents.[23] The relative contributions of pyridine and pyrrole to the total ion intensity (TII) increased with decreasing N contents of the soil samples (Figure 1) and showed a significant negative correlation of $r = -0.92$.

III. EFFECT OF CLAY CONTENT ON TURNOVER RATES OF HUMUS AND ORGANIC COMPOUNDS

Many aspects of this topic have already been discussed by Martin and Haider;[27] therefore, only a few additional aspects will be reported. Several authors have demonstrated that the clay contents in soils have a great effect on the turnover rates of C and on the amounts of C stabilized in soil.[6,28] In a survey of 65 pedons of the Mollisols in the Southern Great

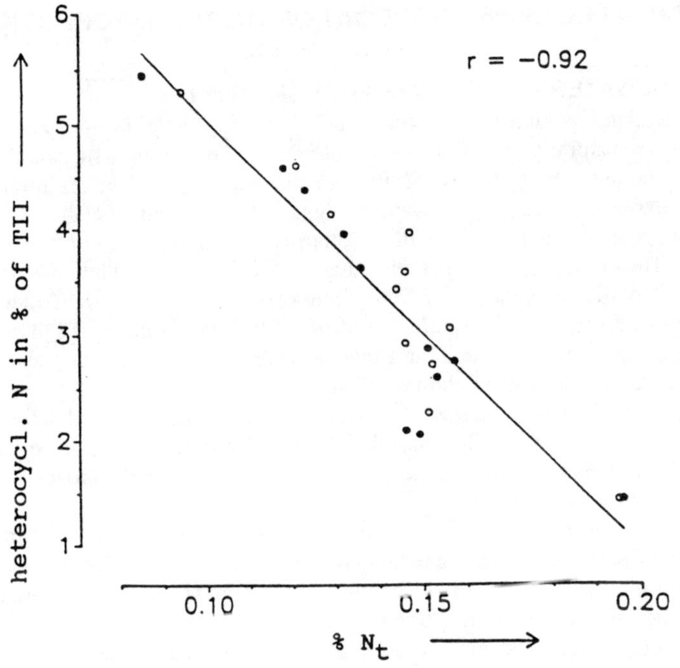

Figure 1 Correlation between total soil N and the relative contribution of heterocyclic nitrogen compounds (pyrrole and pyridine). Pyrolysis data from the Puch and Neuhof sites (○ = 1987, ● = 1988).[23] TII, total ion intensity; N_t, total soil N.

Plains, Nichols[29] showed that organic matter levels were better correlated with clay contents than with precipitation.

The most apparent effects on C turnover and humus formation have been observed in allophanic soils, where humic substances are stabilized by complexation with active Al groups or hydrous oxides of Al and Fe associated with the allophane surfaces. Although allophanic soils are high in organic C, the rate of C loss as CO_2 is slower than from nonallophanic soils. C losses from a wide variety of ^{14}C-labeled substrates in normal agricultural soils of California were compared with those in allophanic soils of southern Chile.[27] Over a 16-week incubation, the presence of allophane in the soil or the addition of 50 to 160 g kg^{-1} of allophane to a sandy soil reduced C losses by 16 to more than 70%.

In contrast to what is known about the effects on turnover rates of sorptive interactions of organic matter with the inorganic matrix, knowledge about direct surface interactions between microbes and clays or other soil components and their effects upon the biological activity in soils *in situ* is relatively less. Stotzky[30] demonstrated that certain clays stimulated growth, respiration, and substrate uptake by bacteria and that the stimulation of fungi was less. Laboratory experiments[31,32] indicated that the addition of montmorillonite to several fungi grown in nutrient solutions increased glucose uptake and the formation of microbial metabolites, in particular the synthesis of humic-type melanins. Anderson and Gray,[33] however, recently indicated that the clay contents of soils were not well correlated with the specific uptake rate of glucose per unit of biomass but were significantly correlated with their metabolic quotients. This means that the loss of CO_2 per biomass unit in a given time period was reduced and the biomass was better stabilized. Ladd et al.[34] noted, in field studies, that soils with higher clay contents generally retained higher proportions of residual ^{14}C and ^{15}N in their microbial biomass.

IV. EFFECTS OF AGGREGATION ON TRANSFORMATION PROCESSES

A. AGGREGATES AND THE FLOW OF NUTRIENTS

Primary particles combine into aggregates of varying sizes in soils. By definition, aggregates are naturally occurring clusters or groups of organomineral particles in which the forces holding the particles together are much stronger than the forces between adjacent aggregates.[35] Tisdall and Oades[36] suggested a hierarchy of aggregates in soils in that microaggregates with sizes of 10 to 250 μm are combined into macroaggregates of >250 μm. They also suggested that microaggregates are combined into macroaggregates by transient, easily decomposable microbial- or plant-derived polysaccharides, by fungal hyphae or fine roots, and by the participation of lipids. The binding agents of particles into microaggregates are more stable and include persistent organic materials, crystalline oxides, and highly disordered aluminosilicates.

Soil aggregation also is an important factor in controlling microbial activity, organic matter turnover, and nutrient flow in soils.[37] Elliot[38] demonstrated that the organic matter associated with macroaggregates is more labile and less highly processed than that associated with microaggregates.

Based on results of mineralization studies, both C and N mineralization in cultivated and native sod soils were always greater in macro- than in microaggregates and increased further in crushed macroaggregates. In cultivated soil, however, total soil organic matter mineralization was less than in native soil (Table 5).

A reduction in humus levels as a result of tillage causes a decrease in macro- and an increase in microaggregates. This reduction in macroaggregation may provide a short-term increase in soil fertility through the release of nutrients from the labile organic matter that binds micro- into macroaggregates. Prolonged cultivation will, therefore, result in a rapid decline in the number of macroaggregates and in labile organic matter.[39]

The importance of polysaccharides, residues of humified plant materials, as well as fine roots and hyphae in stabilizing larger aggregates has been emphasized. More and more, however, several authors[40-43] have become aware of the importance of polar and nonpolar lipids in the improvement of aggregate stability. These lipids can be extracted with benzene, alcohol, or ether, and their addition to soil positively affects the stability of aggregates. Capriel et al.[44] found that in the differently managed Puch and Neuhof plots (Table 3), the amounts of an aliphatic hydrophobic fraction extracted by supercritical n-hexane were well correlated with the soil microbial biomass and the aggregate stability. As indicated in Table 6, the amounts of the n-hexane extracts increased with increases in biomass in both plots. The aggregate stability, determined as the mean weighted diameter (MWD) of 5- to 6-mm aggregates sieved under water, also increased in the same order.

The ^{13}C-NMR spectra of the n-hexane extracts were similar, and signals appeared mostly in the 0- to 50-ppm region and were dominated by a large signal at 29.7 ppm, which is characteristic of CH_2 in polymethylene chains.

A decrease in aggregate stability is generally accompanied by a decrease in the pools of easily decomposable soil organic matter and microbial biomass.[45] Angers et al.,[9] however, reported recently that even when neither total organic C nor biomass C was affected in different cropping systems, soil aggregation was. This adds to the findings of Capriel et al.[44] and others that specific products of microbial biomass, which are rapidly formed or degraded, are of additional importance in contributing to the stability of aggregates.

B. INFLUENCE OF AGGREGATES ON BIOCHEMICAL PROCESSES

Aggregation can cause many microenvironments to rapidly change with time and space in water contents, aeration, pH, or other parameters. The slow movement of oxygen in water-filled pore systems can lead to anaerobiosis in the inner portion of moist aggre-

Table 5 C and N Mineralized from Isolated Macroaggregates and Microaggregates, and Macroaggregates Crushed to Microaggregate Size in Incubations Using Cultivated and Native Sod Soils[38]

Mineralization (d)	C mineralization[a] 0–3	C mineralization[a] 0–20	N mineralization[a] 0–20
Cultivated Soil			
Macroaggregates (0.3–>2 mm)	160	414	27
Crushed macroaggregates	181	493	38
Microaggregates (0.05–0.3 mm)	98	309	16
Native Sod			
Macroaggregates (0.3–>2 mm)	427	1085	68
Crushed macroaggregates	500	1129	94
Microaggregates (0.05–0.3 mm)	255	672	43

[a] In milligrams per kilogram.

gates[46] and thereby modify biological or chemical transformations of chemicals such as nitrate, sulfate, metal derivatives, or organic compounds, including xenobiotics.[47]

1. Aggregates and Denitrification

The anaerobic microsites within aggregates are also the reason for the often observed coincidental occurrence of apparently conflicting processes such as nitrification and

Table 6 Relationships Between Soil Aliphatic Fraction Extracted with Supercritical Hexane, Soil Microbial Biomass, and Soil Aggregate Stability in Soil Samples from Puch and Neuhof[44]

Abbreviation[a]	C_{org} (g kg^{-1})	Biomass (mg C kg^{-1} soil)	Supercritical-hexane extraction (mg C kg^{-1} soil)	Aggregate stability (WMD)[b]
Puch: Luvisol Derived from Loess Loam				
1 BF	7.7	110	140	0.5
2 PM(–)	9.6	220	350	1.0
3 PM(+)	10.4	220	400	1.2
5 CR_2	12.5	320	440	2.8
6 CR_3	13.3	350	670	3.8
7 GL	17.6	970	860	4.3
Neuhof: Gleysol Derived from Loess Loam				
8 CR/N_3	11.2	330	440	2.4
11 CR/N_0	12.9	680	740	4.0

[a] See Table 3 for explanation of abbreviations.
[b] WMD, weighted mean diameter of 5 to 6 mm aggregates after sieving under water through a set of sieves.

denitrification. They are also responsible for the widespread spatial variability of denitrification events in field soils.[48]

Direct measurements of O_2 concentrations in soil aggregates in relation to denitrification rates by Sexstone et al.[49] and Zausig et al.[50] have corroborated the concept of the occurrence of so-called "hot spots" for enhanced denitrification in soils, and mathematical models have been developed that incorporate this idea.[46,51] Measurements using an oxygen microelectrode[49,50] demonstrated that aggregates from cultivated or prairie soils contain expanded anaerobic centers (Figure 2).

The size of aggregates, and even their anaerobic radius, is not always correlated with the observed denitrification rates. Some aggregates with anaerobic centers exhibited no or only little denitrification. As denitrification needs not only nitrate but also easily available carbon sources, the concentrations of either or both may also limit denitrification rates. In general, however, substrate C availability increased with increasing wet-sieved aggregate sizes and was associated with greater rates of denitrification.[52]

2. Influence of Aggregation on Pesticide Degradation

At a pH close to neutrality, denitrification occurs at redox potentials (E_h) of approximately +400 to +300 mV.[47] As the E_h decreases and nitrate is depleted, sulfate reducers become active around –200 mV. After sulfate reduction ceases, methanogenesis occurs at E_h below –200 mV. Those low and long-lasting redox potentials will probably not be reached in the centers of water-saturated aggregates in well-drained field soils but can occur under water-logged conditions.

Aerobic and anaerobic microsites in soils also influence the degradation rates of several xenobiotics, especially those of pesticides.[47,53] Generally, appropriate levels of organic nutrients together with aerobic conditions, which favor microbial growth and activity, increase the rate of pesticide degradation. Grbic-Galic[47] emphasized the enhanced transformation of several aromatic, heterocyclic, and alicyclic compounds, including their halogenated derivatives, under anaerobic conditions. This is particularly true for several pesticides with halogen or nitro groups, which can serve as exogenous electron acceptors for microorganisms under anaerobic conditions. Examples of compounds that are more readily dechlorinated under anaerobic than under aerobic conditions or with nitro groups prone to reduction are shown in Figure 3.

Generally, under anaerobic conditions, lindane (I, γ-isomer of hexachlorocyclohexane) is readily converted into γ-tetrachlorocyclohexene (II) and further into chlorinated benzenes and benzene. Guenzi and Beard[54] and others observed an enhanced conversion of DDT into DDD (IV) in wet and well-manured agricultural soils as compared to unmanured or dry soils.

To determine whether redox potentials, as they can be reached in water-saturated aggregates, are low enough to cause dehalogenation of hexachlorocyclohexane (HCH) isomers, ^{36}Cl-labeled α- or γ-HCH was infused into the centers of moist aggregates (2.5-cm diameter).[93] Oxygen partial pressure in the aggregates, as well as redox potentials, were automatically registered with microelectrodes.[50]

The oxygen partial pressure dropped, within 18 h at a distance of 2 to 5 mm from the surface, to zero, and the redox potential was 200 to 300 mV. Dechlorination, observed by the release of $^{36}Cl^-$, became obvious within 6 d and ranged from 2 to 3% for the α- and 10 to 12% for the γ-isomer of HCH. This was less than the dechlorination observed during anaerobic incubation with a soil slurry or by specific cultures of strict or facultative bacteria,[55] but it was similar to that in submerged or water-saturated soil samples.[56]

Aromatic nitro groups, as in parathion (V) or several dinitroaniline herbicides (VIII-X), are reduced in organic nutrient-rich soils, especially when they are wet. Parathion was found to be most resistant in dry and least resistant in soils with high water contents.[57,58]

Aggregate number[a]	Denitrification rate [ng N g^{-1} h^{-1}]	Aggregate radius [mm]	Anaerobic radius [mm]
CA1	13.0	8.0	1.3
CA2	9.2	12.0	5.5
CA3	2.3	13.0	5.4
CA4	5.3	7.0	3.3
PA1	0.03	9.0	0.3
PA2	1.7	18.0	14.0

[a] Obtained from cultivated silt loam (CA) or an uncultivated native prairie (PA).

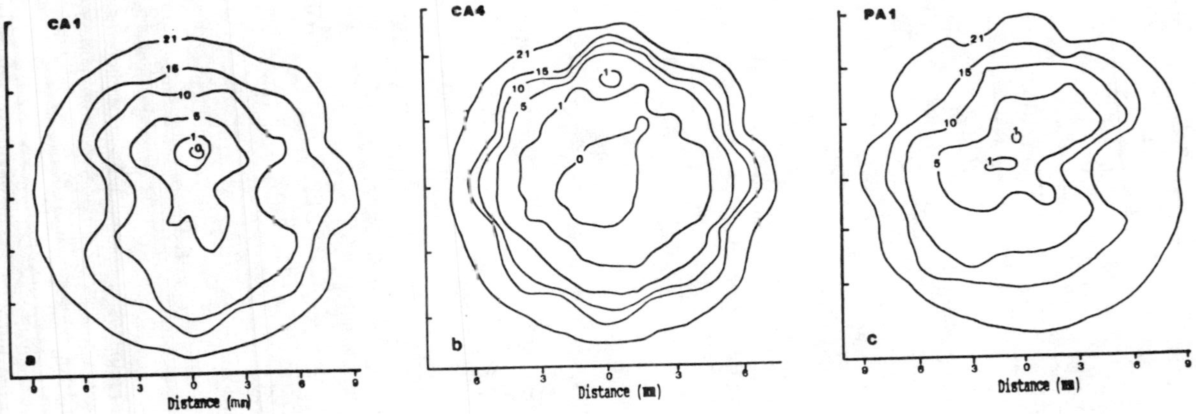

Figure 2 Comparison of denitrification rates, aggregate radius, measured anaerobic radius, and maps of O_2 concentration within aggregates.[49] Figures stand for volume percent of oxygen.

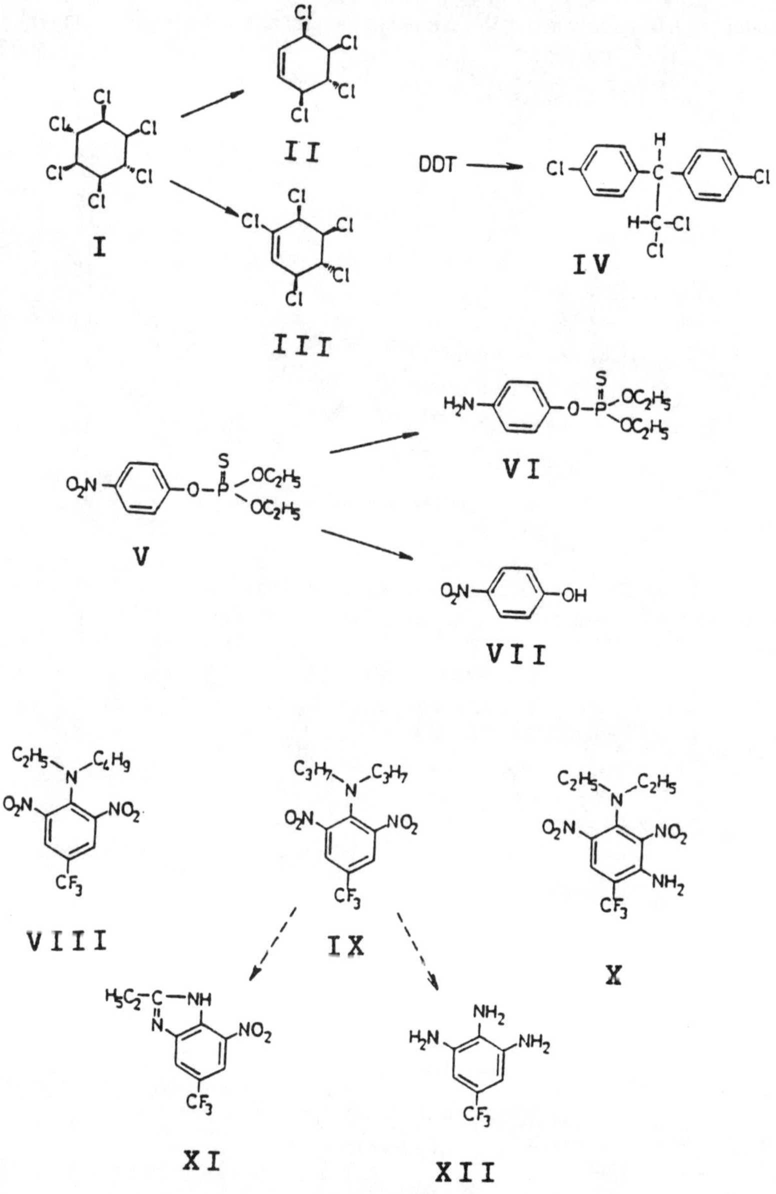

Figure 3 Examples of chlorinated hydrocarbon insecticides or pesticides with nitro groups more easily metabolized under anaerobic than under aerobic conditions.[47,53] Roman numerals are explained in the text.

Similarly, the nitro groups of the herbicide trifluralin (IX) are reduced faster under anaerobic than aerobic conditions. The redox potentials in water-saturated aggregates are apparently low enough to promote these reductions. After reduction of the nitro groups to amino groups, substantial portions of the residues from parathion and other nitro group-containing pesticides are incorporated in soil humus. This formation of bound residues is also promoted under wet soil conditions,[59] indicating that the anaerobic microsites of aggregates are particularly active in such reduction processes and promote the formation of bound residues from pesticides by this manner.

V. INTERACTIONS BETWEEN THE ORGANOMINERAL MATRIX OF SOILS AND LOW-MOLECULAR-WEIGHT ORGANIC COMPOUNDS

The adsorptive reactivity of soil organic and inorganic surfaces with organic molecules depends on the number and types of functional groups at these surfaces and on the chemical characteristics of the organic molecules. The adsorptive interactions are not only important processes in humus formation from naturally occurring precursors, but they are also active in the fixation and detoxification of pesticides and their residues and of other organic xenobiotics. Several recent reports treat these phenomena in detail,[60,61] and they are not addressed here completely. It should be emphasized, however, that sorptive interactions of natural or anthropogenic chemicals in soil occur in a similar manner.

A. INTERACTIONS WITH NATURALLY OCCURRING COMPOUNDS

Phenols and phenolic acids have been identified in extracts from various soil systems and are derived from decomposing plant residues, microbial biosynthesis, or root exudations.[62,63] They have been implicated in various soil processes, including biotic and abiotic humus formation,[64] dissolution of minerals and nutrients,[65] and as possible phytotoxic and allelopathic chemicals in field situations and hydroponic cultures.[66,67] Indigenous soil microbial populations have no difficulty in degrading or transforming exogenously applied phenols or phenolic acids, including the splitting of aromatic rings and their complete assimilation into microbial biomass.[68,69] Microbial or physicochemical reactions may result in their adsorption on the bulk soil material and reduce thereby their concentrations in the free state. Low concentrations of ferulic and protocatechuic acids, and of catechol and its various derivatives, were more readily adsorbed on soil humus and were less rapidly mineralized than higher concentrations. This is exemplified for ^{14}C-labeled ferulic acid added at various concentrations to soils with a pH of 5.6 and 8 (Figure 4).

In both soils, 1 ppm of ferulic acid was less readily decomposed than higher concentrations. Furthermore, the decomposition was slower in the alkaline soil than in the acid soil. The carbon not released as $^{14}CO_2$ was either located in the microbial biomass or became incorporated into humic fractions, and this incorporation was higher in the alkaline than in the acid soil.[70,71] Martin and Haider[27] demonstrated that the pH affected decomposition and incorporation more than the clay content. Lehmann et al.[72] concluded that higher contents of Fe^{3+} and Mn^{4+} compounds diminished degradation rates and increased incorporation into the humus matrix. In both cases, however, abiotic together with biotic reactions led to a rapid disappearance of certain phenolic compounds and to their incorporation into humus. The phytotoxic or allelopathic effects of these and similar compounds, therefore, may not persist for long periods under field conditions, since they are either degraded by microbes or their free toxic form becomes inactivated by sorptive interactions with soil particles. Phenolic compounds, however, may exert toxic influences on seedlings if they are continuously produced in the vicinity of decaying plant residues, as around stubbles, particularly during wet conditions, e.g., in the spring season.[73] Mineralization of lignin and lignin-derived compounds is also retarded by such conditions wherein soil aeration is reduced.[74]

B. INTERACTIONS WITH PESTICIDES AND OTHER XENOBIOTICS

The fate of xenobiotics and pesticides with structural relationships to aromatics is, in many respects, similar to that of naturally occurring phenols. Their bioactivity in soil, e.g.,

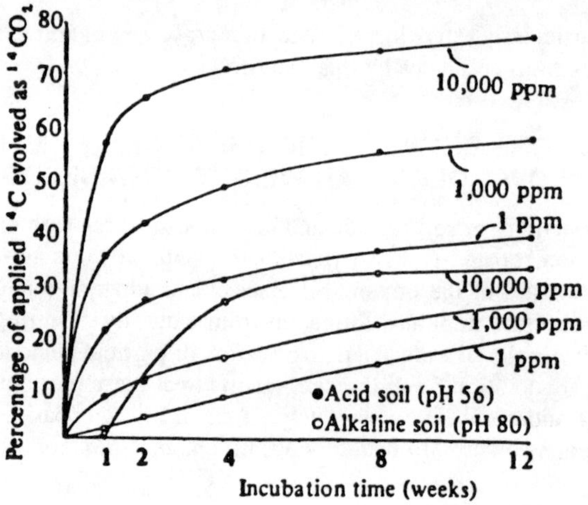

Figure 4 Effect of concentration and pH on decomposition of ^{14}C ring-labeled ferulic acid in soil.[27]

herbicides, can be observed for only a short time, after which it decreases, either by microbial degradation or by sorption on soil colloids or even by "binding" into soil humus fractions (Figure 5).[75]

In many cases, these latter processes are even more effective in detoxification than biodegradation and result in the so-called "bound" or "unextractable" residues. Application of the xenobiotics in a radiolabeled form revealed that their bioavailability was

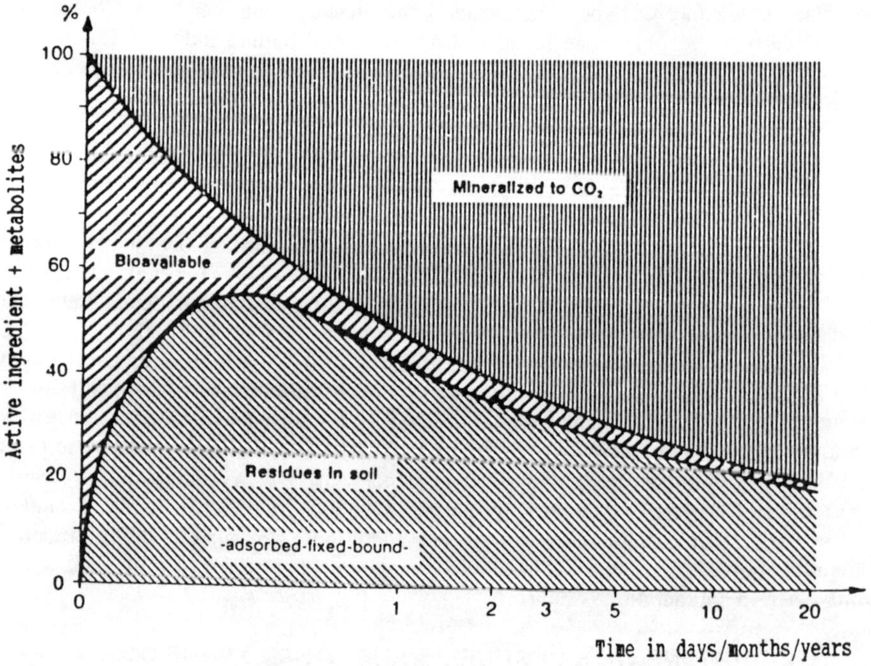

Figure 5 The general fate of pesticides and other xenobiotics in soil.[75]

strongly reduced, although they were still present in soil, primarily combined with the humic matrix.

The amount of nonextractable residues from many ^{14}C-labeled pesticides, their metabolites, or other organic xenobiotics can reach 40 to 60% within days or weeks of soil incubation.[76] These reactions lead to a significant stabilization of the compound against further microbial decay, and it may last many years until it becomes completely mineralized (Figure 5).

The radioactivity is then frequently combined with soil organic matter fractions, such as humic or fulvic acids, or often remains primarily in the humin fraction. This tendency was shown for chlorinated phenols, aromatic amines and their derivatives, carbamates, triazines, and several organophosphates.[77-79] Table 7 shows a selection of ^{14}C-labeled chlorinated phenols, anilines, and catechol derivatives that are known to form bound residues in soil.[80]

During an 8-week incubation period in a Podzolic soil, 20 to 60% of the applied ^{14}C was mineralized to CO_2; thereafter, 8 to 60% could be removed by exhaustive solvent extraction, and the unextractable activity remaining in soil after incubation ranged between 20 and 60% and was highest for 2-chlorophenol and 4-chlorocatechol. Several spontaneous or enzyme-catalyzed coupling reactions between phenols or aromatic amines and humic compounds have been proposed, and plausible models try to emphasize their covalent binding with humic compounds.[77,78,81] The ongoing discussion,[79,82,83] however, indicates that organic xenobiotics can also be sorbed or bound to humic compounds by various types of forces, such as van der Waals forces, hydrogen bonds, charge transfer complexes, and covalent binding. Hydrophobic sorption of nonpolar residues or trapping of molecules in a molecular sieve by humic materials has also been suggested.[84,85] Because the bound residues, however, can be only extracted together with humic fractions or remain in the humin, their status of sorption or binding is still difficult to define.

Recently, attempts have been made, using various labeling techniques in combination with ^{13}C-NMR, to obtain a better characterization of residues remaining in soil after application of the fungicide Dyrene^R, which has anilazine as an active ingredient (Figure 6).

When ^{14}C-labeled anilazine is applied, only 20% or less of the applied radioactivity can be extracted with organic solvents within a few days after contact with the soil. Even during prolonged incubation, only a small portion of the applied radioactivity is evolved as $^{14}CO_2$, and this is observed for the carbons of both the triazine and the aniline moieties. Because small amounts of the mono- and dihydroxy-derivative of anilazine could be detected during incubation, it was hypothesized that strong adsorption occurs after biotic or abiotic removal of one or two chlorine atoms from the triazine ring to form a highly reactive intermediate that reacts with the soil organic matter.[86]

Anilazine with >90% ^{13}C enrichment of the carbons in the triazine ring together with a small amount of the ^{14}C-labeled sample were applied to soil, and the interactions between anilazine residues and humic compounds were investigated by measuring the ^{14}C label and by ^{13}C-NMR spectroscopy.[87] Furthermore, attempts have been made to reduce the ^{13}C signals from the humic matrix by incubating the ^{13}C-labeled anilazine in an artificial soil obtained by humification of ripe corn-stalk material from plants previously grown in a phytotron in which the atmosphere was continuously supplied with $^{12}CO_2$. The milled corn stalks were mixed with quartz-sand and humified aerobically after inoculation with an aqueous compost extract for 20 weeks. Thereafter, the lyophilized samples were mixed with small amounts of the labeled anilazine, remoistened, and incubated.[87]

The ^{13}C-NMR spectra of anilazine, its dihydroxy derivative and dimethylether, as well as the *N*-trimethyl-dihydroxy derivative at natural ^{13}C abundance are shown in Figure 7.

The three carbons of the triazine ring from anilazine have absorptions at $\delta - 164.1$ (C_1), 171.6 (C_2), and 170.6 (C_3), respectively. The chemical shift values of the dihydroxy

Table 7 Decomposition of ^{14}C-Labeled Chlorinated Phenols and Anilines During Aerobic Soil Incubation and Formation of Nonextractable, Bound Residues[80]

	% of ^{14}C applied		
^{14}C compounds ($100\ mg\ kg^{-1}$ soil)	% $^{14}CO_2$ 8 weeks	Solvent extractable	Soil-bound residue
2-Cl-phenol	22	20	58
4-Cl-phenol	44	10	46
2,4-Di-Cl-phenol	56	13	31
4-Cl-catechol	31	8	61
2-Cl-aniline	46	13	41
4-Cl-aniline	13	63	24

derivative at $\delta = 155.5$, 150.5, and 153.7 for C_2, C_3, and C_1, respectively, indicate that this compound is predominantly present in the keto form. C_2 and C_3 of the triazine ring from the dimethoxy derivative coincide to one signal at $\delta = 172.3$ and C_1 with a signal at $\delta = 167.3$. Compared to anilazine, the absorptions of this compound are shifted slightly downfield. The *N*-trimethyl-dihydroxy derivative, another model for sorptive interactions with the humic matrix, is only present in the keto form, as indicated by chemical shift values at $\delta = 150.2$ for C_2 and C_3 and $\delta = 137.8$ for C_1. Other possible metabolites, such as cyanuric acid and 2-chlorophenyl-urea, have shifts for the C=O carbons at $\delta = 149.9$ and 155.2, respectively.[87]

The relatively weak signals from the triazine ring carbons at natural ^{13}C abundance become largely intensified if they are enriched with ^{13}C. The corresponding ^{13}C-NMR spectra for anilazine and its dimethoxy derivative are shown in Figure 8.

Samples from the Podzolic soil or from the sand/humified corn stalk mixture, amended with small amounts of the $^{13/14}C$-anilazine, were incubated for 2 to 6 weeks. After subsequent exhaustive organic solvent extraction of the incubated soils, only about 20%

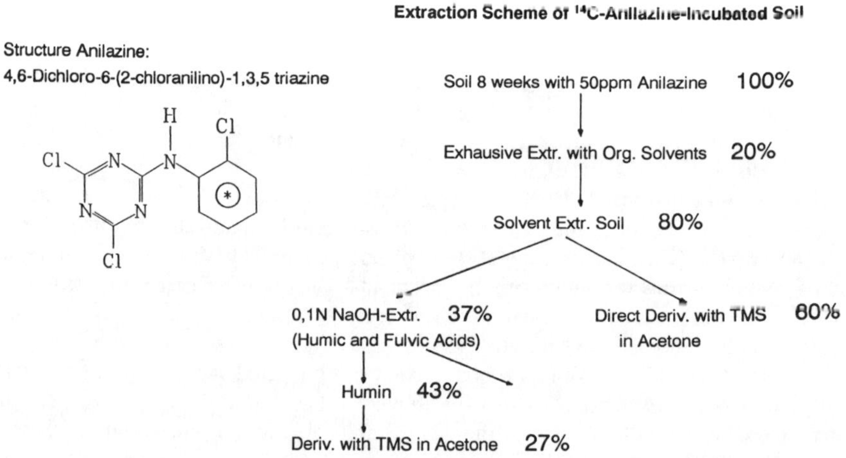

Figure 6 Structure of anilazine, the active ingredient of the fungicide Dyrene^R, and its turnover in soil.[80,87]

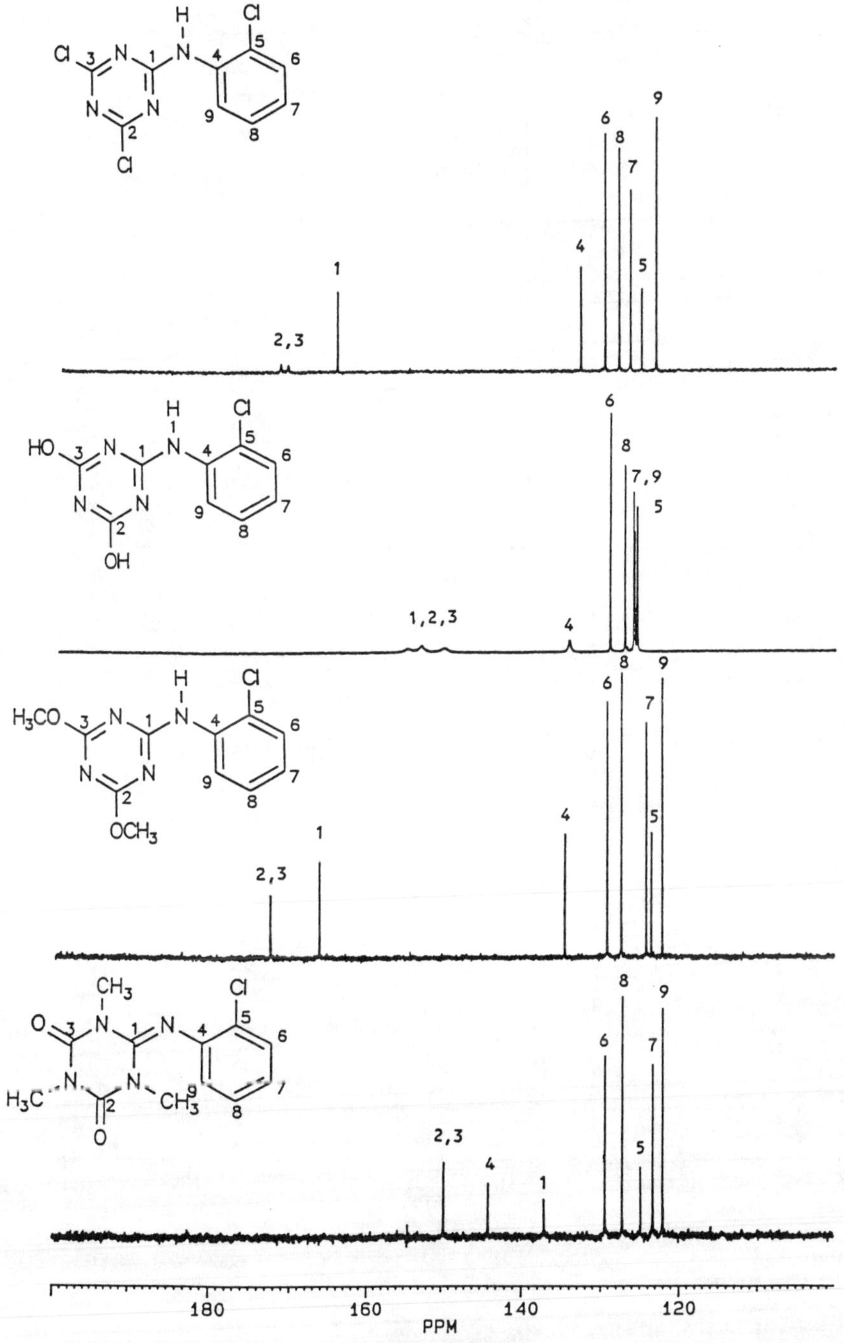

Figure 7 ^{13}C-NMR spectra of anilazine, its dihydroxy derivative, and two other derivatives used as models for the analysis of sorptive interactions. Compounds at natural ^{13}C abundance.[87]

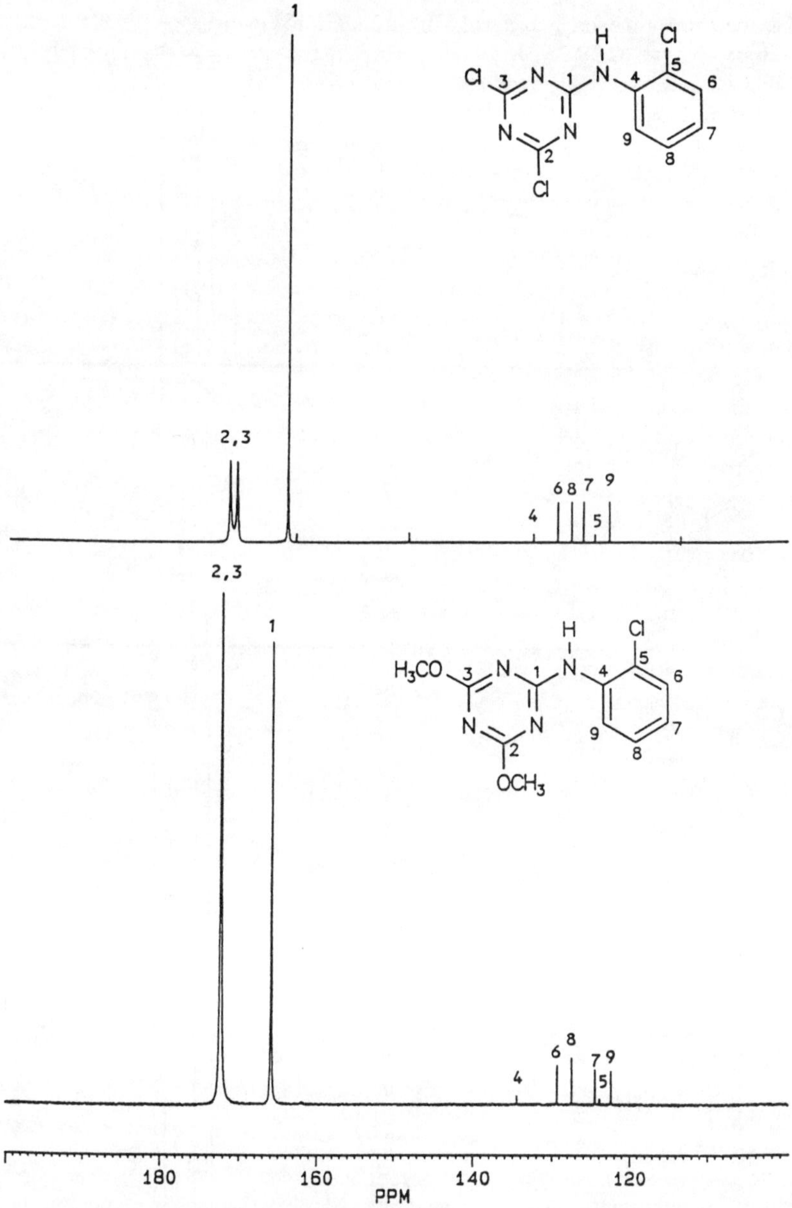

Figure 8 ^{13}C-NMR spectra from anilazine and its dimethoxy derivative with ^{13}C enrichment in the carbons of the triazine ring.[87]

of the remaining ^{14}C activity could be removed, and 70 to 80% remained as nonextractable residues in the soil. Humic acids extracted with NaOH and purified by conventional methods contained 12 to 16% of the initially applied radioactivity, which corresponded to less than 1 mg of anilazine residues in 100 mg of the respective humic acid sample. Their ^{13}C-NMR spectra were measured in deuterated NaOH under inverse-gated decoupling conditions and are shown in Figure 9.

The spectrum of the soil humic acid without anilazine (Figure 9a) has the usual relatively broad signals in the ^{13}C-NMR spectral ranges of humic absorptions. The spectrum from the ^{13}C-anilazine-treated soil (Figure 9b) shows two additional signals which culminate at 172.6 (I) and 166.5 (II) ppm, respectively. The two absorptions have shifts similar to the dimethoxy derivative of anilazine, but they do not correspond with those from anilazine or its dihydroxy or *N*-trimethyldihydroxy derivatives (Figures 7 and 8).

The spectra of humic acids from the humified ^{13}C-depleted corn stalk material, with and without ^{13}C-anilazine amendment, are shown in Figures 9c and d. As a result of the low ^{13}C content of the humic acids, the absorptions are not typical of soil humic acids. The corresponding spectrum of the humic acids that contain anilazine metabolites has two strong peaks at 172.9 and 166.9 ppm, which agree well with the additional peaks of the humic acids from the anilazine-treated soil (Figure 9b) and resemble those of the dimethoxy derivative from anilazine. The results suggest that binding of anilazine metabolites occurs in the form of its dialkoxy derivative by exchange of the two chlorine atoms with functional hydroxy groups in the humic compounds. It was further shown that this binding occurred in the form of ethers and possibly of esters by various hydroxy groups in the humic matrix. Several of the linkages do not seem to be very strong, as a partial release of the free dihydroxy derivative of anilazine could also be observed.

After derivatization of the humic compounds with chlorotrimethylsilane, 60 to 80% of the ^{14}C-labeled bound compounds was dissolved in organic solvents together with the humic compounds.[80,87] High-performance liquid chromatography or thin-layer chromatography revealed that neither the ^{14}C-labeled anilazine metabolites nor any of the aromatic compounds shown in Table 7 could be separated from the coextracted humic compounds. However, covalent bonds, as observed for the compounds just mentioned, may not generally occur with any applied xenobiotic that shows sorptive interactions in soils. A better understanding of the mechanisms involved in such interactions is of great importance, particularly considering the fact that soils are sometimes highly contaminated with organic xenobiotics from various origins and that the behavior and fate of such compounds sorbed on soil is still a "gray" area of research.

VI. SUMMARY AND CONCLUSIONS

This chapter surveyed the considerable impact of sorption phenomena in soils on transformation processes in soil. The organomineral complexes in soil are essential in stabilizing soil organic matter, and seem to be even more important for stabilization than the complex structure of humic compounds. Once these associations are disturbed by changes in soil management practices, the decay rates increase, primarily because of the more easily decomposable soil organic matter fractions that are combined with microbial biomass, microbial products, or partly decomposed plant residues. During such processes, e.g., the numbers of macroaggregates are reduced, whereas those of microaggregates increase. The important role of aggregates and sorptive processes in regulating the flow of carbon and nutrients in soil is becoming more apparent.

Sorptive interactions also influence the location of microbes in soil by connecting them to solid surfaces. According to Hattori,[88] only 10 to 20% of the microbes can be isolated from the soil solution, whereas 80 to 90% are located on solid surfaces, particularly on the surfaces of aggregates.[89] This spatial fixation of microbes not only regulates the access of microbes to certain nutrients, but it also protects microbes from rapid decreases in their biomass. As a result of sorptive interactions, microbial C-loss quotients are significantly negatively correlated with clay and humus contents.[90,91]

Aggregation also influences aerobic and anaerobic microbial processes. The centers of moist aggregates can have anaerobic zones in an otherwise aerobic soil, thus leading to the

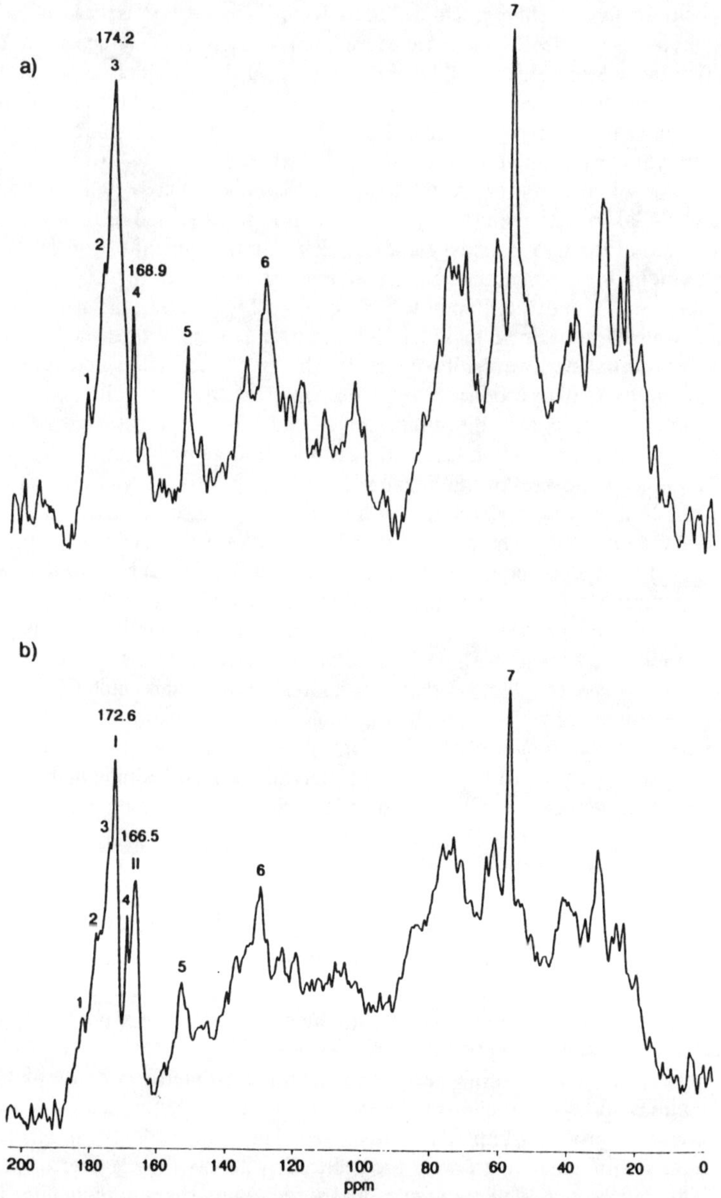

Figure 9 ^{13}C-NMR spectra of purified humic acids in deuterated NaOH from (a) Ap horizon of a Podzolic soil without anilazine amendment; (b) same as a, but with $^{13/14}C$-anilazine amendment after incubation (residues corresponding to 0.6 mg of the originally added anilazine); (c) humified ^{13}C-depleted corn stalk/sand mixture without anilazine amendment; (d) same as c, but with $^{13/14}C$-anilazine amendment after incubation (residues corresponding to 1.6 mg of the originally added anilazine).[87]

coincidental occurrence of apparently conflicting processes such as nitrification and denitrification. It also influences the degradation and transformation of several pesticides by enhancement of microbial reactions that result in dechlorination, reduction of nitro groups, and splitting of aromatic rings, which need anaerobic as well as aerobic conditions.

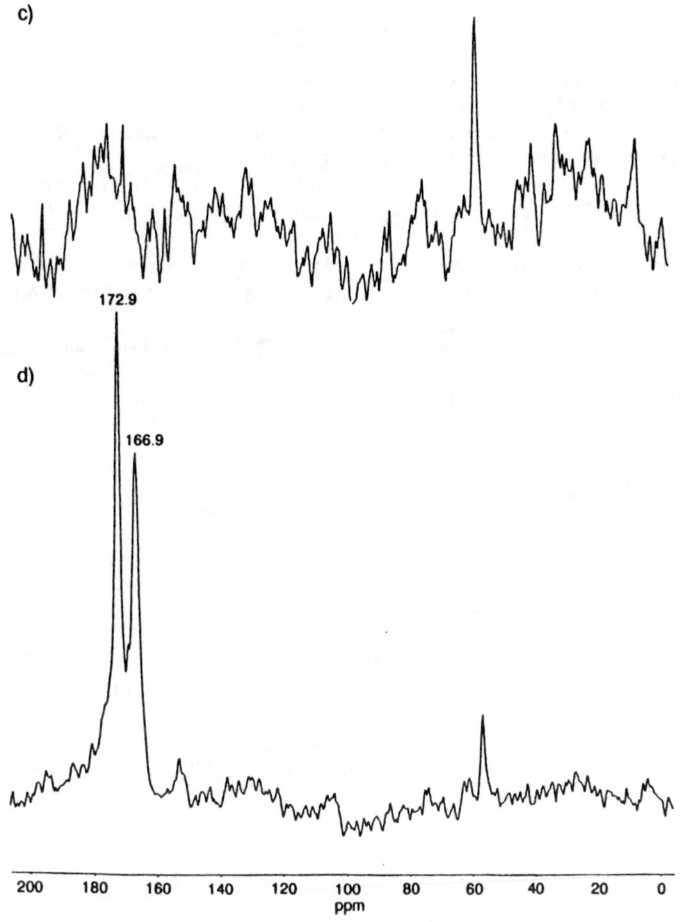

Figure 9 (Continued)

Humic compounds, in addition to inorganic soil constituents, can fix or adsorb low-molecular-weight organic compounds on/in their macromolecular matrix. This is also the reason for the well-known buffering capacity of soils to toxic influences, e.g., by phenols or pesticides and other xenobiotics. Studies have shown that several phenols can be adsorbed or even linked into the humic matrix by either abiotic reactions with Fe^{3+} or Mn^{4+} or through the activity of microbes. This may lead to the formation of bound residues from xenobiotics. This was exemplified for several phenols and anilines and for residues of the fungicide anilazine, showing their covalent linkage into the humic matrix. It must be stressed, however, that the results of these studies cannot be generalized, as other, less strongly sorptive mechanisms may also be operative, leading to nonextractable residues of xenobiotics.

Considerations of the fate of sorbed residues from pesticides and other organic soil contaminants should also include the role of dissolved organic matter as a carrier during leaching and run-off.[92]

REFERENCES

1. **Greenland, D. J.,** Interaction between clays and organic compounds in soils. II. Adsorption of soil organic compounds and its effect on soil properties, *Soils Fert.,* 28, 521, 1965.
2. **Turchenek, L. W. and Oades, J. M.,** Fractionation of organo-mineral complexes by sedimentation and density techniques, *Geoderma,* 21, 311, 1979.
3. **Anderson, D. W., Saggar, S., Bettany, J. R., and Stewart, J. W. B.,** Particle size fractions and their use in studies of soil organic matter. I. The nature and distribution of forms of carbon, nitrogen, and sulfur, *Soil Sci. Soc. Am. J.,* 45, 767, 1981.
4. **Tiessen, H. and Stewart, J. W. B.,** Particle size fractions and their use in studies of soil organic matter. II. Cultivation effects on organic matter composition in size fractions, *Soil Sci. Soc. Am. J.,* 47, 509, 1983.
5. **Campbell, C. A., Paul, E. A., Rennie, D. A., and McCallum, K. J.,** Factors affecting the accuracy of the carbon-dating method in soil humus studies, *Soil Sci.,* 104, 81, 1967.
6. **Jenkinson, D. S. and Rayner, J. H.,** The turnover of soil organic matter in some of the Rothamsted classical experiments, *Soil Sci.,* 123, 298, 1977.
7. **Janssen, B. H.,** A simple method for calculating decomposition and accumulation of "young" soil organic matter, *Plant Soil,* 76, 297, 1984.
8. **Parton, W. J., Schimel, D. S., Cole, C. V., and Ojima, D. S.,** Analysis of factors controlling soil organic matter level in Great Plains Grasslands, *Soil Sci. Soc. Am. J.,* 51, 1173, 1987.
9. **Angers, D. A., Pesant, A., and Vigneaux, J.,** Early cropping-induced changes in soil aggregation, organic matter, and microbial biomass, *Soil Sci. Soc. Am. J.,* 56, 115, 1992.
10. **Hsieh, Y.-P.,** Pool size and mean age of stable soil organic carbon in cropland, *Soil Sci. Soc. Am. J.,* 56, 460, 1992.
11. **Balesdent, J., Wagner, G. H., and Mariotti, A.,** Soil organic matter turnover in long-term field experiments as revealed by carbon-13 natural abundance, *Soil Sci. Soc. Am. J.,* 52, 118, 1988.
12. **Jenkinson, D. S.,** Determination of microbial biomass carbon and nitrogen in soil, in *Advances in Nitrogen Cycling in Agricultural Ecosystems,* Wilson, J. R., Ed., CAB International, Wallingfort, 1988, 368.
13. **Doran, J. W. and Smith, M. S.,** Organic matter management and utilization of soil and fertilizer nutrients, in *Soil Fertility and Organic Matter as Critical Components of Production Systems,* Follett, R. F., Stewart, J. W. B., and Cole, C. V., Eds., SSSA Spec. Publ. No. 19, Soil Science Society of America, 1989, 53.
14. **Doran, J. W.,** Einfluß verschiedener Bewirtschaftungs- und Bearbeitungssysteme auf die organische Bodensubstanz und die Bodenfruchtbarkeit, *Ber. Landw.,* 206, 155, 1992.
15. **Beck, Th.,** Der Einfluß langjähriger Bewirtschaftungsweise auf bodenmikrobiologische Eigenschaften, *Kali Briefe,* 20, 17, 1990.
16. **Schulten, H.-R., Hempfling, R., Haider, K., Gröblinghoff, F. F., Lüdemann, H. D., and Fründ, R.,** Characterization of cultivation effects on soil organic matter, *Z. Pflanzenernähr. Bodenkd.,* 153, 97, 1990.
17. **Fründ, R., Haider, K., and Lüdemann, H. D.,** Influence of soil management practices on the organic matter structure — Investigations by CPMAS 13C NMR-spectroscopy, *Z. Pflanzenernähr. Bodenkd.,* 157, 29, 1994.
18. **Anderson, T.-H. and Domsch, K. H.,** Der Einfluß des Bodengefüges auf mikrobielle Stoffwechselleistungen, *Mitt. Dtsch. Bodenkd. Ges.,* 59/I, 523, 1989.
19. **Springer, U.,** Die Wirkung verschiedener organischer Dünger auf den Humuszustand des Bodens, *Bayer. Landwirtsch. Jahrb.,* 37, 1, 1960.
20. **Kraus, A.,** Pflanzenbauliche und ökologische Aspekte unterschiedlicher Dünge- und Pflanzenschutzintensität, *Bayer. Landwirtsch. Jahrb.,* 58, 259, 1981.
21. **Melian, L.,** Auswirkungen abgestufter Intensitäten im Pflanzenbau auf Lebensgemeinschaften des Ackers, Bodenfruchtbarkeit und Ertrag. *Bayer. Landwirtsch. Jahrb.,* 68, 307, 1991.
22. **Haider, K. and Gröblinghoff, F. F.,** Biochemische Umsetzungen und Humusbildung in Böden unterschiedlicher Bewirtschaftung, *Kali Briefe,* 20, 31, 1990.
23. **Hempfling, R., Schulten, H.-R., and Haider, K.,** Einfluß landwirtschaftlicher Bewirtschaftung auf die organische Bodensubstanz: Identifizierung molekularer Untereinheiten durch Pyrolyse-Massenspektrometrie, *Mitt. Dtsch. Bodenkd. Ges.,* 59, 575, 1989.

24. **Schnitzer, M. and Preston, C. M.,** Supercritical gas extraction of a soil with solvents of increasing polarities, *Soil Sci. Soc. Am. J.,* 51, 639, 1987.
25. **Skjemstad, J. O., Dalal, R. C., and Barron, P. F.,** Spectroscopic investigations of cultivation effect on organic matter in vertisols, *Soil Sci. Soc. Am. J.,* 50, 354, 1986.
26. **Schulten, H.-R. and Schnitzer, M.,** Aliphatics in soil organic matter in fine-clay fractions, *Soil Sci. Soc. Am. J.,* 54, 98, 1990.
27. **Martin, J. P. and Haider, K.,** Influence of mineral colloids on turnover rates of soil organic carbon, in *Interactions of Soil Minerals, with Natural Organics and Minerals,* Huang, P. M. and Schnitzer, M., Eds., SSSA Spec. Publ. No. 17, Soil Science Society of America, Madison, WI, 1986, 283.
28. **Lynch, D. L. and Cotnoir, L., Jr.,** The influence of clay minerals on the breakdown of certain organic substrates, *Soil Sci. Soc. Am. Proc.,* 20, 367, 1956.
29. **Nichols, J. D.,** Relation of organic carbon to soil properties and climate in Southern Great Plains, *Soil Sci. Soc. Am. J.,* 48, 1382, 1984.
30. **Stotzky, G.,** Influences of soil mineral colloids on metabolic processes, growth, adhesion, and ecology of microbes and viruses, in *Interactions of Soil Minerals, with Natural Organics and Minerals,* Huang, P. M. and Schnitzer, M., Eds., SSSA Spec. Publ. No. 17, Soil Science Society of America, Madison, WI, 1986, 305.
31. **Filip, Z., Haider, K., and Martin, J. P.,** Influence of clay minerals on growth and metabolic activity of *Epicoccum nigrum* and *Stachybotrys chartarum, Soil Biol. Biochem.,* 4, 135, 1972.
32. **Haider, K., Filip, Z., and Martin, J. P.,** Einfluss von Montmorillonit auf die Bildung von Biomasse und Stoffwechsel-Zwischenprodukten durch einige Mikroorganismen, *Arch. Mikrobiol.,* 73, 201, 1970.
33. **Anderson, T.-H. and Gray, T. R. G.,** Soil microbial carbon uptake characteristics in relation to soil management, *FEMS Microbiol. Ecol.,* 74, 11, 1990.
34. **Ladd, J. N., Oades, J. M., and Amato, M.,** Microbial biomass formed from ^{14}C, ^{15}N-labeled plant material decomposing in soils in the field, *Soil Biol. Biochem.,* 13, 119, 1981.
35. **Martin, J. P., Martin, W. P., Page, J. B., Raney, W. A., and de Ment, J. D.,** Soil aggregation, *Adv. Agron.,* 7, 1, 1955.
36. **Tisdall, J. M. and Oades, J. M.,** Organic matter and water-stable aggregates in soils, *J. Soil Sci.,* 33, 141, 1982.
37. **Dalal, R. C. and Mayer, R. J.,** Long-term trends in fertility of soils under continuous cultivation and cereal cropping in southern Queensland. III. Distribution and kinetics of soil organic carbon in particle-size fractions, *Aust. J. Soil Res.,* 24, 293, 1986.
38. **Elliot, E. T.,** Aggregate structure and carbon, nitrogen, and phosphorus in native and cultivated soils, *Soil Sci. Soc. Am. J.,* 50, 627, 1986.
39. **Waters, A. G. and Oades, J. M.,** Organic matter in water-stable aggregates, in *Advances in Soil Organic Matter Research: The Impact on Agriculture and the Environment,* Wilson, W. S., Ed., R. Soc. Chem. Spec. Publ. No. 90, 1991, 163.
40. **Dinel, H., Schnitzer, M., and Mehuys, G. R.,** Soil lipids: origin, nature, content, decomposition, and effect on soil physical properties, in *Soil Biochemistry,* Vol. 6, Bollag, J. M. and Stotzky, G., Eds., Marcel Dekker, New York, 1990, 397.
41. **Jambu, P., Coulibaly, G., Bilong, P., Magnoux, P., and Ambles, A.,** Influence of lipids on physical properties of soils, in *Humus Planta Prague,* 1, 46, 1983.
42. **Fustec-Mathon, E., Jambu, P., Joly, G., and Jacquesi, R.,** Analyse et rôle des bitumes dans les sols sableux acides, in *Proc. Symp. IAEA and FAO,* Soil Organic Matter Studies, Vol. 2, Braunschweig, September 6 to 10, 1976, 105.
43. **Schnitzer, M.,** Bedeutung der organischen Bodensubstanz für die Bodenbildung, Transportprozesse in Böden und die Bodenstruktur, *Ber. Landwirtsch.,* 206, 63, 1992.
44. **Capriel, P., Beck, T., Borchert, H., and Härter, P.,** Relationship between soil aliphatic fraction extracted with supercritical hexane, soil microbial biomass, and soil aggregate stability, *Soil Sci. Soc. Am. J.,* 5, 415, 1990.
45. **Powlson, D. S., Brookes, P. C., and Christensen, B. T.,** Measurement of microbial biomass provides an early indication of changes in total soil organic matter due to straw incorporation, *Soil Biol. Biochem.,* 19, 159, 1987.
46. **Smith, K. A.,** A model of the extent of anaerobic zones in aggregated soils and its potential application to estimates of denitrification, *J. Soil Sci.,* 31, 263, 1980.

47. **Grbic-Galic, D.,** Anaerobic microbial transformation of nonoxygenated aromatic and alicyclic compounds in soil, subsurface, and freshwater sediments, in *Soil Biochemistry,* Vol. 6, Bollag, J. M. and Stotzky, G., Eds., Marcel Dekker, New York, 1990, 117.
48. **Parkin, T. B.,** Soil microsites as a source of denitrification variability, *Soil Sci. Soc. Am. J.,* 51, 1194, 1987.
49. **Sexstone, A. J., Revsbech, N. P., Parkin, T. B., and Tiedje, J. M.,** Direct measurement of oxygen profiles and denitrification rates in soil aggregates, *Soil Sci. Soc. Am. J.,* 49, 645, 1985.
50. **Zausig, J., Hell, U., and Horn, R.,** Eine Methode zur Ermittlung der wasserspannungsabhängigen Änderung des Sauerstoffpartial-druckes und der Sauerstoffdiffusion in einzelnen Bodenaggregaten, *Z. Pflanzenernäehr. Bodenkd.,* 153, 5, 1990.
51. **Arah, J. R. M.,** Steady state denitrification: an adequate approximation?, *Mitt. Dtsch. Bodenkd. Ges.,* 60, 413, 1990.
52. **Seech, A. G. and Beauchamp, E. G.,** Denitrification in soil aggregates of different sizes, *Soil Sci. Soc. Am. J.,* 52, 1616, 1988.
53. **Haider, K.,** Anaerobic microsites in soils and their possible effect on pesticide degradation, in *IUPAC Pesticide Chemistry,* Vol. 3, Miyamoto, J. and Kearney, P. C., Eds., Pergamon Press, New York, 1983, 353.
54. **Guenzi, W. D. and Beard, W. E.,** Anaerobic biodegradation of DDT to DDD in soil, *Science,* 156, 1116, 1967.
55. **Jagnow, G., Haider, K., and Ellwardt, P.-C.,** Anaerobic dechlorination and degradation of hexachlorocyclohexane isomers by anaerobic and facultative anaerobic bacteria, *Arch. Microbiol.,* 115, 285, 1977.
56. **Kohnen, R. K., Haider, K., and Jagnow, G.,** Investigations on the microbial degradation of lindane in submerged and aerated moist soil, *Environ. Qual. Safety,* 3, 222, 1975.
57. **Williams, P. P.,** Metabolism of synthetic organic pesticides by anaerobic microorganisms, *Residue Rev.,* 66, 62, 1977.
58. **Lichtenstein, E. P. and Schulz, K. R.,** Persistence of some chlorinated hydrocarbon insecticides as influenced by soil types, rate of application and temperature, *J. Econ. Entomol.,* 52, 124, 1959.
59. **Katan, J., Fuhreman, T. W., and Lichtenstein, E. P.,** Binding of ^{14}C parathion in soil: a reassessment of pesticide persistence, *Science,* 193, 891, 1976.
60. **Huang, P. M.,** Role of soil minerals in transformations of natural organics and xenobiotics in soil, in *Soil Biochemistry,* Vol. 6, Bollag, J. M. and Stotzky, G., Eds., Marcel Dekker, New York, 1990, 29.
61. **Pignatello, J. J.,** Sorption dynamics of organic compounds in soils and sediments, in *Reactions and Movement of Organic Chemicals in Soils,* Sawhney, B. L. and Brown, K., Eds., SSSA Spec. Publ. No. 22, Soil Science Society of America, Madison, WI, 1989, 45.
62. **Whitehead, D. C.,** Identification of p-hydroxybenzoic, vanillic, p-coumaric and ferulic acids, *Nature (London),* 202, 417, 1964.
63. **Whitehead, D. C., Dibb, H., and Hartley, R. D.,** Bound phenolic compounds in water extracts of soils, plant roots and leaf litter, *Soil Biol. Biochem.,* 15, 133, 1983.
64. **Wang, T. S. C., Chen, J.-H., and Hsiang, W.-M.,** Catalytic synthesis of humic acids containing various amino acids and dipeptides, *Soil Sci.,* 140, 3, 1985.
65. **Huang, P. M. and Violante, A.,** Influence of organic acids on crystallization and surface properties of precipitation products of aluminum, in *Interactions of Soil Minerals with Organics and Microbes,* Huang, P. M. and Schnitzer, M., Eds., SSSA Spec. Publ. No. 17, Soil Science Society of America, Madison, WI, 1986, 159.
66. **Yakle, G. A. and Cruse, R. M.,** Effects of fresh and decomposing corn plant residue extracts on corn seedling development, *Soil Sci. Soc. Am. J.,* 48, 1143, 1984.
67. **Blum, U., Dalton, B. R., and Rawlings, J. O.,** Effect of ferulic acid and some of its microbial products on radicle growth of cucumber, *J. Chem. Ecol.,* 10, 1169, 1984.
68. **Haider, K. and Martin, J. P.,** Decomposition of specifically carbon-14 labeled benzoic and cinnamic acid derivatives in soil, *Soil Sci. Soc. Am. Proc.,* 39, 657, 1975.
69. **Stott, D. E., Kassim, G., Jarrell, W. M., Martin, J. P., and Haider, K.,** Stabilization and incorporation into biomass of specific plant carbon during biodegradation in soil, *Plant Soil,* 70, 15, 1983.
70. **Martin, J. P., Haider, K., and Kassim, G.,** Biodegradation and stabilization after 2 years of specific crop, lignin, and polysaccharide carbons in soils, *Soil Sci. Soc. Am. J.,* 44, 1250, 1980.

71. **Kassim, J., Stott, D. E., Martin, J. P., and Haider, K.,** Stabilization and incorporation into biomass of phenolic and benzenoid carbons during biodegradation in soil, *Soil Sci. Soc. Am. J.*, 46, 305, 1982.
72. **Lehmann, R. G., Cheng, H. H., and Harsh, J. B.,** Oxidation of phenolic acids by soil iron and manganese oxides, *Soil Sci. Soc. Am. J.*, 51, 352, 1987.
73. **Mc Calla, T. M.,** Studies on phytotoxic substances from soil microorganisms and crop residues at Lincoln Nebraska, in *Biochemical Interactions among Plants,* National Academy of Science, Washington, D.C., 1971, 39.
74. **Haider, K.,** Problems related to the humification processes in soils of temperate climates, in *Soil Biochemistry,* Vol. 7, Stotzky, G. and Bollag, J.-M., Eds., Marcel Dekker, New York, 1992, 55.
75. **Führ, F., Kloskowski, R., and Burauel, P.-W.,** Bedeutung der gebundenen Rückstände, *Ber. Landwirtsch.*, 198, 106, 1985.
76. **Roberts, T. R. et al.,** Non-extractable pesticide residues in soils and plants, *Pure Appl. Chem.*, 56, 945, 1984.
77. **Bartha, R., You, I.-S., and Saxena, A.,** Humus-bound residues of phenylamide herbicides: their nature, persistence and monitoring, in *Pesticide Chemistry: Human Welfare and the Environment,* Miyamoto, J. and Kearney, P. C., Eds., Pergamon Press, Oxford, 1983, 345.
78. **Parris, G. E.,** Covalent binding of aromatic amines to humates. I. Reactions with carbonyls and quinones, *Environ. Sci. Technol.*, 14, 1099, 1980.
79. **Koskinen, W. C. and Harper, S. S.,** The retention process: mechanisms, in *Pesticides in the Soil Environment,* Cheng, H. H., Ed., SSSA Book Ser. No. 2, Soil Science Society of America, Madison, WI, 1990, 51.
80. **Haider, K., Spiteller, M., Reichert, K., and Fild, M.,** Derivatization of humic compounds: an analytical approach for 'soil bound' organic residues, *Int. J. Environ. Anal. Chem.*, 46, 201, 1992.
81. **Bollag, J.-M. and Liu, S.-Y.,** Biological transformation processes of pesticides, in *Pesticides in the Soil Environment,* Cheng, H. H., Ed., SSSA Book Ser. No. 2., Soil Science Society of America, Madison, WI, 1990, 169.
82. **Hassett, J. J. and Banwart, W. L.,** The sorption of nonpolar organics by soils and sediments, in *Reactions and Movement of Organic Chemicals in Soil,* Sawhney, B. L. and Brown, K., Eds., SSSA Spec. Publ. No. 22, Soil Science Society of America, Madison, WI, 1989, 31.
83. **Pignatello, J. J.,** Sorption dynamics of organic compounds in soils and sediments, in *Reactions and Movement of Organic Chemicals in Soil,* Sawhney, B. L. and Brown, K., Eds., SSSA Spec. Publ. No. 22, Soil Science Society of America, Madison, WI, 1989, 45.
84. **Khan, S. U.,** Distribution and characteristics of bound residues from prometryn in an organic soil, *J. Agric. Food Chem.*, 30, 175, 1982.
85. **Chiou, C. T.,** Theoretical considerations of the partition uptake of nonionic organic compounds by soil organic matter, in *Reactions and Movement of Organic Chemicals in Soil,* Sawhney, B. L. and Brown. K., Eds., SSSA Spec. Publ. No. 22, Soil Science Society of America, Madison, WI, 1989, 1.
86. **Burauel, P., Mittelstaedt, W., and Führ, F.,** Abbau und Fixierung von Anilazine in einer Parabraunerde, *Int. Ber. Inst. Radioagr., KFA Jül.*, 2/83, 1983.
87. **Haider, K., Spiteller, M., Wais, A., and Fild, M.,** Evaluation of the binding mechanism of anilazine and its metabolites in the soil organic matter, *Int. J. Environ. Anal. Chem.*, in press.
88. **Hattori, T.,** *Microbial Life in Soil,* Marcel Dekker, New York, 1973.
89. **Gupta, V. V. and Germida, J. J.,** Distribution of microbial biomass and its activity in different soil aggregate size classes as affected by cultivation, *Soil Biol. Biochem.*, 20, 777, 1988.
90. **Bitton, G., Henis, Y., and Lahav, N.,** Influence of clay minerals, humic acid and bacterial capsular polysaccharides on the survival of *Klebsiella aerogenes* exposed to drying and heating in soil, *Plant Soil,* 45, 65, 1976.
91. **Anderson, T.-H.,** Bedeutung der Mikroorganismen für die Bildung von Aggregaten im Boden, *Z. Pflanzenernaehr Bodenkd.*, 154, 409, 1991.
92. **Kögel-Knabner, I., Knabner, P., and Deschauer, H.,** Dissolved organic matter as carrier for exogenous organic chemicals in soils, in *Advances in Soil Organic Matter Research: The Impact on Agriculture and the Environment,* Wilson, W. S., Ed., R. Soc. Chem. Spec. Publ. No. 90, Royal Society of Chemistry, 1991, 121.
93. **Zausig, J., Horn, R., and Haider, K.,** unpublished results.

Chapter 3

Role of Aluminum and Iron in the Accumulation of Organic Matter in Soils with Variable Charge

K. Wada

CONTENTS

I. INTRODUCTION

Soil organic matter (SOM) plays important roles in alleviating environmental problems as well as nutrient and stress problems in soils. There is a strong concern about factors and mechanisms which determine the SOM content of soils.

A common view on SOM content vs. climate is that tropical soils have lower organic matter contents than temperate soils because of high temperature and decomposition rates. However, Allison[1] and Sanchez,[2] for example, reported that some tropical soils were higher in organic matter than their annual mean temperature would predict. Sanchez[2] compared the SOM contents of randomly chosen Oxisols, Ultisols, and Alfisols from the U.S. He found that there was no difference in this respect between the Oxisols and Mollisols, or between the tropical and temperate Ultisols and Alfisols, and emphasized the importance of a balance between the addition of organic matter to soils and the rate of its decomposition.

Under similar climatic conditions, SOM content is related to soil texture and interaction with inorganic constituents. Sanchez[2] stated that the higher the clay contents and the higher the proportion of oxides and allophane, the lower will be the decomposition rates of SOM. Allison[1] pointed out that the adsorption of SOM on the clay particles, particularly those having the 2:1 layer, and the formation of metal organic complexes, such as Ca, Fe, and Al humates, is important. Theng et al.[3] stated that the chemical structure of humic substances also affects their stability and suggested that alkyl C, as in polymethylene structures, apparently represents a particularly recalcitrant form of soil C and is predominant in the constitution of SOM in many tropical soils.

A review of the interaction of SOM and variable-charge soils by Oades et al.[4] identified mechanisms for maintaining organic matter in variable-charge soils as one major research area deserving immediate attention. The object of this chapter is to illustrate (1) the role of Al and Fe in the determination of SOM content in two sets of soils with variable charge and (2) the effect of Al- and Fe-humus interaction on electric charge and phosphate retention of soils.

0-87371-914-X/95/$0.00+$.50

II. ROLE OF AL AND FE IN THE DETERMINATION OF SOM CONTENT

The first set of soil samples selected for the study were 16 tropical Thai Ultisols and Oxisols and eight temperate Korean Ultisols and Alfisols. They are all A horizons, and the Thai and Korean soils had similar ranges of total C content (0.35 to 3.03 vs. 0.63 to 3.29%). The soils within the respective groups differ not only in soil moisture regime, but also in parent material, vegetation, and management.[5] The Thai soils contain kaolin minerals, whereas the Korean soils contain kaolin minerals, mica, and more or less chloritized vermiculite.[6,7]

The second set of soil samples were A1 and buried A1 horizons of Japanese Andisols collected in the Mt. Aso district[8] and in other districts.[9] Their total C content ranged up to more than 20%, and this accumulation of SOM occurs under humid, temperate climates and mostly grass vegetation and at well-drained sites.

The main approach adopted to determine the role of inorganic constituents which contributed to the retention of SOM was to (1) examine the relationship between the contents of total C and clay and (2) extract the soil with 0.1 *M* $Na_4P_2O_7$ at pH 10 and examine the relationships between the amounts of C, Al, and Fe extracted. Alkali sodium pyrophosphate solutions are good extractants of Al- and Fe-humus complexes in soils,[10] but are poor extractants of Al and Fe in noncrystalline and crystalline oxides and silicates.[12]

A. THAI AND KOREAN SOILS

Figure 1 shows that total C is nearly zero at zero clay and increases with clay content. Carbon extracted by pyrophosphate (C_{pyr}), which represents humic and fulvic acids combined with Al, Fe, and Ca, resembles total C in its relationship with clay content (Figure 2). Linear relationships between total C or C_{pyr} and clay content indicate that the clay contributes to the retention of SOM and extractable humus. Based on these relationships, the soils may be divided into two groups. Each group includes both the Thai and Korean soils, which are different in soil temperature regime (thermic to hyperthermic vs. mesic) and in major clay mineral composition (kaolin minerals vs. kaolin minerals plus mica and chloritized vermiculite), suggesting that these factors do not primarily control the clay-SOM interaction. The C_{pyr}/total C ratio varied from 0.20 to 0.41 in the Thai soils and from 0.21 to 0.37 in the Korean soils, suggesting a similarity in the nature of the SOM in these tropical and temperate soils.

Table 1 shows that C_{pyr} is correlated with Al_{pyr} and Fe_{pyr} but not with exchangeable Ca (ex. Ca). Because there is also a high correlation between Al_{pyr} and Fe_{pyr}, the relationship between C_{pyr} and $(Al + Fe)_{pyr}$ is examined next. As shown in Figure 3, there is a curvilinear relationship common to Thai and Korean soils except for three Korean Ultisols. It suggests that Al_{pyr} and Fe_{pyr} form metal-humus complexes and contribute to humus retention. The amounts of Al_{pyr} and Fe_{pyr} range from 0.5 to 18.9 and from 0.3 to 15.7 m atom (milliatom)/100 g^{-1} soil, respectively. The relationship between C_{pyr} and clay content (Figure 2) and that between C_{pyr} and $(Al + Fe)_{pyr}$ (Figure 3) suggests that Al and Fe are supplied from the clay or from parent rock during soil formation. The atomic ratio of $(Al + Fe)_{pyr}/C_{pyr}$ was in the range of 0.05 to 0.3 for most Thai and Korean soils, indicating a similarity of the Al- and Fe-humus interaction in these soils.

Just as total C and C_{pyr} increased with clay content, so did the residual C (total C – C_{pyr}) (Figure 4). The rate of increase was greater for residual C than for C_{pyr}, suggesting that residual humus, which we may call humin, is retained through enhanced interaction with clay or, more presumably, with larger amounts of Al and Fe. McKeague et al.[11] prepared Al and Fe complexes of fulvic acid from a spodic horizon and found that Al complexes containing >0.37 Al atoms per C atom and Fe complexes containing >0.2 Fe atoms per C atom were not completely pyrophosphate soluble.

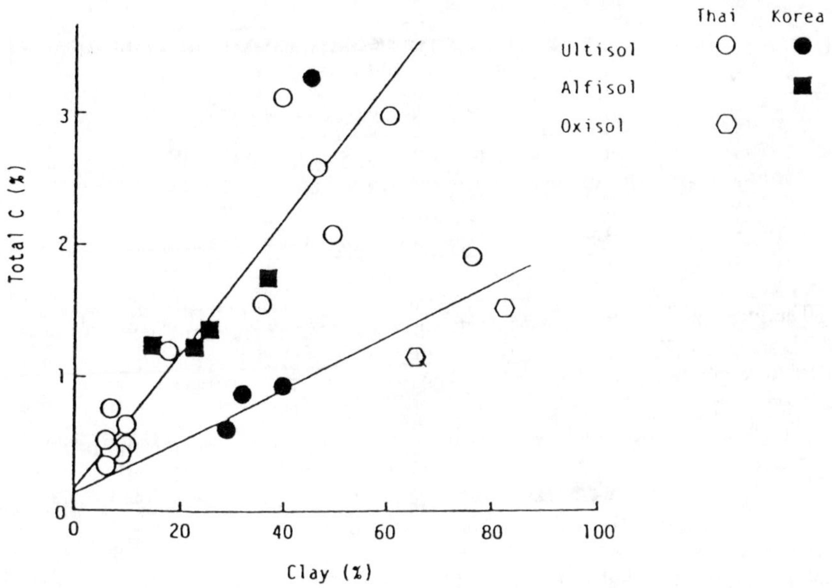

Figure 1 Relationship between total C and clay content in the Thai and Korean soils.[5]

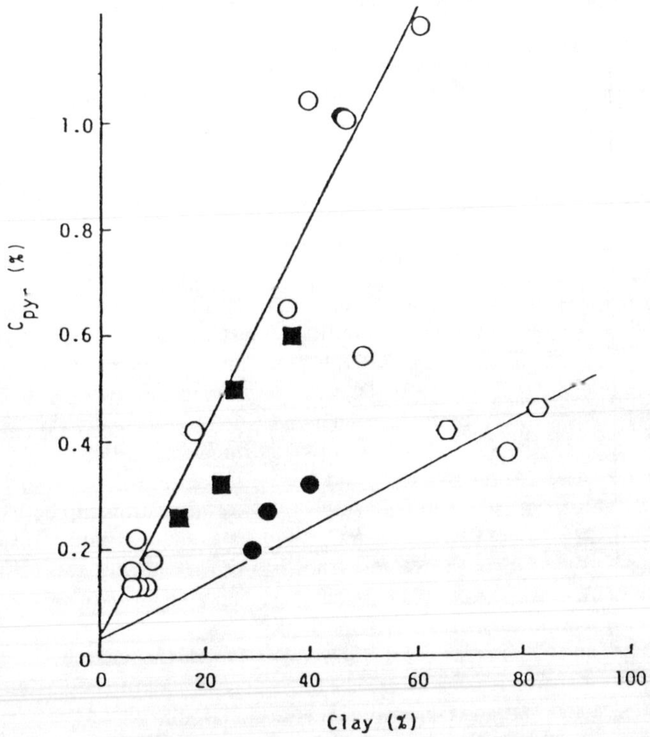

Figure 2 Relationship between pyrophosphate-extractable C and clay content in the Thai and Korean soils.[5] For legend, see Figure 1.

Table 1 Simple Correlation Matrix for C_{pyr}, Al_{pyr}, Fe_{pyr}, and ex. Ca[5]

	Al_{pyr}	Fe_{pyr}	Ex. Ca
C_{pyr}	0.85	0.55	–0.06
Al_{pyr}	1	0.60	–0.30
Fe_{pyr}		1	–0.16
Ex. Ca			1

The major and minor soil groups in Figures 1 and 2 include Thai and Korean soils which are different not only in soil moisture and temperature regimes, but also in parent material, vegetation, and management. The results as a whole indicate, therefore, that the rate of decomposition of humus dominates the organic matter equilibrium in the soils, and that its decomposition rate is primarily determined by Al- and Fe-humus complex formation.

The effect of the clay content on the retention of newly formed humus was examined in a short-time (30-d) incubation study for nine Thai soil samples to which alfalfa meal was added.[13] As shown in Figure 5, the retention of added C was related linearly to clay content as found for total C (Figure 1), but would not be zero at zero clay, indicating that the clay-SOM interaction would be only partial in this case and different in its mechanism from that found for humus present in the soils. The difference between the newly formed humus and the humus present in the soils indicates the absence of such newly formed humus in the Thai and Korean soils studied.

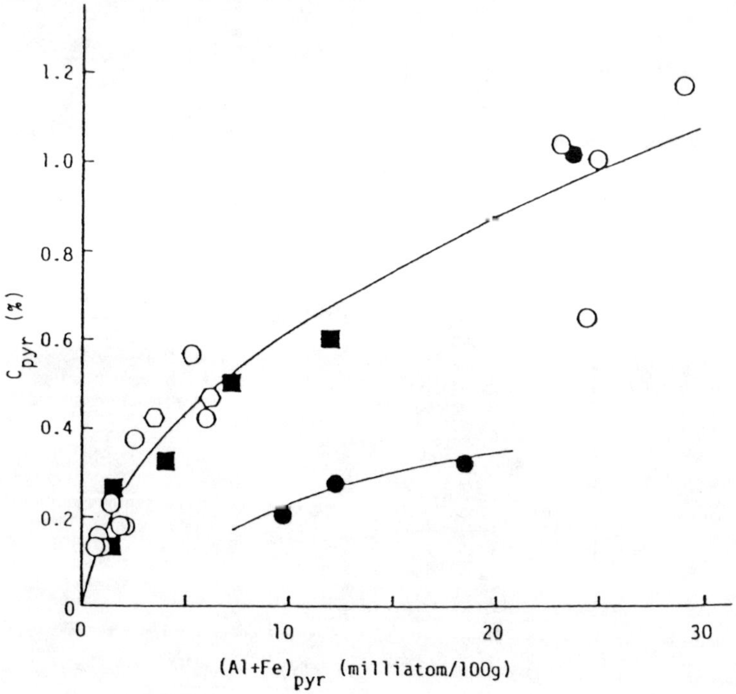

Figure 3 Relationship between pyrophosphate-extractable C and pyrophosphate-extractable (Al + Fe) in the Thai and Korean soils.[5] For legend see Figure 1.

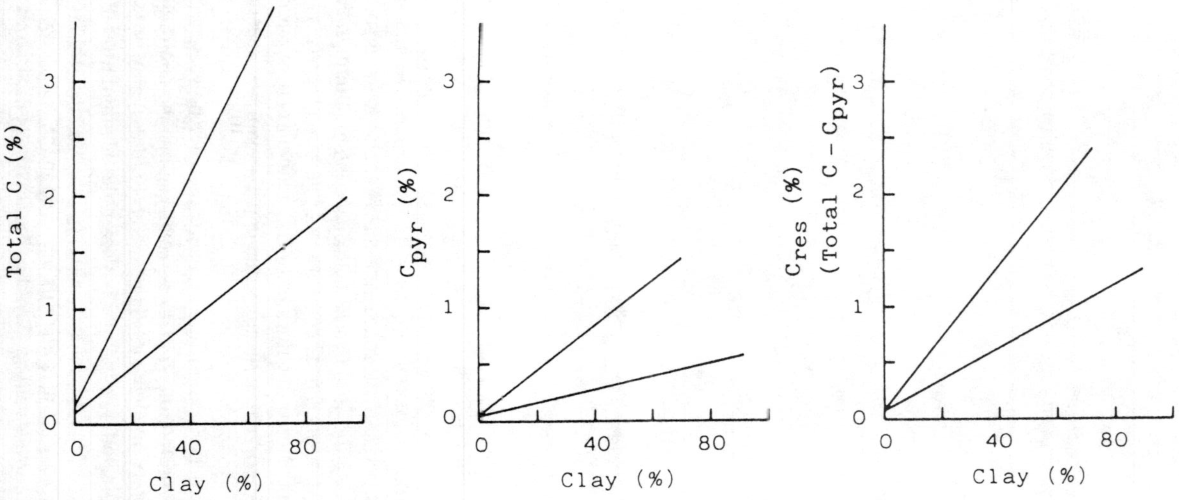

Figure 4 Relationship between total C, pyrophosphate-extractable C and residual C and clay content in the Thai and Korean soils.[5]

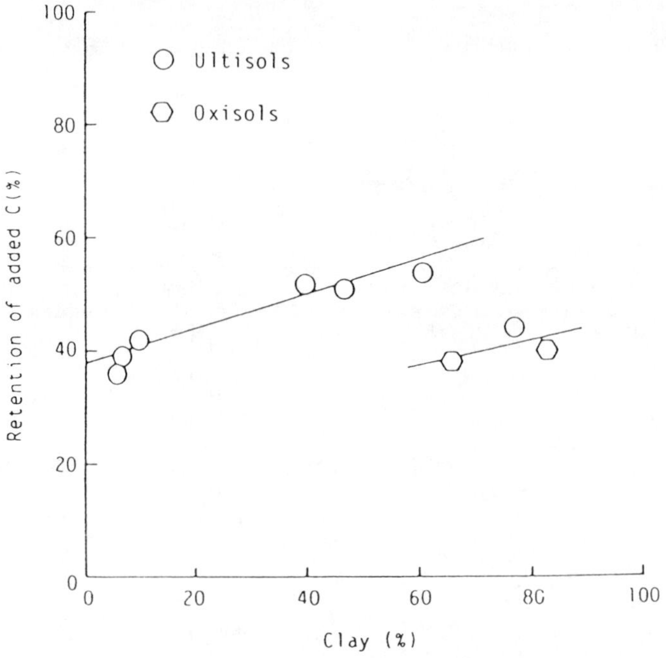

Figure 5 Relationship between the retention rate of added C in a short-term incubation and clay content for selected Thai soils.[5]

B. JAPANESE ANDISOLS

Figure 6 shows the relationship between total C and clay contents for Andisols of the Mt. Aso district in Kyushu. In this district, continuing activities of the central cone result in successive deposition of volcanic ashes. The soils studied are surface soils, subsoils, and their intermediates at depths of 0 to 35, 70 to 100, and 35 to 70 cm, respectively. The subsurface soils are buried surface soils. Like the Thai and Korean soils (Figure 1), the total C content of these Andisols increases with clay content, although the total C content is much higher in the Andisols. The C content would be zero at zero clay for most surface soils but not for subsoils. Generally, the subsoil has a higher clay content than the surface soil at any particular C content, indicating that clay formation proceeds after the burial of the soil.

The accumulation of humus, an important feature of Andisols, was ascribed to its interaction of physical protection by clay, especially allophane and/or sesquioxides. The correlation between total C content and clay content (Figure 6) may be taken as indicating this interaction. As reviewed by Wada,[8-14] however, the accumulation of humus occurred in a number of soils in which allophane and imogolite are absent or are present only in small amounts. Figure 7a illustrates that this occurred in the A1 horizons of soils derived from various ashes in northern Japan.[15] In this figure, Si_{ox} is oxalate-oxalic acid-extractable Si in the soil and represents the amount of Si in allophane and imogolite. The Si_{ox} decreases with increasing total C, indicating that there is an antagonistic relationship between the accumulation of humus and the formation of allophane and imogolite. A further study showed that this antagonistic relationship is due to these two processes competing for Al released by the weathering of volcanic ash (Figure 7b). In Figure 7b, Al_{ox} is oxalate-oxalic acid-extractable Al and represents the sum of Al in allophane and

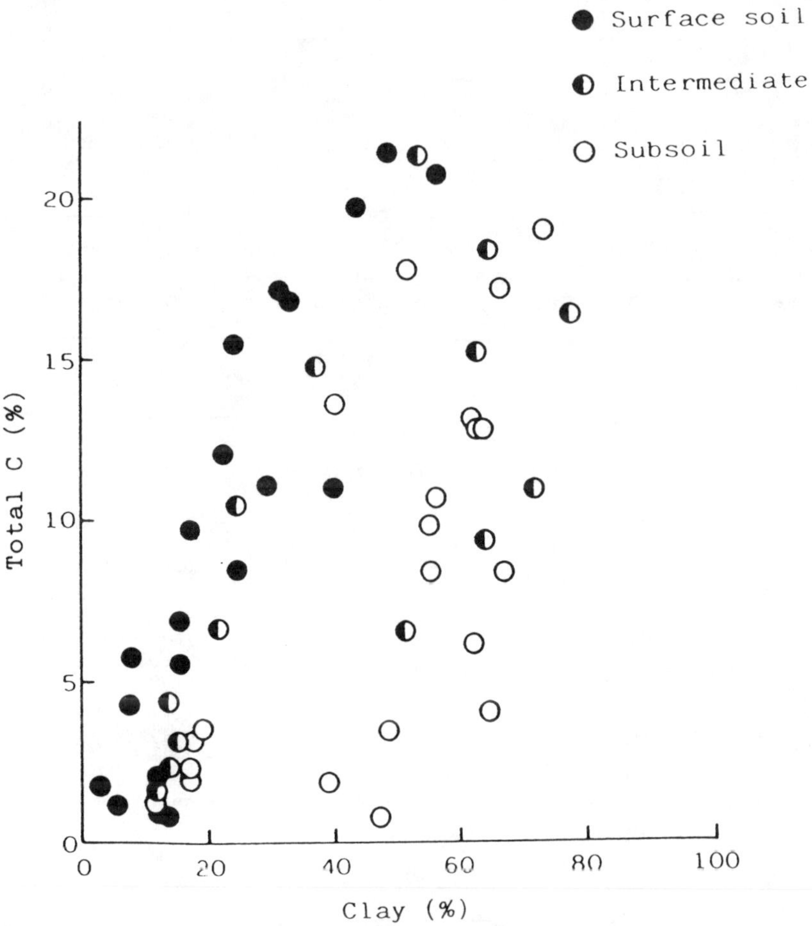

Figure 6 Relationship between total C and clay content for Andisols in Mt. Aso district.[8]

imogolite and Al-humus complexes. The Si_{ox} increases with increasing $(Al_{ox} - Al_{pyr})/Al_{ox}$, indicating that Al humus complex formation limits the formation of allophane and imogolite. Figures 7a and b suggest that the acidity of the soils to which Al-humus contributes can also inhibit the formation of allophane and imogolite.

Figure 8 shows the relationship between C_{pyr} and $(Al + Fe)_{pyr}$ for A1 horizons of Japanese Andisols of different ages. The age of the soil ranges from less than 1000 years to more than 10,000 years, which is indicated by the number attached to each sample plotted. The ratio of C_{pyr}/total C ranged from 0.21 to 0.70, which is similar to, or higher than, that found for the Thai and Korean soils (0.21 to 0.41). C_{pyr} increases almost linearly with $(Al + Fe)_{pyr}$ and is zero at zero $(Al + Fe)_{pyr}$, indicating that Al_{pyr} and Fe_{pyr} form metal-humus complexes with a stoichiometric ratio of metal to C within a fairly narrow range, from 0.17 to 0.23. This value is comparable to that found for most Thai and Korean soils, suggesting that the Al- and Fe-humus interaction is similar for the Thai and Korean soils and the Japanese Andisols. The difference in the amount of humus accumulated in these two groups of soils (Figures 1 and 6) could be due to the difference in the rate of supply of Al and Fe from the clay or from parent rock during the soil formation. The amounts of Al_{pyr} and Fe_{pyr} in the Japanese Andisols ranged from 3.3 to 163 and from 0.9 to 52.1

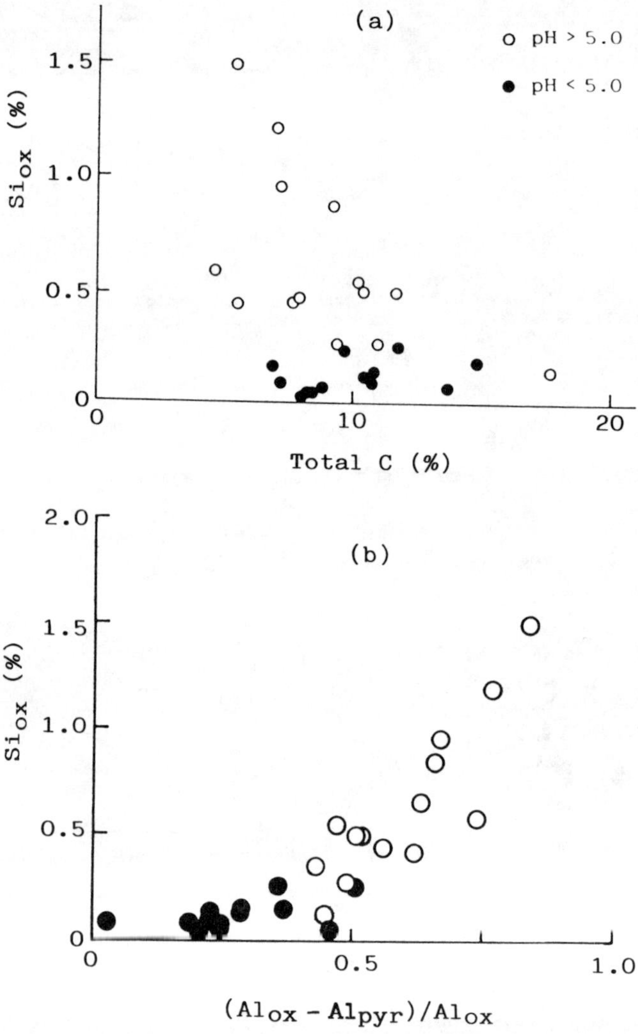

Figure 7 Relationship between Al-humus complex and allophane and imogolite in their formation in Andisols of different pHs from northern Japan.[15] Si_{ox} and Al_{ox}, oxalate-oxalic acid-extractable Si and Al in the soil, respectively; Al_{pyr}, pyrophosphate-extractable Al in the soil.

m atom/100 g soil, respectively, indicating that the interaction with Al mainly contributes to the stability of accumulated humus.

III. EFFECTS OF AL- AND FE-HUMUS INTERACTION ON CHARGE CHARACTERISTICS AND PHOSPHATE SORPTION

Most of the Al and Fe that complex with humus are present as polymeric hydroxy cations which are not extracted by cation exchange, e.g., with 1 *M* KCl. Table 2 shows the difference in the effect of pH on the cation-exchange capacity (CEC) between Andisol and chernozem soil samples in which Al- and partly Fe-humus complexes and Ca-humus complexes are major cation-exchange materials, respectively. A remarkable pH effect for Andisol indicates that, unlike Ca^{2+}, polymeric hydroxy Al and Fe cations with variable

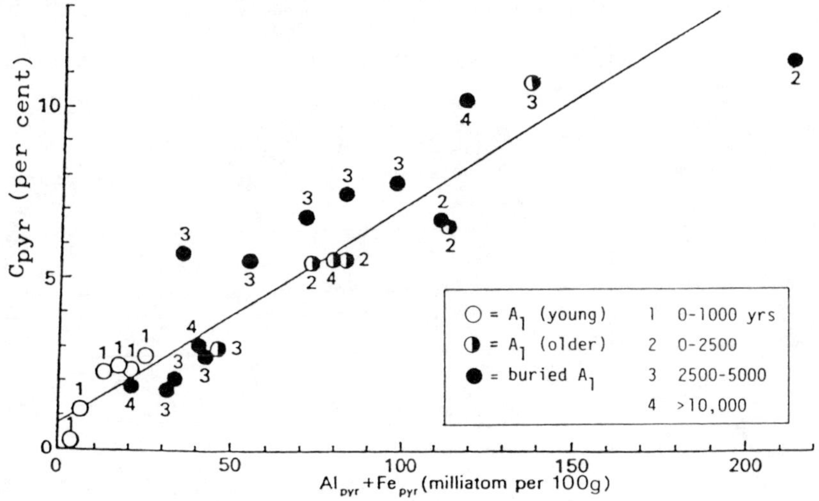

Figure 8 Relationship between pyrophosphate-extractable C and pyrophosphate-extractable (Al + Fe) in Japanese Andisols of different ages.[9]

positive charge interact with and block ionized carboxyl groups in humus, which results in the development of a very variable, pH-dependent negative charge. This is also indicated by comparison of the Andisol with the humic acid, which was extracted from Andisol using NaOH and purified by treatment with a H^+-saturated cation exchanger. In this humic acid, the ionization of most carboxyl groups occurs at pH 4, and the development of negative charge between pH 4 and 7 is small. The difference between the Andisol and the humic acid demonstrates that polymeric hydroxy-Al cations modify the charge characteristics of humic acid. A similar effect resulting from the removal of Al was found for humus in a Podzol B horizon.[19]

The effect of Al- and Fe-humus interactions on phosphate sorption may be different, depending on the nature of the complex formed. Figure 9 shows the relationship between phosphate sorption to attain 0.2 ppm in solution and $(Al + Fe)_{pyr}$ values for the Thai and Korean soils. Although the plots are scattered, possibly reflecting the difference in soil type, there is a tendency for phosphate sorption to increase with increasing $(Al + Fe)_{pyr}$, suggesting that Al and Fe complexed with humus react with phosphate. On the other hand, a study of 14 Oxisol profiles from north Queensland showed that the organic-rich surface soils sorbed much less phosphate than the subsurface soils at any particular sesquioxide content, which was measured by dithionite-citrate extraction.[4] The result was

Table 2 Effect of Metal-Humus Complex Formation on Development of Variable Negative Charge[16-18]

	Relative CEC (pH)[a]				
Sample	**8.0**	**7.0**	**6.0**	**5.0**	**4.0**
Andisol (Al-, Fe-humus)	2.1	1.0	0.5	0.25	—
Chernozem (Ca-humus)	—	1.0	0.9	0.8	0.65
Humic acid (Andisol)	—	1.0	0.9	0.85	0.7

[a] CEC relative to that at pH 7.0.

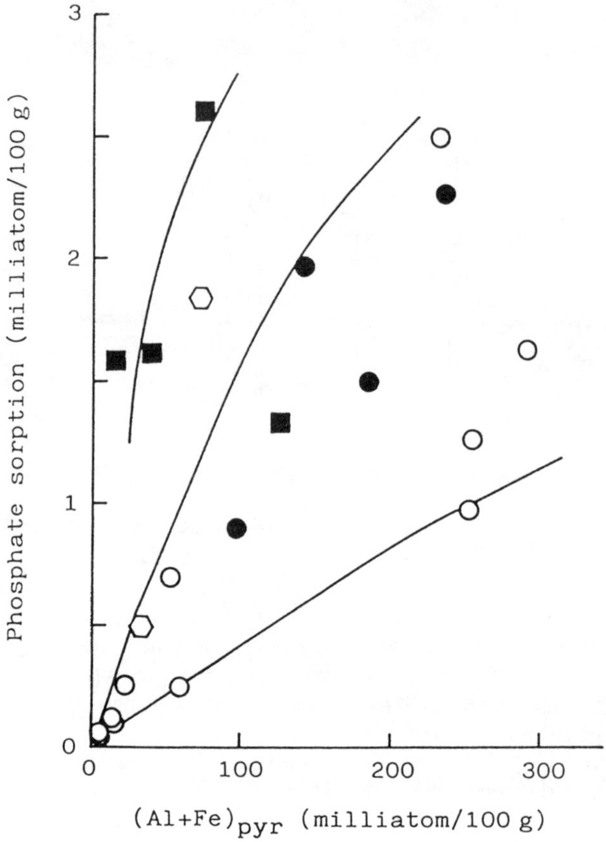

Figure 9 Relationship between phosphate sorption and pyrophosphate-extractable (Al + Fe) for the Thai and Korean soils.[20] For legend, see Figure 1.

interpreted as indicating that when virgin Oxisols are cultivated, the resulting decrease in organic matter content will cause phosphate sorption to increase.

In contrast, phosphate sorption on Andisols used for cultivation of paddy rice and upland crops increases with increasing C content (Figure 10). Phosphate sorption, measured by adding 2.5% ammonium phosphate solution, is larger in soils containing allophane and imogolite than in those containing no or little allophane and imogolite. A large phosphate sorption by soils in which allophane and imogolite are absent or nearly absent, and its increase with increasing C content, demonstrate that polymeric hydroxy-Al and -Fe cations complexed with humus react with phosphate.

IV. SUMMARY AND CONCLUSIONS

1. The interaction of Al and Fe with humus has a primary importance in the determination of SOM content in the tropical Thai Ultisols and Oxisols and the temperate Korean Ultisols and Alfisols as well as in the Japanese Andisols.
2. The Al and Fe that complex and stabilize humus against microbial decomposition are supplied from clay or from parent material during soil formation. Most probably, their

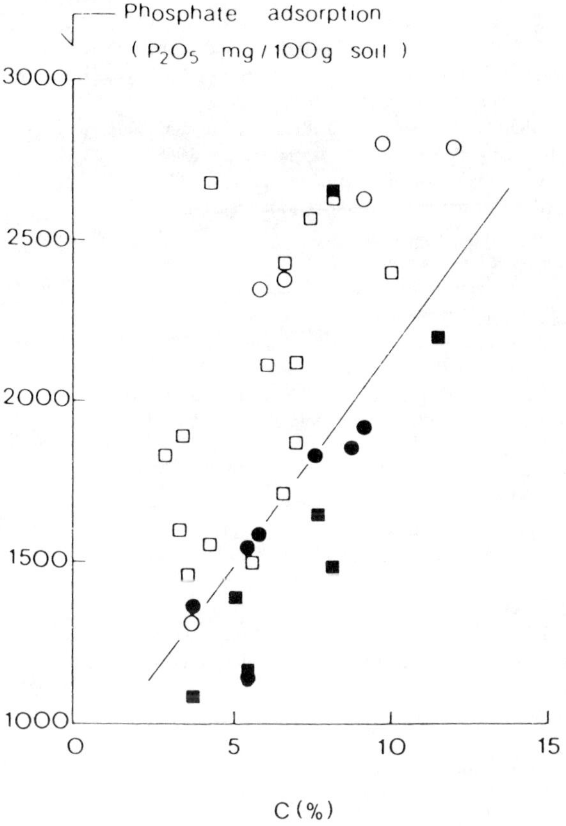

Figure 10 Relationship between C content and phosphate adsorption of Andisols used for paddy rice (circle) and upland crops (square), which differ in containing (open) and not containing (closed) allophane and imogolite.[21]

supply rates determine the SOM content, as demonstrated by the differences in the amounts of humus accumulated in, and Al and Fe extracted with pyrophosphate from, the Thai and Korean soils and Japanese Andisols.

3. Parts of the Al and Fe that complex with humus are present as polymeric hydroxy cations. They block ionized carboxyl groups of humus, resulting in the development of a very variable negative charge and contribute to phosphate sorption.

REFERENCES

1. **Allison, F. E.,** *Soil Organic Matter and Its Role in Crop Production,* Elsevier, Amsterdam, 1973, 128.
2. **Sanchez, P. A.,** *Properties and Management of Soils in the Tropics,* John Wiley & Sons, New York, 1976, 162.
3. **Theng, B. K. G., Tate, K. R., Sollins, P., Moris, N., Nadkarni, N., and Tate, R. T., III,** Constituents of organic matter in temperate and tropical soils, in *Dynamics of Soil Organic Matter in Tropical Ecosystems,* Coleman, D. C., Oades, J. M., and Uehara, G., Eds., University of Hawaii Press, Honolulu, 1989, chap. 1.

4. **Oades, J. M., Gillman, G. P., Uehara, G., Hue, N. V., van Noordwijk, M., Robertson, G. P., and Wada, K.**, Interactions of soil organic matter and variable-charge clays, in *Dynamics of Soil Organic Matter in Tropical Ecosystems,* Coleman, D. C., Oades, J. M., and Uehara, G., Eds., University of Hawaii Press, Honolulu, 1989, chap. 3.
5. **Virakornphanich, P., Wada, S. I., and Wada, K.**, Metal-humus complexes in A horizons of Thai and Korean red and yellow soils, *J. Soil Sci.,* 39, 529, 1988.
6. **Yoshinaga, N., Kato, Y., Nakai, M., Ikeuchi, S., and Maeda, F.**, Clay mineralogy of red- and yellow-colored soils from Korea, *Soil Sci. Plant Nutr.*, 32, 113, 1986.
7. **Yoshinaga, N., Kato, Y., and Nakai, M.**, Mineralogy of red- and yellow-colored soils from Thailand, *Soil Sci. Plant Nutr.*, 35, 181, 1989.
8. **Wada, K.**, The distinctive properties of Andosols, in *Advances in Soil Science,* Vol. 2, Stewart, A. W., Ed., Springer-Verlag, New York, 1985, 173.
9. **Wada, K. and Higashi, T.**, The categories of aluminum- and iron- humus complexes in Ando soils determined by selective dissolution, *J. Soil Sci.*, 27, 357, 1976.
10. **Bremner, J. M. and Lees, H.**, Studies on soil organic matter. II. The extraction of organic matter from soil by neutral reagents, *J. Agric. Sci.,* 39, 274, 1949.
11. **McKeague, J. A., Brydon, J. E., and Miles, N. M.**, Differentiation of forms of extractable iron and aluminum in soils, *Can. J. Soil Sci.*, 48, 27, 1971.
12. **Higashi, T. and Wada, K.**, Size fractionation, dissolution analysis, and infra-red spectroscopy of humus complexes in Ando soils, *J. Soil Sci.,* 28, 653, 1977.
13. **Virakornphanich, P., Wada, K., and Wada, S.-I.**, The amount and nature of humus formed from alfalfa meal in Ultisol, Oxisol, Alfisol and Inceptisol samples in short-term incubation, *Bull. Inst. Trop. Agric., Kyushu Univ.*, 10, 1, 1987.
14. **Wada, K.**, Allophane and imogolite, in *Minerals in Soil Environments,* 2nd ed., Dixon, J. B. and Weed, S. B., Eds., Soil Science Society of America, Madison, WI, 1989, chap. 21.
15. **Shoji, S. and Fujiwara, Y.**, Active aluminum and iron in the humus horizons of Andosols from northeastern Japan: their forms, properties and significance in clay weathering, *Soil Sci.,* 137, 216, 1984.
16. **Wada, K. and Okamura, Y.**, Electric charge characteristics of Ando A1 and buried A1 horizon soils, *J. Soil Sci.,* 31, 307, 1980.
17. **Harada, Y. and Inoko, A.**, Cation-exchange properties of soil organic matter. I. Effects of conditions for the measurement of cation-exchange capacity values of humic acid preparations, *Soil Sci. Plant Nutr.,* 21, 361, 1975.
18. **Yonebayashi, K.**, Studies on organo-mineral complexes in soils, *Sci. Rep. Kyoto Prefect. Univ. Agric., (Kyoto Furitsu Daigaku Gakujutsu Hokoku),* 28, 121, 1976.
19. **Martin, A. E. and Reeve, R.**, Chemical studies of Podzolic illuvial horizons, III. Titration curves of organic-matter suspensions, *J. Soil Sci.,* 9, 89, 1958.
20. **Tokashiki, Y., Kakauto, Y., Wada. S.-I., and Wada, K.**, Forms of active aluminum and iron and phosphorus sorption in Thai and Korean red-yellow soils (continued report), *Abstr. Jpn. Soc. Soil Sci. Plant Nutr.,* 29, 29, 1983.
21. **Mizota, C., Carrasco, M. A., and Wada, K.**, Clay mineralogy and some chemical properties of Ap horizons of Ando soils used for paddy rice in Japan, *Geoderma,* 27, 225, 1982.

Chapter 4

Sorption of Ions by Soil Organic Matter and Clay-Organics at Low Ionic Strength

R. L. Parfitt, L. K. Heng, and M. D. Taylor

CONTENTS

I. INTRODUCTION

The problem of understanding the reaction, mobility, and leaching of ions in soils is extremely complicated but is important because of the environmental pollution caused by unwanted ions and because of losses of nutrient ions through leaching.

In the field, ions react with soils in the presence of water which occupies pores of various dimensions. The water in the small pores is less mobile than the water in the large pores. Therefore, there will be more opportunity for interaction between ions and charged surfaces in the small pores, but because of diffusion between pores, these interactions will influence the concentration of ions in the larger pores.[1]

The charged surfaces are far from clean. The clays may be coated with organic matter and hydrous oxides, influencing both the charge and the size and shape of microaggregates. Soil organic matter is composed of extremely complex molecules which carry a net negative charge in the normal soil pH range. Ions added to the topsoil will be strongly influenced by the charge on the organic matter which is concentrated in this zone.

This review focuses on the sorption of ions on soil organic matter at low ionic strength, which may closely represent the field situation. The interaction of ions with the clay-organic complex in soil samples is also considered.

II. CONCENTRATIONS OF IONS IN SOIL SOLUTION

Under pasture, in soils with udic soil moisture regimes, or wetter, the ionic strength of freshly collected soil solution was found to be quite low, typically 0.005, for much of the

0-87371-914-X/95/$0.00+$.50

year.[2] The ionic strength increased with summer moisture deficit and decreased with winter rainfall, giving a range of 0.003 to 0.016. Calcium was the dominant cation, the concentration ranging from 0.2 to 3.0 mmol l^{-1}; nitrate was the dominant anion, ranging from 0.1 to 6.0 mmol l^{-1}.

In the example given in Table 1 for a soil in an orchard, the calcium concentration in winter was 1.4 mmol l^{-1}. When fertilizer was added in the spring, after 30 d during which time 50 mm of precipitation occurred, the concentration increased to about 12 mmol l^{-1}. The concentration then decreased rapidly as a result of plant uptake and leaching.

For more arid soil moisture regimes, the ionic strength of the soil solution will be higher, and sodium may replace calcium as the predominant cation in the most arid soils.

The negative charge on soils has been found to increase linearly with the square root of the ionic strength of the soil solution in the range 0.001 to 1.[3] Regression analysis has shown that there is a strong relationship between the change in negative charge with ionic strength, organic carbon, and extractable Al.

III. SOIL ORGANIC MATTER AND HUMIC SUBSTANCES

Soil organic matter contains plant debris together with humus, which by definition has none of the cellular organization of plant material.[4] Where soil conditions favor faunal and microbial activity, most of the plant debris is rapidly modified and incorporated into humus. Humus in soils has been difficult to study because of interaction with the cations Ca, Al, and Fe, and with clays and hydrous oxides. Therefore, NaOH has been used to extract soil organic matter, which yields fulvic acid and humic acid fractions; much of our knowledge comes from the study of these fractions.[4]

Molecular weights range from 1000 Da for fulvic acid, up to more than 100,000 for humin. Fulvic acid has the highest oxygen content and the lowest nitrogen content. Humic acids contain 290 to 440 g C kg^{-1} and 41 to 55 g N kg^{-1} and typically have three to four charged carboxylate groups per 1000 Da.[4]

A current model for humic substances in soil, which is gaining acceptance, is a linear polymer which forms a random coil of roughly spherical geometry.[4-6] The size of the coil depends on the molecular weight, the extent of cross-linking on the polymer, the number of charged functional groups on the polymer, and the extent of solvent penetration. In soils, the charge on the functional groups is influenced by the pH of the soil solution, the nature of the cations adsorbed on the humus, and the ionic strength of the soil solution.[6]

Solid-state ^{13}-NMR of soil humic substances shows a prominence of alkyl C in the spectra and a smaller amount of aromatic C,[4,7,8] suggesting that highly aliphatic polymers are the major constituents of soil organic matter.[8] The humic substances found in the clay fraction are also highly aliphatic and appear to be associated with the clay and have a large specific surface area.[7,8] In contrast the O-alkyl C constituents, mainly carbohydrate C, predominate in the larger size fractions.

Since humic substances behave as flexible linear polyelectrolytes at natural pH values, the carboxyl groups probably occur along the length of the polymers.[4,6] Some groups are titrated rapidly and are on the outer shell of the coils, but others are titrated slowly and are within the coils.

At low ionic strength, where monovalent cations (Na, K) balance the charges, the cations will not be in close proximity to each charged site, and the coils will tend to be expanded due to repulsion between charged carboxylate groups. As the ionic strength of the soil solution increases, the ions will be brought closer to each charged site, and the coils will tend to contract.[5] Data from acid-base titrations of humic acids in NaCl solutions (0.001 to 0.100 mol l^{-1}) are consistent with the behavior of charged cylindrical macromolecules with radii ranging from 1.1 nm at 10% ionization to 0.24 nm at 90% ionization.[6] Where weakly dissociating di- or trivalent cations balance the charged sites,

Table 1 Ions in Soil Solution (mmol l^{-1}) Before and After Adding Fertilizer, Apple Orchard, 0 to 10 cm, Hapludult

	NO_3	SO_4	Cl	Ca	Mg	K
Winter	2.0	0.3	0.8	1.4	0.1	0.4
Spring	17.2	4.2	4.4	12.6	1.8	1.3
Summer	1.3	0.6	1.5	1.4	0.1	0.2

Note: N and K, 80 kg ha^{-1}; S, 110 kg ha^{-1}.

there will be less intramolecular repulsion, and the polymers will be more tightly coiled and tend to have spherical geometry. However, because cation exchange can occur rapidly, there is probably a degree of openness in the molecules.

IV. ACID FUNCTIONAL GROUPS

The most distinctive feature of the acid-base properties of "clean" fulvic acids is the high proportion of carboxylate groups with log K_i (protonation constant) values grouped about a weighted mean of 2.6 to 2.7 (50 to 70%).[9] These groups will be substantially dissociated in dilute aqueous solution. It was suggested[9] that a high proportion of these groups belong to isolated monoprotic entities such as amino acids and salicylates or diprotic moeities such as malonates. The results[9] indicated a small proportion (10 to 20%) of polyprotic carboxyl moieties in fulvic acids.

For humic acids, the pH values at half neutralization, pK(a = 0.5), decreased from 5.6 to 4.9 as ionic strength increased from 0.001 to 0.1 mol l^{-1}.[6] Extrapolation of acid-base titration data to zero ionic strength gave values too high for simple carboxylic acids and indicated electrostatic effects arising from the macromolecular nature of the humic acid. When the electrostatic effects were subtracted, the results indicated the presence of more than a single weakly acidic group.[6] This is consistent with numerous chemical degradation studies that show a variety of types of carboxyl groups in humic acids.[10]

The total concentration of acidic groups in humic substances has recently been measured at 350 to 380 cmol[–] kg^{-1} for humic acids[6] and >400 cmol[–] kg^{-1} for fulvic acids.[11] The charge at pH 7 was close to 200 cmol[–] kg^{-1} for humic acids.[6]

Most of these data, however, have been gathered using Na-saturated humic substances, and there are little data for Ca-saturated humic substances. The adsorption of triazines and paraquat has been studied on Ca-humate and was found to be less than on Na-humate.[12] The data were consistent with the lower accessibility of carboxylate groups when Ca was present.

V. INTERACTIONS OF SOIL ORGANIC MATTER AND SOIL CLAYS

The topic of the interaction between organic matter and clays has been thoroughly reviewed by several authors (e.g., References 4, 8, and 13). The charge on either organic matter or clays can be modified when they are close enough to interact in soils.

The different clay-sized minerals fall into two main groups: those with permanent negative charge and those with variable charge.[8] The clays with permanent negative charge, such as smectite, are more likely to interact with humic substances in soils if Ca or Al ions are present since these cations can form bridges between clay and humic molecules.[4,13] Both entropy and enthalpy factors favor the formation of these complexes, and humic substances are stabilized against decomposition. The interaction probably depends on the specific surface area, the charge, and the composition of the clay, but this is a field requiring further research.

The clay-sized minerals having variable charge include allophane, iron oxides, and clays with broken AlO and FeO bonds.[8] These clays react readily with the carboxyl groups on humic substances, forming bonds to Al and Fe atoms on the surfaces of the clays. However, because of steric factors on the humic molecules, not all of the carboxyl groups can react with the surface.[8]

The presence of organic matter in Andisols and Oxisols has the effect of lowering the point of zero charge of the clay-sized minerals with variable charge.[4,8] This may be because of the preponderance of negative charge on the humic substances, or it may arise because the positive sites on the clays are partly blocked by interaction with negatively charged carboxylate groups.

Further research is needed as to the nature of the adsorbed humic molecules, but there is evidence that the adsorbed molecules in soils may have molecular weights of no more than a few thousand Da and that aliphatic components and carbohydrates are dominant.[8]

VI. CALCIUM ADSORPTION

Calcium is the cation which usually predominates in soils, but there have been few studies of calcium adsorption at low ionic strength on the clay-organic complex and at a range of pH values. Early investigations with humic substances and with peat have shown that the preference for alkaline-earth ions is Mg < Ca < Sr < Ba (see Reference 14). Recent results gave the order Mg < Sr < Ca < Ba and suggested that exchange behavior occurred at widely separated carboxylate sites and that $(RCOOM)^+$ ion pairs were formed.[14]

Where experiments have been carried out on soil samples from A and B horizons with a similar clay mineral content but different organic matter content, the effect of the charged (exchange) sites of organic matter on Ca adsorption can be observed.[8,15,16]

In these experiments, at pH values below 5, aluminum occupies some of the exchange sites, and the sum of calcium plus aluminum gives an estimation of the negative charge of the sample. No exchangeable aluminum was detected above pH 5, probably because of stronger adsorption and/or precipitation of hydroxy-Al species at higher pH.

For an allophanic soil, adsorption increased with pH for both the A horizon and the B horizon (Figure 1a). Although the allophane contents were both about 200 g kg^{-1}, more calcium was adsorbed by the A horizon than by the B horizon. This results from the greater amount of organic matter (129 g C kg^{-1}) in the A than in the B horizon (24 g C kg^{-1}). There is a gradual increase in Ca adsorption as the pH increases. The groups which are active in adsorbing calcium are the carboxylate groups in organic matter together with the AlO^- groups in allophane,[17] and more groups dissociate as the pH increases.[6] At pH 7, there are about 250 $cmol_c$ kg^{-1} of carboxylate sites on organic matter in New Zealand soils and 50 $cmol_c$ kg^{-1} on allophane.[42] At pH 5.5, about half of these sites are dissociated.

For a recent soil which contained smectite, the adsorption of calcium changed little with pH for the B horizon (7 g C kg^{-1}) (Figure 1b). However, for the A horizon, the adsorption increased with pH. The increase may also be ascribed to the dissociation of carboxyl groups on the organic matter in the A horizon (31 g C kg^{-1}). Thus, at low ionic strength, adsorption of cations such as calcium increases with pH when organic matter is present in soils. The increase is noticeably steeper between pH 6 and 7.

VII. POTASSIUM ADSORPTION

Allophane generally has a low preference for potassium; therefore, a study of potassium adsorption in A and B horizons of allophanic soils can give information about adsorption on humic substances.[18] At low ionic strength, more potassium was adsorbed by the A horizon than by the B horizon for any one equilibrating solution (Table 2, Figure 2)

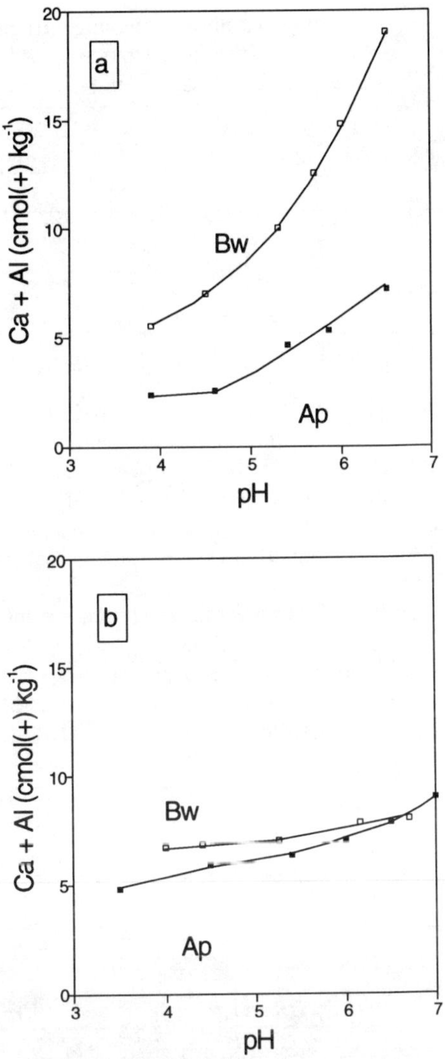

Figure 1 Calcium and aluminum adsorption measured at different pH for (a) an allophanic soil and (b) a recent soil, using 2 mmol l^{-1} $CaCl_2$. No exchangeable aluminum was detected above pH 5.[15]

because more sites, arising from the presence of humic substances, were present in the A horizon. However, the potassium made up a smaller proportion of the total adsorbed cations in the A horizon.

When the preference of the soil for potassium is calculated from these data using the Gaines and Thomas convention,[18] the results show that both A and B horizons have a preference for potassium over calcium at typical soil solution concentrations (Figure 3). The preference decreases with increasing amount of potassium, and the preference in the A horizon is less than the preference in the B horizon. This probably arises from a greater proportion of potassium-selective sites in the B horizon[18] which may be present on minerals such as chlorite and vermiculite.[19]

When potassium is added to topsoils, it will be held initially in the topsoil by the potassium-selective sites, and calcium will be leached from the soil. If potassium is added in large amounts so that most of the selective sites become occupied, potassium will move lower in the soil profile. Some potassium may be lost by leaching during intense rain events by the flow of soil solution through larger pores.[20]

Table 2 Ions in Soil Solution and on Exchange Sites for a Column of Allophanic Soil at pH 5.7 Equilibrated for 16 h[18]

		Solution (mmol l^{-1})			Exchange (cmol(+) kg^{-1})		
Horizon	Depth (cm)	Ca	Mg	K	Ca	Mg	K
Ap	0–10	1.4	0.5	0.4	16.0	2.4	0.6
Bw	20–30	1.4	0.5	0.4	5.3	1.0	0.4

VIII. ALUMINUM ADSORPTION

There is strong evidence that Al is complexed with humic substances not only in Spodosols but also in acid Inceptisols and Andisols.[21,22] The Al can be estimated by extraction with Na-pyrophosphate reagent[23,24] and only a small proportion of this Al is exchangeable with KCl (1 mol l^{-1}), suggesting that the Al is strongly complexed within the coiled molecules. For some Andisols, the total C/Al-pyrophosphate atomic ratio decreases from 30 to 15 as the age of the organic matter increases,[25] indicating that more Al is complexed with humic substances with time.

When Al ions are added to humic substances, the adsorbed Al does not seem to react with base during titrations; the Al ions also reduce the cation exchange capacity of the humic substances,[26,27] probably as a result of the formation of complexes.

The complexes involve Al ions which carry a positive charge smaller than 1.0 and which are bonded to OH ions. The OH/Al molar ratio is in the range 0.7 to 2.5 for

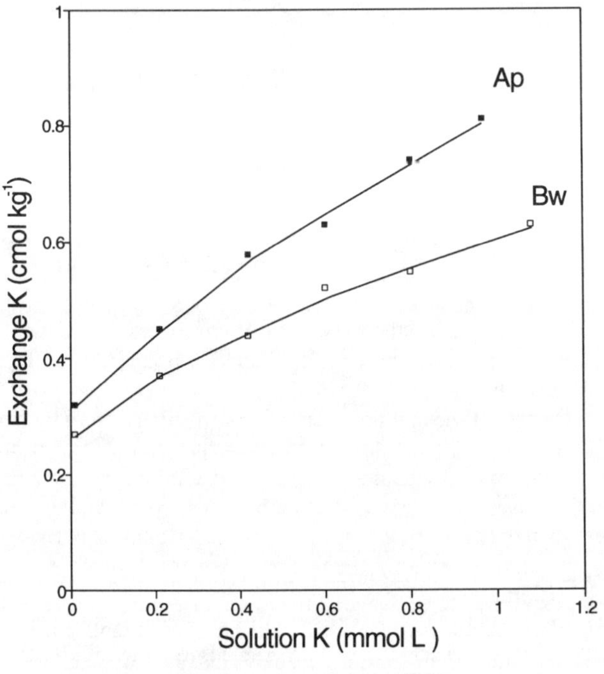

Figure 2 Exchange K adsorbed in a soil column equilibrated with solutions containing 4.2 mmol l^{-1} nitrate and increasing K/Ca ratio for an allophanic soil.[18]

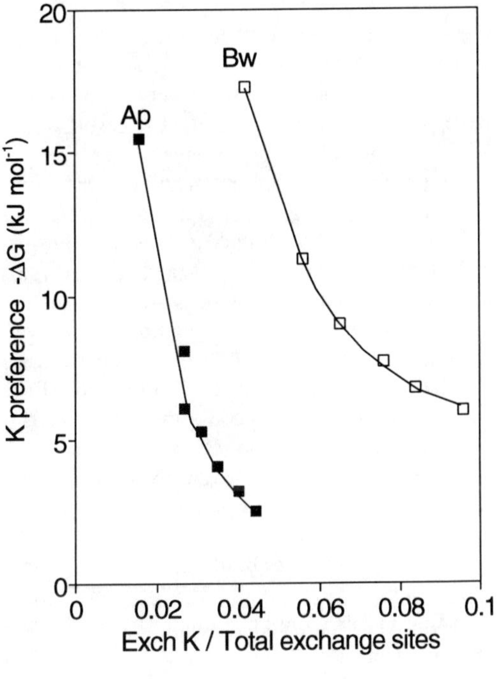

Figure 3 K preference plotted against fraction of K on exchange sites for an allophanic soil.[18]

complexes in Andisols, and the polyhydroxy-Al species appear to increase as the soils became older.[28] Al-humus complexes have high phosphate adsorption capacity, which is consistent with the presence of exchangeable OH ligands.

A recent model, which incorporates complexation and nonspecific ion-exchange, accounts for acid-base titration data and solution Al data.[29] The model also uses two types of carboxyl groups dissociating at pH <7, weak acid groups dissociating at pH >7, which would include weak carboxylic acids and phenolic groups, and a single equilibrium constant for Al reacting with humic samples ($K = 10^{-3.3}$ to $10^{-3.8}$) similar to those for simple diprotic acids. The model gives reasonable predictions of solution pH and concentration of Al^{3+}.

Because Al is strongly complexed with humic substances and is present in all soils, some Al is likely to be adsorbed on most soil humic molecules.[29] In most soils, the amounts may be small since pyrophosphate-extractable Al values are often <2 g kg^{-1} where the organic carbon is about 30 g kg^{-1}. Some carboxylate groups, however, will be blocked by Al and will not be available for exchange with other cations.

IX. FUNCTIONAL GROUPS AND METAL ADSORPTION, INCLUDING HEAVY METALS

Different metals react with different humic substances in their own characteristic manner because of different amounts and types of functional groups, and variations in stability constants.[30] The reaction is also influenced by the degree of metal loading on the humic substance and the pH of the soil solution.[31]

Some metals adsorb by ion exchange, but others form inner-sphere complexes where the metal ions replace the water molecules adjacent to the functional groups.[29]

For fulvic acids, it is thought that metals such as Cu react with ligands with salicylate , phthalate-, malate-, catechol-, or acetylacetone-like arrangements.[11,30,32] However, when metallic impurities were removed from fulvic acids, at low ionic strength and in the pH range 4.0 to 6.5, complexing of Cu and Cd was best modeled by carboxylic acids such as citric acid and malic acid.[33]

The adsorption of Cu by smectite was reduced when fulvic acid or carboxylic acids were present.[34] This was thought to result from the lowered activity of Cu in solution when organic complexing agents were present.

Since N and S are present in humic substances, it is important to consider a wide range of ligands. For any given complexing ligand, the metals form complexes that increase in stability as the metal acidity increases (radius decreases), according to the Irving-Williams series: Ba, Sr, Ca, Mg, Mn, Co, Ni, Cu, and Zn.[31] Experimental results for metal adsorption on humic substances are consistent with this series.[35]

The relative stability of the complexes also varies with the nature of the ligand and the presence of O, N, or S.[31] The "softer" acids, those with greater numbers of d-electrons (Ni, Cu), have a stronger preference for the softer ligands (containing N), but the harder acids (Ca, Mg) have a preference for the hard ligands (containing O). By the Pearson classification of Lewis acids, the relatively soft acids such as Cd and Pb should prefer the soft S-containing ligands such as sulfhydryl groups in humic substances. In contrast, the hard acids (Ca, Mg) will bond preferentially to carboxylate ligands. Therefore, at low metal loading, Cd is strongly preferred to Ca through reaction at specific sites containing S ligands. Cu and Zn are borderline acids with intermediate behavior.

For humic acids, Cu and VO form inner-sphere complexes, but most of the first-row transition metals (Mn, Fe(II), and Co) and Ca and Mg appear to form outer-sphere complexes.[31] Thus, the nonchelating polycarboxylic acids may be better models of the metal bonding sites on humic substances than chelating molecules. Exceptions are the trivalent metals such as Fe(III), which has a behavior consistent with the Fe complexes of polyphenols. The bonding of Fe(III) is further complicated, however, because the formation of solid hydroxides such as goethite or ferrihydrite is both thermodynamically favorable and rapid over a wide range of conditions of pH, temperature, and ionic strength.[4,8,31,32]

Metals can generally be desorbed from humic substances, but when iron oxides are present, the desorption of metals is reduced.[35] This may arise from precipitation reactions or from the diffusion of metals into the pores of the oxides. These are complicated systems, and further research is required.

X. CADMIUM ADSORPTION AT LOW LOADING

Low loading of cadmium to soils occurs when phosphate fertilizers containing cadmium are added. If fertilizer is added at the rate of 500 kg ha^{-1} for 40 years and the fertilizer contains 50 mg kg^{-1} of cadmium, the concentration of total cadmium in the soil (0 to 10 cm) will increase by 1 mg kg^{-1}, assuming there are no losses. Increases of this magnitude have been measured on high-producing farms in New Zealand where phosphate fertilizer has been applied annually for many years (Table 3). The cadmium in topsoils is higher than in subsoils (Table 4), and at these relatively low loadings, the cadmium is probably strongly adsorbed at specific sites, such as the sulfhydryl groups in humic substances, and it does not move to the subsoil.

Total cadmium in samples (0 to 15 cm) of an allophanic soil was correlated ($r^2 = 0.68$) with total carbon, and this is consistent with cadmium being adsorbed on sites in organic matter (Figure 4).

A large proportion of the cadmium was desorbed from the allophanic soil when shaken with 0.05 mol l^{-1} sodium chloride for 3 h at pH 2.5, which are the conditions present in the digestive tract of grazing animals (Figure 5). The dissolution of cadmium with decreasing pH followed the desorption of aluminum, but not iron, in this soil, and this is consistent with cadmium also being adsorbed on the surface of allophane by forming inner-sphere complexes.[31]

Table 3 Total Cd in Two Archived Soils and Soils Sampled in 1991 (0 to 15 cm)

	Gley soil		Allophanic soil	
	1941	**1991**	**1960**	**1991**
Cd (mg kg^{-1})	0.5	0.9	0.7	1.8

Table 4 Total Cd Measured at Different Soil Depths

Horizon	Depth (cm)	Allophanic soil (mg kg^{-1})	Gley soil (mg kg^{-1})
Ap	0–2	1.5	0.9
	2–7	1.3	0.8
	7–15	1.5	0.6
Bw	15–25	1.0	0.5
	25–35	0.9	0.4

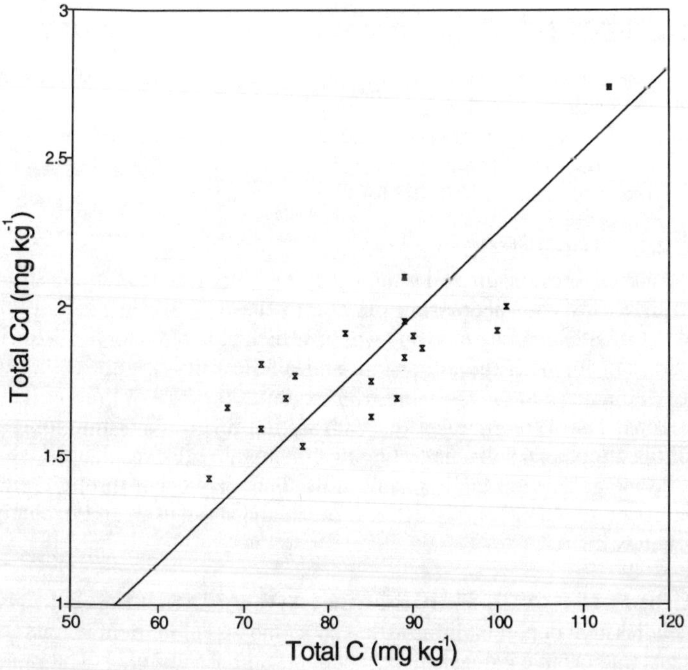

Figure 4 Total Cd in samples of allophanic soil (0 to 15 cm) containing different amounts of organic C in one field.

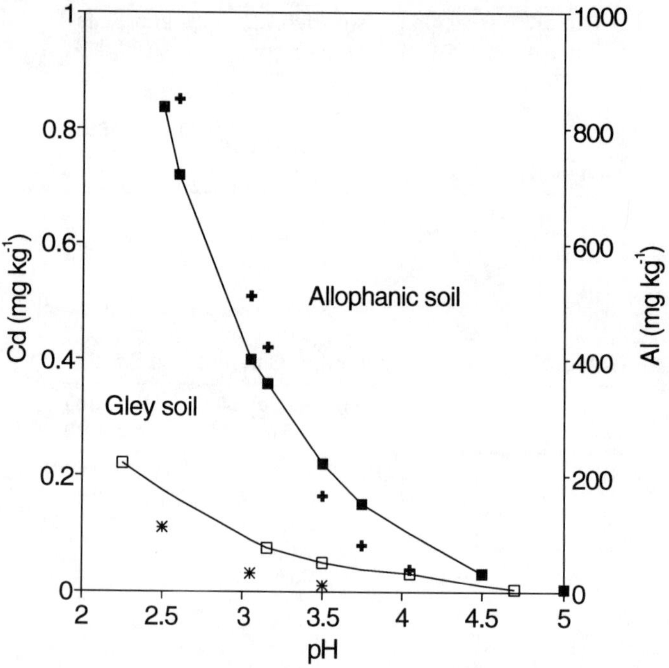

Figure 5 Cd (square symbols) and Al (cross symbols) dissolved from an allophanic soil and a gley soil at different pH levels using 0.05 mol l^{-1} NaCl.

XI. EFFECT OF ORGANIC MATTER ON ANION SORPTION

Anions adsorb on AlOH and FeOH sites in soil, and there is evidence that these sites may be blocked by organic matter, thereby reducing anion adsorption.[36] Chloride and nitrate are weakly adsorbed, forming outer-sphere complexes. On the other hand, phosphate is strongly adsorbed because it forms an inner-sphere complex. Sulfate shows intermediate behavior, and there is debate about the form of the complex.

A. CHLORIDE ADSORPTION

When chloride adsorption on soil samples is measured at low ionic strength, the adsorption increases with decreasing pH due to the increase in positive charge on AlOH and FeOH sites.[15] More chloride was adsorbed at low pH on the B horizon than on the A horizon for both the allophanic and oxidic soils (Figure 6), although the amounts of allophane (200 g kg^{-1}) and iron oxide (200 g kg^{-1}) were similar in the A and B horizons. The difference between adsorption on A and B horizons was most marked for the allophanic soil, suggesting that the humic substances block some of the sites for chloride adsorption in allophanic soils. This may occur through outer-sphere interaction between $-COO^-$ and $-AlOH_2^+$ groups and through ligand exchange where –COOAl- bonds form.

B. CHLORIDE, NITRATE, FLUOROACETATE, AND SULFATE LEACHING

Leaching studies give important information about the loss of nutrient anions (NO_3, SO_4) and pollutants (radioactive iodine and technetium, fluoroacetate, etc.). The losses may be accelerated in the presence of organic matter.

When a 5-ml pulse containing 50 μmol_c of anion was added to the top of a small soil column and then leached with 5-ml pulses of distilled water, fluoroacetate and chloride

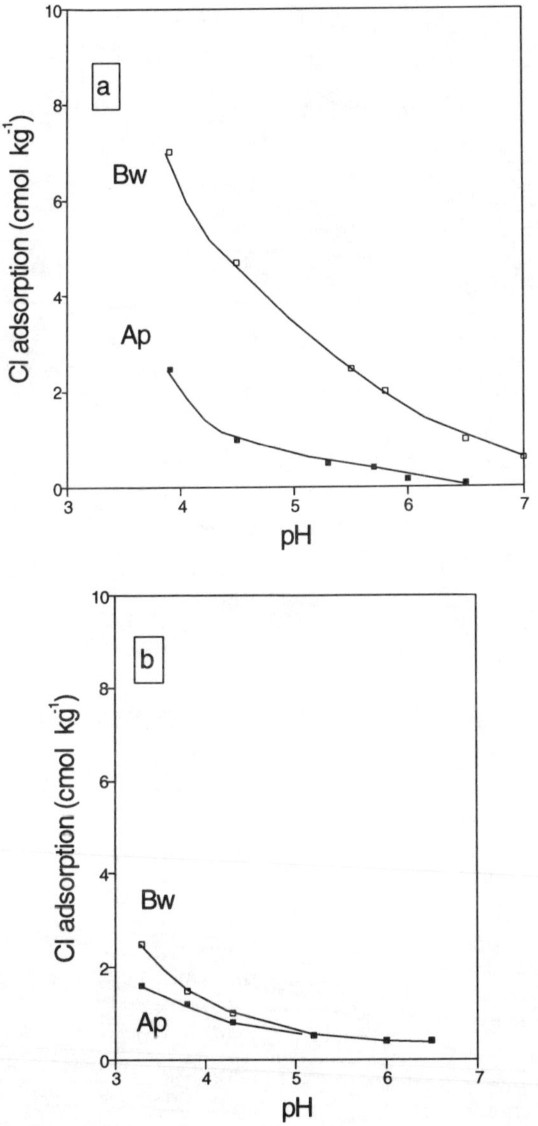

Figure 6 Cl adsorption measured at different pH for (a) an Allophanic soil and (b) an Oxidic soil using 2 mmol l^{-1} $CaCl_2$.[15]

gave breakthrough curves which were very close to those of nitrate (Figure 7). Nitrate was eluted completely from the column in 2.5 pore volumes for a brown topsoil, 3.1 pore volumes for an allophanic topsoil, and 3.5 pore volumes for an allophanic subsoil (data not presented) in the pH range of 5 to 6. In these experiments, three pore volumes is equivalent to 150 mm of rainfall. The allophanic soil adsorbed more nitrate than the brown soil, and the subsoil adsorbed more than the topsoil, due to the greater positive charge (see Figure 6). When these data were modeled using the convective-dispersive equation,[37] there was good agreement between the data and the output from the model (Figure 7). The rate of leaching was fastest in the brown soil, which showed slight

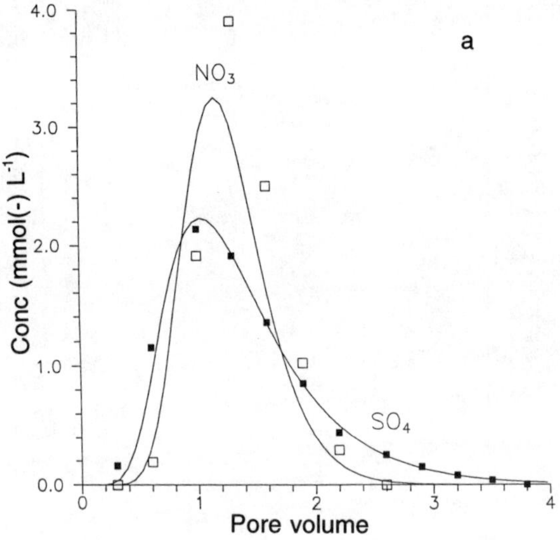

Figure 7a

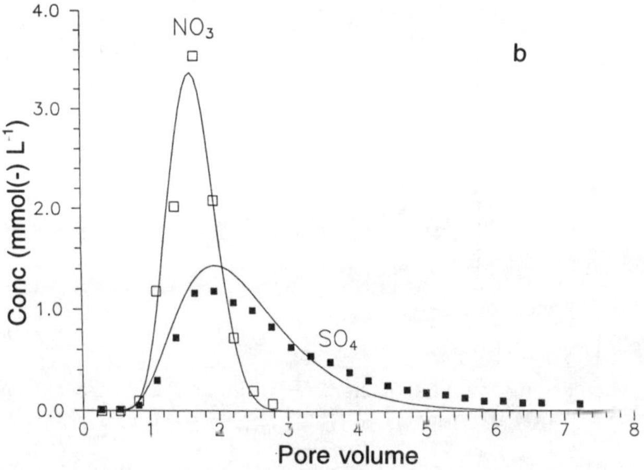

Figure 7b

negative adsorption for nitrate, and was slowest for the allophanic subsoil, which showed some adsorption in this pH range. Since, in these experiments, all other variables were very similar, the rate was probably influenced by the presence of minerals, such as allophane, which adsorb anions. The rate of leaching of these anions, which form outer-sphere complexes, can be slightly reduced, particularly at low pH where there is an increased positive charge in Andisols and Oxisols, and in the absence of humic substances which can block positive charge sites.[38]

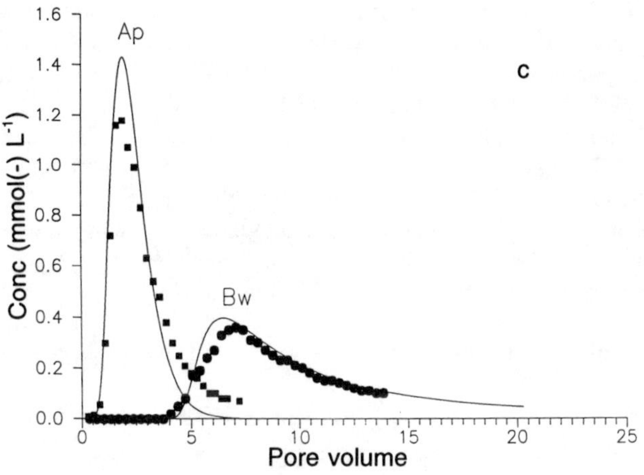

Figure 7 Breakthrough curves for anions using 8-cm columns for (a) Brown soil Ap, (b) Allophanic soil Ap, and (c) SO_4^{-2} on Allophanic soil Ap and Bw. Solid lines are modeled using the convective-dispersive equation.[37]

The elution of sulfate was slower than for nitrate (Figure 7). The breakthrough curves all have a tail, which was greatest for the allophanic subsoil (Figure 7) and least for the brown soil. van Genuchten and Wierenga[39] have listed some possible chemical and physical phenomena giving rise to tailing. The peak in the breakthrough curve occurred at two pore volumes for the allophanic topsoil and at seven pore volumes for the subsoil (Figure 7). This was due to adsorption on positively charged sites, there being more sites in the subsoil and fewer sites in the topsoil where organic matter probably blocked some of the sites (Figure 6). This is consistent with data showing that surface horizons of Oxisols adsorbed less sulfate than subsurface horizons.[8] All of the sulfate was leached from the A horizon, but for the B horizon, a part was released slowly (Figure 7). This may arise from stronger binding of sulfate by allophane in the absence of organic matter. Bolan et al.[40] found that less sulfate was leached from soils when the sulfate was incubated with soil prior to leaching, and they suggested a change in the nature of adsorption with time which indicates that the desorption of sulfate is not reversible. Experiments with lysimeters, however, show that some leaching occurs through macropores, and considerable losses of sulfate have been observed.[41]

XII. SUMMARY AND CONCLUSIONS

Soil organic matter contains humic substances, which are extremely complicated linear polymers and which carry negative charges due to the dissociation of many different carboxyl groups. When Ca or Al ions are present, the polymers form coils of spherical geometry. Al ions and hydroxy-Al ions may be present largely within the coils as inner-sphere complexes which are not readily exchangeable. Ca ions, on the other hand, probably form outer-sphere complexes which are exchangeable. In the pH range of 5 to 7, the negative charge increases markedly with pH, and the number of outer-sphere complexes with Ca will also increase. The carboxyl groups appear to be related largely to aliphatic acids.

The adsorption of many other metals depends on the loading as well as the pH. Thus, small amounts of K are adsorbed with a high preference by the organic-clay complex in A horizons. As the loading increases, the preference decreases.

At low loading, heavy metals are also adsorbed strongly by humic substances and thus are retained in A horizons. The soft metals Cd and Pb probably interact with the soft S-containing ligands. Cu probably forms inner-sphere complexes with N-containing ligands and polycarboxylic ligands.

The adsorption of anions by soils is reduced by the presence of organic matter, which blocks AlOH and FeOH adsorption sites. This is most clearly observed for sulfate in Andisols, where leaching is more rapid through A horizons than through B horizons.

It is expected that the effects of organic matter on ion adsorption will increase as the content of the organic matter increases. The effect will be particularly important in sandy soils where there are few other exchange sites, and in Oxisols and Andisols where the charge characteristics of the oxides are modified by the presence of organic molecules. Clearly, more research is required so that field conditions can be modeled, taking account the interactions of Al with humic substances and using realistic loadings of ions at low ionic strength.

ACKNOWLEDGMENTS

The authors are grateful to soil survey and soil analysis colleagues for their assistance, and to Bob Lee and Benny Theng for comments on the manuscript.

REFERENCES

1. **Schulin, R., Papritz, A., Fluhler, H., and Selim, H. M.,** Calcium and magnesium transport in aggregated soils at variable ionic strength, *Geoderma,* 44, 129, 1989.
2. **Edmeades, D. C., Wheeler, D. M., and Clinton, O. E.,** The chemical composition and ionic strength of soil solutions from New Zealand topsoils, *Aust. J. Soil Res.,* 23, 151, 1985.
3. **Black, A. S. and Campbell, A. S.,** Ionic strength of soil solution and its effect on charge properties of some New Zealand soils, *J. Soil Sci.,* 33, 249, 1982.
4. **Oades, J. M.,** An introduction to organic matter in mineral soils, in *Minerals in Soil Environments,* Dixon, J. B. and Weed, S. B., Eds., Soil Science Society of America, Madison, WI, 1989, chap. 3.
5. **Hayes, M. H. B., MacCarthy, P., Malcolm, R. L., and Swift, R. S.,** *Humic Substances II,* Wiley-Interscience, Chichester, 1989.
6. **Barak, P. and Chen, Y.,** Equivalent radii of humic macromolecules from acid-base titration, *Soil Sci.,* 154, 184, 1992.
7. **Wilson, M. A.,** *NMR Techniques and Applications in Geochemistry and Soil Chemistry,* Pergamon Press, Oxford, 1987.
8. **Coleman, D. C., Oades, J. M., and Uehara, G.,** *Dynamics of Organic Matter in Tropical Ecosystems,* University of Hawaii, Honolulu, 1989.
9. **Gregor, J. E. and Powell, H. K. J.,** Protonation reactions of fulvic acids, *J. Soil Sci.,* 39, 243, 1988.
10. **Stevenson, F. J.,** *Humus Chemistry,* Wiley-Interscience, New York, 1982.
11. **Ephraim, J. H., Reddy, M. M., and Marinsky, J. A.,** Ion binding by humic substances: considerations based on the solution chemistry and heterogeneity of humic substances, in *Humic Substances in the Aquatic and Terrestrial Environments,* Allard, B., Boren, H., and Grimvall, A., Eds., Springer-Verlag, 1991, 263.
12. **Hayes, M. H. B. and Mingelgrin, U.,** Interactions between small organic chemicals and soil colloidal constituents, in *Interactions at the Soil Colloid — Soil Solution Interface,* Bolt, G. H., DeBoodt, M. F., Hayes, M. H. B., and McBride, M. B., Eds., Kluwer, Dordrecht, 1991, 375.
13. **Theng, B. K. G.,** Clay-humic interactions and soil aggregate stability, in *Soil Structure and Soil Aggregate Stability,* Rengasamy, P., Ed., Institute for Irrigation and Salinity Research, Tatura, Australia, 1986, 32.
14. **Baes, A. U. and Bloom, P. R.,** Exchange of alkaline earth cations in soil organic matter, *Soil Sci.,* 146, 6, 1988.

15. **Parfitt, R. L.,** Surface charge in some New Zealand soils measured at typical ionic strength, *Aust. J. Soil Res.,* 30, 331, 1992.
16. **Gillman, G. P. and Sumpter, A. S.,** Surface charge characteristics and lime requirements of soil from basaltic, granitic, and metamorphic rocks in high rainfall tropical Queensland, *Aust. J. Soil Res.,* 24, 173, 1986.
17. **Parfitt, R. L.,** Allophane in New Zealand — a review, *Aust. J. Soil Res.,* 28, 343, 1990.
18. **Parfitt, R. L.,** Potassium - calcium exchange in some New Zealand soils, *Aust. J. Soil Res.,* 30, 145, 1992.
19. **Goulding, K. W. T.,** Thermodynamics and potassium exchange in soils and clay minerals, *Adv. Agron.,* 36, 215, 1983.
20. **Williams, P. H., Gregg, P. E. H., and Hedley, M. J.,** Mass balance modelling of potassium losses from grazed dairy pasture, *N. Z. J. Agric. Res.,* 33, 661, 1990.
21. **Parfitt, R. L. and Claydon, B.,** Andisols — a view of the new order in Soil Taxonomy, *Geoderma,* 49, 181, 1991.
22. **Stevenson, F. J. and Vance, G. F.,** Naturally occurring aluminum-organic complexes, in *The Environmental Chemistry of Aluminum,* Sposito, G., Ed., CRC Press, Boca Raton, FL, 1989.
23. **Shoji, S. and Yamada, H.,** Comparisons of mineralogical properties between tephra-derived Spodosols from Alaska and nontephra-derived Spodosols from New England, *Soil Sci.,* 152, 162, 1991.
24. **Parfitt, R. L. and Childs, C. W.,** Estimation of forms of Fe and Al: a review and analysis of contrasting soils using dissolution and Moessbauer methods, *Aust. J. Soil Res.,* 26, 121, 1988.
25. **Wada, K. and Higashi, T.,** The categories of Al- and Fe-humus complexes in Ando soils determined by selective dissolution, *J. Soil Sci.,* 27, 357, 1976.
26. **Hargrove, W. L. and Thomas, G. W.,** Titration properties of Al-organic matter, *Soil Sci.,* 134, 216, 1982.
27. **Shinagawa, A., Miyuauchi, N., and Higashi, T.,** Preparation of Al-humates and their aluminum content and cation-exchange capacity, *Soil Sci. Plant Nutr.,* 28, 1, 1982.
28. **Higashi, T.,** Characterisation of Al/Fe-humus complexes in Dystrandepts through comparison with synthetic forms, *Geoderma,* 31, 277, 1983.
29. **Tipping, E. and Hurley, M. A.,** A model of solid-solution interactions in acid organic soils, based on the complexation properties of humic substances, *J. Soil Sci.,* 39, 505, 1988.
30. **Mountney, A. W. and Williams, D. R.,** Computer simulation of metal ion-humic and fulvic acid interactions, *J. Soil Sci.,* 43, 679, 1992.
31. **McBride, M. B.,** Reactions controlling heavy metal solubility in soils, *Adv. Soil Sci.,* 10, 1, 1989.
32. **Murray, K. and Linder, P. W.,** Fulvic acids: structure and metal binding. II. Predominant metal binding sites, *J. Soil Sci.,* 35, 217, 1984.
33. **Cressey, P. J., Monk, G. R., Powell, H. K. J., and Tennent, D. J.,** Fulvic acid studies: evidence for a polycarboxylate co-ordination mode at soil pH, *J. Soil Sci.,* 34, 783, 1983.
34. **McBride, M. B.,** Forms and distribution of copper in solid and solution phases of soil, in *Copper in Soils and Plants,* Loneragan, J. F., Robson, A. D., and Graham, R. D., Eds., Academic Press, New York, 1981, 25.
35. **Swift, R. S. and McLaren, R. G.,** Micronutrient adsorption by soils and soil colloids, in *Interactions at the Soil Colloid — Soil Solution Interface,* Bolt, G. H., DeBoodt, M. F., Hayes, M. H. B., and McBride, M. B., Eds., Kluwer, Dordrecht, 1991, 266.
36. **Parfitt, R. L.,** Anion adsorption by soils and soil materials, *Adv. Agron.,* 30, 1, 1978.
37. **Biggar, J. W. and Nielsen, D. R.,** Miscible displacement and leaching phenomena, in *Irrigation in Agricultural Lands,* Hagen, R. M., Ed., Agronomy Monograph, American Society of Agronomy, Madison, WI, 1967, 254.
38. **Ishoguro, M., Song, K.-C., and Yuita, K.,** Ion transport in an allophanic Andisol under the influence of variable charge, *Soil Sci. Soc. Am. J.,* 56, 1789, 1992.
39. **van Genuchten, M. TH. and Wierenga, P. J.,** Mass transfer studies in sorbing porous media. I. Analytical solutions, *Soil Sci. Soc. Am. J.,* 40, 473, 1976.
40. **Bolan, N. S., Scotter, D. R., Syers, J. K., and Tillman, R. W.,** The effect of adsorption on sulfate leaching, *Soil Sci. Soc. Am. J.,* 50, 1419, 1986.
41. **McLay, C. D. A., Cameron, K. C., and McLaren, R. G.,** Influence of soil structure on sulfate leaching from a silt loam, *Austr. J. Soil Res.,* 30, 443, 1992.
42. **Parfitt, R. L.,** unpublished data.

Chapter 5

Water Potential, Soil Microhabitats, and Microbial Development

M. Robert and C. Chenu

CONTENTS

I. INTRODUCTION

Water potential is one of the main factors that govern both the survival and activity of microorganisms in soils and the competition or predation among them.

Plants and microbes take their water from soil, and the water potential represents the energy with which water is retained in soils. It is composed of two components: the matric potential and the osmotic potential, which becomes important when the concentration of the solutes is above 10^{-2} *M*. If we exclude saline soils, the solute concentration will be below this limit; therefore, matric potential is the main factor.

If we consider the literature, water potential is often not taken into account in laboratory and field studies. There are two possible explanations: (1) water potential is difficult to monitor, especially in experiments where there are microbes, (2) the water potential units are very often confusing, which does not help!

However, water potential has a critical role in soils. It may either directly affect the microorganisms themselves or indirectly modify the microhabitats within the soil.

The direct effect is physiological: the internal water potential of microorganisms is in equilibrium with the outside water potential; hence, low water potentials induce plasmolysis of the cells. Different physiological responses of microorganisms consist of (1) osmoregulation, i.e., the production of internal solutes to adjust internal water potential, (2) forms of resistance (spores, etc.), and (3) production of extracellular polysaccharides.

The physiological effects on, or responses of, microorganisms are not discussed here. Instead, we discuss the indirect effects of soil water potential related to soil fabric. The first section is devoted to the effect of water potential on microbes through clay interactions. The second part, addresses the different soil microhabitats for microbes, as they are of major importance for the ecology of microbes in soils.

0-87371-914-X/95/$0.00+$.50

II. EFFECTS OF WATER POTENTIAL ON MICROBIAL ACTIVITIES IN SOIL

A. DEFINITION AND EXPRESSION OF WATER POTENTIAL

Table 1 presents[1] the different units currently used to express water potential (Ψ), which is the pressure or tension of the water retained in the soil. In the first column are pF values or "potential of free energy," which is the logarithm of the pressure or suction [pF = log (Ψ)]. In the second column are the equivalent pressures in mega Pascals, which are the new units to replace bars (third column). Soil water potential is always less than atmospheric pressure, and thus is negative. High water potentials correspond to suctions close to the value of zero, which represents saturation, to –0.01 MPa (pF0 to pF2); low water potentials are from –1 to –100 MPa (pF4 to 6).

In the fourth column, the relative humidity of air is presented. This last expression does not describe well the variations of potential in soils; for example, a variation from pF1 to pF4 corresponds to a variation in relative humidity from 0.9999 to 0.9927. The last column represents the maximum diameter of water-saturated pores (micrometers) for a corresponding pressure. This important characteristic is obtained by Laplace's law.

The normal water regime in soil during the year ranges from saturation (pF0) to pF4. Thus, air-dried samples subjected to very high pressures (100 MPa) represent an artifact which never occurs in soil except within the upper few centimeters, and rewetting an air-dried sample is even more disturbing.

There are strong interactions between soil fabric, specifically clay behavior, and water potential. For this subject, we refer to Robert and Chenu.[2]

B. GENERAL EFFECT

The general effect of water potential is well illustrated by classical curves.[3,4] There is a general correlation between the number and activity of microbes and water potential (Figure 1). There is an optimum close to pF2 (0.01 MPa); then, the number of organisms or their activity decreases with the water potential.

At pF5 or 5.6 (–10 to –40 MPa), microbial activity is limited.[5] This is a general tendency, but there are also large differences among organisms, bacteria being the most sensitive to decreasing water potential and actinomycetes and fungi being the most resistant (Figure 1).[4]

C. HIGH WATER POTENTIAL

At high water potentials, most pores are water saturated. In water, the diffusion of oxygen is 1/10,000 of that in air; thus, oxygen may become limiting. Oxygen limitation occurs at very different soil water contents, according to the texture or structure of soils. Doran et al.[6-8] have introduced a new parameter, the water-filled pore space (WFPS), which is determined on the basis of apparent volume and water content. A comparison of different soils is thus possible on the same curves. In Figure 2A, respiration increases with WFPS in the 20 to 60% range because available water increases and air content remains sufficient. Beyond 60 to 70% WFPS, there is a decrease in respiration due to saturation of the pore space.

For the denitrification process, which is anaerobic, an effect is illustrated (Figure 2B), denitrification increases as the water saturation of pores increases.

D. INTERMEDIATE TO LOW WATER POTENTIALS

There is a progressive desaturation of soil pores as water potential decreases. The subsequent decrease in the continuity of water films is responsible for restricted diffusion in the soil with both water and nutrient limitations.

The main effect on microbial activities in this range of water potential concerns soil shrinkage and the deformation of clay fabric as water potential decreases, which can

Table 1 Water Potential Expression With Different Units and Maximum Size of Water-Saturated Pores[1,12]

pF	Ψ (MPa)	Ψ (bars)	Relative humidity (%)	Maximum diameter of water-saturated pores (μm)
1	–0.001	–0.01	99.999	300
2	–0.01	–0.1	99.99	30
3	–0.1	–10	99.93	3
4	–1	–10	99.27	0.3
5	–10	–100	92.70	0.03
6	–100	–1000	50.00	0.003

provoke mechanical constraints on soil microorganisms. The best example for illustrating the mechanical action of clay concerns the survival of *Pseudomonas solanacearum* in soils.[9] *Pseudomonas* is an important soil-borne pathogen in the tropical region, and field studies in the West Indies have attributed the suppressiveness of Vertisols to a physical problem.

Studies were performed in the laboratory with two strains of *Pseudomonas* with both reference clays and clays extracted from the soils of the West Indies (Vertisols, Oxisols).[10,11] Figure 3 shows the main results. At constant high water potential, there is no change in the number of bacteria. On the contrary, lower water potential (pF4 or 4.4) causes a rapid decline of both strains (Figure 4).

Using a control with bacteria alone, it was shown that water potential alone could not explain the mortality of bacteria observed in the presence of clay. With TEM and cryo-SEM, the bacteria have been located in the clay matrix of the Vertisol, within the pores between clay particles.

In order to interpret the results concerning *Pseudomonas* mortality, some reference to the changes in clay organization[1,12] with fluctuations of water potential is needed (Figure 4). Between pF2 and 4 (–0.01 and –1MPa), there is a considerable decrease in porosity. The 2 μm-sized pores, where the bacteria are mostly situated, close. A mechanical compression may thus explain the bacterial mortality.

As can be seen in Figure 4, such a deformation of the pores with water potential exists for all the clay minerals having 1- or 2-μm particles. Some explanations can be put

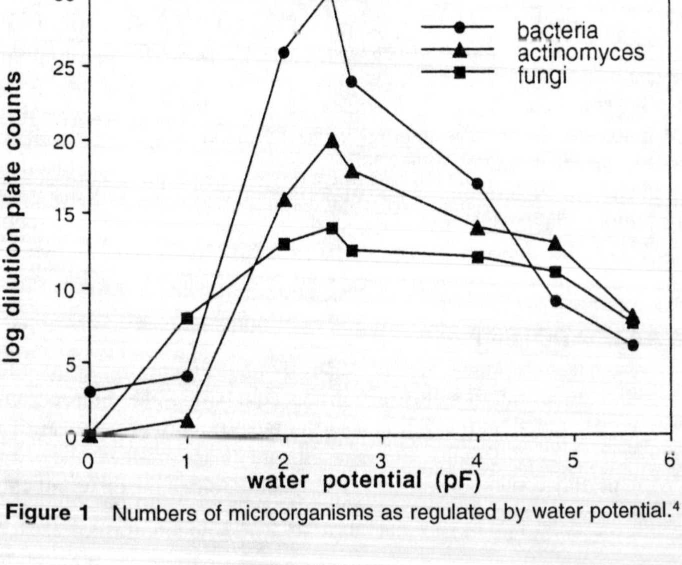

Figure 1 Numbers of microorganisms as regulated by water potential.[4]

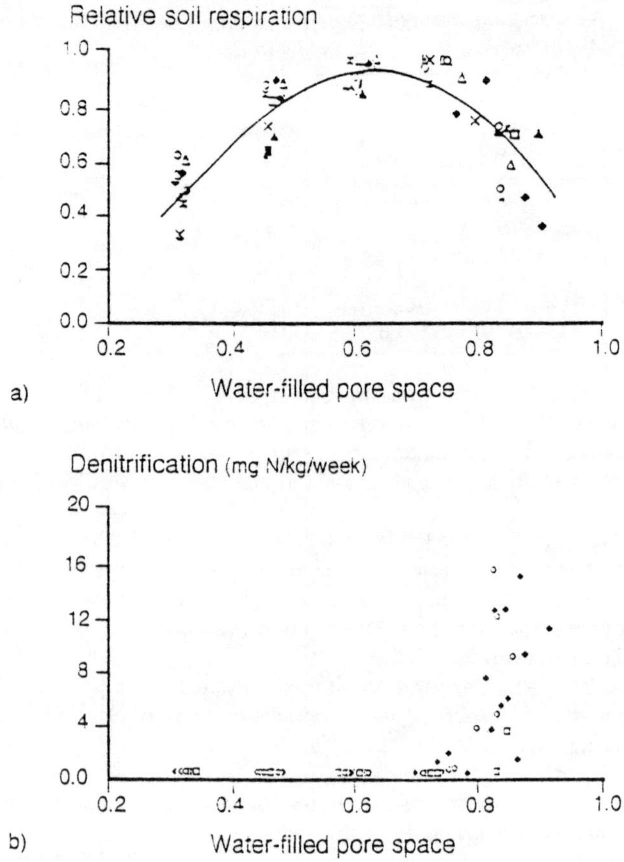

Figure 2 Microbial activities as regulated by water-filled pore space (A) respiration, and (B) denitrification. The different symbols stand for different soils.[6]

forward on what occurs in natural conditions: Vertisols of the West Indies are suppressive, as opposed to the Oxisols.[9] The Vertisol organization is characterized by two main levels of porosities: a few big fissures and numerous small pores (2 μm) typical of smectites. The bacteria are located in the latter and would be subjected to shrinking.

In Oxisols, kaolinite or halloysite particles are associated into aggregates.[2] The main porosity available for bacteria is interaggregate porosity, where the mechanical effect does not occur, and this may explain why disease can develop in these soils.

The application of such results can be important in remedying this disease in tropical areas. For example, the introduction of irrigation in Guadeloupe Vertisols seems to favor the development of the disease. This could be explained by the fact that water potential fluctuations during the dry season are decreased or even suppressed.

E. LOW WATER POTENTIAL

Many studies have been conducted on the survival of bacteria during desiccation in order to understand the fate of microorganisms introduced to soils. The microorganism which has been the most studied and which is also the most important because of its current inoculation into soils is *Rhizobium*. Some results are clear, and they show that survival increased with the clay content of soils.[13,14] The specific role of the different clay minerals

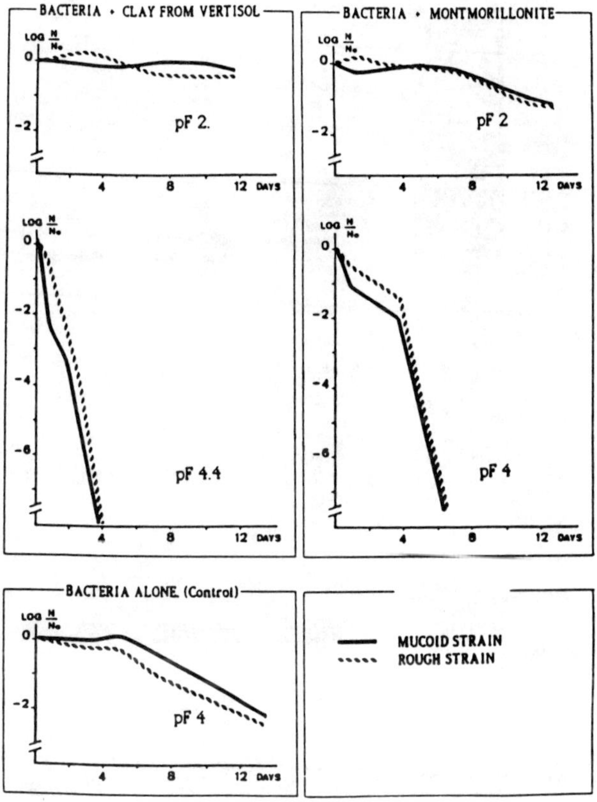

Figure 3 Survival of *P. solanacearum* in montmorillonite and in clays extracted from a West Indies Vertisol.[11]

is less clear, as added clay affected survival in different ways: kaolinite decreased survival,[13] illite had no effect,[13] and montmorillonite increased survival.[13,15,16]

How does clay affect the survival of bacteria? Different hypotheses can be put forward: the clay could act either by lowering the rate of desiccation, which was shown by Zechman and Casida[17] and Hartel and Alexander,[18] or by increasing the final water content of soils. Some results demonstrated that the final water content of soils increased with clay content, following the order: kaolinite < illite < montmorillonite.[13,15]

Increased survival of the microorganism may be ascribed to the fact that (1) more water is available for microorganisms with clays[15] and (2) the cells are more thoroughly dried due to suction exerted by clay.[14,16] It is difficult at present to favor one mechanism more than the other.

F. CONCLUSIONS

We have briefly reviewed the role of water potential mediated through clay or soil organization. At high water potential, there are anaerobic conditions with limiting O_2; then, for intermediate values, we pointed out a mechanical effect which is mediated directly by clay particles. If pathogenic microbes are considered, such an effect is beneficial. At low water content (low Ψ), clays provide protection against desiccation.

An important observation is that in soils, the most rapid changes in water potential occur when a dry soil is wetted rather than during desiccation.[19,20] This sudden wetting

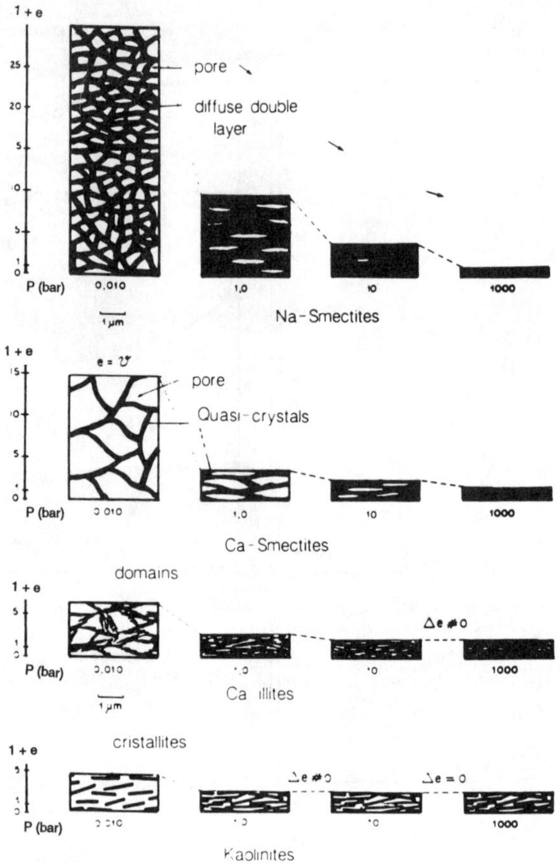

Figure 4 Clay microorganizations as affected by water potential.[1,8]

provokes a release of solutes by the cells and an influx of water within the cell, with subsequent bursting. For example 70% of total C biomass is released by wetting a dry soil.[19] This phenomenon has not yet been sufficiently studied, and the role of clays is unknown.

Another comment concerns the complexity of soil organization. For a given value of water potential, there is a heterogeneity of microsites in soils which may be illustrated by the occurrence of both aerobic and anaerobic regions within a soil at a given time (Figure 5).[21] Furthermore, a correlation has been established between aggregate size distribution and the extent of limitation of denitrification in soils under aerobic and anaerobic conditions.[22,23]

III. SOIL MICROHABITATS AS A FUNCTION OF WATER POTENTIAL AND SOIL ORGANIZATION

A. THE PORE SIZE APPROACH

Kilbertus[24] used TEM to demonstrate, by enumeration of bacteria on thin sections of different soil samples, that most bacteria were located in 1- to 2-μm pores.

On laboratory samples, Postma et al.[25,26] introduced *Rhizobium* bacteria into loamy soils maintained at different water potentials. These soil samples were prepared at different initial moisture contents in which water-filled pores were of different diameters.

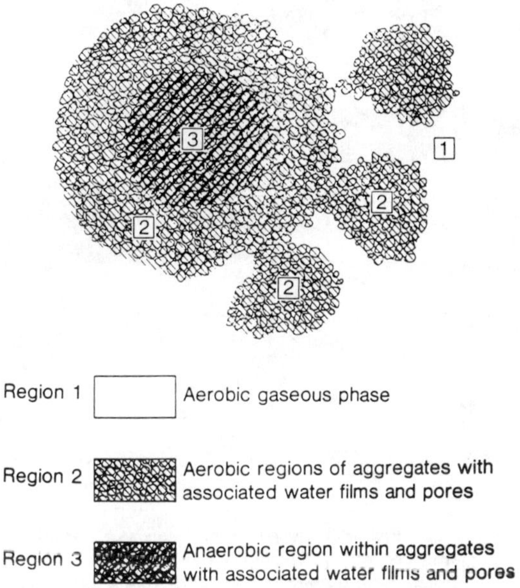

Figure 5 Aerobic and anaerobic regions in aggregates.[21]

After incubation at pF2 (–0.01 MPa), the bacteria were washed out, and it was observed that the numbers of rhizobia retained by the soil decreased with a drop in the initial moisture content of the samples. The less saturated the porosity was initially, the more places were vacant for inoculated bacteria. Retention of the cells occurred for water potentials less than –0.01 or –0.1 Pa, which, using Laplace's law, corresponded to pore necks between 0.6 and 9 μm.

Heidjnen and van Veen[27-29] introduced rhizobia into soils nonamended or amended with 10% kaolinite or bentonite. The survival of rhizobia was clearly increased by the addition of clay, especially bentonite (Figure 6A). On the moisture release curve of soil samples (Figure 6B), it can be seen that the addition of bentonite increased the water retention at –0.1 MPa (pF2). This corresponds to an increase of the pores with a diameter less than 6 μm, where bacteria would be protected against predation by protozoans that are bigger than these pore necks.

Since then, it was demonstrated using scanning electron microscopy and a cryoapparatus[30] that (1) such pores existed and had sizes ranging from 2 to 10 μm, (2) increasing the amount of bentonite added to soil notably increased the number of these pores by unit volume of soil, and (3) bacteria were indeed located in these pores.

Postma and van Veen[26] defined different types of pores among the total pore space (Figure 7): (1) accessible pore space (>0.8 μm), which corresponds to the smaller size of most bacteria, (2) habitable pore space, which is water saturated and ranges from 0.8 μm to a certain value, x, depending on the water potential, and (3) protective pore space that is not accessible to predation and would be even more restricted in size (0.8 to 6 μm).

Clays, especially the smectites, are extremely important in soils because they provide these protective and habitable pores of less than 6 μm which would correspond to the suitable microhabitats for bacteria.

B. THE INNER-OUTER AGGREGATE APPROACH

The inner-outer aggregate approach is better known and has been defined by Hattori.[31,32] Hattori enumerated microorganisms after either a mild washing of aggregates (microbes

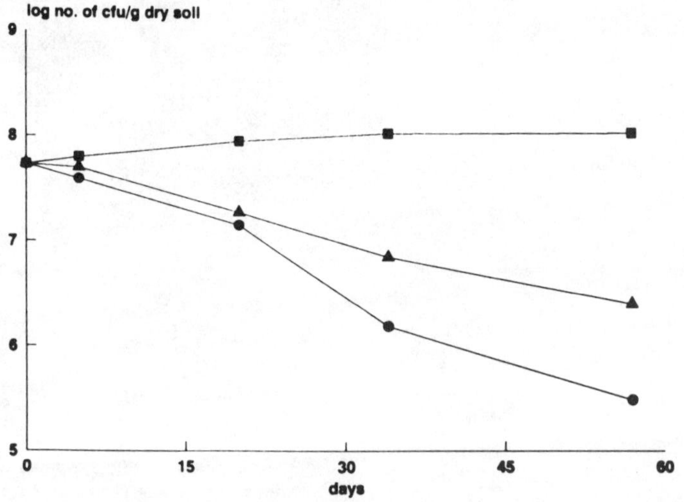

Population dynamics of *R. leguminosarum* biovar *trifolii* in loamy sand (●), loamy sand amended with 10% kaolinite (▲) and loamy sand amended with 10% bentonite (■), determined using plate count techniques. cfu = colony forming units. The largest difference on a log scale found between two duplicate survival levels was 0.20.

A

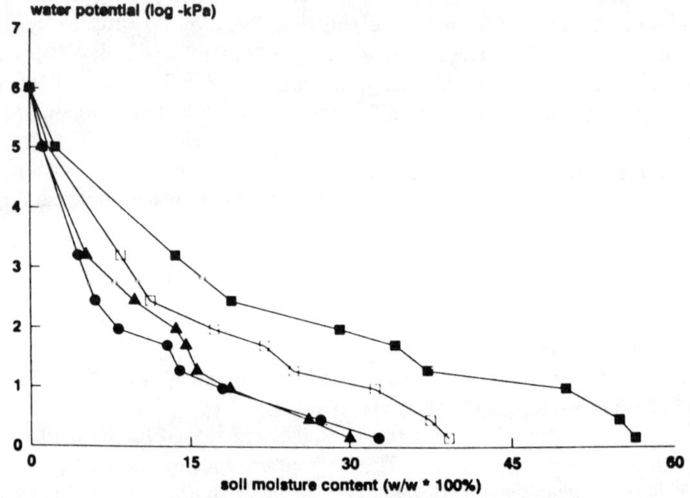

Moisture characteristic curves of loamy sand (●), loamy sand amended with 10% (■) or 5% bentonite (□) and loamy sand amended with 10% kaolinite (▲). The bulk density was 1.2 g/cm³.

B

Figure 6 Survival of *Rhizobium* bacteria in (A) clay-amended soils and (B) moisture release curves of the clay-amended soils.[27]

presumably located on outer surfaces of aggregates) or a dispersion of aggregates (microorganisms presumably located within aggregates). Gram-negative bacteria are found mainly within the aggregates, and fungi, actinomycetes, or sporulated bacteria mainly on external surfaces.[31] Vargas and Hattori[33] established a correlation between microbial habitats and predation (Figure 8): in the absence of protozoa, both inner and outer bacteria

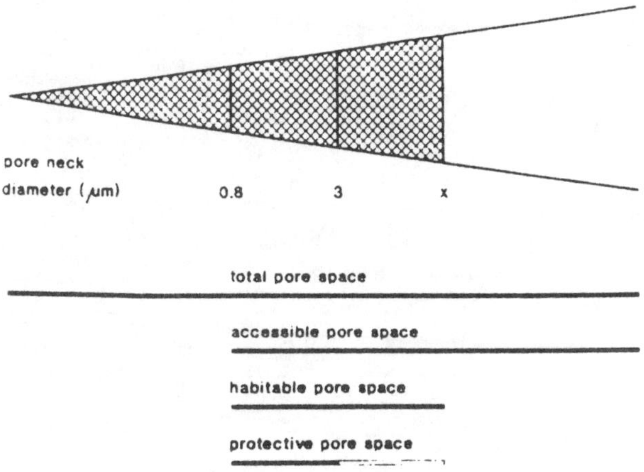

Figure 7 Schematic representations of the total accessible habitable and protective pores space in soil. Hatched areas represent pores filled with water and x is the pore neck diameter that is water-filled at the water potential used (e.g., x is 30 µm at –0.01 MPa).[26] ■ and □ indicate if pore volume or pore surface area, respectively, of a certain pore diameter class are expected to be important.

are abundant. If protozoa are present, only the inner bacteria are protected and can persist[33] (Figure 8A). A correlation was also established between predation and water content: as the saturation of pores increased, protozoa could reach a greater volume of pores and could feed on more bacteria[33] (Figure 8B).

A similar approach was used by Jocteur-Monrozier et al.[34] concerning the mineralization of organic matter in Alfisols, which have no stable aggregates and thus only external positions, as compared to Vertisols, which have stable aggregates with both internal and external surfaces colonized by bacteria.

C. MICROBIALLY DERIVED MICROHABITATS

The existence of different microhabitats as defined in the above studies depends, as demonstrated by Jocteur-Monrozier et al.,[34] on the existence of stable aggregates and the fact that microbes can penetrate them.

In addition, it should be stated that the microorganisms themselves contribute to the organization and aggregation of soil constituents. They modify their immediate environment, and therefore build their own microhabitats. In experimental studies conducted in the laboratory with pure clays,[35-38] it was shown that all microorganisms and even roots have the same effects on clay organization: orientation of clay particles around the cell, adhesion of clay particles to cells, and secretion of polysaccharides that bind clay particles to the cells. In some cases (fungi, roots), there is a strong physical entanglement of clay particles. By such mechanisms, each microorganism may form an aggregate around its cell, and this aggregate becomes its microenvironment or microniche.

The existence of such aggregates was demonstrated in soils by several authors,[39] and it was noted that bacteria are in many cases surrounded by clay coatings,[40] the overall structures being stable microaggregates.[2]

D. MICROBIAL HABITATS AND WATER POTENTIAL

The three approaches provide insights into the possible microhabitats for soil organisms. It is certain that clay micropores are very important protective microhabitats; the same applies to the inner pores of aggregates when they are accessible, which is not

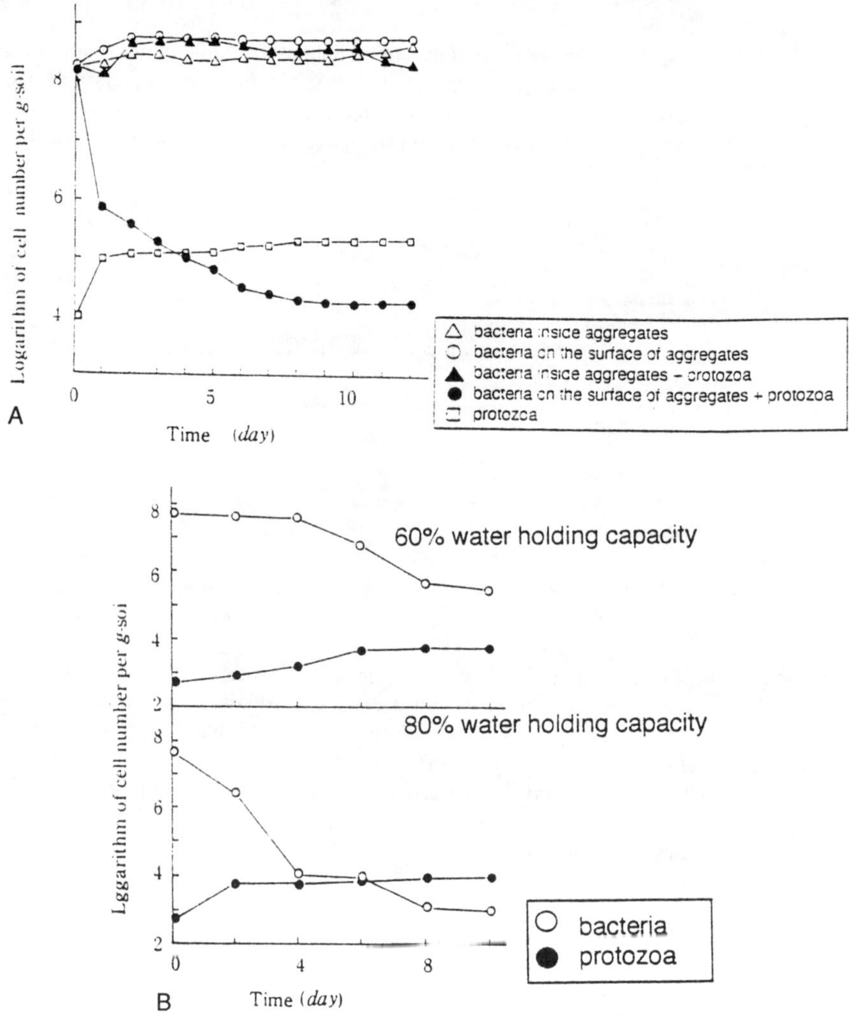

Figure 8 Predation of the bacterium *Klebsiella aerogenes* by protozoa as affected by location of the microorganism. (A) Accessibility of the bacteria to protozoa; (B) effect of water content on predation.[33]

always the case. This problem of accessibility is reduced if we consider that microbes are the main actors in soil microaggregation and that they build their own "inner" pores inside these microaggregates. The different approaches to microbial habitats may be compared, taking into consideration the water potential. There are similarities between small pores (for example, the clay porosity) and the inner surfaces in aggregates. Such microhabitats provide good protection against predation. The pores are saturated by water most of the time and are subjected to few hydric variations, but O_2 becomes limiting, which induces anaerobic conditions, and the accessibility of substrates may also be limited.

In the case of large pores, for example, in sandy soils or on the external surfaces of aggregates, there is extensive predation. Pores are frequently unsaturated, and drastic water potential fluctuations occur. These sites are mainly aerobic, and more substrates are presumably accessible due to water flow.

It is not possible to define a "perfect" microhabitat for soil microorganisms. For example, small pores located in clay or within aggregates may allow survival of microorganisms but also present some disadvantages for activity, such as lack of available O_2. It remains to be investigated whether bacteria are active or dormant in such microhabitats.

IV. SUMMARY AND CONCLUSIONS

Water potential determines several important phenomena in soil: (1) the availability of water, oxygen, and substrates, (2) competition and predation relationships among microorganisms, and (3) the movement of microorganisms.

Besides a direct physiological effect, water potential acts through both soil organization, especially aggregation, and the clay fabric. Clays have a major and very complex role in the survival of microorganisms that can stem from mechanical compression, leading to bacteria mortality or protection from predation, desiccation, etc. One of the main roles of clay is the building up of microhabitats that have roughly the size of bacteria (1 to a few micrometers), with water available in a wide water potential range, and which give good protection against desiccation and predation.

If we want to apply the results to the solution of environmental problems, it is important to recognize the significance of oxidation-reduction reactions in microbial metabolism.[41] These reactions govern the nitrification/denitrification processes, the oxidation of organic matter and xenobiotics, the geochemical cycling of S, Fe, Mn, etc., and a part of heavy metal dynamics.[42] The occurrence of such oxidative vs. reductive conditions is directly controlled by water potential.

REFERENCES

1. **Tessier, D.,** *Etude Expérimentale de l'Organisation des Matériaux Argileux. Hydration, Gonflement et Structuration au Cours de la Dessiccation et de la Réhumectation,* Ph.D. thesis, University of Paris, 1984.
2. **Robert, M. and Chenu, C.,** Interactions between soil minerals and microorganisms, in *Soil Biochemistry,* Vol. 7. Stotzky, G. and Bollag, J. M., Eds., Marcel Dekker, New York, 1992, 307.
3. **Seifert, T. J.,** Influence of the size of soil structural aggregates on the degree of nitrification. II. The role of aeration, *Folia Microbiol.,* 9, 363, 1964.
4. **Williams, S. T., Shameemullah, M., Watson, E. T., and Mayfield, C. I.,** Studies on the ecology of actinomycetes in soil. VI. The influence of moisture tension and growth and survival, *Soil Biol. Biochem.,* 4, 215, 1972.
5. **Dommergues, Y. R.,** Contribution à la dynamique microbienne des sols en zone semi-aride et en zone tropicale sèche, *Ann. Agron.,* 4, 265, 5, 391, 1962.
6. **Doran, J. W., Mielke, L. N., and Power, J. F.,** Microbial activity as regulated by soil water-filled pore space, in *Trans. Int. Congr. Soil Science,* Vol. 3, Kyoto, Japan, 1990, 94.
7. **Linn, D. M. and Doran, J. W.,** Aerobic and anaerobic microbial populations in no-till and plowed soils, *Soil Sci. Soc. Am. J.,* 48, 794, 1984.
8. **Linn, D. M. and Doran, J. W.,** Effect of water-filled pore space on carbon dioxide and nitrous oxide production in tilled and non-tilled soils, *Soil Sci. Soc. Am. J.,* 48, 1267, 1984.
9. **Béreau, M. and Messian, C. M.,** Réceptivités commparées des sols à l'infestation par Pseudomonas solanacearum. *Ann. Phytopathol.,* 7, 191, 1975.
10. **Schmit, J. and Robert, M.,** Action des argiles sur la survie d'une bactérie phytopathogene Pseudomonas solanacearum E.F.S., *C. R. Acad. Sci.,* 299, 733, 1984.
11. **Schmit, J., Prior, P., Quiquampoix, H., and Robert, M.,** Studies on survival and localization of Pseudomonas solanacearum in clays extracted from vertisols, in *Plant Pathogenic Bacteria,* Klement, Z., Ed., *Proc. 7th Int. Conf. Plant Pathogenic Bacteria,* Akademiai Kaido, Budapest, 1990, 1001.

12. **Tessier, D.,** Behavior and microstructure of clay minerals, in *Soil Colloids and Their Associations in Aggregates,* De Boodt, M. F., Hayes, M. H. B., and Herbillon, A., Eds., NATO ASI Ser. B, Vol. 215, 1991, 387.
13. **Osa-Afiana, L. O. and Alexander, M.,** Differences among cowpea rhizobia in tolerance to high temperatures and desiccation in soil, *Appl. Environ. Microbiol.,* 53, 435, 1982.
14. **Al-Rashidi, R. K., Loynachan, R. K., and Frederick, L. R.,** Desiccation tolerance of four strains of *Rhizobium japonicum, Soil Biol. Biochem.,* 14, 489, 1982.
15. **Bitton, G., Henis, Y., and Lahav, N.,** Influence of clay minerals, humic acid and bacterial polysaccharide on the survival of *Klebsiella aerogenes* exposed to drying and heating in soils, *Plant Soil,* 45, 65, 1976.
16. **Bushby, H. V. A. and Marshall, K. C.,** Some factors affecting the survival of root-nodule bacteria on desiccation, *Soil Biol. Biochem.,* 9, 143, 1977.
17. **Zechman, J. M. and Casida, L. E.,** Death of *Pseudomonas aeruginosa* in soil, *Can. J. Microbiol.,* 28, 788, 1982.
18. **Hartel, P. G. and Alexander, M.,** Role of extracellular polysaccharide production and clays in the desiccation tolerance of cowpea Bradyrhizobia, *Soil Sci. Soc. Am. J.,* 50, 1193, 1986.
19. **Kieft, T. L., Soroker, E., and Firestone, M. K.,** Microbial biomass response to rapid increase in water potential when dry soil is wetted, *Soil Biol. Biochem.,* 19, 119, 1987.
20. **Salema, M. P., Parker, C. A., Kigby, D. K., Chatel, D. L., and Armitage, T. M.,** Rupture of nodule bacteria on drying and rehydration, *Soil Biol. Biochem.,* 14, 15, 1982.
21. **Klein, A. D. and Thayer, J. S.,** Interactions between soil microbial communities and organometallic compounds, in *Soil Biochemistry,* Vol. 6., Bollag, J. M. and Stotzky, G., Eds., Marcel Dekker, New York, 1990, 431.
22. **Stanford, G. and Epstein, E.,** Nitrogen mineralization-water relations in soils, *Soil Sci. Soc. Am. Proc.,* 38, 103, 1974.
23. **Sexstone, A. J., Reusbech, N. P., Parkin, T. N., and Tiedje, J. M.,** Direct measurement of oxygen profiles and denitrification rates in soil aggregates, *Soil Sci. Soc. Am. J.,* 49, 645, 1985.
24. **Kilbertus, G.,** Etude des microhabitats contenus dans les agrégats du sol relation avec la biomasse bactérienne et la taille des protocaryotes présents, *Rev. Ecol. Biol. Sol,* 17, 543, 1980.
25. **Postma, J., van Veen, J. A., and Walter, S.,** Influence of different initial moisture on the distribution and population dynamics of introduced *Rhizobium leguminosarum* biovar trifolii, *Soil Biol. Biochem.,* 21, 437, 1989.
26. **Postma, J. and van Veen, J. A.,** Habitable pore space and survival of *Rhizobium leguminosarum* biovar *trifolii* introduced into soil, *Microb. Ecol.,* 19, 146, 1990.
27. **Heijnen, C. E. and van Veen, J. A.,** A determination of protective microhabitats for bacteria introduced to soils, *FEMS Microbiol. Ecol.,* 85, 73, 1991.
28. **Heijnen, C. E., van Elsas, J. D., Kuikman, P. J., and van Veen, J. A.,** Dynamics of *Rhizobium leguminosarum* biovar *trifolii* introduced into soil; the effect of bentonite clay on predation by protozoa, *Soil Biol. Biochem.,* 20, 483, 1988.
29. **van Elsas, J. D., Dijkstra, A. F., Govaert, J. M., and van Veen, J. A.,** Survival of *Pseudomonas fluorescens* and *Bacillus subtilis* introduced into two soils of different texture in field microplots, *FEMS Microbiol. Ecol.,* 38, 151, 1986.
30. **Heijnen, C. E., Chenu, C., and Robert, M.,** Micro-morphological studies on clay- and unamended loamy sand, relating survival of introduced bacteria and soil structure, *Geoderma,* 56, 195, 1993.
31. **Hattori, T. and Hattori, R.,** The physical environment in soil microbiology: an attempt to extend principles of microbiology to soil microorganisms, *Crit. Rev. Microbiol.,* 4, 423, 1976.
32. **Hattori, T.,** Soil aggregates as microhabitats for microorganisms, *Rep. Inst. Agric. Res. Tohoku Univ.,* 37, 23, 1988.
33. **Vargas, R. and Hattori, T.,** Protozoan predation of bacterial cells in soil aggregates, *FEMS Microbiol. Ecol.,* 38, 233, 1986.
34. **Jocteur-Monrozier, L., Ladd, J. N., Fitzpatrick, R. W., Foster, R. W., and Raupach, M.,** Physical properties, mineral and organic components and microbial biomass content of size fraction in soils of contrasting aggregation, *Geoderma,* 49, 37, 1991.
35. **Dorioz, J. M. and Robert, M.,** Etude expérimentale de l'interaction entre champignons et argiles: conséquences sur la microstructuration des sols, *C. R. Acad. Sci.,* 295, 511, 1982.

36. **Dorioz, J. M. and Robert, M.,** Aspects microscopiques des relations entre microorganismes ou végétaux et les argiles. Conséquence sur la microorganisation et microstructuration des sols, in *Soil Micromorphology,* Fedoroff, N., Bresson, L. M., and Courty, M. A., Eds., Proc. 7th Int. Working Meeting on Soil Micromorphology (Paris, 1985), 1987, 353.
37. **Guckert, A., Breisch, H., and Reisinger, O.,** Interface sol-racine. I. Etude au microscope électronique des relations mucigel — argiles — microorganismes, *Soil Biol. Biochem.,* 7, 241, 1975.
38. **Dormaar, J. F. and Foster, R. C.,** Nascent aggregates in the rhizosphere of perenial ryegrass *(Lolium perenne* L.), *Can. J. Soil Sci.,* 71, 465, 1991.
39. **Tisdall, J. M. and Oades, J. M.,** Stabilization of soil aggregates by the root system of ryegrass, *Aust. J. Soil Res.,* 17, 429, 1979.
40. **Foster, R. C.,** Microenvironments of soil microorganisms, *Biol. Fertil. Soils,* 6, 189, 1988.
41. **Stotzky, G.,** Influence of soil minerals on metabolic processes, growth, adhesion, and ecology of microbes and viruses, in *Interactions of Soil Minerals with Organics and Microbes,* Huang, P. M. and Schnitzer, M., Eds., SSSA Spec. Publ. No. 17, Soil Science Society of America, Madison, WI, 1986, 305.
42. **Bourquin, A. W.,** Les microorganismes au secours des sites pollués, *Biofutur,* 93, 24, 1990.

Chapter 6

Effect of Citric Acid on Interlayer Adsorption of Hydroxy-Aluminosilicate Ions by Montmorillonite

G. Q. J. Lou and P. M. Huang

CONTENTS

I. INTRODUCTION

Renewed interest in Al chemistry can be attributed to the acid rain-induced mobilization of Al and its impact on pedogenesis, soil properties, and Al toxicity to living organisms.[1,2] Of soluble Al species, hydroxy-Al ions have received ample attention.[3] Laboratory studies indicated that monosilicic acid could readily react with hydroxy-Al ions to form "protoimogolite" or hydroxy-aluminosilicate (HAS) ions,[4,5] and the formation of such Al silicate species has been suggested to reduce the toxicity of soluble Al to fish in acid waters.[6] Hydroxy-aluminosilicate ions are also the precursor of imogolite and protoimogolite allophane, the poorly crystalline and noncrystalline aluminosilicates commonly found in Spodosols, Andisols, and some other acidic soils.[7-9] The natural occurrence of soluble HAS species was indicated by the synthesis study in which imogolite was formed upon heating of the soil solutions collected from shattered sandstone underlying a brown forest soil (Inceptisol).[10] A recent study predicted that Al silicate species accounted for up to 95% of the total inorganic mononuclear Al in natural waters.[11] A thorough understanding of the factors controlling the transformation of HAS species in acidic soils and natural waters is, therefore, essential.

Recent studies revealed that expandable layer silicates played an important role in affecting the transformation of HAS species.[12,13] Lou and Huang[12,14] provided evidence that the interlayer surfaces of expandable layer silicates were able to adsorb a substantial amount of HAS species, and such a process leads to the formation of siliceous Al interlayers in the expandable clays over the range of soluble Al and Si concentrations commonly found in acidic soils and natural waters. The interlayer adsorption of HAS species by smectite was shown to inhibit the formation of noncrystalline aluminosilicates by preventing the HAS species from forming a discrete solid phase.[15] The interlayer

0-87371-914-X/95/$0.00+$.50

adsorption mechanism thus provides an interpretation for the observation that imogolite and the related noncrystalline aluminosilicates usually occur in spodic horizons that are low in layer silicate clays[16] and that larger amounts of noncrystalline aluminosilicates are present in sandy than in loamy Spodosols.[17]

Several factors have been shown to affect the interlayer adsorption of HAS species by expandable layer silicates. Hydroxy-aluminosilicate interlayers in montmorillonite can equilibrate with soluble Si and Al at levels of 10^{-5} to 10^{-4} *M*, which is the range of solution Si and Al concentrations in acid soils of temperate regions.[14] A soluble Si concentration of $>6.6 \times 10^{-5}$ *M* is required for the formation of HAS species and their adsorption by montmorillonite.[13] Vermiculite, which has a higher layer charge and lower interlayer expansibility than montmorillonite, has been shown to adsorb an appreciable amount of HAS species but with a lower Si/Al ratio and smaller size than those adsorbed by montmorillonite.[13] However, little information is available on the effect of natural organic ligands on the interaction of HAS species with expandable layer silicates.

Complexing organic acids such as citric acid are constantly introduced into soil through vegetation and microbial activity. The total concentration of complexing organic acids in soil solutions is estimated to be 10^{-5} to 10^{-4} *M* and may reach 10^{-3} *M* or higher in the soil rhizosphere.[18] The complexation of organic ligands with HAS species, the precursor of noncrystalline aluminosilicates, has been shown to perturb the structure and promote the surface reactivity of the noncrystalline mineral colloids.[19,20]

This study was undertaken to investigate how the interactions of citric acid with HAS species would affect the interlayer adsorption of the HAS by montmorillonite and the extent to which the adsorption process was affected.

II. MATERIALS AND METHODS

A. PREPARATION OF HYDROXY-ALUMINOSILICATE SPECIES AND CITRATE-HYDROXY-ALUMINOSILICATE COMPLEXES

A suitable volume of 6×10^{-2} *M* $AlCl_3$ solution was added to 1.6×10^{-3} *M* $Si(OH)_4$ solution, which was obtained by passing a Na-metasilicate solution through a H^+-form Dowex 50W-X8 ion-exchange resin column, to give a Si/Al molar ratio of 0.46. The mixed solution was titrated with 0.12 *M* NaOH with continuous stirring at a rate of 0.33 ml min^{-1} to an OH/Al molar ratio of 2.0. The resulting HAS solution had an initial pH of 4.38, which decreased to 4.00 after 4 months of aging. The concentration of soluble Si and Al in the aged HAS solution was 1.44×10^{-3} and 3.10×10^{-3} *M*, respectively.

A 3×10^{-2} *M* Na citrate solution was added to a citric acid solution of the same concentration until the pH of the mixture reached 4.00. Suitable volumes of the citrate-citric acid mixture were added to 150-ml aliquots of HAS solutions such that the citrate/Al molar ratio of HAS-citrate mixtures was 0, 0.01, 0.05, 0.1, 0.5, and 1.0, respectively. The pH of these mixtures was 4.00, 3.96, 3.71, 3.50, 3.22, and 3.35, respectively. The decrease in solution pH upon mixing was an indication of the reaction between citrate and HAS species, which is discussed in the next section.

B. REACTION OF HYDROXY-ALUMINOSILICATE SPECIES AND CITRATE-HYDROXY-ALUMINOSILICATE COMPLEXES WITH MONTMORILLONITE

Montmorillonite (SWy-1, Crook, WY) was supplied by the Clay Minerals Society. The <2-µm fraction of the montmorillonite was obtained by sedimentation and then saturated with Ca^{2+} ions. The montmorillonite had a CEC of 76 cmol kg^{-1}.

A suitable volume of montmorillonite suspension was added to the HAS-citrate mixtures so that the clay/solution ratio was maintained at 0.375 g l^{-1}. The suspension pH in the system without citrate was 4.03, and the pH of the systems containing various amounts of citrate was adjusted to the same value by adding 0.1 *M* NaOH. The initial concentration of citrate in HAS-citrate-clay suspensions was 0, 2.5×10^{-5}, 1.25×10^{-4}, 2.5×10^{-4}, 1.25×10^{-3}, and 2.5×10^{-3} *M*, respectively, and the initial concentration of soluble Si and Al in the six treatments was 1.20×10^{-3} and 2.58×10^{-3} *M*, respectively. The resulting suspensions were then agitated at 25°C for 1 week. The final pH of the six treatments was 3.94, 3.95, 3.93, 3.91, 4.01, and 3.75, respectively.

C. EXAMINATION OF REACTION PRODUCTS

At the end of the reaction period, the solution and solid phases were separated by filtration through a Sartorius cellulose nitrate membrane filter of 0.01-μm pore size. The Si and Al in solution were determined by the Aluminon and Molybdate-blue methods,[21] respectively, after the destruction of citrate with H_2SO_4 and HNO_3.[22] The amounts of Si and Al adsorbed were calculated by taking the difference between the initial and final Si or Al concentrations.

The clay complexes collected by filtration were saturated with K^+ ions by washing four times with 0.5 *M* KCl solution. The procedure also removed occluded solutions of HAS and hydroxy-Al species or $Si(OH)_4$. X-ray diffraction analysis of the resultant clay complexes was carried out on a Rigaku D/Max-RBX X-ray diffractometer, using parallel-oriented specimens air dried on glass slides at 25°C or heated at 110°C for 24 h and at 350°C for 2 h, respectively. A portion of the clay complexes was examined by infrared spectroscopy after freeze drying. Infrared (IR) spectra were recorded on a Perkin-Elmer 983 IR spectrometer, using a KBr (200-mg) disc containing 1 mg of the freeze-dried sample.

III. RESULTS

A. ADSORPTION OF HYDROXY-ALUMINOSILICATE SPECIES BY MONTMORILLONITE AS AFFECTED BY CITRATE

The reaction between HAS species and citrate was indicated by the decrease in solution pH upon the addition of citrate to HAS solutions. The pH of the HAS solutions shifted from 4.00 to 3.22 – 3.96 after the addition of a citrate-citric acid mixture of pH 4.00. The magnitude of the pH change generally increased with solution citrate/Al ratio (0.01 to 1.0). The decrease in solution pH suggested that protons were released into solution upon the reaction between citrate and the Al in HAS and/or hydroxy-Al species (Figure 1).

The complexation of citrate with Al in HAS and hydroxy-Al species was likely to result in one or more of the following changes: (1) a decrease in positive charge density of the Al-containing species, (2) an increase in their ionic sizes, and (3) the competition of citrate with monosilicate and hydroxy groups, causing the decomposition or partial decomposition of HAS and even hydroxy-Al species. These changes are expected to affect the interlayer adsorption of HAS species by montmorillonite.

Figure 2a shows the adsorption of soluble Al and Si by montmorillonite from HAS solutions of different citrate/Al ratios. The amount of Al adsorbed increased initially with the increasing solution citrate/Al ratio (Figure 2a). The increase in Al adsorption can be attributed to the decrease in positive charge on HAS and/or hydroxy-Al species due to their complexation with citrate. Hence, a larger amount of the Al-containing species had to be adsorbed to satisfy the negative layer charge of montmorillonite. The amount of soluble Si adsorbed also increased initially with the increasing citrate/Al ratio (Figure 2a). Since montmorillonite adsorbs Si in the form of HAS species rather than monosilicic acid,[12] the increase in Si adsorption is an indication of the enhanced adsorption of HAS species as a result of their complexation with citrate. However, the Si/Al ratios of the species adsorbed by montmorillonite decreased in the systems containing citrate (Figure 2b), which can be attributed to the competition between citrate ligand and $Si(OH)_4$ for

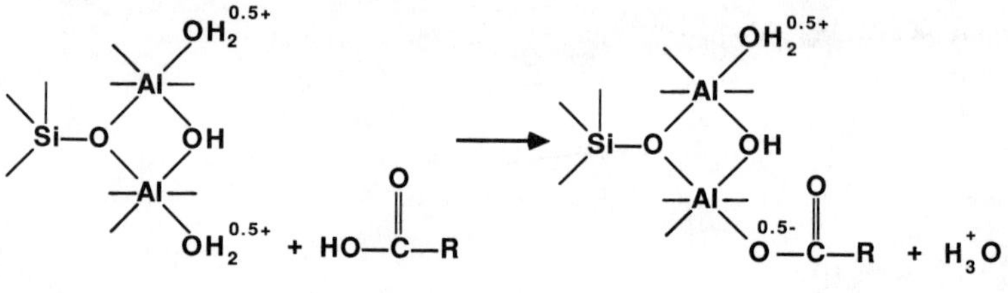

Figure 1 Proposed reaction between hydroxy-aluminosilicate species and citrate.

hydroxy-Al. The adsorption of soluble Al and Si by montmorillonite from HAS solutions was drastically reduced when the solution citrate/Al molar ratio increased to 0.5, and little adsorption of Al and Si was observed when the citrate/Al molar ratio reached 1.0.

B. FORMATION OF HYDROXY-ALUMINOSILICATE INTERLAYERS IN MONTMORILLONITE AS AFFECTED BY CITRATE

Infrared spectroscopy revealed the formation of HAS-montmorillonite and citrate-HAS-montmorillonite complexes in the systems studied. The unreacted montmorillonite exhibited IR absorption bands at 3636 cm^{-1} (Al-OH stretching), 3440 cm^{-1} (OH– stretching of adsorbed water), 1046 cm^{-1} (Si-O stretching), and 917 cm^{-1} (Al-OH bending),[23] respectively (Figure 3a). The HAS-montmorillonite complexes showed additional IR absorption at ~940 cm^{-1} (major Si-O band of adsorbed HAS species[11]) and at 3440-cm^{-1} region (stretching of structural OH) (Figure 3b). The IR absorption at these two regions increased with the citrate/Al molar ratio from 0 to 0.1 (Figures 3b to e), which was consistent with the increased absorption of HAS species by the montmorillonite (Figure

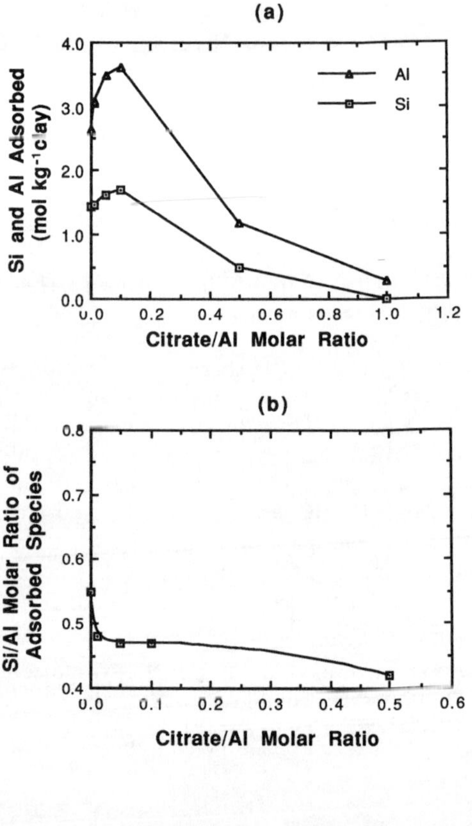

Figure 2 (a) Adsorption of Si and Al by Ca-montmorillonite from solutions containing hydroxy-aluminosilicate species and various amounts of citrate; (b) Si/Al molar ratio of the adsorbed species as a function of the citrate/Al ratio.

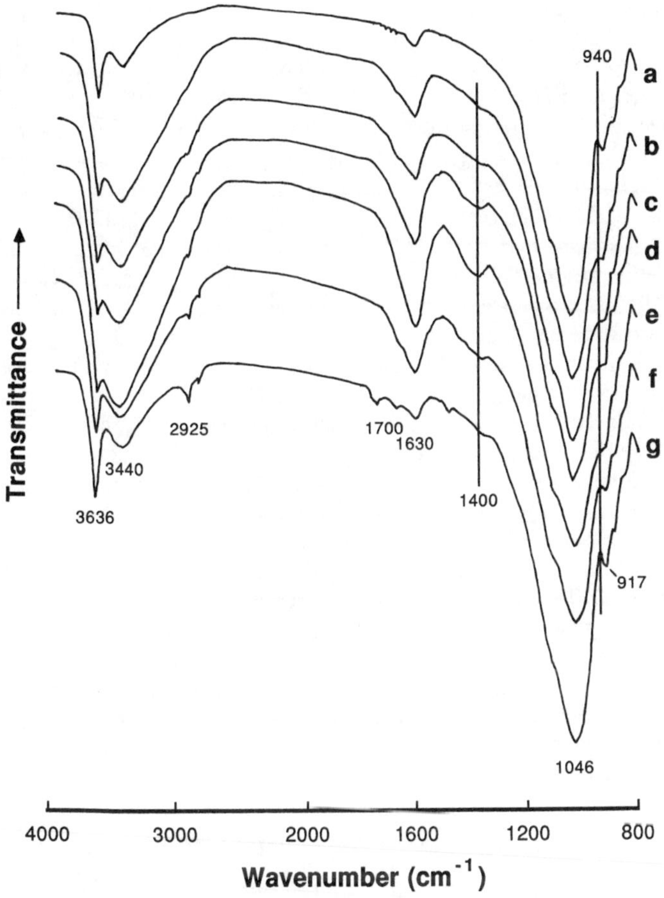

Figure 3 Infrared spectra of (a) montmorillonite and the montmorillonite reacted with the mixture of hydroxy-aluminosilicate species and citrate at citrate/Al molar ratios of (b) 0, (c) 0.01, (d) 0.05, (e) 0.1, (f) 0.5, and (g) 1.0. Clay complexes were K saturated.

2a). The IR study, therefore, confirmed the enhanced adsorption of HAS species by montmorillonite in the presence of citrate at 10^{-5} to 10^{-4} *M*. A further increase in the citrate/Al ratio to 0.5 and 1.0 resulted in a decrease in IR absorption in these regions (Figures 3f and g), which was also supported by the data obtained from the adsorption study (Figure 2a).

The adsorption of citrate-HAS complexes by montmorillonite was confirmed by the IR absorption bands at 1400 cm^{-1} (carboxylate groups) and the increased IR absorption at 1630 cm^{-1} (carboxylate groups and adsorbed water molecules) (Figures 3c to e). Since the adsorption of citrate alone by montmorillonite was not detectable, the citrate had to be adsorbed in the form of complexes. The fact that the IR absorbance at 1400 and 1630 cm^{-1} increased with the increase in the amount of HAS species adsorbed suggested that citrate was adsorbed in the form of citrate-HAS complexes. At higher citrate/Al ratios (0.5 and 1.0), citrate was more likely adsorbed as its complexes with Al and hydroxy-Al. At a citrate/Al molar ratio of 1.0, for example, the presence of citrate in clay complexes was clearly indicated by the IR bands at 2925 cm^{-1} (–CH), 1700 cm^{-1} (–COOH), and 1630 cm^{-1} ($–COO^-$) (Figure 3g), but no adsorption of Si was observed (Figure 2a). The decomposition of HAS species at a citrate/Al molar ratio of 1.0 was indicated by the predominance of 1080- and 1200-cm^{-1} bands (silicic acid) in the IR spectrum of the freeze-dried solution prepared by mixing HAS solution with a citrate solution (not shown).

The effect of citrate on the formation of HAS interlayers in montmorillonite was evidenced by X-ray diffraction analysis. The adsorption of HAS species in the interlayer space of montmorillonite resulted in a basal spacing of 18.0 Å under air-dried condition (Figure 4, 25°C, a), which collapsed to 17.2 Å after heating at 110°C (Figure 4, 110°C, a). The 17.2-Å spacing corresponded to an interlayer thickness of 7.6 Å that resulted from the intercalation of a monolayer of HAS species or protoimogolite.[12,24] The shoulders at ~9°(2θ) of the X-ray diffractograms can be attributed to a second-order diffraction of the major diffraction peaks (Figure 4).

The coadsorption of citrate with HAS species was confirmed by the increase in basal spacing of the clay complexes from 18.0 to 20.9 Å (under air-dried condition) as the initial citrate/Al molar ratio increased from 0.01 to 0.1 (Figure 4, 25°C, b to d). Upon heating at 110°C, these interlayers collapsed more easily than the HAS interlayers without citrate (Figure 4, 110°C, a to d). It is noteworthy that the X-ray diffraction patterns of the clay complexes formed at the citrate/Al ratio of 0.1 was very diffuse (Figure 4, 110°C, d), indicating the presence of disordered materials in the interlayer of montmorillonite. A further increase in the citrate/Al ratio (0.5 and 1.0) resulted in the deposition of little interlayer materials as a result of the decomposition of HAS species at the high citrate concentration (Figures 4e and f, 25, 110, and 350°C). Upon heating at 350°C, all the interlayer complexes containing citrate exhibited lower basal spacings (14.8 to 16.5 Å) than the HAS interlayer without citrate (16.6 Å), in spite of the fact that larger amounts of Si and Al were adsorbed in the citrate treatments. The lower thermal stability of the HAS-citrate interlayers in montmorillonite indicates the perturbation of the structure of HAS polymers as a result of their complexation with the organic ligand.

IV. SUMMARY AND CONCLUSIONS

Citrate can either enhance or reduce the interlayer adsorption of HAS species by montmorillonite, depending on the concentration of the organic ligand. In the presence of citrate at low concentrations (10^{-5} to 10^{-4} *M*, citrate/Al molar ratios of 0.01 to 0.1), the adsorption of HAS species by montmorillonite increased as a result of the complexation of HAS with citrate and hence the loss of some positive charges on the Al-containing species. A similar effect was observed in our previous studies which compared the effect of the Si/Al molar ratio on the adsorption of hydroxy-Al species and HAS species (lower positive charge density) by smectite.[12,13]

Although citrate at low concentrations enhanced the interlayer adsorption of HAS species, the citrate-HAS interlayers exhibited a lower thermal stability than the HAS interlayers without citrate, suggesting the perturbation of the structure of HAS species by citrate through a complexation reaction. The strong complexing ability of citrate was also demonstrated in the study, which indicated that citrate at a citrate/Al molar ratio as low as 0.005 was able to perturb the crystallization of imogolite.[20]

The present study showed that at citrate/Al molar ratios of 0.5 and 1.0 (COOH/Al molar ratios of 1.5 and 3.0), citrate substantially reduced the formation of HAS interlayers in montmorillonite due to the partial decomposition (COOH/Al = 1.5) and decomposition (COOH/Al = 3.0) of HAS species. Farmer et al.[9] reported the decomposition of HAS (protoimogolite) when fulvic acid was present in excess (COOH/Al molar ratio ≥8). The higher COOH/Al ratio required for the decomposition of HAS species in the presence of fulvic acid may be related to the distribution of –COOH groups in the humic polymers and hence steric factors.

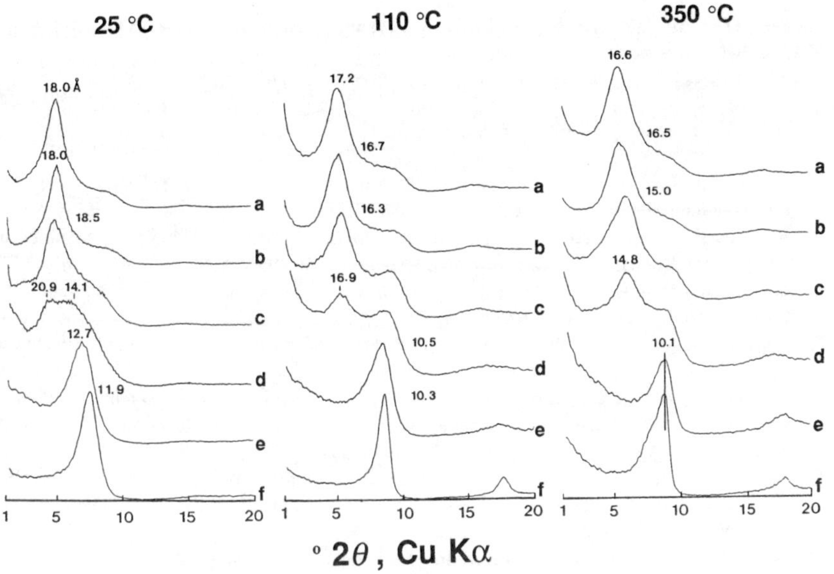

Figure 4 X-ray diffractograms of the HAS-montmorillonite complexes formed at citrate/Al molar ratios of (a) 0, (b) 0.01, (c) 0.05, (d) 0.1, (e) 0.5, and (f) 1.0. K-saturated, parallel-oriented specimens were air dried at 25°C and heated at 110°C for 24 h and at 350°C for 2 h, respectively.

This study suggested that complexing organic ligands in the concentration range commonly found in bulk soils (10^{-5} to 10^{-4} *M*) perturbed but did not prevent the formation of siliceous Al interlayers in expandable layer silicate through the interlayer adsorption of HAS species. However, in the localized zones where biological activities are intense, such as in the soil rhizosphere and near decomposing plant residues, the organic acids produced (10^{-3} *M* or higher) may hamper the interlayer adsorption of the soluble Al silicate species by expandable clays. The perturbation effect of organic ligands and the variation in soluble Al and Si levels[14] and in interlayer expandability[13] often result in the formation of partial rather than complete interlayers in expandable layer silicates. These factors are, therefore, important in controlling the composition, structure, and surface properties of the interlayer materials pertaining to the transformation of nutrients and pollutants in acidic environments.

REFERENCES

1. **Havas, M. and Jaworski, J. F.,** *Aluminum in Canadian Environment,* National Research Council of Canada, Ottawa, 1986, chap. 3–5.
2. **Stevenson, F. J. and Vance, G. F.,** Naturally occurring aluminum-organic complexes, in *The Environmental Chemistry of Aluminum,* Sposito, G., Ed., CRC Press, Boca Raton, FL, 1989, chap. 5.
3. **Huang, P. M.,** Ionic factors affecting aluminum transformation and the impact on soil and environmental sciences, *Adv. Soil Sci.*, 8, 1, 1988.
4. **Farmer, V. C., Fraser, A. R., and Tait, J. M.,** Characterization of the chemical structures of natural and synthetic aluminosilicate gels and sols by infrared spectroscopy, *Geochim. Cosmochim. Acta*, 43, 1417, 1979.
5. **Wada, S.-I. and Wada, K.,** Formation, composition and structure of hydroxy-aluminosilicate ions, *J. Soil Sci.*, 31, 457, 1980.
6. **Birchall, J. D., Exley, C., Chappell, J. S., and Phillips, M. J.,** Acute toxicity of aluminium to fish eliminated in silicon-rich acid waters, *Nature (London)*, 338, 146, 1989.

7. **Farmer, V. C., Fraser, A. R., and Tait, J. M.,** Synthesis of imogolite: a tubular aluminium silicate polymer, *J. Chem. Soc. Chem. Commun.,* 1977, 462, 1977.
8. **Wada, S.-I., Eto, A., and Wada, K.,** Synthetic allophane and imogolite, *J. Soil Sci.*, 30, 347, 1979.
9. **Farmer, V. C., Russell, J. D., and Berrow, M. L.,** Imogolite and proto-imogolite allophane in spodic horizons: evidence for a mobile aluminium silicate complex in Podzol formation, *J. Soil Sci.*, 31, 673, 1980.
10. **Yoshinaga, N., Makai, M., and Minagawa, T.,** Formation of imogolite and allophane in shattered sandstone underlying brown forest soil, *Soil Sci. Plant Nutr.*, 30, 555, 1984.
11. **Browne, B. A. and Driscoll, C. T.,** Soluble aluminum silicates: stoichiometry, stability, and implications for environmental geochemistry, *Science,* 256, 1667, 1992.
12. **Lou, G. and Huang, P. M.,** Hydroxy-aluminosilicate interlayers in montmorillonite: implications for acidic environments, *Nature (London),* 335, 625, 1988.
13. **Lou, G.,** Interactions of Soluble Aluminum Silicate Species with Expandable Phyllosilicates and Citrate, Ph.D. thesis, University of Saskatchewan, Saskatoon, 1991.
14. **Lou, G. and Huang, P. M.,** Hydroxy-aluminosilicate (HAS) interlayering in montmorillonite as influenced by soluble Si and Al concentrations, in *Agronomy Abstracts*, 82nd Annu. Meet. American Society of Agronomy, San Antonio, TX, 1990, 351.
15. **Lou, G. and Huang, P. M.,** Effect of montmorillonite on the formation of noncrystalline aluminosilicates, in *Trans. 14th Int. Congr. Soil Science,* International Society of Soil Sciences, Kyoto, Japan, Vol. 7, 1990, 402.
16. **Wang, C., McKeague, J. A., and Kodama, H.,** Pedogenic imogolite and soil environments: case study of Spodosols in Quebec, Canada, *Soil Sci. Soc. Am. J.*, 50, 711, 1986.
17. **Childs, C. W., Parfitt, R. L., and Lee, R.,** Movement of aluminium as an inorganic complex in some Podzolised soil, New Zealand, *Geoderma,* 29, 139, 1983.
18. **Huang, P. M. and Violante, A.,** Influence of organic acids on crystallization and surface properties of precipitation products of aluminum, in *Interaction of Soil Minerals with Natural Organics and Microbes*, Huang, P. M. and Schnitzer, M., Eds., SSSA Spec. Publ. No. 17, Soil Science Society of America, Madison, WI, 1986, chap. 6.
19. **Inoue, K. and Huang, P. M.,** Influence of selected organic ligands on the formation of allophane and imogolite, *Soil Sci. Soc. Am. J.,* 50, 1623, 1986.
20. **Lou, G. and Huang, P. M.,** Nature and charge properties of X-ray noncrystalline aluminosilicates as affected by citric acid, *Soil Sci. Soc. Am. J.*, 53, 1287, 1989.
21. **Jackson, M. L.,** *Soil Chemical Analysis Advanced Course,* published by the author, Madison, WI, 1969, 59.
22. **Higashi, T. and Ikeda, H.,** Dissolution of allophane by acid oxalate solution, *Clay Sci.*, 4, 205, 1974.
23. **van der Marel, H. W. and Beutelspacher, H.,** *Atlas of Infrared Spectroscopy of Clay Minerals and their Admixtures,* Elsevier, Amsterdam, 1976, 37.
24. **Lou, G., Huang, P. M., and Ripmeester, J. A.,** Structural studies of hydroxy-aluminosilicate polymers intercalated in hectorite by solid-state ^{29}Si and ^{27}Al NMR, in *Abstracts*, 28th Annu. Meet. Clay Minerals Society, Houston, TX, 1991, 103.

Chapter 7

Microbial Oxidation of Pyrites in Relation to its Efficiency in Alkali Soil Reclamation

P. Sharma and A. Swarup

CONTENTS

I. INTRODUCTION

Alkali soils constitute about 2.8 million hectares in the indo-Gangetic alluvial plains of northern India.[1] Similar soils have also been reported in several other parts of the world.[2] These soils are unproductive due to high pH, high levels of exchangeable sodium, and carbonate and bicarbonate deposits detrimental to the physicochemical properties of the soil.[3] Reclamation of alkali soil essentially requires the replacement of sodium by calcium ion on the exchange complex and leaching of sodium out of the root zone. Calcium is usually supplied by amending these soils with a calcium salt such as gypsum ($CaSO_4$) or an acidifier such as iron pyrite (FeS_2) which is capable of dissolving the calcium carbonate invariably present in these soils. Organic matter, because of its influence on nutrient availability and soil improvement, is also recommended for reclamation of alkali soils.[4] Both gypsum and pyrite have successfully been used as amendments for alkali soil reclamation with almost equal efficiency.[5-8] However, according to some authors, pyrite is inferior to gypsum in its efficiency for alkali soil reclamation.[9,10] Since the reclamative efficiency of pyrite is mainly dependent on its oxidation and consequent acid production, it is important to study its oxidation rate under varying conditions. Chemical oxidation of pyrite is a slow process, but some iron and sulfur oxidizing bacteria such as *Thiobacillus ferrooxidans* remarkably hasten the oxidation and solubilization of pyrite.[11,12] The soluble sulfur content of different supplies of agricultural grade (AG) pyrite can be variable due to differential microbial oxidation caused by different physicochemical properties of pyrite and storage conditions. The present study examined the effect of *T. ferrooxidans* inoculation on the efficiency of pyrite and compared it with that of gypsum, with or without FYM, in alkali soil reclamation. In a pot experiment, the effects of three different supplies of AG pyrite, varying in soluble sulfur contents, on rice growth and alkali soil reclamation were also studied.

0-87371-914-X/95/$0.00+$.50

II. MATERIALS AND METHODS

A. ORGANISM, CULTURE MEDIUM, AND CULTURE CONDITIONS

T. ferrooxidans 2580 was obtained from the National Culture Collection Centre, Pune, and was maintained and cultured in MS medium with $FeSO_4 \cdot 7H_2O$ as an energy source.[13] However, in pyrite oxidation studies, pyrite was used as an energy source in MS medium. The cultures were grown in a shaking device maintained at 30°C.

B. AG PYRITE — SOURCE AND COMPOSITION

Three different supplies of AG pyrite were obtained from Pyrites, Phosphates and Chemicals Ltd., Government of India. Their physical and chemical properties were tested and are shown in Table 1. The surface area of pyrite was determined according to Gee and Bauder.[14] Soluble sulfur in pyrite was determined by a turbidimetric method with barium chloride precipitation.[15] Total pyrite sulfur was determined according to Vogel.[16] Soluble Fe and Mn were determined by atomic absorption of the filtrates obtained by dissolving 1 g of pyrite in 100 ml of distilled water. Total Fe and Mn of pyrite were determined after acid digestion of pyrite according to Jackson.[17]

C. OXIDATION OF PYRITE AT DIFFERENT pH

MS medium containing 4% pyrite (PIII) was autoclaved in conical Erlenmeyer flasks and adjusted to pH 2, 4, 6, 8, and 10 with sterilized NaOH. The flasks were then inoculated with *T. ferrooxidans* 2580 (0.14 mg N), keeping an uninoculated control for each pH value. The flasks were incubated at 30°C and the pH of each flask noted after every 24 h for 7 d. The total decline in pH after 7 d of incubation was compared in the various treatments.

D. COLUMN LEACHING EXPERIMENT

A laboratory column (11-cm diameter, 41-cm length) experiment was conducted with barren alkali soils collected from the Gudha Experimental Farm of the Central Soil Salinity Research Institute, Karnal, India. These soils are formed under the influence of sodium carbonate and are highly sodic throughout the soil profile. The surface texture (0 to 15 cm) varies from sandy loam to loam, the clay fraction being predominantly illitic with a moderate amount of intergrade minerals. The soils have been classified as Aquic Natrustalfs according to the Soil Taxonomy. The initial characteristics of the soil used for the studies are presented in Table 2. The soil was ground, sieved through a 2-mm sieve, and packed uniformly so that a 30-cm length of the soil with a bulk density of 1.4 to 1.5 mg cm^{-3} was attained. Treatments consisted of: T_1, control; T_2, pyrite (equivalent to 50% gypsum requirement, GR, on a total sulfur basis); T_3, inoculated pyrite (50% GR); T_4, gypsum (50% GR); T_5, FYM (farmyard manure) (1g 100 g^{-1} soil); T_6, pyrite + FYM; T_7, inoculated pyrite + FYM; and T_8, gypsum + FYM. The required amounts of AG gypsum, pyrite (PIII, Table 1), and FYM, as per treatments, were applied and mixed in the top 10-cm layer of the soil. Inoculation of pyrite was done by spraying a 10-ml culture (0.28 mg N) of *T. ferrooxidans* onto the pyrite in the columns. Pyrite-treated soils were kept moist on the surface for 1 week for maximum pyrite oxidation. The soils were then leached with tap water (electrical conductivity [EC], 0.3 dS m^{-1}) under a constant water head for 2 months. Leachates were collected periodically and analyzed for pH, EC, cations, and anions. After termination of the experiment, the soils were taken out of the columns, cut into two parts (upper 0 to 15-cm and lower 15 to 30-cm depths), and analyzed for various soil properties according to the standard analytical procedure.[17] Fe and Mn were determined immediately after sampling in wet samples. Fe was extracted with 1 *N* NH_4OAc, pH 3.0, and analyzed by the orthophenanthroline method.[17] Mn was extracted with 1 *N* NH_4OAc, pH 7.0, and determined colorimetrically.[18] Reducible forms of Mn were deter-

Table 1 Physical and Chemical Properties of Three Different Supplies of Pyrite

Pyrite supply	pH (1:10)	Surface area ($m^2\ g^{-1}$)	Sulfur ($g\ kg^{-1}$) total	Sulfur soluble	Iron ($g\ kg^{-1}$) total	Iron soluble	Manganese ($g\ kg^{-1}$) total	Manganese soluble
PI	2.0	1.36	280	110	200	110	2	0.4
PII	5.1	0.87	316	16	200	5	2	0.1
PIII	3.5	1.50	236	40	162	11	1	0.1

mined by using 0.2% hydroquinone as reducing agent.[18] Active iron was determined according to Assami and Kumuda.[19]

E. POT EXPERIMENT USING THREE SUPPLIES OF PYRITE

Alkali soil (pH 10.4, ESP 87) was ground, sieved, and filled in pots with 20 kg of soil per pot. Three different supplies (PI, PII, PIII; Table 1) of pyrite were applied at a rate of 150 g per pot (50% GR) on the surface of the soil. The pyrite was kept moist for 1 week, to allow for its oxidation. The pots were then flooded and 30-d-old rice seedlings were transplanted. A basal dose of 5 ppm Zn, through $ZnSO_4$, was applied. Urea at a rate of 100 ppm N was applied in three equal splits (one third at transplanting, one third each at 21 and 42 d after transplanting). The rice yield was recorded in terms of grain and straw weight at the end of the experiment. Soil pH was tested at the end of the experiment to assess the extent of reclamation under different treatments.

III. RESULTS AND DISCUSSION

T. ferrooxidans 2580 showed increased acid production from pyrite with increasing pH from 2 to 8, with a maximum at pH 8, as shown by the decline in pH in the culture flasks (Figure 1). However, at pH 9 and 10, its activity declined. Uninoculated pyrite also produced acid as a result of chemical oxidation, but to a lesser extent as compared to inoculated pyrite. The optimum pH for *T. ferrooxidans* activity ranges from 2.0 to 3.0, and suboptimal pH levels result in suppression of bacterial activity.[20] It is possible that at higher pH levels, bacteria exist in micronitches of low pH on the pyrite surface and gradually decrease the pH of the culture medium. At pH 9 to 10, the bacterial activity is completely suppressed due to the absence of such micronitches. However, in the culture

Table 2 Initial Characteristics of the Experimental Soil Used for Column and Pot Experiments

Soil characteristic	Value
Particle size distribution (%)	
Sand	62.1
Silt	20.2
Clay	17.7
pH (1:2, soil:water)	10.4
EC (ds m^{-1} (1:2, soil:water)	4.0
Exchangeable sodium percentage (ESP)	87.0
Cation-exchange capacity (cmol kg^{-1})	10.3
Organic carbon (g kg^{-1})	1.4
$CaCO_3$ (g kg^{-1})	27.5
Gypsum requirement (cmol kg^{-1})	11.7

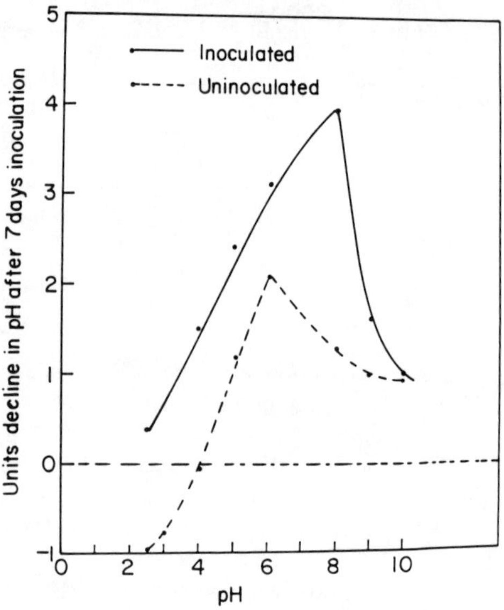

Figure 1 Acid production from pyrite (PIII) with or without inoculation with *T. ferrooxidans,* as seen in terms of total decline in medium pH after 7-d incubation of pyrite with MS medium at different initial pH.

flasks with low starting pH levels (2.0 to 3.0), a rapid microbial oxidation may cause a sharp decline in pH, reaching suboptimal pH levels which suppress further bacterial activity.

In the column-leaching experiment, the leachates appeared first in the inoculated pyrite treatments, followed by uninoculated pyrite and gypsum treatments (Figure 2). Alkali soils usually have low infiltration rates due to the dispersion of soil colloids, and their reclamation by replacement of Na^+ with Ca^{2+} on the clay complex results in aggregation and improved infiltration rates.[21] The oxides of ferrous and ferric ions released from pyrite also cause aggregation of clays and improve the infiltration of soils.[22]

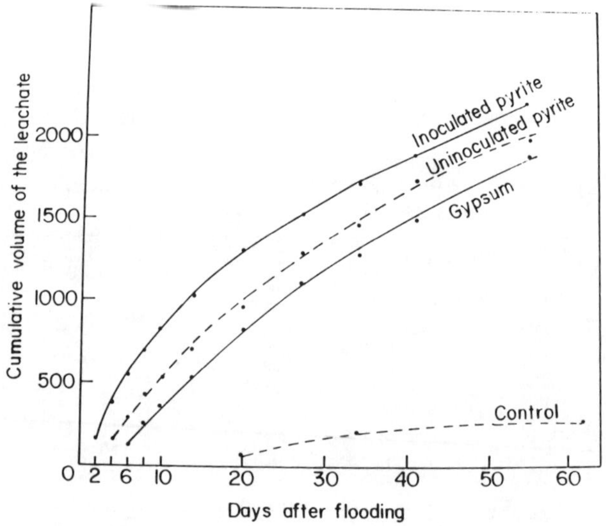

Figure 2 Cumulative volume of the leachates collected periodically from soil columns.

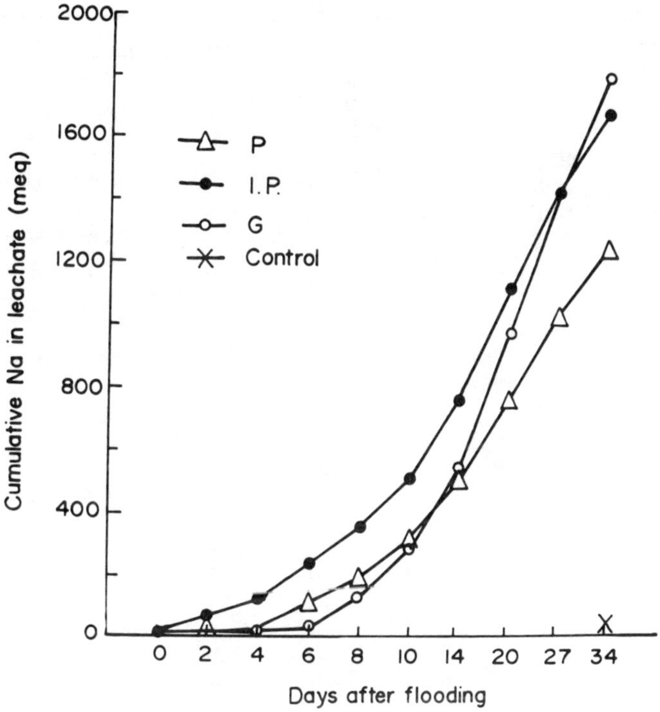

Figure 3 Cumulative sodium of the leachate collected periodically from soil columns.

The leachates, when tested for cumulative sodium, showed the highest level in gypsum-treated soils, followed by inoculated pyrite, pyrite alone, and control (Figure 3). Since the reclamation of sodic soil is brought about by replacement of sodium on the exchange complex by calcium, the efficiency of the various amendments can be assessed by the amount of sodium removed and leached from the columns. The reclamation efficiency of various amendments, as judged by the reduction in pH and ESP of the soils (Table 3), was in the order: gypsum + FYM > gypsum > inoculated pyrite > inoculated pyrite + FYM > uninoculated pyrite > uninoculated pyrite + FYM > FYM > control. Inoculation of pyrite resulted in an appreciable reduction in pH and ESP, thereby bringing it close to that of gypsum. FYM increased the efficiency of gypsum, while it decreased that of pyrite, in reclaiming the sodic soil. FYM had earlier been reported to increase the efficiency of gypsum under field conditions.[23] In the case of pyrite treatments, it seems that FYM created reducing conditions[24] which might have impeded the oxidation of pyrite, thereby reducing its reclamation efficiency. Inoculated pyrite resulted in a substantial increase in extractable and active iron in the soil (Table 3), as compared to all other amendments. Pyrite had earlier been reported to increase the levels of iron and sulfur in soil.[7] Inoculation of pyrite thus appears to improve its nutrient value for the soil. FYM alone or in conjunction with pyrite or gypsum proved effective in enhancing the extractable (water-soluble + exchangeable) iron and manganese in the soil (Table 3). However, corresponding decreases in the easily reducible form of manganese were noticed. The enhanced availability of Fe and Mn following application of the amendments is attributed to intensified microbial and chemical reduction releasing Fe^{2+} and Mn^{2+} from oxides of iron and manganese.[25] In the pot experiment using three different supplies of pyrite, grain and straw yields were highest in the pyrite with 11% soluble sulfur, followed by pyrite with

Table 3 Effect of Amendments on the Changes in pH, ESP, and Forms of Iron and Manganese in Alkali Soil After 2 Months of Column Leaching Experiment

Treatment	Soil depth (cm)	pH (1:2)	ESP	Fe (mg kg^{-1}) NH$_4$OAc extractable	Fe (mg kg^{-1}) Active	Mn (mg kg^{-1}) NH$_4$OAc extractable	Mn (mg kg^{-1}) Reducible
T_1 Control	0–15	10.1	68.7	10.8	260	11.5	108
	15–30	10.4	87.2	13.0	252	12.8	107
T_2 Pyrite (PIII)	0–15	9.5	43.6	28.8	395	19.5	102
	15–30	10.2	72.7	32.1	388	20.1	100
T_3 Inoculated	0–15	9.1	29.7	42.9	496	19.1	104
Pyrite (PIII)	15–30	10.2	79.7	45.8	480	19.8	103
T_4 Gypsum	0–15	8.6	23.7	14.2	258	15.9	104
	15–30	10.0	70.2	15.6	245	18.6	100
T_5 FYM	0–15	10.0	67.0	65.0	210	69.1	55
	15–30	10.4	35.0	42.8	235	51.8	69
T_6 Pyrite (PIII)	0–15	9.7	48.5	85.2	345	65.8	55
+ FYM	15–30	10.2	71.9	70.6	360	48.9	71
T_7 Inoculated	0–15	9.5	42.0	98.0	475	67.3	53
Pyrite (PIII)	15–30	10.1	71.0	76.6	492	50.1	70
+ FYM							
T_8 Gypsum +	0–15	8.6	22.3	66.8	218	68.0	54
FYM	15–30	10.0	71.1	45.6	242	51.2	68

4 and 1.6% soluble sulfur (Table 4). This shows that the reclamation efficiency of pyrite is directly related to its soluble sulfur content. The contradictory reports of various authors with regard to the efficiency of pyrite in alkali soil reclamation may be due to the quality of pyrite used in the experiments. For a judicious use of pyrite, it is necessary to determine its oxidation level and the amount of soluble iron and sulfur present at the time of application. Since the oxidation rate of pyrite is very slow and more so at high pH levels of sodic soils, a separate method needs to be devised to calculate the gypsum requirement of soil when pyrite is to be used as an amendment. It may be concluded from the present study that pyrite can be used as an efficient alternative amendment to gypsum for alkali soil reclamation if its soluble sulfur level can be increased by inoculation with *T. ferrooxidans*. The efficiency of pyrite has been found to be directly related to its soluble

Table 4 Effect of Different Supplies of Pyrite (PI, PII, PIII) on Alkali Soil Reclamation

Treatment	Soil depth (cm)	pH	Grain yield (g per pot)	Straw yield (g per pot)
Control	0–15	10.0	0.0	0.0
	15–30	10.1		
PI	0–15	8.8	27.8	21.0
	15–30	9.9		
PII	0–15	9.8	1.4	11.3
	15–30	9.9		
PIII	0–15	9.5	31.0	23.0
	15–30	9.9		
LSD at $p = 0.05$		—	7.1	—

sulfur content. Therefore, pyrite should be used on the basis of water-soluble sulfur content instead of total sulfur in alkali soil reclamation.

IV. SUMMARY AND CONCLUSIONS

The efficiency of AG pyrite in alkali soil reclamation depends largely on its oxidation and consequent acid production. To increase the rate of oxidation of pyrite and soluble sulfur content, inoculation of pyrite with *T. ferrooxidans* was carried out. In a laboratory column (11-cm diameter, 41-cm length), highly alkaline soil (initial pH, 10.4; exchangeable sodium percentage,87) was treated with different soil amendments, including inoculated pyrites. The soil treatments consisted of control, pyrite (50% GR on sulfur basis), inoculated pyrite, gypsum (50% GR), FYM (1 g 100 g^{-1} soil), pyrite + FYM, inoculated pyrite + FYM, and gypsum + FYM. The amendments were applied at the surface of the soil columns. The pyrite-treated soils were kept moist for 1 week for maximum pyrite oxidation. The soils were then leached with tap water (EC, 0.3 dS m^{-1}) under a constant water head for 2 months. Leachates were collected periodically and analyzed for cations and anions. The reclamation efficiency, as judged by the reduction in pH and ESP of the soils, was in the order: gypsum + FYM > gypsum > inoculated pyrite > inoculated pyrite + FYM > uninoculated pyrite > uninoculated pyrite + FYM > FYM > control. Inoculated pyrite resulted in a substantial increase in active iron in the soil as compared to all other amendments. FYM alone or in conjunction with pyrite and gypsum increased the amount of NH_4OAc-extractable iron and manganese in the soil. Pot experiments were also conducted to evaluate the efficiency of pyrite containing varying soluble sulfur content (1.6, 4, and 11%) on the yield of rice and alkali soil reclamation. The results showed that the efficiency of pyrite in reclaiming alkali soil and improving crop yield was directly related to water-soluble sulfur.

ACKNOWLEDGMENTS

The technical assistance provided by H. Tomar and S. Singh is gratefully acknowledged.

REFERENCES

1. **Singh, N. T.,** Salt-affected soils: some reflections on the state of knowledge, *J. Indian Soc. Soil Sci.*, 37, 620, 1989.
2. **Kovda, V. A.,** Alkaline soda-saline soils, *Agrochem. Talajt.*, 14 (Suppl.), 15, 1965.
3. **Gupta, R. K. and Abrol, I. P.,** Salt-affected soils: their reclamation and management for crop production, *Adv. Soil Sci.*, 11, 223, 1990.
4. **Swarup, A.,** Influence of organic matter and flooding on chemical and electrochemical properties of sodic soils and rice growth, *Plant Soil,* 106, 135, 1988.
5. **Singh, N. T., Hira, G. S., and Singh, R.,** Comparative effects of pyrites and gypsum on physico-chemical properties of a sodic soil, in *Seminar on Use of Sedimentary Pyrites in Reclamation of Alkali Soil,* Fertiliser Association of India-Phosphates and Pyrites Chemical Limited, New Delhi, 1978, 121.
6. **Jaggi, T. N.,** Use of amjhore pyrites as alkali soil amendment, in Proc. Semin. Use of Sedimentary Pyrites in Reclamation of Alkali Soil, Lucknow, 1978, 34.
7. **Tewari, K. N., Dwivedi, B. S., Upadhyay, G. P., and Pathak, A. N.,** Sedimentary pyrites as amendment for sodic soils and carrier of sulfur and iron — a review, *Fert. News,* 29, 31, 1984.
8. **Sharma, O. P. and Gupta, R. K.,** Comparative performance of gypsum and pyrite in sodic vertisols, *Indian J. Agric. Sci.*, 56, 423, 1986.
9. **Mehta, K. K. and Abrol, I. P.,** Pyrites for reclamation of alkali soils, *Indian Farm.*, 26, 20, 1976.

10. **Verma, K. S. and Abrol, I. P.,** Effects of gypsum and pyrites on soil properties in a highly sodic soil, *Indian J. Agric. Sci.,* 50, 844, 1980.
11. **Hutchins, S. R., Davidson, M. S., Brierley, J. A., and Brierley, C. L.,** Microorganisms in reclamation of metals, *Annu. Rev. Microbiol.,* 40, 311, 1986.
12. **Sharma, P. and Varma, A. K.,** Microbial reclamation of metals from ores and industrial waste waters, *Indian J. Microbiol.,* 31, 1, 1991.
13. **Harrison, A. P., Jr.,** The acidophilic, thiobacilli and other acidophilic bacteria that share their habitat, *Annu. Rev. Microbiol.,* 38, 265, 1984.
14. **Gee, G. W. and Bauder, J. W.,** Particle size analysis, in *Methods of Soil Analysis,* Klute, A., Ed., American Society of Agronomy and Soil Science Society of America, Madison, WI, 1986, 383.
15. **Snell, F. D. and Snell, C. T.,** *Calorimetric Methods of Analysis,* Vol. 2, D. Van Nostrand, Toronto, 1949.
16. **Vogel. A. I.,** *A Text Book of Quantitative Inorganic Analysis,* 3rd ed., The English Language Book Society and Langman Group, Ltd., London, 1961.
17. **Jackson, M. L.,** *Soil Chemical Analysis,* Prentice Hall, New Delhi, 1967.
18. **Sherman, G. D., McHargue, J. S., and Hodkins, W. S.,** Determination of active manganese in soil, *Soil Sci.,* 54, 253, 1942.
19. **Assami, T. and Kumuda, K.,** A new method for determining free iron in paddy soils, *Soil Sci. Plant Food,* 5, 141, 1959.
20. **Karavaiko, G. I.,** *Microbiological Processes for Leaching of Metals—State of Art Review,* Centre for International projects, GKNT, Moscow, 1985.
21. **Frenkel, H., Goertzen, J. O., and Rhoades, J. D.,** Effect of clay type and content, exchangeable sodium percentage and electrolyte concentration on clay dispersal and soil hydrolic conductivity, *Soil Sci. Soc. Am J.,* 42, 32, 1978.
22. **Shanmuganathan, R. T. and Oades, J. M.,** Modification of soil physical properties by addition of calcium compounds, *Aust. J. Soil Res.,* 21, 285, 1983.
23. **Dargan, K. S.,** Agronomic and cultural practices for crop production in salt-affected soils, *Curr. Agric.,* 3, 1, 1979.
24. **Swarup, A.,** Effect of chemical and electrochemical changes in submerged sodic soils on yield and nutrition of rice, *Fert. News,* 37, 25, 1992.
25. **Swarup, A.,** Transformation and availability of iron and manganese in submerged sodic soils in relation to yield and nutrition of rice, *Fert. News,* 34, 2, 1989.

Chapter 8

Modification of Gelation Properties of Colloidal Solids from Oil Sands: Extraction Impact on Fine Tailings Formation

L. S Kotlyar, B. D. Sparks, H. Kodama, and R. Schutte

CONTENTS

I. INTRODUCTION

The hot water extraction process is presently used in Alberta to extract bitumen from oil sands. The tailings remaining after bitumen extraction are predominantly a slurry of sand, silt, clay, and a minor amount of bitumen in water. The slurry is transported to tailings or sedimentation ponds, where the coarse solids settle rapidly. Fine solids and almost all the bitumen remain in suspension and flow to the center of the pond to form a sludge (fine tails) containing substantial amounts of the process water. The toxicity, loss of bitumen, diluent naphtha, and process water are among the most serious problems associated with the buildup of these clay sludges.[1,2]

In previous work,[3,4] the intractable behavior and high water-holding capacity of these sludges was attributed to the presence of colloidal solids material, or so-called "ultrafines" (UF), which show a tendency for gel formation. The purpose of this work was threefold: (1) to determine the chemical and mineralogical characteristics of UF solids, (2) examine the effect of particle size and water chemistry on gelation by using ^{2}H-NMR, and (3) demonstrate that structure formation, and therefore the environmental hazards associated with sludge ponds, can be prevented by changing the properties of the water during hot water extraction.

II. EXPERIMENTAL

The fine tails samples were supplied by Suncor Research Ltd. Ultrafines were separated in the form of an aqueous colloidal suspension layer (A), using the procedure described in detail elsewhere.[4] Pond water in layer A was replaced with distilled water by high-speed (366,000 g) centrifugation. The suspension of the UF in distilled water was identified as A-I. A similar procedure was used to replace distilled water in the A-I sample by pond water; this latter suspension was designated A-II.

All three suspensions (A, A-I, and A-II) were then centrifuged sequentially at increasing speeds. Solids from the A-I, sedimented at 500 g, 1500 g, and 91,000 g, were

0-87371-914-X/95/$0.00+$.50

subjected to flotation in the presence of toluene to separate bitumen and the associated biwetted solids from the hydrophilic components (AS), which remained in the aqueous phase. All AS samples were used for X-ray diffraction (XRD), transmission electron microscopy (TEM) and X-ray photoelectron spectroscopy (XPS) analysis.

XRD analysis was performed on powder specimens by random mounting for the determination of whole-mineral composition, and in preferred orientation for detecting small amounts of smectite minerals. The samples were analyzed using a Scintag PAD II diffractometer with Co radiation and a graphite monochromator; the current and voltage used were 40 mA and 40 kV, respectively. The morphology of clay particles was investigated by TEM with a Carl Zeiss transmission electron microscope Model 902, operated at 80 kV. XPS spectra were recorded with a PHI 5500 Instrument® (PHI Electronics, Eden Prairie, MN) with an Al Kα source of X-rays.

To prepare samples for the gelation study, the A-I suspension was centrifuged at 500 g; supernatant was decanted and centrifuged at 91,000 g. Dilutions were prepared from the concentrated suspensions sedimented at the above speeds. Deuterium nuclear magnetic resonance spectra were recorded with a Bruker MSL 300 spectrometer (magnetic field, 7.1 T) at a frequency of 46.07 MHz. In typical experiments, 16 transients were acquired in 8 K points, with a spectral width of 1000 Hz.

III. RESULTS AND DISCUSSION

Accumulated knowledge allows us to define sludge as "colloidal matter which has a tendency to form a highly porous three-dimensional network structure and has, as a result, very slow deliquoring characteristics."[5] Such a network is capable of entrapping other sludge components such as coarser solids, free and emulsified bitumen, etc. The structure is thixotropic, that is, it undergoes viscosity thinning under shear, and this allows most of the entrapped coarse solids and bitumen to be separated by using mechanical agitation, followed by mild centrifugation (Figure 1, bottom and top layers). The remaining tails (Figure 1, intermediate layer, A) occur in a flocculated state, where the fundamental units are flocs, with an associated immobilized water phase, rather than discrete particles.[6] The state of aggregation may be changed dramatically by substitution of pond water in suspension A with distilled water. The above inference on degree of aggregation is evident from the results of fractionation tests carried out under different degrees of centrifugation on both A and A-I suspensions (Figure 2). In the case of suspension A in pond water, nearly 90% of the solids settled at 1500 g as opposed to only 56% for the colloid A-I in distilled water. It can also be seen from Figure 2 that this effect of different water chemistry was completely reversible by simple replacement of one suspending medium for another (compare the results for the A and A-II suspensions).

A. PROPERTIES OF ULTRAFINES

The mineralogical properties of the ultrafines were examined by the XRD technique. The results of both random mount and orientated sample analyses showed that in all cases, the main phyllosilicates were mica and kaolinite. The XRD on orientated samples detected only a trace amount of smectite only in the AS-91,000 sample. This component had not been identified in the random mount technique because its concentration was below the detection limit for that method. The interstratified minerals identified appeared to be partially hydrated micaceous types or vermiculitic minerals; the amounts detected were in the minor to trace range.

Results of transmission electron microscopy show a decrease in average particle size, ranging from about 200 nm for the coarsest fraction, AS-500, to 50 nm for the finest fraction, AS-91,000. A typical TEM micrograph is shown in Figure 3, which illustrates

Figure 1 Separation of sludge by mechanical agitation followed by centrifugation. Sample on the left is the original sludge, whereas the right-hand sample has been treated (agitated and centrifuged). In the treated sample, the top layer is bitumen containing some associated solids, the intermediate layer is an aqueous colloidal suspension (A), and the bottom layer is a compact sediment containing the coarser solids.

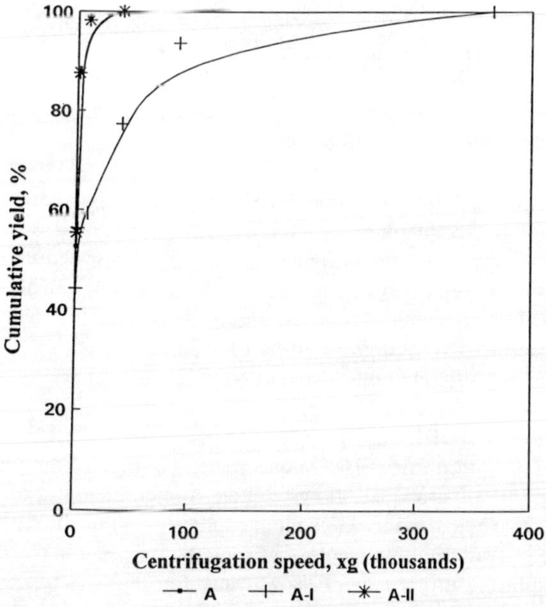

Figure 2 Effect of water chemistry on the state of aggregation of ultrafines.

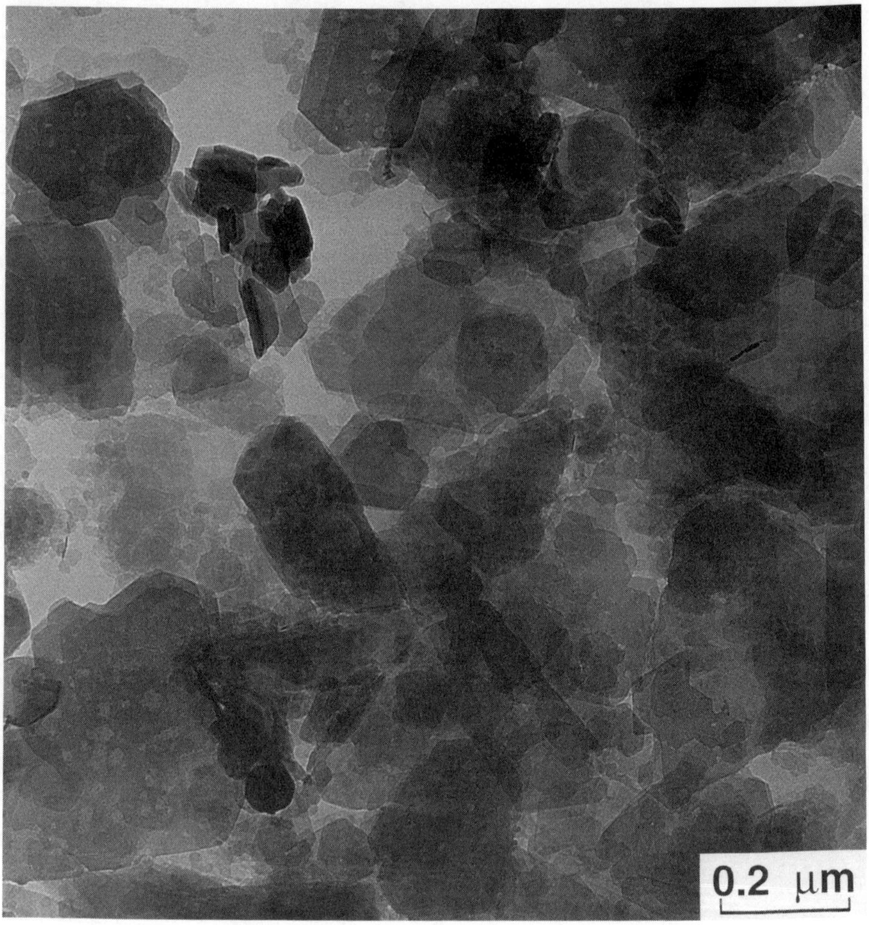

Figure 3 Transmission electron microscopy (TEM) of subfraction of hydrophilic portion of colloidal solids sedimented at 500 g (AS-500).

the results for AS-500. It can be seen that the particles were small and platy; a few were hexagonal, although the morphology of most was irregular.

A typical XPS survey spectrum of the AS solids (AS-500) is shown in Figure 4. The atomic concentration (atomic %) of the observed elements for all three fractions are tabulated in Table 1. The chemical composition of the surface for each of the fractions was similar. The surfaces were dominated by oxygen atoms, with a contribution from carbon, aluminum, and silicon atoms.

B. GEL-FORMING PROPENSITY

A study of sol-gel transition using ^{2}H-NMR as probe was carried out on subfractions of the A-I sedimented at 500 g and 91,000 g with an average diameter of 200 and 50 nm, respectively. The ^{2}H-NMR method relies on the splitting (△) of the deuterium peak due to rapid exchange of deuterium molecules between the bulk and surface environments.[7] Previously,[8] the splitting effect was observed only for smectite-type clays. However, XRD identification of kaolinite and mica in the current samples demonstrates that these minerals also exhibit this behavior, providing that the particles are ultrafine, anisotropic, charged, and nearly free of paramagnetic species such as iron (see Table 1).

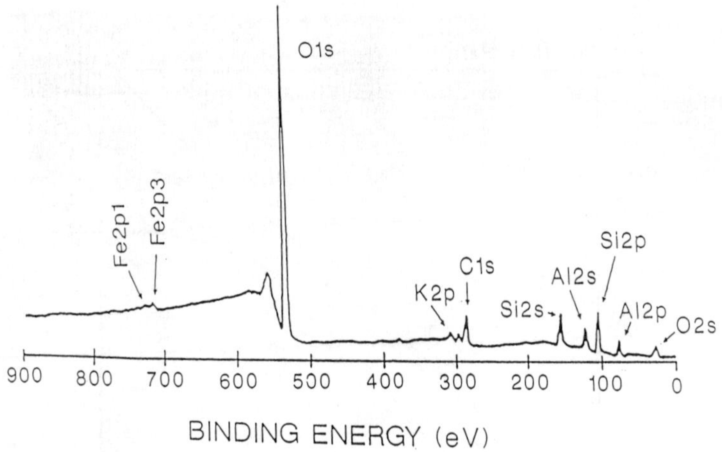

Figure 4 Typical X-ray photoelectron spectroscopy (XPS) survey of the hydrophilic portion of colloidal solids, AS (results for subfraction sedimented at 500 g, AS-500).

If aggregation is absent (or limited), then, with an increase in particle concentration, the amount of orientated water and therefore residual splitting increases linearly. As the gel structure forms, it inhibits free rotation of the particles, and the splitting gradually disappears. This is illustrated by the results shown in Figures 5a and b (experiment with pond water) for splitting Δ plotted vs. concentration after various periods of time. When Δ approaches 0, a spanning cluster, which extends throughout the sol, is formed and a stiff gel results; this last link is no different from the others which form before and after this point,[9] except that it is responsible for the creation of a continuous solid network and therefore for complete loss of the freedom of particles to orientate in the magnetic field. For 50 nm ultrafines, the minimum concentration required to form a stiff gel ($\Delta = 0$) is 0.8 vol%, as opposed to 7.8 vol% for the dispersion of 200-nm particles (Figures 5a and b). These results demonstrate that particle size of ultrafines is an important parameter governing the sol-gel transition. It is interesting that corresponding plots of Δ for both fractions in distilled water, measured at different times, remain close to that measured at zero time, suggesting that both sols are relatively stable under these conditions (Figures 5c and d).

C. SLUDGE PREVENTION

It follows from the above observations that in order for sludge to form, two components are absolutely essential: ultrafines and "bad" water containing significant amounts of electrolytes, such as those present in sludge. It is known that UF are an integral part of oil-sands ores and cannot be avoided. There is, however, an indication that sludge

Table 1 X-Ray Photoelectron Spectroscopy of the Subfractions of Hydrophilic Colloidal Solids (AS) Sedimented at 500 g (AS-500), 1500 g (AS-1500), and 91,000 g (AS-91,000)

	Atomic %					
Fraction	**C**	**O**	**Si**	**Al**	**K**	**Fe**
AS-500	11.1	62.4	14.8	10.7	1.0	0.2
AS-1500	13.7	59.1	15.2	11.1	0.9	0.2
AS-91,000	13.5	60.7	16.1	9.0	0.7	0.3

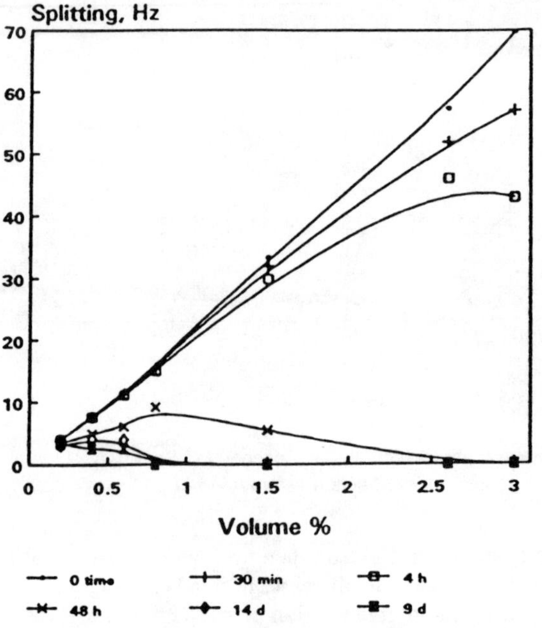
Splitting, Hz
70
60
50
40
30
20
10
0
0
0.5
1
1.5
2
2.5
3
Volume %
0 time
30 min
4 h
48 h
14 d
9 d

A

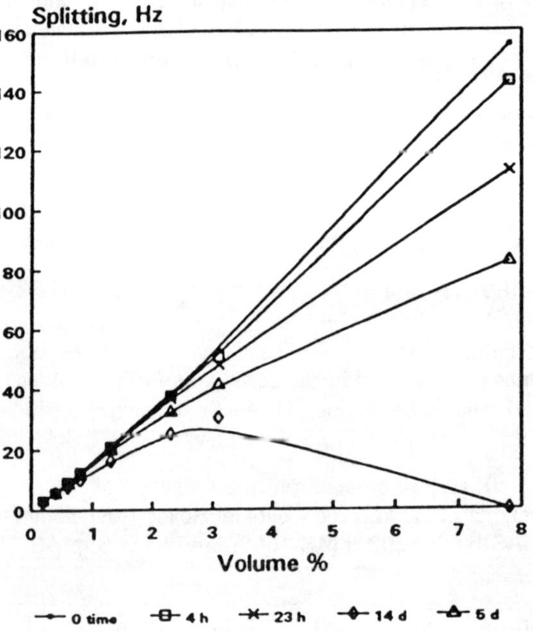
Splitting, Hz
160
140
120
100
80
60
40
20
0
0
1
2
3
4
5
6
7
8
Volume %
0 time
4 h
23 h
14 d
5 d

B

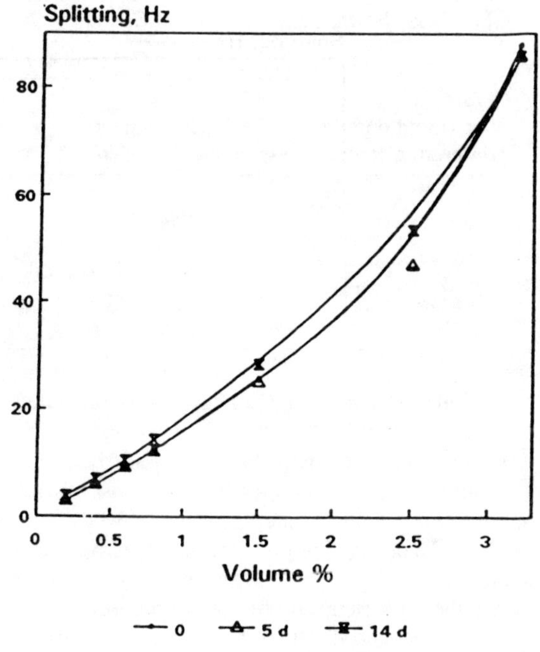

C

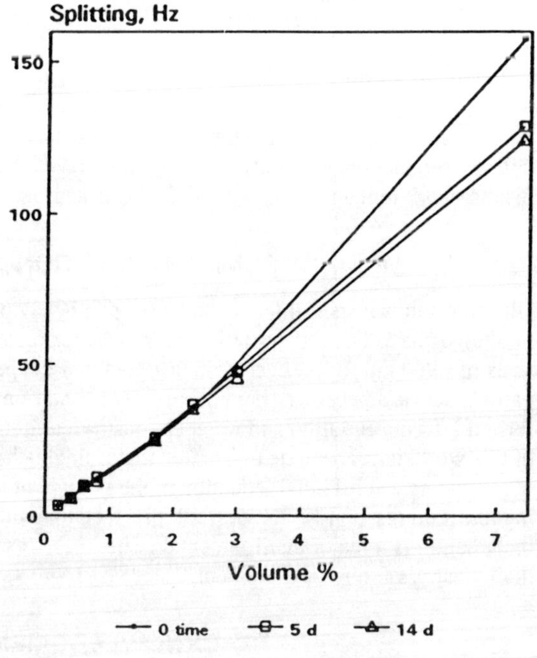

D

Figure 5 Concentration dependence of ^{2}H-NMR splitting (△) for (a) 50-nm and (b) 200-nm ultrafines (UF) in pond water, and (c) 50-nm, and (d) 200-nm UF in distilled water measured at different times. Note: at zero time, measurements are made immediately after shaking.

Table 2 Separation of Colloidal Solids by Centrifugation

Fractions separated under different g forces	% of total dry colloidal solids		
	Tailings produced with sodium silicate	Tailings produced with caustic addition	Sludge
200	22.7	42.9	50.7
500	17.4	39.7	29.2
1500	None	17.4	17.8
91,000	34.1	Traces	2.3
Unsedimented at 91,000 g	25.8	None	None

formation can be prevented by changing the properties of the water by, for example, using a dispersing agent such as sodium silicate instead of sodium hydroxide for pH modification during hot water extraction.[10] It should be mentioned that the use of sodium silicate can cause formation of amorphous iron oxide;[11] therefore, special attention should be paid to the amount of the latter component. In the case of tailings produced in the presence of sodium silicate, most of the ultrafines were well dispersed (Table 2). This allowed them to be separated by centrifugation almost solely on the basis of their particle sizes. In the case of the blank sample from conventional separation using caustic addition, the ultrafines were in an advanced state of flocculation. Size fractionation of these solids, by centrifugation, produced fundamental "flow units"[6] comprised of flocs, with their associated immobile liquid, as well as entrapped coarse emulsified droplets and larger colloidal solids. There was a striking similarity between the behavior of colloidal solids from both the blank test and mature sludge. Therefore, the results show that a change in water chemistry raises the possibility of recycling a greater proportion of the tailings water, while avoiding the adverse effect of sludge viscosity on bitumen separation. The end result of this approach would be a significantly smaller tailings pond. A preliminary evaluation of this concept has been carried out using small-scale batch extraction equipment. Process water was recycled at least nine times without adverse effects on bitumen recovery. A larger scale, continuous pilot test has also been conducted in which fine tailings production was only 30% of the expected level. Despite these encouraging results, work remains to be done before this approach can be commercialized.

IV. SUMMARY AND CONCLUSIONS

Ultrafine solids responsible for the intractable behavior and high water-holding capacity of sludges have been identified as aluminosilicate clays (kaolinite, mica) with particle sizes of <200 nm. It has been demonstrated that the presence of UF with an appropriate particle size is a necessary but not sufficient condition for gelation. The sludge will only form if UF concentration and water chemistry requirements are met. Based on the results of this work, it has been demonstrated that a single change in process additive is capable of producing a substantial reduction in the volume of sludge-forming components entering the tailings ponds. In view of the fact that oil-sands tailings disposal and site reclamation is a major environmental issue, these results could have a major impact on the industry's future development.

ACKNOWLEDGMENTS

We are grateful to Dr. Y. Deslandes for TEM and XPS analysis.

REFERENCES

1. **Camp, F. W.,** Processing Athabasca tar sands tailings disposal, in *Proc. Symp. on Tar Sands*, 26th Can. Chem. Eng. Conf., Toronto, 1977, pap. 9a.
2. **Kessick, M. A.,** Clay slimes from the extraction of Alberta oil sands. Florida phosphate matrix and other mined deposits, *Can. Inst. Min. Bull.*, 71, 80, 1978.
3. **Sparks, B. D., Kotlyar, L. S., and Majid, A.,** A study of factors affecting the stability of tailings sludge produced by the hot water bitumen extraction process, Paper #91-117 presented at Petroleum Society of Canadian Institute of Mining and Alberta Oil Sands Technology and Research Authority, Banff, Alberta, Canada, 1991.
4. **Kotlyar, L. S., Sparks, B. D., Schutte, R., and Capes, C. E.,** Gel forming propensity of colloidal solids from fine tails formed during extraction of bitumen from Athabasca oil sands by the hot water process, *AOSTRA J. Res.*, 8, 55, 1992.
5. **Somasundaran, P.,** in *Fine Particle Processing: Principles of Flocculation, Dispersion and Selective Flocculation,* Somasundaran, P., Ed., American Institute of Mechanical Engineers, New York, 1980, 947.
6. **Michaels, A. S. and Bolger, J. C.,** Settling rates and sediment volumes of flocculated kaolin suspensions, *Ind. Eng. Chem. Fundam.*, 1, 25, 1962.
7. **Ripmeester, J. A., Kotlyar, L. S., and Sparks, B. D.,** ^{2}H NMR and the sol-gel transition in suspensions of colloidal clays, *Colloids Surf.,* 78, 57, 1993.
8. **Grandjean, J. and Laszlo, P.,** Deuterium nuclear magnetic resonance studies of water molecules restrained by their proximity to clay surface, *Clays Clay Miner.,* 37, 403, 1989.
9. **Brinker, C. J. and Scherer, G. W.,** *Sol-Gel Science. The Physics and Chemistry of Sol-Gel Processing,* Academic Press, New York, 1990, chap. 5.
10. **Schutte, R.,** personal communication, 1992.
11. **Referee,** private communication, 1993.

Chapter 9

Position Paper of Part I

M. Schnitzer, K. Haider, A. Herbillon, R. L. Parfitt, M. Robert, and K. Wada

The chapters presented in this section deal with general soil quality as influenced by the interactions of soil minerals with organic components, and microorganisms. Highlights of the papers presented at the workshop are summarized below.

Humic substances, especially fulvic acids, are important ligands in the formation in soils of metal- and clay-organic matter (OM) complexes (M. Schnitzer). The formation of these complexes is one of the most fundamental reactions in soil chemistry, which has profound physical, chemical, and biological effects on soil quality. Both water-soluble and water-insoluble complexes are formed. The chemical structures of humic acids, fulvic acids, and humins are essentially similar except that fulvic acids have lower molecular weights and contain, per unit weight, less carbon but more oxygen and CO_2H groups than the other two fractions. In agricultural soils, humic substances, bonded to clay particles through complexation with Fe and Al, play a major role in soil aggregation, and so assist in establishing satisfactory drainage and aeration and providing protection against erosion. Other soil organic matter (SOM) components that may act as binding agents are polysaccharides and wax-like materials.

The formation of organomineral complexes in soil protects humic substances even more effectively against microbial decay than their complex chemical structure (K. Haider). Enhanced degradation of soil organic matter after changes in management or climate is mainly due to the combination of the fractions with microbial biomass, microbial products, or partly decomposed plant residues. Aggregation of soil particles influences the flow of carbon and other nutrients in the soil, the location of microbes, and ongoing aerobic and anaerobic microbial processes. Organomineral complexes can fix or adsorb organic or inorganic compounds, and are the reason for the buffering capacity of soils to toxic influences of pesticide residues or other xenobiotics.

In some soils, hematite-goethite assemblages may change spontaneously to goethite-bearing ones (A. Herbillon). This process is accompanied by a change in color from reddish to yellowish, and appears to be due to the preferential reduction of ferric iron in hematite by ferri-reducing bacteria and subsequent dissolution of hematite. Goethite is less affected by reduction than hematite because it is richer in Al. This process may be responsible for the transformation of Oxisols, with good physical characteristics, to Ultisols, which have less favorable physical properties. It may also play an important role in the bauxitization of laterites.

In a study on retention of cations and anions by soils, soil solutions were extracted from topsoil and subsoil samples (R. L. Parfitt). All samples showed a preference of K over Ca, although the preference was less in the topsoil than in the subsoil. Thus, Ca rather than K is likely to be leached from the soils tested. Cd accumulates in the topsoil, probably in the clay-organic complex. NO_3, Cl, and pesticides are readily leached from a topsoil. SO_4 is leached from the topsoil but retained by the subsoil if it contains allophane. The presence of OM appears to reduce the adsorption of anions on positively charged mineral surfaces, possibly by competition for sites.

Drying (low water potential) as well as rewetting (high water potential) exert stresses on microbes (M. Robert and C. Chenu). At high water potential, the ratio of water to air

0-87371-914-X/95/$0.00+$.50

in the soil determines the occurrence of aerobic vs. anaerobic processes. At low water potential, desiccation occurs which affects microbial activity and survival. Water potential is important with regard to pollution and bioremediation.

In Al horizons of Japanese Andisols, which contain up to more than 20% C, young humus has a low complexing capacity for Al and Fe (K. Wada). With time and soil formation, the complexing capacity increases so that Al and Fe released from volcanic ash are eventually complexed by SOM in the form of polymeric hydroxy ions. In old, buried Andisol Al horizons, SOM also reacts with Al and Fe in allophane and clays. The polymeric hydroxy-Al and -Fe ions can also react with phosphate. The blocking of ionized CO_2H groups in the SOM by hydroxy-Al and -Fe ions results in a variable negative charge which contributes to stable aggregate formation.

Relatively high concentrations of citric acid reduce the interlayer adsorption by montmorillonite of hydroxy-aluminosilicates in acidic environments (J. G. Q. Lou and P. M. Huang). The surface activity of Andisols is affected by the interaction of allophane with SOM (M. L. Mora et al.). Microbial oxidation of pyrite in relation to its efficiency in alkaline soil reclamation is reported (P. Sharma and A. Swarup). Acid rain affects the release of cations from forest soils; the major cation released from mineral soil horizons is Ca^{2+}, followed by Mg^{2+} and K^{+}, but the release of Al is not apparent (T. Sakuma et al.). Colloidal, chemical, and mineralogical characteristics of fine separates from oil sands are reported and the implications in environmental quality are discussed (L. S. Kotlyar et al.; G. S. R. Krishnamurti and P. M. Huang).

SUMMARY AND RECOMMENDATIONS

1. The papers presented at this workshop are a mixture of applied and more fundamental studies. It is likely that future research in this area will be done along similar lines. It will be necessary to do applied investigations to solve problems of immediate urgency. At the same time, it seems that there is a great need for more fundamental research on inorganic-organic interactions in soils.
2. The more fundamental studies could be done on humins, which are naturally occurring metal-OM and clay-OM complexes, about which we know relatively little.
3. Research along these lines could also be done on laboratory-prepared metal-humic and clay-humic complexes or on specific size fractions of soils which could then serve as models for what is occurring in natural soils.
4. The formation of metal-humic and clay-humic complexes exerts physical, chemical, and biological effects on soil quality. More seems to be known at this time about the physical and chemical effects, so that future research should pay more attention to the biological effects of these complexes.
5. It is incumbent upon us to be vigilant about maintaining, and possibly improving, soil quality in order to maintain our ability to feed an ever increasing population.

Part II: Transformations of Natural and Anthropogenic Organic Compounds as Affected by Soil Minerals and Microorganisms

A. Natural Organics

Chapter 10

Recent Advances in Organomineral Interactions: Implications for Carbon Cycling and Soil Structure

J. M. Oades

CONTENTS

I. INTRODUCTION

Since the publication of *Interactions of Soil Minerals with Natural Organics and Microbes*,[1] advances in organomineral interactions relevant to carbon cycling and soil structure have come from (1) application of solid-state ^{13}C-NMR to soils and soil fractions, (2) further studies of calcium and aluminum bridges between organic and mineral components, (3) studies of the spatial distribution of organic materials in soil aggregates of various sizes, and (4) catchment studies of the release of organomineral complexes from soils to water and implications for water quality.

This review summarizes recent information obtained from solid-state CPMAS ^{13}C-NMR on soils and soil fractions. The results show the presence of considerably more aliphatic materials than previously considered from wet chemistry and tend to confirm that organic matter added to soils which are dominantly aerobic are degraded oxidatively to polymethylene end products. Stabilization of aromatic components occurs in some soils by interactions with calcium and aluminum. General stabilization of a range of organic chemicals occurs in soils by interactions with the inorganic matrix. The spatial distribution of organic debris within larger aggregates has been illustrated by scanning electron microscopy and in smaller aggregates by chemical methods and sectioning, followed by transmission electron microscopy using a range of electron-dense materials which react specifically with certain groups of chemicals. Finally, the integration of various organomineral interactions on a catchment scale is reflected in the colloidal components released from soils into water. The quality of water harvested from catchments is controlled by the sorptive capacities of the soils and the accessibility of soil surfaces to moving water.

0-87371-914-X/95/$0.00+$.50

II. APPLICATIONS OF SOLID-STATE ^{13}C-NMR TO SOILS

The application of solid-state CPMAS ^{13}C-NMR to soils has developed substantially in the last decade. Salient features have been described by Wilson.[2] Developments in equipment such as larger probes, higher spinning speeds, and magnetic fields up to 200 MHz allow chemical characterization of organic materials which represent less than 5% by weight of a soil or sediment. Interfering paramagnetic materials such as iron can be removed by reductive procedures. While there are still questions to be answered with respect to some quantitative aspects of the technique, it is now possible to describe the overall chemistry of complex mixtures of biopolymers. The spectrum in Figure 1 illustrates the major chemical groupings of organic materials obtained from a nonwetting sand and is the basis for the interpretation of other NMR spectra in this review.

A. ^{13}C-ENRICHED SUBSTRATES AND MICROBIAL SYNTHESIS

This technique can be applied to fungi and bacteria and, with the aid of ^{13}C enrichment of substrates, can be used to follow the development of microbial tissues and metabolites during the decomposition of a ^{13}C-enriched substrate such as glucose.[3,4] In Figure 2, a direct comparison can be made of the chemistry of fungal and bacterial isolates with the microbial biomass utilizing glucose in an Alfisol. The comparative chemistry indicates the predominance of fungi in the microbial biomass and shows clearly the dominance of the polymethylene peak around a 30-ppm chemical shift in the newly formed organic materials. In these experiments, little if any aromatic material was synthesized by the dominantly fungal biomass. Substrates enriched with ^{13}C are currently very expensive, but this approach has potential for studies of syntheses and degradation of ^{13}C-labeled organic materials in soils.

B. PARTICLE SIZE FRACTIONS

A second approach where solid-state NMR has been successful is the investigation of the comparative chemistry of particle size fractions in soils. It was shown some time ago[5] that there were substantial differences in the organic chemistry of different particle size fractions obtained from soil after disaggregation by ultrasonic energy. More recently, these differences have been clearly illustrated using NMR. The coarser particle size fractions usually require a density separation of organics from inorganics, which has the major advantage for solid-state NMR studies of concentrating the organic materials from about five- to 20-fold. Resolution of spectra is markedly improved, even after short scanning times.

An example of spectra covering the range of particle sizes from a Mollisol is shown in Figure 3.[6] From these spectra, it is clear that:

1. Carbohydrate contents are least in fine particles.
2. Tannin or phenolic groupings can be found only in coarse fractions.
3. Aliphatic or polymethylene materials increase in relative importance in fine fractions.
4. Aromatic components show the greatest relative concentrations in particles with diameters from about 2 to 50 μm.

These and other data show that the organic chemistry in clay fractions is dominated by aliphatic or polymethylene components and not aromatics. Moreover, Baldock et al.[6] have made calculations which indicate that the accumulation of polymethylene materials is most likely to be a result of both the recalcitrant nature of $-CH_2$-chains and synthesis by the microbial population. By comparison the concentration of aromatic components in particles with diameters of 2 to 50 μm was calculated to be due to the relative concentration of components from lignin by oxidation of other components such as carbohy-

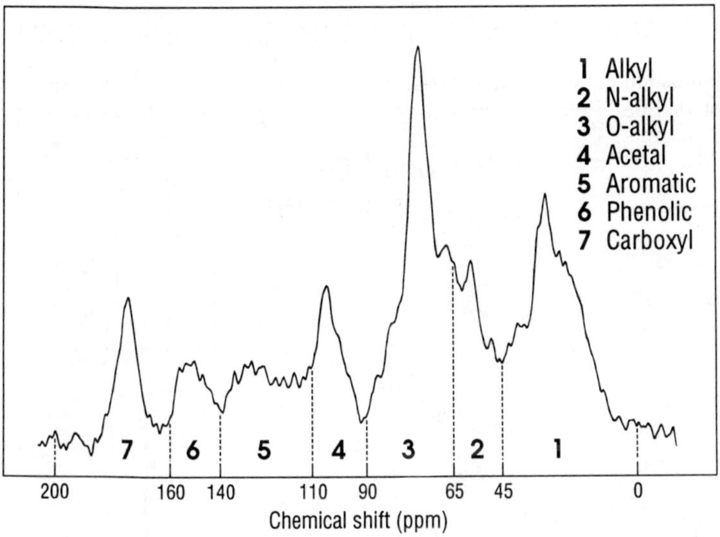

Figure 1 CPMAS ^{13}C-NMR of organic matter from a nonwetting sand. Chemical shifts were used to delineate the chemical groups.

drates and proteins. Synthesis of aromatics in soils was considered to be insignificant compared with the legacies from lignin.

Baldock et al.[6] suggested that in aerobic soils, the changes in organic matter from coarse to fine particles represents the progressive oxidation of plant materials, and developed the schemes of Oades[7] and Hatcher and Spiker[8] to illustrate a simple oxidative degradation system for plant materials (Figure 4). Factors which slow down the rate of decomposition or which selectively stabilize specific chemicals are discussed later.

C. WHOLE SOILS

NMR data on whole soils can be obtained directly or by summing the organic chemistry of individual fractions. Too few data are currently available to draw many conclusions, but the organic chemistry of Oxisols compared with Alfisols and Mollisols indicated more materials in later stages of decomposition, perhaps due to the tropical environment in which Oxisols have developed.[6]

Comparisons have been made of the organic chemistry of an Alfisol under an old pasture and under alternate wheat-fallow rotation and a virgin site. The pasture soil had good water-stable aggregation in the 0 to 10-cm layer and almost 3% organic carbon. The wheat-fallow soil with about 1% C was devoid of water-stable aggregates >250 μm diameter. Differences have been sought in the chemistry of the organic matter in these soils to explain the differences in structural stability and the composition of the old "resistant" organic matter. Differences could not be detected between the Alfisol under extremes of management with respect, for example, to the detailed descriptions of polysaccharides.[9] Similarly, Stevenson[10] was unable to show effects of management on the protein composition of soils.

Comparisons of the composition of the two soils under very different management regimes resulting in almost a threefold difference in organic matter contents revealed that there were some differences in the coarse fractions related to the composition of the plant materials being added but that overall, and particularly within the clay fractions, the composition of the organic matter was identical within the limits of the NMR techniques used[11] (Figure 5).

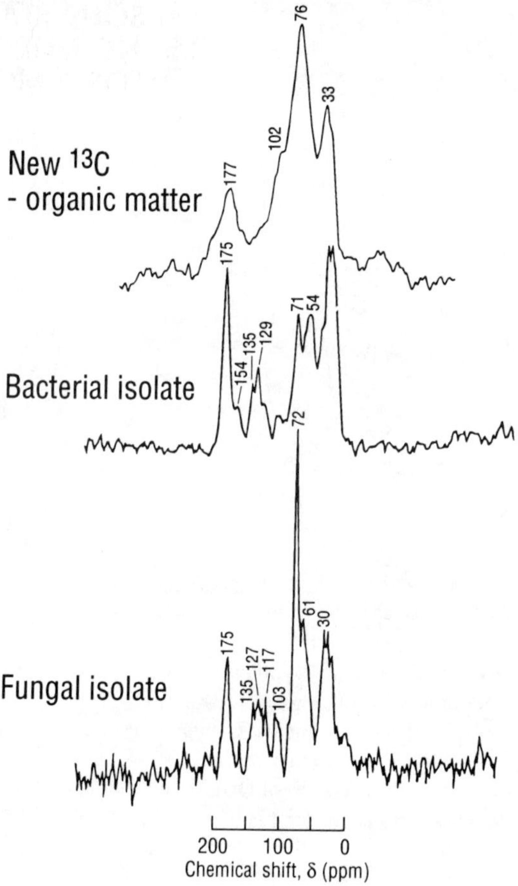

Figure 2 Comparison of fungal and bacterial tissues with new organic matter synthesized from ^{13}C glucose in an Alfisol.[3]

This raises questions about the chemistry of the "inert" pool and "recalcitrant" chemicals with long half-lives. It is suggested from the point of view of carbon cycling that such pools will not be defined chemically because they are controlled by organomineral interactions, including the spatial distribution of organic materials within the inorganic soil matrix. Thus, although within limits the photosynthate entering a particular soil may be different under different management regimes, once it has been decomposed by the microbial population, the products have the same composition which is characteristic for that soil.

Further data which help to substantiate this concept compare the composition of organic chemicals extracted from soils by NaOH with the composition of the nonextracted materials by solid-state NMR, even after HCl-HF treatments to dissolve silicates (Figure 6).

Newman and Tate[12] suggested that the ^{13}C-NMR of alkaline extracts of soils gave a good indication of the chemistry of soil organic matter, and the results in Figure 6 confirm this with some minor adjustments. Clearly, NaOH-soluble materials are similar in composition to the insoluble materials except that NaOH has extracted carbohydrates more efficiently than other chemical groups. Even the organic materials remaining after NaOH extraction and dissolution of silicates using HF-HCl mixtures were similar to the compositions of organic matter in the whole soil and that extracted by NaOH. Some O-alkyl groups survived the two chemical treatments, and it may be that these materials are polyhydroxy acids firmly associated with clays and not true carbohydrates.

The organic chemistry of different particle size fractions was clearly different, and yet the chemical extraction procedures do not appear to fractionate organic matter into its

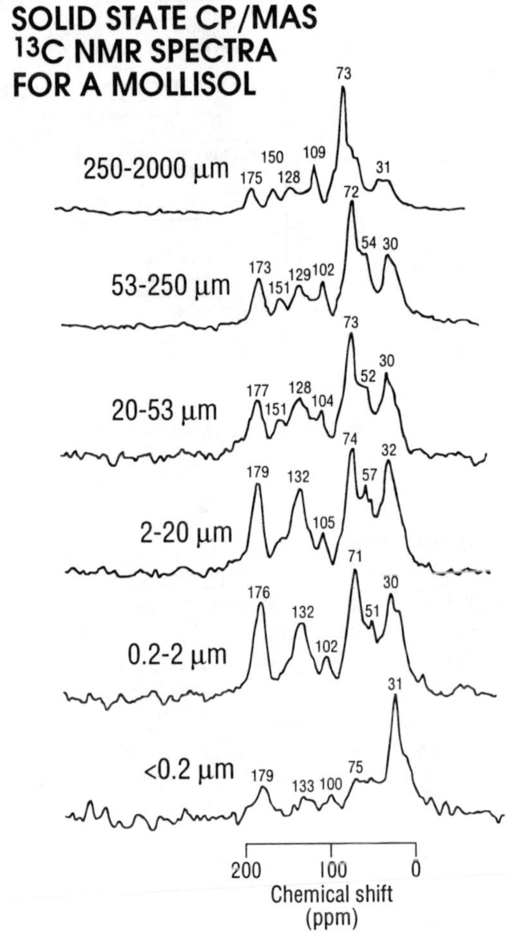

Figure 3 CPMAS ^{13}C-NMR spectra of particle size fractions from a Mollisol.[6]

chemical constituents. We must then explore the possibility that the extractability of soil organic matter is primarily controlled by interactions with the inorganic matrix. The resulting stabilization of the organic materials is not specific for a particular group of chemicals. The protection of organic materials is nonspecific, presumably by adsorption through functional groups and polyvalent cations, and applies to the whole mixture of biological chemicals. This concept helps to explain the similar organic chemistry for old and new materials and the fact that the content of any particular chemical group has not been correlated more strongly with soil aggregation than total soil carbon. It also explains why "humin fractions" have been shown to take up the least amounts of ^{14}C from labeled substrates and, conversely, have low quantities of natural ^{14}C. These properties are controlled by organomineral interactions and not organic chemistry. We will see shortly that the protection of organic materials may be explained not only by adsorption, but also by the spatial distribution of organic particles in the soil matrix.

III. NONWETTING SANDS

In sands, particularly but not exclusively in semiarid environments, the interaction of polymethylene components with the surfaces of sand grains causes the sand grains to become hydrophobic. The nonwetting sands are a problem across southern Australia. The

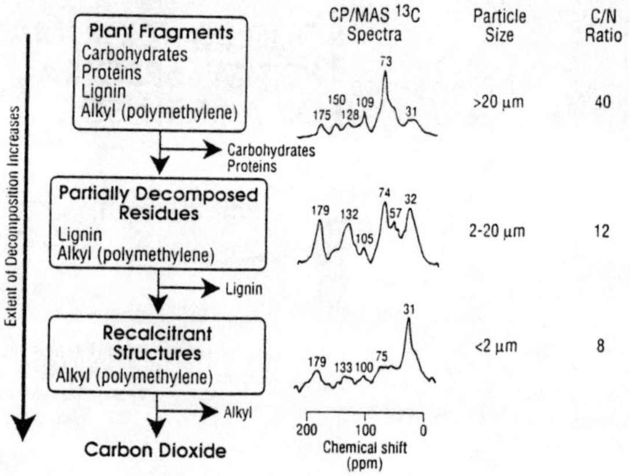

Figure 4 A scheme for the oxidative decomposition of organic materials in aerobic soils.[6]

patches of nonwetting sand, which are often meters in diameter, prevent germination of seeds and, because the sands are not cohesive, become trouble spots which initiate wind erosion. The materials responsible have similarities to the waxes in plant cuticles.[13] The waxes involved are complex, containing both straight and branched polymethylene chains. To interact with the mineral surfaces of sand grains and clays, the waxes must contain substantial numbers of functional groups.

The presence of the waxes on sand grains can be shown by measurements of the molarity of ethanol droplets (MED test) which just penetrate the sand surface in less than 10 s. In sands, the values range from 0 (wettable) to 4 (severely repellent). Calculations of the amounts present have shown that the hydrophobic components probably occur on

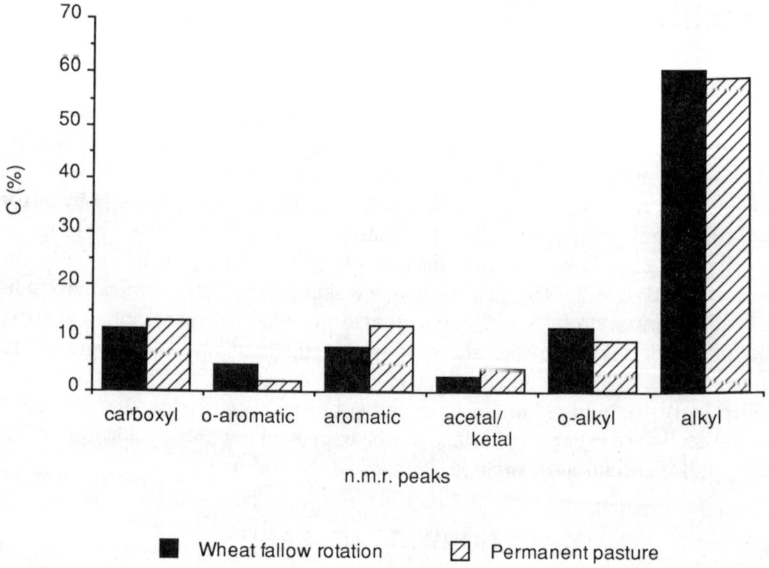

Figure 5 Comparison of organic matter in an Alfisol under pasture and under a wheat-fallow rotation.[11]

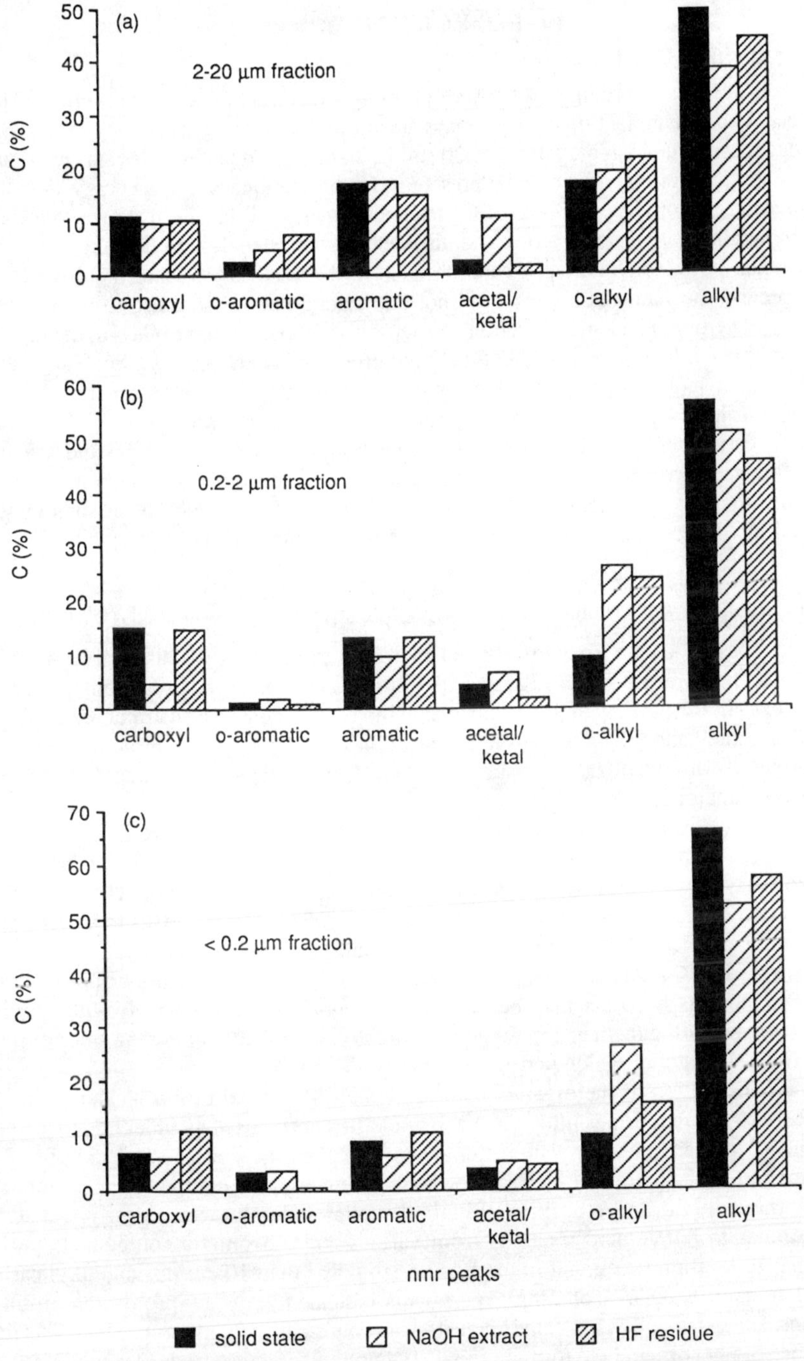

Figure 6 Comparison of organic matter in various fractions from an Alfisol.[11]

the silica surfaces of sand grains as fine globules, but much of the water repellency is also caused by fine particulate matter with the overall chemistry illustrated in Figure 1. The water repellency can be decreased by the addition of dispersive clays,[14] which is being practiced on some farms.

IV. POLYCATION BRIDGES

A. CALCIUM

The role of Ca as a binding agent in soils of high-base status has been reviewed.[15,16] There appears to be an interaction between clays and aromatic substances in Ca-rich soils which produces very stable aggregates 1 to 20 μm in diameter. The aggregates survive limited ultrasonic dispersion and would be present in larger aggregates if less energy were used for disaggregation. The removal of Ca from an organic soil has been shown to increase the mineralization of both carbon and nitrogen,[17] while the addition of Ca to some soils decreased the mineralization of carbon added to the soils as ^{14}C-labeled substrates.[18,19] The precise mechanisms involved are not understood but are assumed to be the cross-linking of carboxyl and other functional groups in a manner similar to Ca binding in clays. The linking could also be between functional groups and clay surfaces. Such aggregates also contain large proportions of the microbial biomass. The organic matter in these aggregates have been shown to be very stable in Vertisols and Mollisols.[20,21] Since Ca is the only cation which can be added to soil economically to improve both pH and structure, it would seem worthy of further study in leached soils. The improvements in structural stability may be incremental rather than dramatic, but Ca availability in soils may be crucial to conservation of organic matter and sustainable management systems.

B. ALUMINUM

The process of Podzolization has challenged soil scientists for decades because the interactions of Al and Fe with organic materials produces very distinctive A and B horizons. In a recent paper, Skjemstad[22] emphasized the role of aluminum-organic complexes in the profiles of giant Podzols on the eastern coast of Australia. Examination of the organic materials in various horizons of these Podzols led to a modified model for the processes of Podzolization in subtropical and tropical regions in which Al rather than Fe plays a major role.[23]

Older profiles on the coastal dune systems of eastern Australia are dominantly siliceous sands with A horizons sometimes exceeding 20 m in depth. The bleached Al horizon may be several meters deep with no colloidal or organic matter. There is a black Bh horizon dominated by C and Al with some Fe grading through a Bhs to a Bs horizon containing disordered aluminosilicates (protoimogolite) and/or ferrihydrite. The Bhs horizon has less C than Bh horizons but contains maximum accumulations of iron oxides. The Bh and Bhs horizons may be only a few centimeters in thickness with no clear boundary separating them and may run horizontally or vertically along the sides of pipes of A horizon material which penetrate to the B horizon.

The composition of the organic matter in a well-developed profile is shown in ^{13}C-NMR spectra of the surface litter, the A1, Bh, and Bhs horizons of an older Podzol (Figure 7). The litter was typical for plant material showing phenolic C at 143 and 153 ppm. The A1 horizon was dominated by aromatic and polymethylene components, presumably concentrated by the decomposition of both phenolics and polysaccharides. The material was similar to humic acid fractions from various soils. Aromatic components clearly dominated the Bh horizon, but in the Bhs horizon, less than 10 cm distant, alkyl carbon was dominant. Dipolar dephasing experiments indicated very few protonated aromatic carbons, and calculations indicated about 1.9 carboxyls per aromatic ring. There was little evidence of phenols, and the formula shown in Figure 8 was proposed. This model is very similar to that proposed by Schnitzer and co-workers as a basis for "fulvic acid" from the Bh horizon of the Armadale Podzol and subsequently extrapolated to other humic substances.[24,25] Whether such materials represent a final stage of lignin degradation or are synthesized by fungi is not known.

The organic materials in the Bhs and Bs horizons were dominated by alkyl materials, and the contents of carboxyl C exceeded those of aromatic carbons. Most carboxyl groups

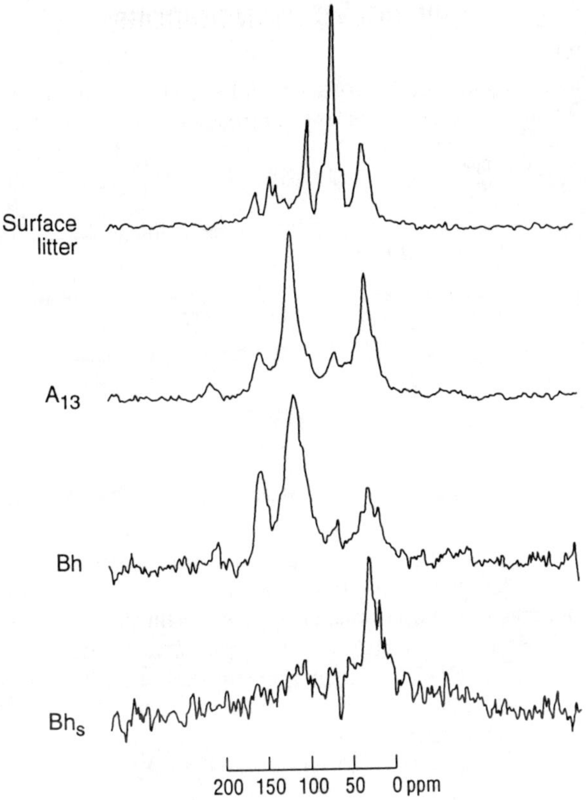

Figure 7 Organic matter in the horizons of an old giant Podzol.[23]

existed in alkyl-carboxylic structures similar in properties to the polymaleic acid model proposed by Anderson et al.[26] and Farmer.[27] Infrared and other data indicated that the material giving rise to O-alkyl signals may be hydroxycarboxylic acids rather than polysaccharides. The data indicate a change from aryl to alkyl in the sequence Bh-Bhs-Bs. This may also apply to other Podzols.[27,28]

The data can be summarized to construct a standard profile for giant Podzols (Figure 9). None of the theories on Podzolization will explain the development of these profiles satisfactorily. An explanation may be that initially, weathering releases Al and Si which interact to form disordered aluminosilicates, which are leached from the A horizon to form a B horizon. These active protoimogolite components adsorb leached organic

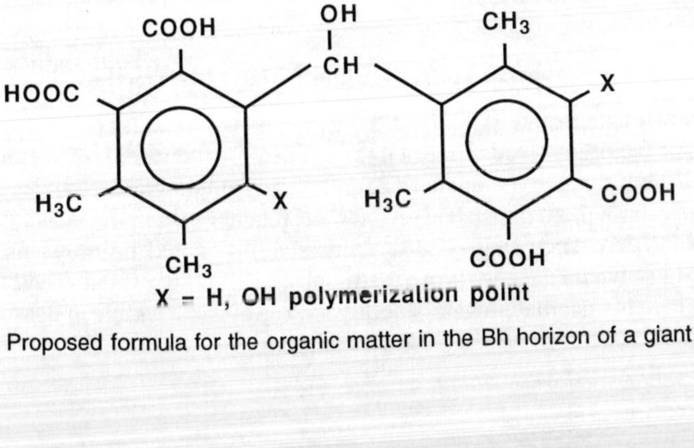

Figure 8 Proposed formula for the organic matter in the Bh horizon of a giant Podzol.[23]

Horizons	Inorganic fine material	Organic material
Litter	-	Plant fragments carbohydrate rich
A1	Virtually absent	Plant debris and roots Aromatic and acidic
A2	Virtually absent (Bleached quartz sand)	Very little
Bh	Extractable Al Crystalline Fe oxides	Acidic aromatic
Bhs	Proto-imogolite and oxyhydroxides	Acidic aliphatic/aromatic
Bs	Proto-imogolite and oxyhydroxides	Highly acidic aliphatic
C	Some weatherable minerals	Very little

Figure 9 Generalized profile characteristics of old giant Podzol.[23]

materials to form a new Bhs horizon. Al and Fe released by further weathering would accumulate in the Bhs horizon and aid in trapping further organic materials. Eventually, sufficient organic matter would accumulate to form a Bh horizon which would then produce a redox environment capable of reducing iron, which would move down the profile and become separate from the main accumulation of Al. The Bh horizon would continue to develop and move downward, preceded by the movement of iron. Thus, the A1 horizon would slowly deepen.

This mechanism depends very much on the association of Al with acidic aromatic components to produce the black color of the Bh horizon and to stabilize the aromatic components against decomposition and further movement. Iron does not appear to be associated with these materials, and most of the organic matter associated with iron oxides appears to be aliphatic in character.

This is illustrated in Figure 10, which shows the solution ^{13}C-NMR of dithionite extracts of the Bh and Bhs horizons of the giant Podzol. The extracts were freed from salts including ferrous iron by ultrafiltration to obtain materials >1000 Da. The organic materials present were dominantly aliphatic, with some carbohydrate and carboxyl groups. These spectra are similar to those published for polymaleic acids and polyuronides.[29,30] Similar spectra have been obtained from organic materials left when natural ferrihydrite was dissolved by reduction. The conclusion is that the organic materials associated with iron oxide surfaces are aliphatic rather than aromatic in character.[31] The reason for this may be that the oxidation of iron is not normally an abiotic chemical process, but is mediated by microorganisms. This may explain why natural ferrihydrites contain significant quantities of organic materials.

V. PROTECTION OF ORGANIC MATERIALS

In their review on aggregate stability and organic matter, Tisdall and Oades[32] concluded that the major factor involved in the stability of macroaggregates, >250-μm diameter in soils, was the enmeshment of smaller aggregates and single-grain particles by roots and hyphae, particularly those dependent on the roots for energy, i.e., the vesicular arbuscular mycorrhiza (VAM). They suggested that this stability would be transient because the roots and hyphae would die and form a ready substrate for the soil fauna and flora. In fact, the rate of loss of macroaggregate stability in a grassland which undergoes tillage is

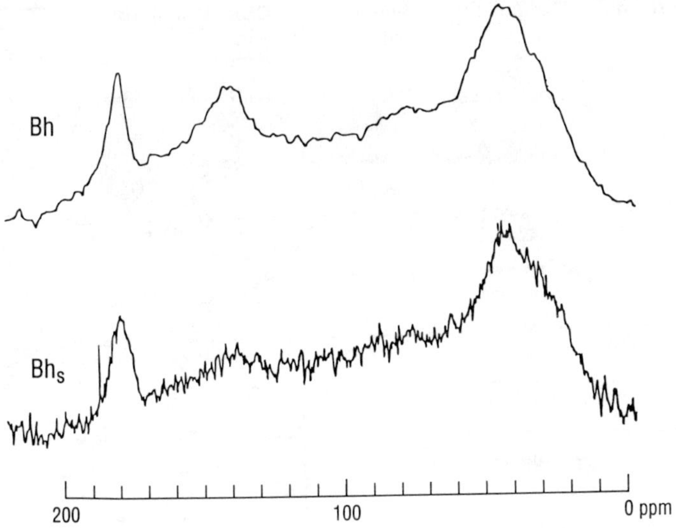

Figure 10 ${}^{13}C$-NMR spectra of desalted dithionite extracts of the organic horizons of a giant Podzol.[23]

similar to the rate of decay of plant material in soils. The decomposition is initially rapid and becomes increasingly slower with time, although the bulk chemistry of the organic materials does not appear to change significantly after the rapid phase of decomposition.

The decomposition of roots has been studied by scanning electron microscopy of aggregates obtained from soils by particle-size fractionations after limited disaggregation of soils by wetting of air-dry soils.[33,34] In larger aggregates, root fragments were clearly seen encrusted to some extent by inorganic materials, presumably through the agency of the rhizosphere and VAM. In aggregates of diameter less than about 200 µm, elongate shapes could be seen, and careful observation showed that many of these elongated aggregates were remnants of roots (and stems) almost completely encapsulated by inorganic coatings. The centers of these aggregates could be cleaned by ultrasonic dispersion of aggregates and separated from inorganic material using heavy liquids. The particulate cellular debris indicated an array of plant remnants which probably represent the lignified cellular skeletons of plant tissues. Examination of smaller aggregates (50- to 100-µm range) revealed the presence of equidimensional aggregates occasionally showing vestiges of cellular material and often cavities presumed to be pores left after complete decomposition of plant debris. The surrounding inorganic particles were presumably held together by organic materials derived from the rhizosphere and metabolic products from the organisms which utilized the roots (or stems) as energy sources.

The encapsulation of pieces of organic material will clearly slow down decomposition processes, but unfortunately we have almost no information on aggregate turnover. The overall view of the protection of organic matter by coatings of inorganic materials is illustrated in Figure 11.

Similar protection of organic materials occurs in particles <20 µm where organic cores have been demonstrated by sectioning and transmission electron microscopy.[35] Recently, Skjemstad et al.[39] have shown that organic materials in microaggregates <20-µm diameter were protected from high-energy ultraviolet photooxidation. The protection was greater in silt fractions than in clay fractions. In some clay fractions, about one fifth of the organic matter was shown to be "internal" and inaccessible to UV light. In

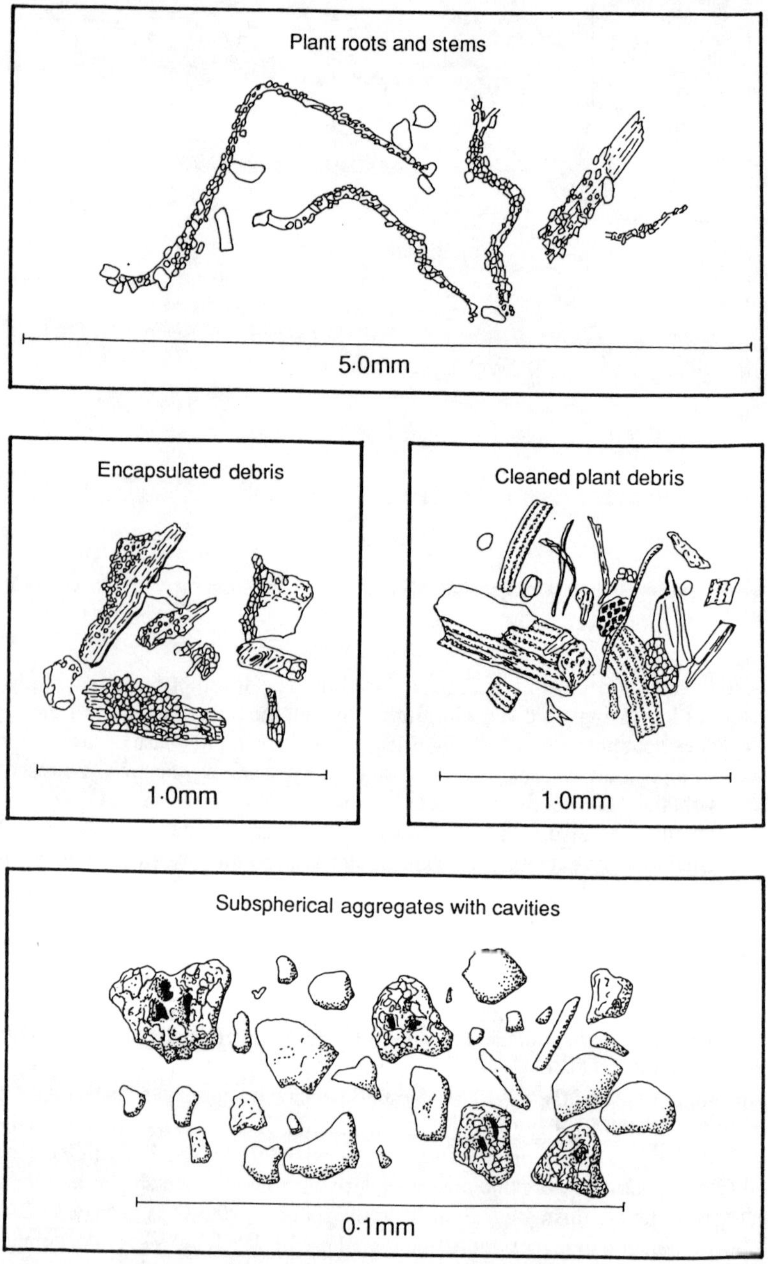

Figure 11 Encapsulation of plant debris by inorganic materials during decomposition of plant materials. (Drawn by Paul Dangerfield.)

silt-sized aggregates, the protected organic matter was about one third of the total material present. By use of a ^{14}C accelerator mass spectrometer, it was demonstrated that both clay and silt fractions contained modern carbon which was presumably associated with the outsides of aggregates. After 4 h of photooxidation, the internal organic material was shown to be old, with mean residence times of 200 and 300 years

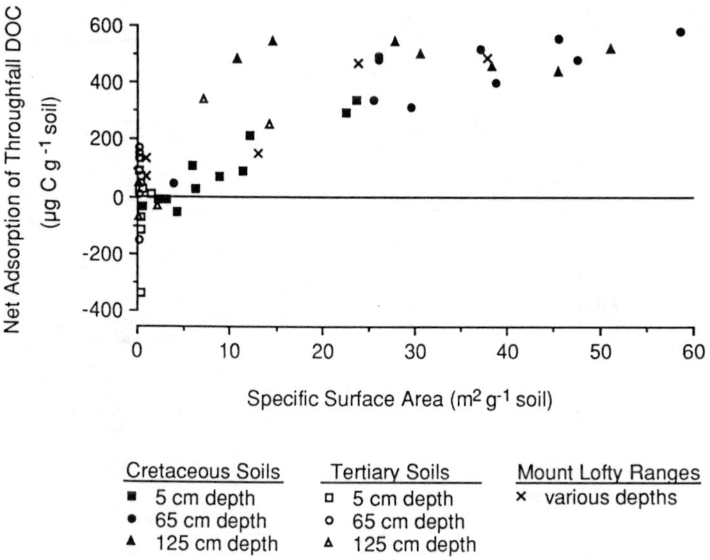

Figure 12 Relation of the dissolved organic carbon content of water to the texture of A horizons.[37,38]

BP, respectively. If the accessibility of organic material to UV light is similar to the accessibility to enzymes, then perhaps this method could form a basis for the measurement of the "inert pool" of organic matter in soils.

A biological approach to protection mechanisms in soils has been described by Amato and Ladd.[36] ^{14}C-labeled glucose and ^{14}C-labeled legume materials were incubated in soils in the field for 44 and 66 weeks respectively. The soils ranged in texture from 2 to 73% clay (CEC of 20 to 804 mmol kg^{-1}). Residual ^{14}C from ^{14}C glucose incubation (15 to 30%) increased with clay content due entirely to increases in the residual ^{14}C in the microbial biomass, i.e., residual ^{14}C in nonbiomass was 14% of the input in all soils. Residual ^{14}C from ^{14}C legume ranged from 20 to 36% and was greater for mildly acidic soils. The differences in residual ^{14}C between glucose and legume substrates could initially be explained by the slow decomposition of the more complex biopolymers such as lignin, but subsequently by the protection methods described above.

VI. ORGANOMINERAL INTERACTIONS AND WATER QUALITY

In the Mediterranean climate of South Australia, the high rainfall areas represent the major catchment areas for metropolitan Adelaide reservoirs. Some of these reservoirs contain water with dissolved organic carbon (DOC) contents up to 20 mg l^{-1}, while others contain only several parts per million of DOC. Recent investigations have shown that water quality was controlled by the ability of soils to retain organic matter and prevent it from entering the aqueous phase for transport to reservoirs.[37]

The adsorption capacity of the soils was shown to be related to the clay content of A horizons or the surface area of the soils (Figure 12). The data initially established for a pair of catchments were extended to a range of catchments in the Mount Lofty ranges and another pair of catchments in the Otway Ranges Victoria.[38] The combined data (Figure 13) show that the DOC of water from catchments decreased from about 20 mg l^{-1} in catchments with sandy A horizons to about 5 mg l^{-1} in catchments with loamy A horizons. The quality of the water with respect to organic matter was related to the adsorptive capacity of the clays. It was presumed initially that the water quality was controlled by

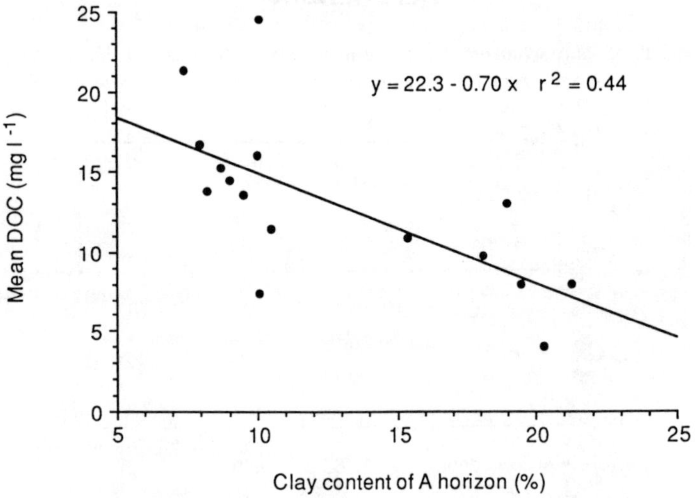

Figure 13 The influence of surface area (clay content) on adsorption of organic materials.[37]

the different management regimes in the catchments, but the results show clearly that the major factor is the adsorption capacity of the A horizons.

The adsorption of organic material on the surfaces of the A horizon materials is also influenced by the salt contents and the sodium adsorption ratios (SAR) of the water.[40] As would be predicted from the theories of clay dispersion and adsorption mechanisms, the release of fine colloidal material, including organic carbon from soils, is greatest at low electrolyte concentrations, especially when SAR values are high. The data show the control of the release of carbon and clay from soils by electrical conductivity and SAR of soil water. Turbidity, a measure of clay content, was strongly correlated with total organic carbon. Thus, the desorption which resulted was of a clay-organic complex and not simply organic matter from clay. Examination of the organic matter in streams by ^{13}C-NMR showed the usual complex of chemical groups present in soils and leads to a reiteration of the view that the organoclay complexes in soils are the result of nonspecific adsorption of complex mixtures of biopolymers. Desorption releases the whole complex. Both P and N contents were correlated with turbidity and carbon, indicating the importance of organomineral interactions in catchment soils in controlling the quality of water.

VII. SUMMARY AND CONCLUSIONS

1. The use of solid-state CPMAS ^{13}C-NMR enables characterization of organomineral associations not previously possible.
2. There is a strong association of polymethylene materials with clay fractions.
3. There may be specific stabilization of aromatic acidic materials by Ca and Al.
4. There is a general stabilization of organic materials by occlusions of organic particles in the soil inorganic matrix, i.e., the control of C turnover in the longer term is by organomineral associations and not by organic chemistry.
5. The colloidal components released from soils and entering the aqueous phase are an organomineral complex. The dispersed phase contains C, N, and P and impacts on water quality.

REFERENCES

1. **Huang, P. M. and Schnitzer, M.,** *Interactions of Soil Minerals with Natural Organics and Microbes,* SSSA Spec. Publ. No. 17, Soil Science Society of America, Madison, WI, 1986.
2. **Wilson, M. A.,** *NMR Techniques and Applications in Geochemistry and Soil Chemistry,* Pergamon Press, Oxford, U.K., 1987.
3. **Baldock, J. A., Oades, J. M., Vassallo, A. M., and Wilson, M. A.,** Solid state CP/MAS ^{13}C N.M.R. analysis of bacterial and fungal cultures isolated from a soil incubated with glucose, *Aust. J. Soil Res.,* 28, 213, 1990.
4. **Baldock, J. A., Oades, J. M., Vassallo, A. M., and Wilson, M. A.,** Significance of microbial activity in soils as demonstrated by solid-state ^{13}C NMR, *Environ. Sci. Technol.,* 24, 527, 1990.
5. **Turchenek, L. W. and Oades, J. M.,** Fractionation of organo-mineral complexes by sedimentation and density techniques, *Geoderma,* 21, 311, 1979.
6. **Baldock, J. A., Oades, J. M., Waters, A. G., Peng, X., Vassallo, A. M., and Wilson, M. A.,** Aspects of the chemical structure of soil organic materials as revealed by solid-state ^{13}C NMR spectroscopy, *Biogeochemistry,* 16, 1, 1992.
7. **Oades, J. M.,** Organic matter in the Urrbrae soil, in *Red Brown Earths of Australia,* Oades, J. M., Lewis, D. G., and Norrish, K., Eds., Waite Institute, University of Adelaide, and CSIRO Division of Soils, 1982, 63.
8. **Hatcher, P. G. and Spiker, E. C.,** Selective degradation of plant biomolecules, in *Humic Substances and Their Role in the Environment,* Frimmel, F. H. and Christman, R. F., Eds., John Wiley & Sons, Chichester, 1988, 59.
9. **Oades, J. M.,** Studies on soil polysaccharides. III. Composition of polysaccharides in some Australian soils, *Aust. J. Soil Res.,* 10, 113, 1972.
10. **Stevenson, F. J.,** *Humus Chemistry, Genesis, Composition, Reactions,* John Wiley & Sons, New York, 1982.
11. **Oades, J. M., Water, A. G., Vassallo, A. M., Wilson, M. A., and Jones, G. P.,** Influence of management on the composition of organic matter in a red-brown earth as shown by ^{13}C nuclear magnetic resonance, *Aust. J. Soil Res.,* 26, 289, 1988.
12. **Newman, R. H. and Tate, K. R.,** Use of alkaline soil extracts for ^{13}C NMR characterization of humic substances, *J. Soil Sci.,* 35, 47, 1984.
13. **Ma'shum, M., Tate, M. E., Jones, G. P., and Oades, J. M.,** Extraction and characterization of water-repellent materials from Australian soils, *J. Soil Sci.,* 39, 99, 1988.
14. **Ma'shum, M., Oades, J. M., and Tate, M. E.,** The use of dispersible clays to reduce water-repellency of sandy soils, *Aust. J. Soil Res.,* 27, 797, 1989.
15. **Oades, J. M.,** The retention of organic matter in soils, *Biogeochemistry,* 5, 35, 1988.
16. **Oades, J. M.,** Associations of colloids in soil aggregates, in *Soil Colloids and Their Associations in Aggregates,* De Boodt, M. F., Hayes, M.H.B., and Herbillon, A., Eds., Nato AS1 Ser. B, Vol. 215, Plenum Press, NY, 1990, 463.
17. **Gaiffe, M., Duqyet, B., Tavant, H., and Bruckert, S.,** Stabilite biologique et comportement physique d'un complexe argilo-humique place dans differentes conditions de saturation en calcium ou en potassium, *Plant Soil,* 77, 271, 1984.
18. **Muneer, M. and Oades, J. M.,** The role of Ca-organic interactions in soil aggregate stability. I. Laboratory studies with ^{14}C-glucose, $CaCO_3$ and $CaSO_4 \cdot 2H_2O$, *Aust. J. Soil Res.,* 527, 389, 1989.
19. **Muneer, M. and Oades, J. M.,** The role of Ca-organic interactions in soil aggregate stability. II. Field studies with ^{14}C-labelled straw, $CaCO_3$ and $CaSO_4 \cdot 2H_2O$, *Aust. J. Soil Res.,* 27, 401, 1989.
20. **Dalal, R. C. and Mayer, R. J.,** Long term trends in fertility of soils under continuous cultivation and cereal cropping in southern Queensland. III. Distribution and kinetics of soil organic carbon in particle size fractions, *Aust. J. Soil Res.,* 24, 293, 1986.
21. **Anderson, D. W. and Paul, E. A.,** Organo-mineral complexes and their study by radiocarbon dating, *Soil Sci. Soc. Am. J.,* 48, 298, 1984.
22. **Skjemstad, J. O.,** Genesis of Podzols on coastal dunes in southern Queensland. III. The role of aluminium-organic complexes in profile development, *Aust. J. Soil Res.,* 30, 645, 1992.
23. **Skjemstad, J. O., Waters, A. G., Hanna, J. V., and Oades, J. M.,** Genesis of Podzols on coastal dunes in southern Queensland. IV. Nature of the organic fraction as seen by ^{13}C nuclear magnetic resonance spectroscopy, *Aust. J. Soil Res.,* 30, 667, 1992.

24. **Schnitzer, M. and Khan, S. U.,** *Humic Substances in the Environment,* Marcel Dekker, New York, 1972.
25. **Schnitzer, M.,** Humic substances: chemistry and reactions, in *Soil Organic Matter,* Schnitzer, M. and Khan, S. U., Ed., Elsevier, New York, 1978.
26. **Anderson, H. A., Berrow, M. L., Farmer, V. C., Hepburn, A., Russell, J. D., and Walker, A. D.,** A reassessment of Podzol formation processes, *J. Soil Sci.,* 33, 125, 1982.
27. **Farmer, V. C.,** Significance of the presence of allophane and imogolite in Podzol Bs horizons for Podzolization mechanisms: a review, *Soil Sci. Plant Nutr.,* 28, 571, 1982.
28. **De Coninck, F.,** Major mechanisms in formation of spodic horizons, *Geoderma,* 24, 101, 1980.
29. **Spiteller, M. and Schnitzer, M.,** A comparison of the structural characteristics of polymaleic acid and a soil fulvic acid, *J. Soil Sci.,* 34, 525, 1983.
30. **Martin, F. and Gonzalez-Vila, F. J.,** About the similarity between polymaleic acid and water soluble humic fractions, *Z. Naturforsch.,* 39, 244, 1984.
31. **Oades, J. M., Gillman, G. P., and Uehara, G.,** Interactions of soil organic matter and variable-charge clays, in *Dynamics of Soil Organic Matter in Tropical Ecosystems,* Coleman, D. C., Oades, J. M., and Uehara, G., Eds., University of Hawaii, Honolulu, 1989.
32. **Tisdall, J. M. and Oades, J. M.,** Organic matter and water-stable aggregates in soils, *J. Soil Sci.,* 33, 141, 1982.
33. **Waters, A. G. and Oades, J. M.,** Organic matter in water-stable aggregates, in *Advances in Soil Organic Matter Research: The Impact on Agriculture and the Environment,* Wilson, W. S., Ed., Royal Society of Chemistry, Cambridge, 1991, 163.
34. **Oades, J. M. and Waters, A. G.,** Aggregate hierarchy in soils, *Aust. J. Soil Res.,* 29, 815, 1991.
35. **Jocteur-Monrozier, L., Ladd, J. N., Fitzpatrick, R. W., Foster, R. C., and Raupach, M.,** Components and microbial biomass content of size fractions in soils of contrasting aggregation, *Geoderma,* 49, 37, 1991.
36. **Amato, M. and Ladd, J. N.,** Decomposition of ^{14}C-labelled glucose and legume material in soils: properties influencing the accumulation of organic residue C and microbial biomass C, *Soil Biol. Biochem.,* 24, 455, 1992.
37. **Nelson, P. N., Cotsaris, E., Oades, J. M., and Bursill, D. B.,** The influence of soil clay content on dissolved organic matter in stream waters, *Aust. J. Mar. Freshwater Res.,* 41, 761, 1990.
38. **Nelson, P. N., Baldock, J. A., and Oades, J. M.,** Concentration and composition of dissolved organic carbon in streams in relation to catchment soil properties, *Biogeochemistry,* 19, 27, 1993.
39. **Skjemstad, J. O., Janik, L. J., Head, M. J., and McClure, S. G.,** High energy ultraviolet photo-oxidation: a novel technique for studying physically protected organic matter in clay- and silt-sized aggregates, *J. Soil Sci.,* 44, 485, 1993.
40. **Skene, T. M. and Oades, J. M.,** The effects of SAR and electrolyte concentration on water quality: laboratory studies. *Soil Science* (in press) 1995.

Chapter 11

The Role of Short-Range Ordered Mineral Colloids in Abiotic Transformations of Organic Components in the Environment

P. M. Huang

CONTENTS

I. INTRODUCTION

Short-range ordered (SRO) mineral colloids in soils include a wide range of inorganic constitutes. Metal oxides[1,2] and other minerals such as aluminosilicates (e.g., allophane, imogolite)[3] and carbonates[4] may exist in soils in short-range ordered forms. These mineral colloids are characterized by a relatively high surface area and charge density and, thus, interact readily with organic molecules. Furthermore, the SRO mineral colloids often coat crystalline minerals in soil environments. Therefore, besides their direct contribution to the abiotic transformations of organics in soils, these mineral colloids may affect the transformations by altering the surface properties of crystalline minerals. The effect of these mineral colloids on the abiotic degradation of organics can be considerably larger than indicated by their contents in the soil. Ill-defined and coated mineral particles may contribute to most of the reactive surface area in many soils. Much of the more detailed research on surface-related abiotic transformations of organic components was conducted with pure crystalline mineral components of soils, however.

The importance of the association between mineral colloids and organic materials was recognized about 6 decades ago.[5,6] Most of the natural organic components in soils are

0-87371-914-X/95/$0.00+$.50

associated with a series of mineral colloids. The data presented by Greenland[7] show that between 52 and 98% of the organic carbon in a wide range of soils is associated with the clay fraction. It is likely that most of the remaining organic carbon is linked to metal oxides and hydroxides.[8] Therefore, abiotic transformations of organic components should be closely associated with the surface reactivity of SRO mineral colloids of soils.

The relatively high specific surface and disordered network of the SRO mineral colloids would influence the make-up of the solution. The dissolved metal ions can complex with organics in the soil solutions. These solution complexation reactions[9] may influence subsequent transformations of organics in soil environments. Furthermore, dissolved metals such as Fe and Mn can influence the redox potential of the liquid phase,[4] which would, in turn, affect the abiotic degradation of organics in the soil.

Well-defined organic compounds and humic substances can interact with Al and Fe and perturb their transformations.[10-17] These reaction products are short-range ordered in nature. Organic components are part of the disordered network of these reaction products, and their dynamics in soils would thus be influenced by the formation of SRO mineral colloids.

The high specific surface and charge density of the SRO mineral colloids[2,3,18] contribute to the retention of organic components on the surfaces of these materials. The retention should control chemical and biological transformation processes of organic matter in soils. Therefore, the dynamics of turnover of organic components would be affected by the retention process.

The reactive surfaces of the SRO mineral colloids also contribute to the abiotic catalytic degradation of natural organic components in soils.[19-22] The research data indicate that these mineral colloids have a role similar to enzymatic catalysis. Therefore, the catalytic power of SRO mineral colloids of soils in the dynamics of organic transformations via abiotic processes in the pedosphere deserves increasing attention.

II. FORMATION OF METAL-ORGANIC COMPLEXES IN SOIL SOLUTIONS AND OTHER NATURAL WATERS

The organic components in the soil solution can be divided into two groups: (1) well-defined biochemicals and (2) a series of acidic, yellow- to black-colored polyelectrolytes referred to as fulvic and/or humic acids (FAs, HAs). These soluble organics may form stable complexes with metals released from weathering of the SRO mineral colloids.

Biochemical compounds are ubiquitous in soils. These biochemicals are present in microbial metabolites, root exudates, canopy drip, and decay products of plant and animal tissues. They have a transitory existence in soils due to microbial destruction[9] and abiotic transformations.[22] The amounts found at any one time represent a balance between the formation and the degradation. High amounts of these biochemicals would be expected to be produced in localized zones where biological activity is intense, such as in the rhizosphere,[23] near decomposing plant residues,[24] and in the organic layers of Alfisols and Spodosols.[9] Soils amended with manures and other organic wastes may also be relatively rich in biochemicals.

Soil biochemical compounds include simple organic acids, amino acids, sugar acids, phenols, phenolic acids, hydroxamate siderophores, and complex polymeric phenols.[9] The kinds and amounts of these biochemicals are variable in most agricultural soils, depending on the plant species, the soil type on which the plants are growing, and land management.[25] These compounds are being constantly added to soils through natural vegetation, microbial metabolism, and farming practices.

The SRO mineral colloids can influence the chemistry of the soil solution, which would, in turn, affect the transformations of soil biochemicals. For instance, poorly ordered carbonates can act as a buffer, and hydrous oxides, hydroxides, and oxyhydroxides containing redox active metals such as Fe and Mn can modify redox processes of certain biochemicals such as phenols and phenolic acids. Furthermore, dissolved metals such as Al and Fe form stable complexes with many biochemical compounds in the soil solution.[9,18] Complexation of these biochemicals with dissolved metals occurs predominantly with organic groups containing oxygen. The organic groups containing nitrogen generally have weak interactions with the metals. Therefore, transformations of soil biochemicals that contain COOH, phenolic-, enolic-, and aliphatic-OH groups, and possibly ketonic C=O and ester functional groups would likely be affected by dissolved metals in the soil solution.

Besides well-defined biochemical compounds, FAs are important soil solution organic ligands. The results obtained with natural waters[26,27] have shown that the bulk of the soluble organic matter consists of FAs. The ability of FAs to form stable complexes with metal ions can be attributed to their high content of O-containing functional groups. Amino groups may also be involved to some extent. The total acidities of FAs range from 640 to 1420 cmol (H^+) kg^{-1}; COOH contents of FAs range from 520 to 1120 cmol (H^+) kg^{-1}.[28] The results of infrared spectroscopy studies have confirmed that COOH groups, or more precisely carboxylate (COO^-), play a prominent role in the complexing of FAs by metal ions. Some evidence indicates that OH, C=O, and NH groups are also involved.[29-31] The conditional stability constants of metal-FAs have been reported.[9]

The stability constant is an important characteristic of a metal-organic complex, because it provides a quantitative measure of the affinity of the ligand to the metal.[9,28] Therefore, stability constants of metal-organic complexes are essential for a full understanding of the factors affecting the abiotic transformations of organics. Furthermore, the stability constants should be related to the biodegradation of organics as influenced by complexation of metals released from the SRO mineral colloids and the dynamics of carbon turnover.

The stability constants of the complexes formed between Al and selected low-molecular-weight organic acids are shown in Table 1. Among the five organic acids listed, citric acid gives rise to the most stable complexes with Al, and *p*-hydroxybenzoic acid has the least affinity for Al as indicated by the stoichiometric stability constants. More research is needed to advance knowledge on the biodegradability of soil biochemical compounds as influenced by the formation of soluble organometal complexes in the soil solution. Furthermore, these complexes may contain mixed ligands. The stability of these mixed ligand complexes and their subsequent transformations in the soil solution are even less understood.

III. INCORPORATION OF ORGANICS INTO THE STRUCTURAL NETWORK OF SHORT-RANGE ORDERED MINERAL COLLOIDS

Organic components are an integral part of soils. They vary in their chemical composition, structure, nature of functional groups, molecular and particle size, and other physicochemical properties. Therefore, they also vary in their ability to perturb the crystallization of Al and Fe precipitate[2,18] and influence the formation of SRO aluminosilicates, leading to the formation of ill-defined mineral colloids.[17] Both well-defined biochemical compounds and humic substances influence the neoformation of the SRO mineral colloids.

The positively charged edges of polynuclear Al ions, in the absence of organic ligands, readily undergo hydrolysis in the pH range common in soil environments. If organic ligands are present in the system, the coordination sites of polynuclear Al are occupied by organic ligands instead of H_2O molecules, imposing a restraint on the hydrolysis of

Table 1 Stability Constants of the Complexes Formed Between Al and *p*-Hydroxybenzoic, Aspartic, Tannic, Malic, and Citric Acids at 25°C[32]

Organic acid	Stability constants of the complexes log K_1	log K_2
p-Hydroxybenzoic acid	1.66	—
Aspartic acid	2.60	—
Tannic acid	3.78	—
Malic acid	5.14	8.52
Citric acid	7.37	13.90

Al.[32] If more organic ligands are present in the system, the greater would be the replacement of H_2O molecules and blocking of the coordination sites of Al, and the greater the restraint on the subsequent hydrolysis of polynuclear Al ions. If the concentration of organic ligands is held constant, the greater the affinity of the organic ligands for Al, the more extensive would be their occupation of the coordination sites of Al, and the more efficient the restraint imposed on the hydrolysis of polynuclear Al ions. The occupation of the coordination sites of Al by organic ligands disrupts the hydroxyl bridging mechanism indispensable for the polymerization of Al ions. The hydrolytic products of Al formed in the presence of citric, malic, and aspartic acids at 10^{-4} *M* in the system are noncrystalline to poorly crystalline to X-rays (Figure 1b to d). *p*-Hydroxybenzoic acid, which forms the least stable complex with Al (Table 1), does not inhibit the crystallization of hydrolytic precipitation products of Al (Figure 1a). The critical molar ratio of organic acid to Al in inhibiting the crystallization of Al hydroxides varies with the affinity of organic acids toward Al.

Higher-molecular-weight organic acids such as tannic acid also perturb the crystallization of precipitation products of Al.[11] Kodama and Schnitzer[14] reported that FAs significantly hamper the crystallization of Al hydroxides. As the ratio of FA to Al reaches 0.1, only noncrystalline precipitates are formed. Fulvic acid resembles low-molecular-weight aliphatic acids such as citric and malic acids in that it contains COOH and aliphatic OH groups. It also resembles quercetin and tannic acid as it contains phenolic hydroxyl and ketonic C=O groups. The crystallization of Al hydroxides is inhibited by the presence of very small amounts of flavone quercetin.[12] Through these functional groups, FAs form stable complexes with Al and retard the crystallization of Al hydroxides.

During the formation of the SRO Al hydroxides, organic ligands are incorporated into the network of these mineral colloids.[34] The residual C contents of the precipitation products of Al formed in the presence of citric and tannic acids after NaOCl and H_2O_2 treatments range from 22 to 60% and from 55 to 90% of the original C contents, respectively (Table 2). A substantial proportion of C still remains in the Al precipitation products after the treatments. The contents of organic C which remained after the NaOCl treatment were less than those remaining after H_2O_2 treatment. The greater efficiency of NaOCl than H_2O_2 treatment is in agreement with results obtained from soils by Anderson[35] and Lavkulich and Wiens.[36] Since citric and tannic acid in the absence of Al precipitation products are completely oxidized by two to three treatments of H_2O_2 or NaOCl, organic ligands coprecipitated with Al are apparently adsorbed and incorporated in internal structures of Al precipitation products. Consequently, the organic C was protected and not easily oxidized.

Infrared (IR) spectra of Al precipitation products formed at tannic acid/Al molar ratios (R) of 0.1 and 0.02 (Figures 2a and b) show many bands (1700, 1600, 1500, 1443, 1361, 1305, and 1205 cm^{-1}) characteristic of tannate. After oxidation by NaOCl, spectra of the

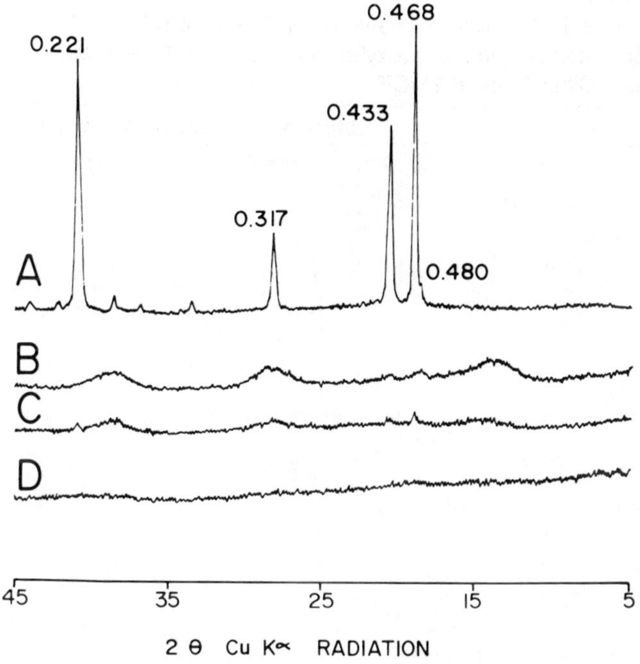

Figure 1 X-ray diffraction patterns of hydrolytic products of Al precipitated at the initial Al concentration of 1.10×10^{-3} *M,* OH/Al molar ratio of 3.0, and aged for 40 d at room temperature in the presence of 10^{-4} *M.* (a) *p*-Hydroxybenzoic acid; (b) aspartic acid; (c) malic acid; and (d) citric acid. d-Spacings are in nanometers.[33]

sample formed at R = 0.1 (Figure 1c) is practically unchanged (only a slight reduction of bands at 1500 and 1200 cm^{-1}), evidently due to the remaining large organic C content (133 g kg^{-1}) of the precipitate (Table 2). The sample formed at R = 0.02 shows all the characteristic bands of tannate before and after oxidation (Figures 2b and d), but some of them (1500, 1443, 1305, and 1205 cm^{-1}) appear much weaker or broader after oxidation. This is attributed to relatively low residual organic C content (42 g kg^{-1}) after oxidation. After the NaOCl treatment, the precipitation products of Al remain unchanged to X-ray

Table 2 Effects of H_2O_2 and NaOCl Treatments on Organic C Contents of Al Precipitation Products Formed in the Presence of Selected Organic Acids[34]

		Organic C (g kg^{-1})[b]		
Ligand	**R**[a]	**No treatment**	**After H_2O_2 treatment**	**After NaOCl treatment**
Citric acid	0.01	10	9	6
Tannic acid	0.01	100	55	22
Tannic acid	0.02	130	79	42
Tannic acid	0.10	302	ND[c]	133

[a] R indicates the initial ligand/Al molar ratio.
[b] Average error, 2.8%.
[c] Not determined.

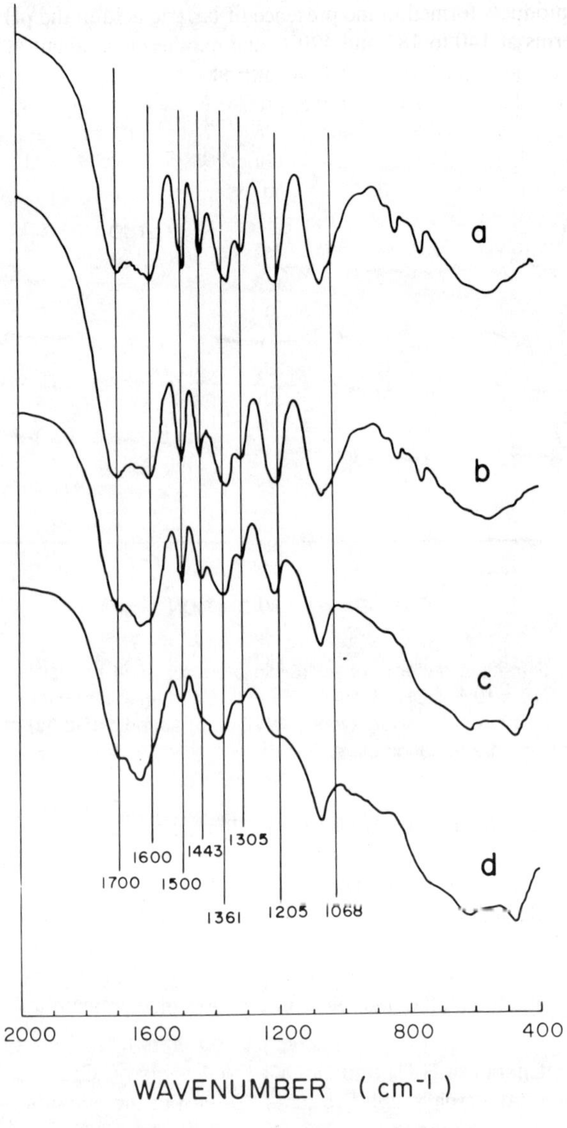

Figure 2 The IR spectra of Al precipitation products showing the presence of organic ligands (tannate) coprecipitated with Al before (a, R = 0.1; b, R = 0.02) and after (c, R = 0.1; d, R = 0.02) NaOCl treatment (the R indicates the initial ligand/Al molar ratio). Tannic acid has strong bands at 1725, 1620, 1545, 1445, 1330, and 1220 cm^{-1}.[34]

diffraction analyses. The IR and X-ray data of the precipitate formed in the presence of aspartate and citrate as influenced by the oxidation treatment show the same trend. The IR spectra coupled with the evidence obtained from the organic C contents of the poorly crystalline Al precipitation products before and after oxidation treatment of NaOCl or H_2O_2 further indicate that organic ligands are present on the surfaces and in the structural network of Al precipitation products.

Recently, Violante and Huang[37] reported that differential thermal analyses of Al precipitation products formed in the presence of tartaric acid in the pH range of 6 to 10 show endotherms at 140 to 185 and 420°C and exotherms at about 400, 490, 750, and 850°C. The endotherms at 140 to 185°C are attributed to the sorbed water. The exotherms at 400 to 850°C are attributed to the oxidation of organic ligands present in the solids. The exotherms at about 750 and 850°C appear clearly only in the samples containing high C contents (≥45 g C kg^{-1}). The exotherm at about 850°C is present in the samples with a C content ≥90 g kg^{-1}. These results show that some organic ligands are decomposed only at very high temperatures, indicating that these organic ligands are strongly bound to Al ions or present in the interior of the solid phases and thus protected from oxidation. Figure 3 shows the influence of temperature on the C content in the samples formed at pH 8.0 and at an initial tartaric acid/Al molar ratio of 0.1 and 0.25. At 200°C, the C content slightly increases, evidently because mainly water was removed at this temperature from the samples. The loss of weight of these materials at 200°C was, respectively, 83 g kg^{-1} (R = 0.1) and 61 g kg^{-1} (R = 0.25). At 300°C, there was a considerable decrease of C content in both the samples (31 g kg^{-1} at R = 0.1 and 70 g kg^{-1} at R = 0.25). At 400°C, the C content in the sample at R = 0.1 remains virtually unchanged (30 g kg^{-1}), whereas that of the sample at R = 0.25 is drastically reduced (28 g kg^{-1}). Finally, at 600°C, the C contents in both samples are very close (12 and 16 g kg^{-1}). It is evident that at high temperatures (≥400°C), the amounts of the organic residues, which still remain in the network of the aluminum oxides, are similar regardless of the quantity of the organic ligands initially present. The organic molecules which are adsorbed on the external surfaces or are less protected by the oxides are apparently oxidized at lower temperatures. Small amounts of C still remain after heating samples at 600°C for 12 h.

Nonhumified and humified materials are also associated with SRO aluminosilicates. Recent experimental evidence[15,38,39] has substantiated the hypothesis based on field observations[40,41] that soil organic components may retard the formation of allophane and imogolite. While allophane and imogolite are nearly absent in organic matter-rich surface horizons of some Inceptisols,[40,41] a fraction of poorly ordered aluminosilicate gel materials has been detected in these horizons. The aluminosilicate gel materials detected in soil are termed "allophane-like constituents"[42] or "ill-defined fraction of the allophane-imogolite complex."[43] These materials are X-ray noncrystalline and have IR spectra similar to those of the "ill-defined aluminosilicate complex"[38,39] formed in the presence of low-molecular-weight organic acids or humic acids. This indicates that some of these aluminosilicate gel materials in soil are stabilized by soil organic components.

Lou and Huang[44] reported that as the citrate/Al molar ratio increased from 0 to 0.025, the amount of imogolite in the precipitates decreased and that of X-ray noncrystalline aluminosilicate materials increased, which was indicated by the gradual disappearance of the IR absorption doublets at 995 and 937 cm^{-1}, increasing absorption at 965 cm^{-1} (Figures 4a to f), and the broadening of the X-ray diffraction peaks (Figures 5a to f). The ignition loss increases with the initial citrate/Al molar ratio which is attributed to the increased incorporation of citrate ligands in the precipitates (Table 3) as indicated by the increasing IR absorption of COO^- at 1400 cm^{-1} (Figures 4b to f). The electron micrographs (Figure 6) also reveal that the perturbing effect of citrate ligand increases with the increasing citrate/Al molar ratio. The occupation of some of the coordination sites of Al in the hydroxy-aluminosilicate ions by citrate ligands apparently perturbs the regular linkage of the hydroxy-aluminosilicate fragments into the imogolite tube, leading to the formation of structurally disordered materials (Figure 7).

Lou and Huang[45] further reported that citrate ligands at a low concentration (10^{-5} *M*, citrate/Al molar ratio of 0.01) during the formation of allophane did not affect the unit particle size (42 Å) of the allophane formed, but did modify its surface properties by promoting the formation of aluminosilicate gel materials which apparently blocked some

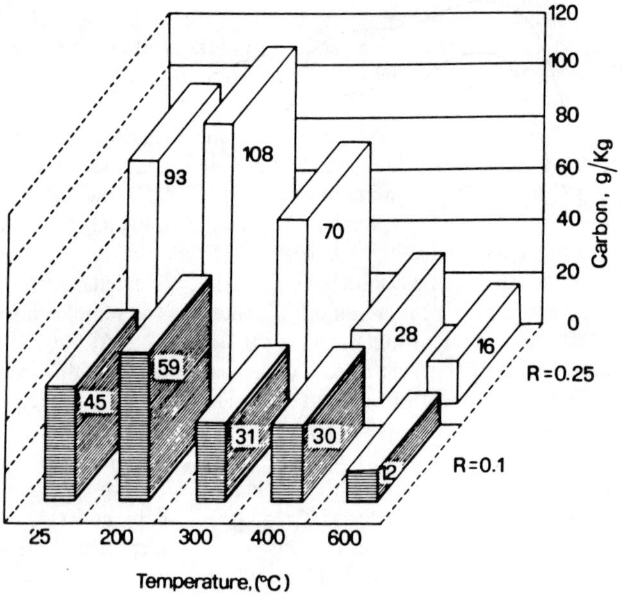

Figure 3 The carbon content in Al precipitation products formed at pH 8.0 and at R = 0.1 and 0.25 before (25°C) and after heating at 200, 400, and 600°C. R stands for the initial tartaric acid/Al molar ratio.[37]

of the intersphere channels and defect pores on allophane shells. The products formed at the initial citrate/Al ratio of 0.01 appear to be the mixture of allophane and the aluminosilicate gel materials. As the initial citrate/Al ratio of the parent solutions increases, the SiO_2/Al_2O_3 molar ratio of the allophanic materials decreases and the amount of citrate ligands incorporated increases. The increase in the incorporation of citrate ligands accounts for the increasing IR absorption of COO^- at 1400 and 1634 cm^{-1}. The IR spectra suggest that nearly all of the COOH groups of the incorporated citric acid are in the form of COO^- (1400 and 1634 cm^{-1}) since the IR absorption band of COOH groups of citric acid at 1700 to 1750 cm^{-1} is not apparent in the IR spectra. Citrate ligands at the citrate/Al ratio of 0.10 distort the formation of allophane structure by complexing with Al in the hydroxy-aluminosilicate fragments, thus preventing the normal linkage of these fragments. The data obtained from the small-angle X-ray scattering intensity curves show that in contrast to the allophane formed in the absence of citrate ligands and the allophane-containing materials formed at the low citrate/Al ratio (0.01), the aluminosilicate gel materials formed at the high citrate/Al ratio (0.1) have very low scattering intensities. The resultant aluminosilicate gel materials consist of "unit particles" or clusters of the randomly arranged hydroxy-aluminosilicate fragments, with a gyration radius of 10.4 Å or an average diameter of 27 Å. The aluminosilicate gel materials exhibit much higher net surface charge densities than those of allophanes, which is attributable to the extensive exposure of the edge sites of hydroxy-aluminosilicate fragments.

Inoue and Huang[39] reported that the SiO_2/Al_2O_2 ratio of the precipitates formed from solutions containing hydroxy Al ions, orthosilicic acid, and humic substances decreases with increasing organic C content (Figure 8). The strong competition of such humic substances with orthosilicic acid for the coordination sites of Al is considered to account for the marked impedance of the humic substances on the formation of allophane and the incorporation of organic C in the network of SRO mineral colloids.

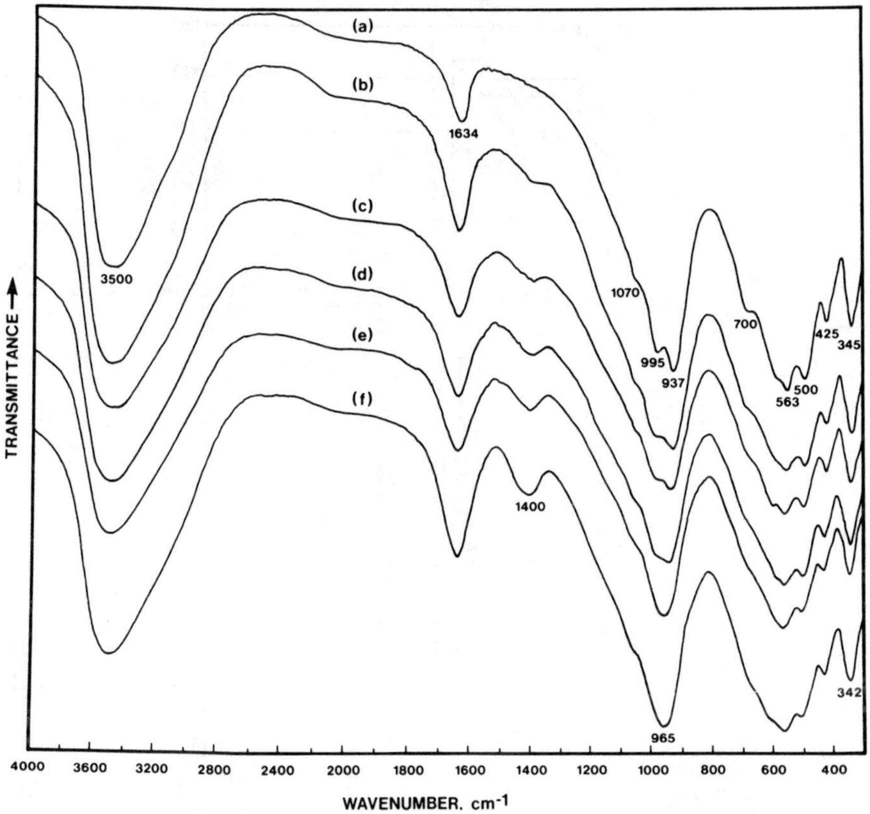

Figure 4 Infrared spectra of precipitates formed both in the absence and presence of citric acid. Initial citrate/Al molar ratio: (a) 0, (b) 0.005, (c) 0.01, (d) 0.015, (e) 0.02, and (f) 0.025.[44]

Organic components are also very intimately associated with SRO Fe oxides. Many soil profile analyses have shown that the ratio of the oxalate extractable Fe (Fe_o) to dithionite extractable Fe (Fe_d) is higher in horizons richer in organic carbon such as surface soils and B horizons of Spodosols (Figure 9). The Fe_o is mainly attributed to the SRO iron oxide, ferrihydrite, whereas Fe_d can reasonably well be set equal to the total amount of secondary iron oxides in soils. Therefore, in carbon-rich soil horizons, Fe, which is mobilized through weathering of iron silicates and/or through reduction of iron oxides, is preferentially precipitated as ferrihydrite, rather than crystalline Fe oxides such as goethite, hematite, and lepidocrocite. For example, in placic horizons from various countries,[47] those with a high Fe_o/Fe_d ratio contain mainly ferrihydrites, whereas in those with a low Fe_o/Fe_d ratio, goethite and lepidocrocite dominate (Figure 10). The significant positive correlation between the Fe_o/Fe_d ratio and the organic carbon content supports the idea that organic matter, through its interaction with surface Fe, may be the main reason for inhibiting the formation of goethite and lepidocrocite and promoting the formation of ferrihydrite in these and similar carbon-rich horizons. Organic matter may be associated with ferrihydrite as surface-bound organic molecules. Organic matter may also be associated with Fe as Fe-organic complexes. Therefore, the association of organic matter with SRO Fe oxides should affect the dynamics of transformations of organic components in soils.

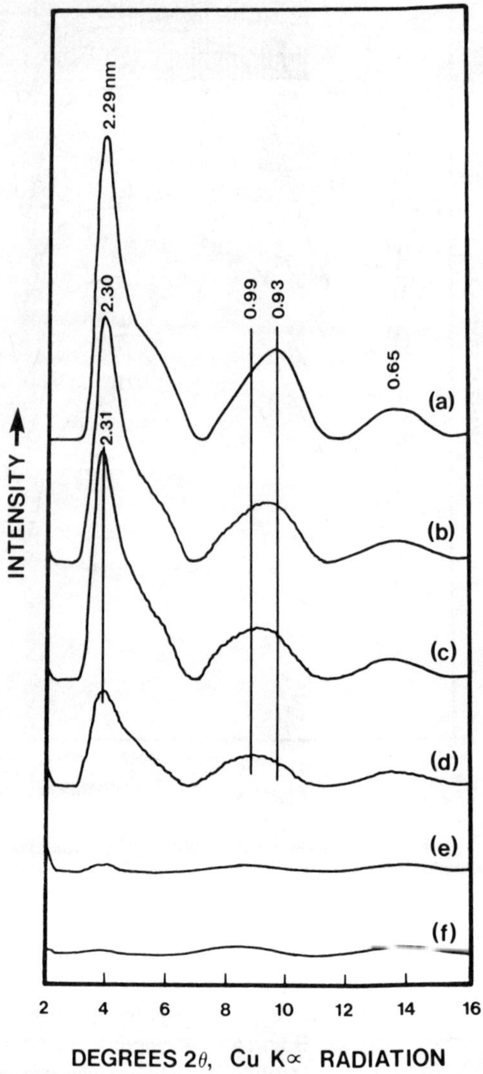

Figure 5 X-ray diffractograms of the precipitates formed both in the absence and presence of citric acid. Initial citrate/Al molar ratio: (a) 0, (b) 0.005, (c) 0.01, (d) 0.015, (e) 0.02, and (f) 0.025. Air-dried, parallel-oriented specimens.[44]

Table 3 Composition of the Precipitates Formed Both in the Absence and Presence of Citric Acid[44]

Citrate/Al molar ratio	SiO_2	Al_2O_3	Ignition loss	Citrate	SiO_2/Al_2O_3 molar ratio
			(g kg^{-1})		
0	257	514	253	0	0.85 ± 0.02
0.01	220	523	270	36.3	0.72 ± 0.01
0.02	230	500	296	51.6	0.79 ± 0.01

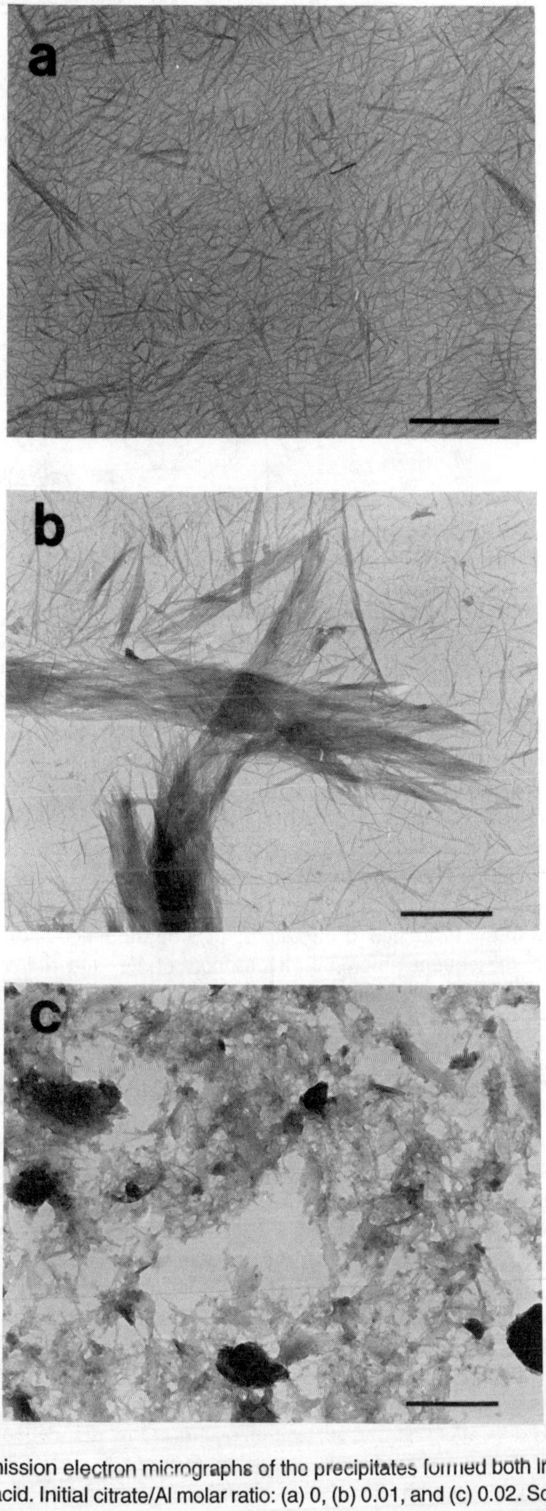

Figure 6 Transmission electron micrographs of the precipitates formed both in the absence and presence of citric acid. Initial citrate/Al molar ratio: (a) 0, (b) 0.01, and (c) 0.02. Scale bar: 500 nm.[44]

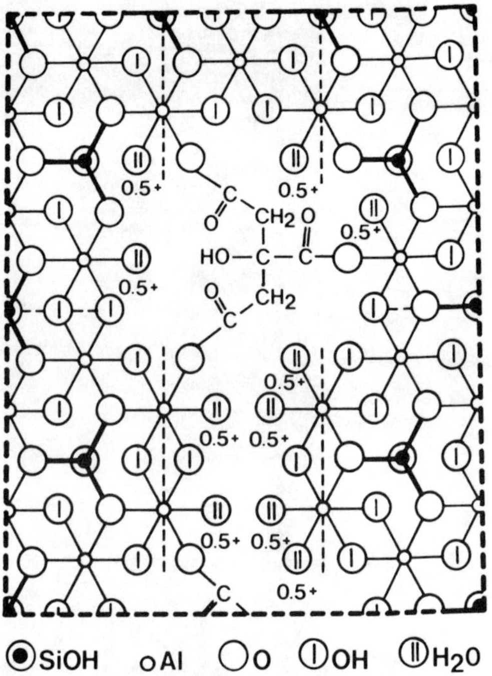

Figure 7 Proposed model for the structural distortion and exposure of $Al\text{-}OH_2^{0.5+}$ groups due to the incorporation of citrate ligand into the precipitates.[44]

Krishnamurti and Huang[48] show that organic ligands have a retarding influence on the rate of Fe(II) oxidation. The decrease in the rate constant is related to the stability constant of the Fe(II)-ligand complex. The kinetics of Fe(II) oxidation and the complexation of Fe(III) with the ligands influence the kinetics of hydrolysis of Fe(III) and modify the crystallization processes of the oxidation products of Fe(II), resulting in the preferential formation of lepidocrocite in the presence of oxalate, goethite in the presence of acetate, and X-ray noncrystalline Fe oxides in the presence of tartrate. At a citrate/Fe molar ratio of 0.1, no precipitate is formed. Therefore, the stability constants of Fe(II)-ligand complexes are related to the influence of organic ligands on the kinetics of Fe(II) oxidation, the nature of the subsequent precipitation products of Fe, and the associated abiotic transformations of organic components. This area of research is emerging and should receive increasing attention.

IV. RETENTION OF ORGANICS ON SURFACES OF MINERAL COLLOIDS

Retention is an important abiotic process controlling transformations of organic components in the environment. This process should influence the turnover times of natural organic components.

Most of the work on the retention of organic components on soil minerals has been conducted on crystalline clay minerals. Relatively less information is available on the retention of organic components on SRO mineral colloids. However, these materials are very reactive because of their high specific surface and exposed edge sites.

Short-range ordered Al hydroxides have been shown to retain many organic compounds, including phthalate, oxalate, salicylate, and benzoate.[49-54] Retention of organic acids is highly pH dependent; strong adsorption occurs at low pH. Bidentate functionality of organic molecules is an important factor in forming a strong bond with singly coordinately aluminol sites on the edge faces of Al hydroxides. The dominant crystal faces of gibbsite are unreactive. The retention of these organics is attributed to ligand-

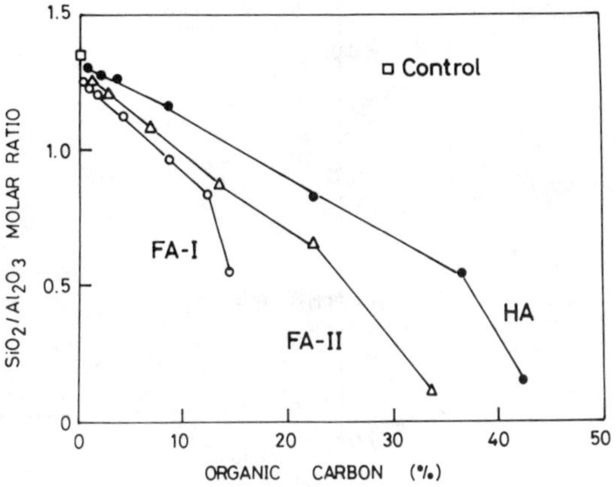

Figure 8 Relationship between SiO_2/Al_2O_3 molar ratios and organic carbon content of precipitates formed from solutions containing hydroxy Al ions, orthosilicic acid, and humic substances.[39]

exchange reactions involving a single aluminol group and the deprotonated or singly protonated anion. The general pH dependence for retention of these organic acids is in good agreement with observations of FA retention on Al oxides and hydroxides.[50,52,53]

McBride and Wesselink[55] recently compared the adsorption of catechol on SRO Al hydroxide, gibbsite, and boehmite. They found that only singly coordinated aluminol groups on the edge faces of Al hydroxides are reactive, similar to observations made on phosphate and oxalate adsorption.[49] Infrared spectroscopy suggests that the surface complex formed is a 1:1 bidentate complex of deprotonated catecholate with Al ions at the surface. However, compared with organic acids, catechol retention by SRO Al hydroxides exhibits a different pH-dependent pattern, increasing with increasing pH. The opposite trend in pH dependence in the retention behavior is likely due to the catalytic effect of SRO Al hydroxide on oxidative polymerization of catechol and the subsequent retention of catechol-derived polymers on the surface of the hydroxides.

In addition to the foregoing mechanisms, organic components may be retained on the surfaces of SRO Al hydroxides through hydrogen bonding between aluminol groups and organic functional groups. The retention of organics may also proceed by adsorption of the metal-ligand complexes; this effect has been observed with FAs.[52] Few studies of competitive adsorption of organics have been conducted. The effect of changes in surface charge and potential is even less understood.

Organic components can also be retained on the surfaces of hydroxy-Al-clay complexes. Phenolic compounds, which are widely distributed in soils, have been shown to be adsorbed on hydroxy-Al-montmorillonite and hydroxy-Al-vermiculite complexes.[56] Goh and Huang[57,58] reported that the formation of hydroxy-Al-montmorillonite complexes in the acidic pH range that

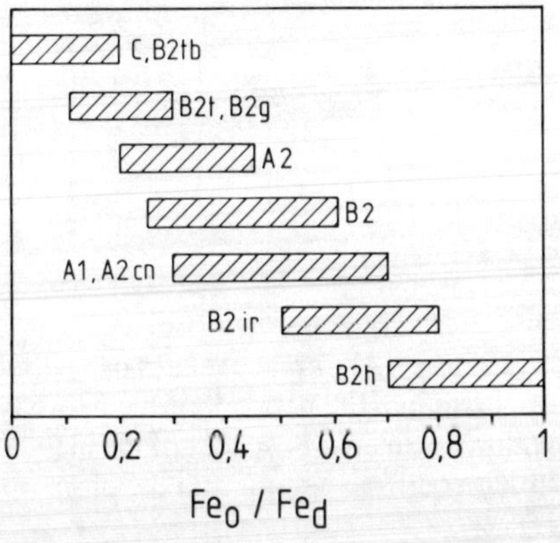

Figure 9 Range of Fe_o/Fe_d ratios in horizons of soils in the humid temperate area.[46] cn, concretions.

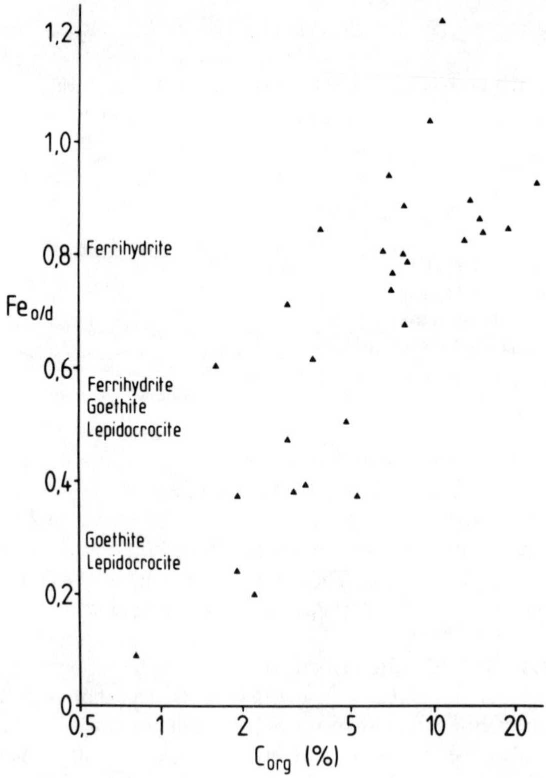

Figure 10 Relationship between Fe_o/Fe_d and iron-oxide mineralogy and total organic carbon in placic horizons from various countries.[47]

is most conducive to chloritization[59] is significantly perturbed by low-molecular weight organic acids such as citric and tannic acids. Once hydroxy-Al ions interact with organic acids to form hydroxy-Al-organic ligand complexes, the adsorption of these complexes by interlayer spaces of montmorillonite can take place only if the net attractive forces between the positively charged sites of the complexes and clay surfaces can overcome the repulsive forces between the anionic ligands of the complexes and clay surfaces and any steric hindrances that may be present. The organic ligand/Al molar ratio has an important influence on the formation of hydroxy Al interlayers. The polynuclear Al ion of double-ring structure[1] can be chelated by organic ligands such as citrate,[32] and its structure can be depicted as:

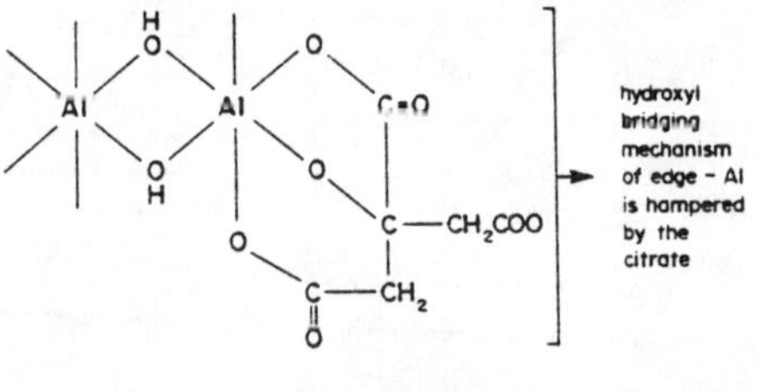

(Structure A)

Hydroxy-Al-citrate chelates can be retained on interlayer spaces.[57] The incorporation of citrate into the network of the hydroxy-Al interlayer causes structural distortion from within the interlayer spaces of the clay, as indicated by the broadening of the (001) peak in X-ray diffractograms and the obscuring of the IR absorption bands at 3695 and 3570

cm^{-1} representing the presence of hydroxy-Al interlayers. In addition, the dehydroxylation endotherm is spread over a wide range of temperatures ranging from 435 to about 520°C as the citrate/Al molar ratio is increased from 0.1 to 0.5. The perturbation to the interlayer hydroxy-Al by citrate has resulted in more gradual loss of structural water upon thermal treatment. The complexation of Al ions with tannic acid also takes place, since the stability constant of the complex formed between Al and tannic acid is $10^{3.78}$ (Table 1). The hydroxy-Al-tannate complex may enter the interlayer space of smectite. Further growth of the hydroxy-Al-tannate complex to a large hydroxy-Al-tannate complex may be possible through polymerization of both Al and tannate:[58]

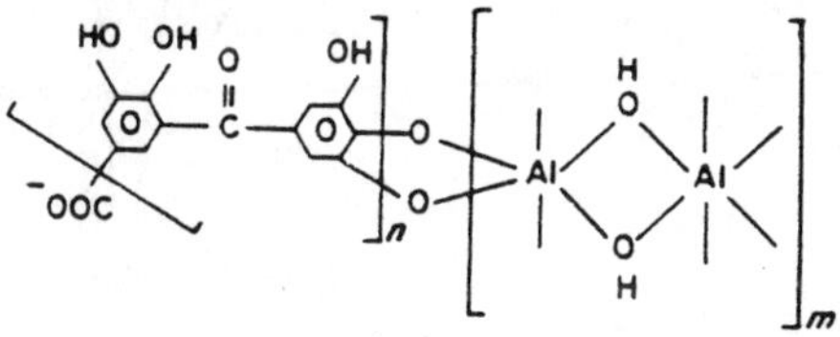

(Structure B)

The high molecular weight of the hydroxy-Al-tannate complexes apparently causes them to precipitate as a discrete phase, which acts as a cementing agent between clay particles:

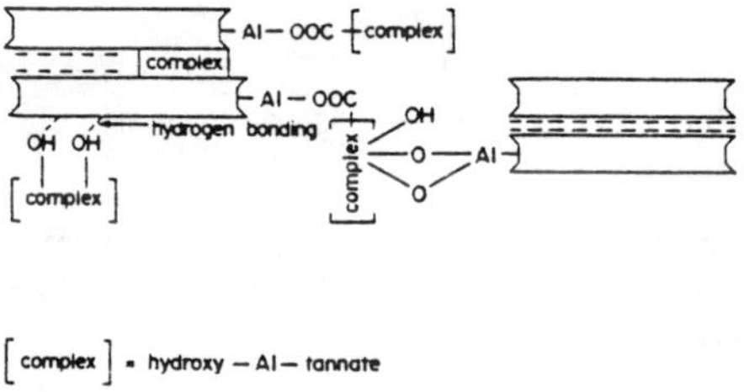

(Structure C)

In addition to the chemical bonding between hydroxy-Al-tannate complexes and clay particles, the forces of physical adsorption are additive in high-molecular-weight complexes. The aggregation of hydroxy-Al-clay particles is thus promoted by tannic acid, which is then retained in the aggregate.

Recent studies of Singer and Huang[60] show that with increasing concentration of HA, hydroxy-Al interlayering is perturbed, as indicated by X-ray diffraction, infrared absorption, and CEC data. Hydroxy-Al-HA complexes can be adsorbed by expansible clay mineral colloids because some of them may be positively charged. This would lead to the perturbation of Al interlayering due to the complexation of Al by HA. This supports the observation of Rich[59] that organic matter-rich soils are not likely to have hydroxy-Al interlayering. Also, oxidized conditions and frequent wetting and drying cycles, cited as favoring Al interlayering,[59,61] might be acting indirectly by controlling organic matter levels. Compared with Ultisols and Alfisols, where levels of organic matter are low, interlayering is relatively less common in Mollisols and Spodosols, which are rich in organic matter.[62] The data obtained from the treatment of the complexes with pyrophosphate indicate an intimate association of HA with hydroxy-Al-clay complexes. Only partial removal of organic C from the complexes by sodium hypochlorite treatment suggests that some of the adsorbed HA was very strongly bound with the hydroxy Al-clay complexes. The shift to a higher-temperature endotherm (575°C) in thermogravimetric analysis (Figure 11), diffuse X-ray diffractograms of the hydroxy-Al-HA-montmorillonite, decreased adsorption of HA during complex formation with collapsed montmorillonite, and decreased d-spacing after NaOCl treatment[60] suggest that some of the HA in the complex is located in the interlayers of montmorillonite.

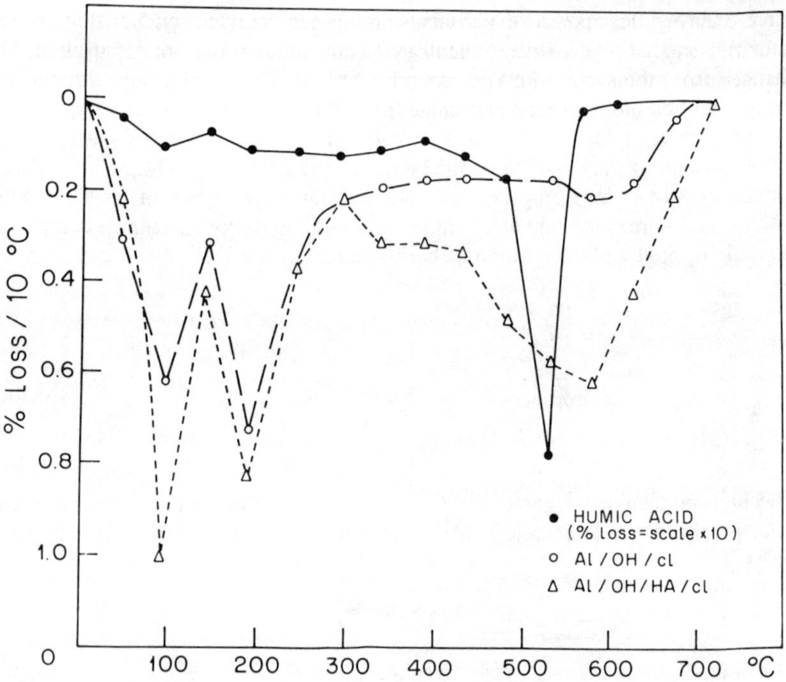

Figure 11 Differential thermogravimetric curves of pure HA, hydroxy Al-montmorillonite (Al/OH/cl) complex, and hydroxy Al-HA-montmorillonite complex (Al/OH/HA/cl) formed at pH 4.8 in the presence of 4 mg HA.[60]

In the hydroxy-Al-HA montmorillonite complexes formed when the clay is added first and HA last, perturbation of Al interlayering by HA is minimal, although the adsorption of HA by the hydroxy-Al-montmorillonite complex is substantial. The data thus indicate that the formation of hydroxy-Al-HA complexes in solution and their subsequent adsorption into the interlayer space of montmorillonite perturb the Al interlayering. The subsequent transformations of HA in the hydroxy-Al interlayer network remain obscure, however.

The short-range ordered aluminosilicates, allophane and imogolite, are effective in both nonspecific and specific adsorption of organic anions.[3] For instance, acetate has been shown to be adsorbed by coulombic interaction on these mineral colloids, whereas citrate, which has a strong affinity toward Al, is more strongly adsorbed and in a greater amount. In addition, van der Waal interactions contribute to the adsorption of high-molecular-weight organic compounds. Inoue and Wada[63] demonstrated much greater adsorption of humified clover extract by allophane and imogolite than by montmorillonite and halloysite. The importance of the protective action of SRO aluminosilicate colloids against biodegradation of organic components which become incorporated in soils deserves increasing attention.

The adsorption of organic components on the surfaces of iron oxides is of great significance for the transformations and migration of soluble natural organic molecules in soils. The adsorption of humic components by iron oxides is considered one of the processes by which Spodosol Bh horizons are formed. Reeve and Fergus[64] demonstrated the decoloration of "black" waters derived from the organic layers of a Spodosol in Queensland by passing through the ochreous sand of the C horizon. Adsorption experiments using synthetic iron oxides and HA and FA extracted from soils substantiate this concept. Many of the rigorous experiments on this subject were conducted in crystalline iron oxides and oxyhydroxides.[50,65] Infrared data indicate that, on adsorption, OH is exchanged from the surface Fe-OH group by organic functional groups through ligand exchange, and carboxyl groups are ionized:

$$-Fe-OH + HOOC-R \rightarrow -Fe-OOC-R + H_2O$$

In addition, H bonding may also significantly contribute to the adsorption of organic molecules on iron oxide surfaces, especially at a higher concentration of organic components. The ligand exchange and H bonding cause multiple contacts to form a strong association between organic molecules and iron oxides.

Humic acid and FA have been shown to be adsorbed on freshly precipitated iron oxide (protoferrihydrite). The extent of adsorption varies with the types of soils: Mollisol-HA > Alifsol-HA > Spodosol-B-FA.[66] This SRO iron oxide retains between 0.22 and 0.68 mg C (FA extracted from a Spodosol Bhir) per gram of Fe.[67] The retention of humics on iron oxides is also of great importance in lakes and rivers. On a per unit weight basis, the SRO iron oxide (ferrihydrite) retains more humic substances than goethite and hematite.[68] In the presence of divalent cations (Ca and Mg), adsorption increases, possibly because divalent cations reduce the electrostatic repulsion between adsorbed humic molecules by coadsorption through functional groups of the humic substances. There is a slight competition in the adsorption on iron oxide surfaces between humic substances and phosphate.

Little is known about the kinetics and mechanisms of the interactions of a series of SRO mineral colloids, which differ in structural configuration and surface properties, with natural biomolecules, humic substances, enzymes, and antibiotics.

V. ABIOTIC CATALYSIS OF TRANSFORMATIONS OF ORGANICS

The significance of biological degradation in the transformations of organics in soil is well established. The catalytic role of soil minerals, especially SRO mineral colloids, in the abiotic transformations of organics is at the emerging stage. Natural organics can undergo chemical breakdown through catalysis of SRO mineral colloids in soil. Abiotic catalysis of these colloids can also lead to oxidative degradation of polyphenols and other organics and the subsequent formation of humic substances. Short-range ordered metal oxides and hydroxides, and aluminosilicates are reactive in promoting abiotic degradation of organic compounds. Therefore, the dynamics of organics in soils and associated environments should be influenced by the nature and properties of these mineral colloids.

Manganese oxides (birnessite, cryptomelance, and pyrolusite), which occur in soils and sediments,[69] can act as Lewis acids, which accept electrons from diphenols, leading to their oxidative degradation.[19] The rate-determination step in the formation of HAs from simple phenols and phenolic acids by oxidative degradation seems to be the formation of a semiquinone radical involving a one-electron transfer reaction.[70] Semiquinones will couple with each other to form a stable HA polymer. In contrast to electron transfer reactions, the coupling of radicals requires little activation energy.[71] Coupling of semiquinones rather than formation of quinones should kinetically be the preferred reaction in the transformation of diphenols to humic macromolecules.

Birnessite (δ-MnO_2) promotes the abiotic generation of CO_2 through its ability to cleave the ring structure of pyrogallol (1,2,3-trihydroxygenzene) under ambient temperature and pressure in a system free of microbial activity.[21] At the end of a 90-h reaction period, the release of CO_2 in the birnessite-pyrogallol system was 18.9 times higher than that in the pyrogallol system (Table 4) The absorbances of the supernatants of the birnessite-pyrogallol system are 10.8 and 12.5 times at 472 and 664 nm, respectively, higher than those of

Table 4 Promoting Effect of Birnessite on the CO_2 Release and Absorbances of the Supernatant in the Birnessite-Pyrogallol System at the End of a 90-h Reaction Period[21]

Reaction condition		Absorbance		
Birnessite	Pyrogallol	472 nm	664 nm	CO_2 release (μmol)
P[a]	P	2.91 ± 0.01[b]	0.25 ± 0.01	463.8 ± 0.3
A[c]	P	0.27 ± 0.01	0.02 ± 0.00	24.5 ± 0.3

[a] In the presence.
[b] The average deviation from the mean.
[c] In the absence.

the pyrogallol system. The results thus indicate that the formation of pyrogallol-derived polymers is strongly promoted by birnessite. Furthermore, the total yields of synthesized humic polymers formed in the birnessite-pyrogallol system is 10.5 times higher than those formed in the pyrogallol system. The IR data show that unpolymerized pyrogallol is strikingly different from its derived polymers.[21] The IR spectra of pyrogallol-derived polymers[21] resemble those of natural HA and FA.[72] The titration of pyrogallol-derived polymers to pH 7.0 resulted in the virtual disappearance of the absorption band at 1714 to 1721 cm^{-1}, the concomitant appearance of a band at 1583 to 1591 cm^{-1} and intensification of the band at 1370 to 1377 cm^{-1}.[21] On the basis of the existing information on the nature of soil humic substances,[73-75] such shifts in the absorption bands indicate that the ~1715-cm^{-1} band is due mainly to carboxyl groups in the synthesized humic polymers. Furthermore, it is noteworthy that pure pyrogallol has a strong absorption band of aromatic ring structure at 1610 cm^{-1}, while the pyrogallol-derived humic polymers have strong absorption bands of carboxyl groups at ~1715 cm^{-1}. The band positions observed for the NMR spectrum of HA formed in the pyrogallol solution in the absence of birnessite (Figure 12a) are 180.9, 152.2, 149.8, 137.3, 133.7, 133.1, 125.4, 118.8, 113.9, and 111.4 ppm, whereas those of HA formed by the promoting effect of birnessite (Figure 12b) are 196.3, 170.7, 144.6, 131.9, ~121.3, 82.9, and 43.0 ppm. The spectrum of HA (Figure 12b) resembles those of natural HAs.[76,77] The band positions are listed in Table 5. The NMR spectra (Figure 12) corroborate the IR evidence and the data on CO_2 release (Table 4), revealing that the ring cleavage of pyrogallol and the associated formation of aliphatic fragments occur through the promoting effect of birnessite. The origin of the aliphaticity of humic substances in soils is still not clear.[72,78] The data (Table 4, Figure 12) indicate that the abiotic ring cleavage of polyphenols promoted by soil inorganic components such as birnessite may in part account for recent findings of the high aliphaticity of natural humic substances.[76,79] Furthermore, ring cleavage of polyphenols abiotically promoted by Mn oxide can be a pathway of C turnover and a mechanism contributing to the aliphatic nature of humic substances in the environment.

More recent data[80] show that the abiotic release of CO_2 in the birnessite-polyphenol and polyphenol systems increases as light intensity is increased from 0 to 500 μE $m^{-2}s^{-1}$ and as temperature is increased from 5 to 45°C at a light intensity of 0 μE $m^{-2}s^{-1}$ (dark). On the other hand, the absorbances of the supernatants in the birnessite-polyphenol and polyphenol systems decrease as light intensity is increased from 0 to 500 μE $m^{-2}s^{-1}$, despite an increase in absorbances as temperature is increased from 5 to 45°C in the dark. This is attributed to a consequence of the photofragmentation of polyphenolics and the nature of their degradation products. The data indicate that photochemical and thermal effects on the abiotic degradation of polyphenolics via SRO inorganic components in soils and sediments merit attention.

Shindo and Huang[81] compared the catalytic effects of Mn oxide and tyrosinase on the oxidative polymerization of diphenols over the pH range of 4 to 8. Mn oxide influences

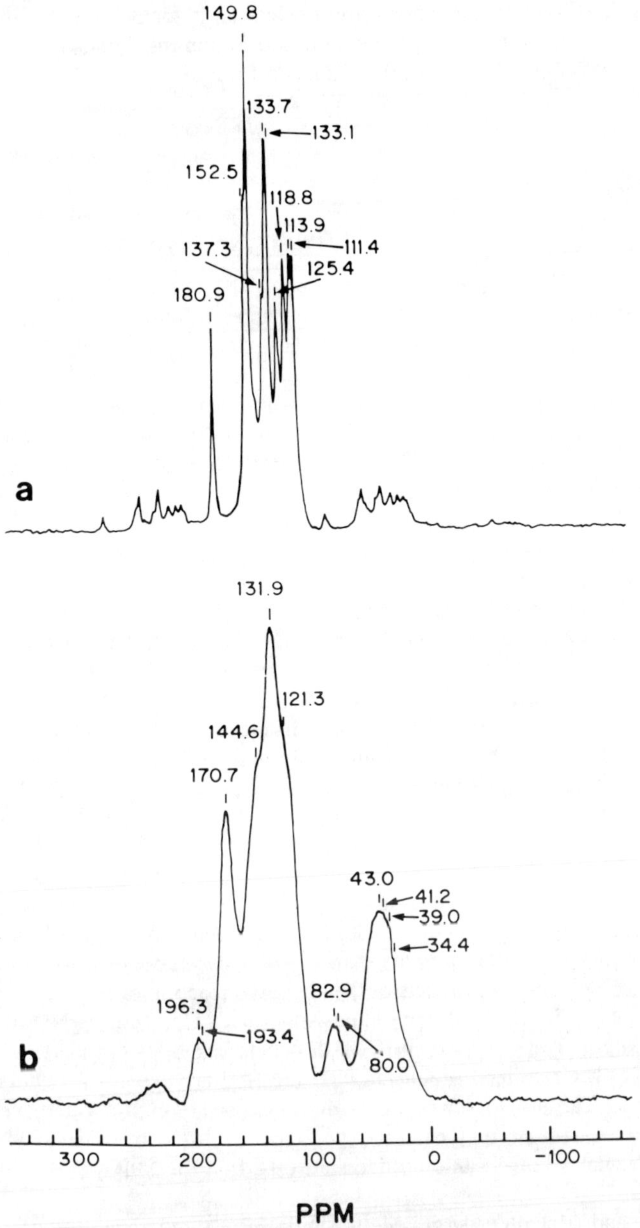

Figure 12 Solid-state ^{13}C-NMR spectra of dry, powdered HA formed in the reaction systems with cross-polarization and magic-angle spinning (CPMAS), proton decoupling, contact time 2.00 ms, recycle time 1.00 s, spectral width 20,000.0 Hz, spinning speed 3700 Hz, number of scans 60,000, and spectrometer frequency 37.7 MHz. Synthesized HA: (a) pyrogallol and (b) birnessite-pyrogallol systems.[21]

the darkening of hydroquinone and resorcinol to a larger extent than does tyrosinase, whereas the reverse is true for catechol. In the tyrosinase system, the darkening of catechol is more efficient compared with hydroquinone and resorcinol. In both the Mn

Table 5 The Band Positions and Assignments for Solid-State CPMAS ^{13}C-NMR Spectra of Humic Polymers Formed in the Reaction[21]

Chemical shift (ppm)	Assignment
196.3	Carbons in aldehydes and ketones
180.9–170.7	Carboxyl carbon in carboxyl and ester groups
152.5–111.4	Aromatic carbons and olefinic carbons
82.9	CO carbons — alcohols, esters, ethers, carbohydrates
43.0	Alkyl carbons

oxide and tyrosinase systems, the degree of the darkening of resorcinol is generally much lower than for the other phenols studied. The yields of HAs are also significantly influenced by the kind of catalyst and diphenol used. In the Mn oxide system, the yield of HAs is in the order: hydroquinone > catechol > resorcinol. In the tyrosinase system, catechol produces the highest yield of HA, followed by hydroquinone and resorcinol. These findings indicate that the relative catalytic effects of Mn(IV) oxides and tyrosinase would vary with the type of diphenols in natural environments.

Manganese oxide also promotes the abiotic formation of NH_3-N and nitrogenous polymers in hydroquinone-glycine systems in the common pH range of soils.[20] Manganese oxide acts as a Lewis acid by accepting electrons from hydroquinone, which is thus oxidized and polymerized. The products of the reaction of glycine with hydroquinone polymerize to form nitrogenous polymers, thereby incorporating glycine into the polymers during the oxidative polymerization of hydroquinone; a partial deamination of glycine occurs during the process. Mn oxide also catalyzes the polycondensation of pyrogallol and glycine and the subsequent formation of N-bearing polycondensates (Figure 13), which resemble natural humic substances in IR and ESR spectra. The data also reveal that Mn oxide promotes the deamination and decarboxylation of glycine. Deamination and decarboxylation of amino acids as catalyzed by Mn oxide may be a pathway of C turnover and N transformation. Mineralization of N via abiotic processes should not be overlooked.

Phenolic acids have been shown to be oxidized rapidly in the presence of MnO_2 to form a number of soluble products.[83] Some of the soluble products of oxidation have somewhat higher molecular weights than the parent compounds, as shown by mass spectrometry. The soluble products of the chemical reaction of ferulic acid and MnO_2 do not contain any ferulic acid-derived polymers. This is in contrast to the products of enzymatic polymerization that range from dimers to hexamers found by Liu et al.[84] and Bollag et al.[85] The oxidized products of ferulic acid are evidently rapidly sorbed on MnO_2 surfaces. These adsorbed products appear to exist as high-surface-area polymers, which could be similar to the humic polymers reported by Shindo and Huang.[20] However, the nature of organic products adsorbed on the MnO_2 surface needs to be established.

Pohlman and McColl[86] described the initial and rapid redox processes between polyhydroxyphenolic acid and soil or Mn oxide suspensions. The rates of oxidation of polyhydroxybenzoic acid by soil and Mn oxides are similar in magnitude in both suspensions. Polyhydroxyphenolic acids with *para*- and *ortho*-OH groups are rapidly oxidized by Mn oxides, with spectral evidence suggesting that the reaction leads to polymeric humic products probably via semiquinone or benzoquinone derivatives. The rapid disappearance of polyhydroxyphenolic acid is accompanied by the formation of Mn^{2+} and colored humic products in both soil and Mn oxide suspensions. On the other hand, polyhydroxyphenolic acids with *meta*-oriented phenolic-OH groups are not oxidized by soil or Mn oxide suspensions within a 120-min reaction period.

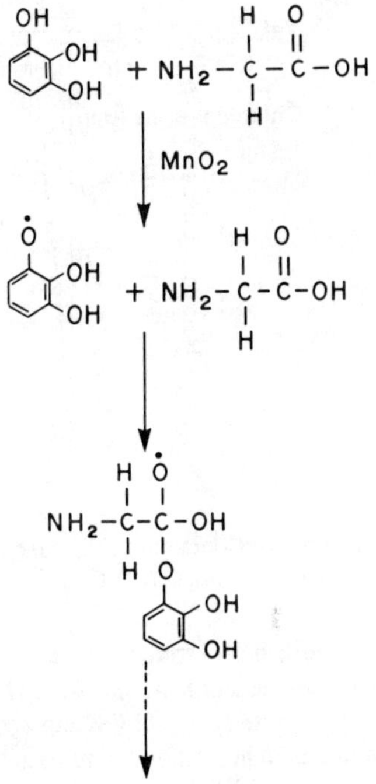

Figure 13 Reaction of pyrogallol-derived free radicals with glycine.[82]

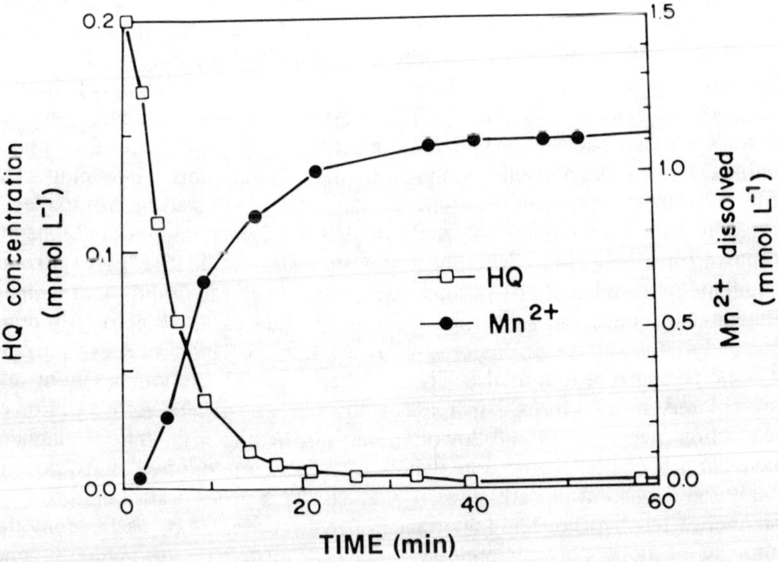

Figure 14 Change in solution concentration of hydroquinone (HQ) and Mn^{2+} due to the oxidation reaction. Initial concentrations were 2×10^{-4} *M* hydroquinone and 600 nm of hausmannite (Mn_3O_4) at pH 6.[89]

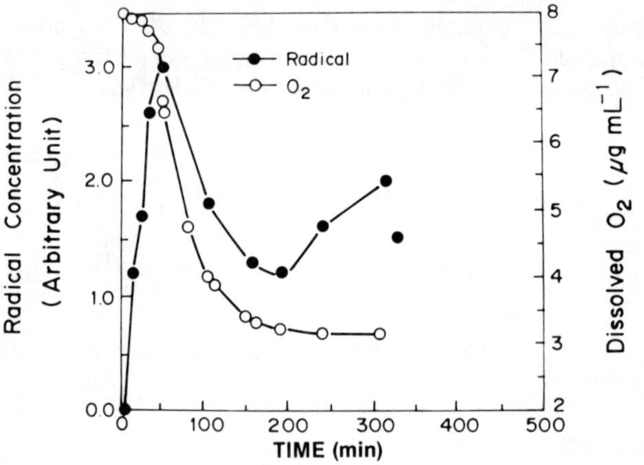

Figure 15 Concentrations of *p*-benzosemiquinone radical (arbitrary units) and dissolved O_2 in solution as a function of time.[89]

McBride[87,88] studied the factors affecting the oxidation of dihydroxybenene by birnessite and the extent of Mn^{2+} released. The kinetics of the oxidation of hydroquinone by housmannite at pH 6 was studied by Kung and McBride.[89] Electron transfer from hydroquinone to the oxide is indicated by the release of Mn^{2+} accompanying the loss of hydroquinone solution (Figure 14). Electrons extracted from hydroquinone reduces surface Mn and thus solubilize the Mn. The ESR data obtained during the electron transfer process establish the presence of a quintet of hyperfine lines arising from four equivalent protons (hyperfine coupling constant = 2.4 G, g value = 2.005), which confirms the *p*-benzosemiquinone anion radical spectrum.[90,91] The radical concentration initially increases, but then decreases simultaneously with the consumption of dissolved oxygen (Figure 15). Once O_2 is depleted, the concentration of free radical gradually increases again. The rapidly generated semiquinone anion radical is apparently slowly oxidized by dissolved O_2 in solution.

Many organic compounds other than polyphenolics have also been shown to solubilize Mn from Mn oxides by reduction.[92,93] These organics include salicylic, pyruvic, and oxalic acids. Organics such as malate[94] and glutamate[95] are oxidatively decomposed by Mn oxides. Even monophenolic compounds, particularly those containing electron-donating substituent groups on the aromatic ring, can be oxidized by Mn oxides.[96,97]

The ability of iron oxides to catalyze the oxidative polymerization of polyphenols has been known for a long time. Scheffer et al.[98] studied the oxidative polymerization of hydroquinone in the presence of various iron oxides at pH 3 to 7 and room temperature in an attempt to model the formation of humic substances. Dark-colored products in solution and on the surface of the iron oxides are formed. This process is substantially faster in the presence of iron oxides. The absorbance at 472 nm, which is a measure of the dark-colored products formed, increases with increasing pH. The effect of iron oxides on the reaction increases in the following order: ferrihydrite > goethite > maghemite > lepidocrocite > hematite. Ferrihydrite, which is an SRO iron oxide, is most powerful in promoting the formation of dark-colored products. This may be attributed to the large surface area of ferrihydrite. The importance of hydrous oxides of Fe in promoting the formation of phenolic polymer appears to be also related to the nature of phenolic compounds.[99]

Short-range ordered Al hydroxides catalyze the oxidative polymerization of polyphenols. The formation of humic substances from the catechol and pyrogallol is substantially greater in the presence of Al hydroxide than in its absence.[100] Aluminum forms complexes with polyphenols and the semiquinone radical.[101] The Al^{3+} ion appears to promote delocalization of electrons from phenolic oxygen atoms into the π-orbitol system through displacement of protons from the phenolic groups. This would make the polyphenols more susceptible to oxidation.

Precipitated SRO silica and soluble silicic acid in aqueous solution have the ability to oxidize hydroquinone, catechol, and pyrogallol.[102] The surface of quartz has a disturbed layer which is short-range ordered in nature.[103,104] Similar surficial features are present on natural grains of quartz in soils and geological deposits.[105] Therefore, polymerization of polyphenols may be catalyzed by the disturbed surface of quartz in soils and associated environments.

Shindo and Huang[19] reported that in the systems free of microbial activity, among the metal oxides studied, Mn(IV) oxide is most effective in the abiotic oxidation of a hydroquinone solution over the pH range of 4 to 8 which is common in soils. In the nonsterilized systems in which abiotic and biotic processes were not differentiated, Larson and Hufnal[106] also reported that the Mn(IV) oxides are the most efficient among the metal oxides in oxidative polymerization of dissolved phenols. Wang[107] critically compared the catalytic power of Mn(IV), Fe(II), Al, and Si oxides in pyrogallol transformations. The sequence for the CO_2 released from the reaction systems through the ring cleavage of pyrogallol under air is: Mn(IV) oxide-pyrogallol >>> Fe(III) oxide-pyrogallol > Al oxide-pyrogallol > Si oxide-pyrogallol > pyrogallol. The amounts of CO_2 released from the systems under an N_2 atmosphere are drastically reduced. The IR spectra of synthesized FA (Mw <1000) formed in the oxide-pyrogallol systems show the development of the absorption band at around 1715 cm^{-1} which may be attributed to the presence of carboxyl as well as aldehydic and ketonic carbonyl groups in polymers. The IR spectra of these polymers formed in Si oxide- and Al oxide-pyrogallol systems are still very close to that of unpolymerized pyrogallol, whereas those of the polymers formed in Fe(III) oxide- and Mn(IV) oxide-pyrogallol systems are very similar to those of humic substances of soils. The IR spectra of the HA (MW >1000) and FA (MW >1000) formed in the Mn oxide-, Fe oxide-, Al oxide-, and Si oxide-pyrogallol systems resemble those of natural humic substances. The data indicate that the catalytic power of Mn(IV), Fe(III), Al, and Si oxides in the ring cleavage of pyrogallol and the development of carboxyl and aliphatic groups in the resultant humic polymers is greatly influenced by the nature of the oxides.

Short-range ordered aluminosilicates such as allophane exhibit a catalytic effect on the oxidative changes of polyphenols.[108,109] The catalytic power of other SRO aluminosilicates in the oxidative degradation of polyphenols remains obscure. Furthermore, the formation of SRO aluminosilicates is significantly influenced by inorganic ligands, nonhumified organic acids, humic substances, metallic cations, and expansible-layer silicates.[17] These ionic factors perturb the genesis of allophanes and imogolite to various extents, resulting in the formation of ill-defined aluminosilicate complexes (allophane-like constituents, or an ill-defined fraction of the allophane-imogolite complex) and hydroxyaluminosilicate-interlayered expansible-layer silicates. The catalytic ability of these SRO mineral colloids in the degradation of polyphenolics and other organics in the environment remains to be uncovered.

VI. IMPACT OF ABIOTIC TRANSFORMATIONS OF ORGANICS ON THE PEDOSPHERE

Abiotic transformations of natural organic components have significant impacts on soil physical, chemical, and biological properties and related environmental quality, as summarized below.

A. DYNAMICS OF TURNOVER OF ORGANIC CARBON

Abiotic transformations of organic components through the formation of soluble metal-organic complexes, incorporation of organics into the structural network of SRO mineral colloids during neogenesis, and retention of organics on the surfaces of these mineral colloids should have a protective effect on the biodegradation of organic components. Allophanic soils, which are commonly termed Andisols, are especially active in stabilizing organic C and commonly contain 4 to 30% or more humus.[110-112] The mean residence time of humus in allophanic soils range from 2000 to 5000 years.[113] The humic substances in allophanic soils are believed to be stabilized by complexation with active Al groups or hydrous oxides of Al and Fe associated with the allophanic surfaces.[114] The FA fraction from allophanic soils is usually high in silica, even after repeated purification.[115] Silica-humus complexation may account for some of the resistance of the humus to biodegradation.

The rate of loss of soil organic C as CO_2 is slower from allophanic soils than from nonallophanic soils.[116-118] Zunino et al.[112] studied the loss of C from a wide variety of ^{14}C-labeled, readily degradable organic substrates in the normal agricultural soils of California and Chile and the allophanic soils of southern Chile. The presence of allophane in the soil or the addition of 5 to 16% allophane to a very sandy soil decreases C losses by 16 to 73%. The loss of C over a 1-year period from the additions of more resistant polymer substrates, such as plant lignins and fungal melanins, is generally decreased to a much greater extent in the allophanic soils than in the normal agricultural soils.[118] However, the protective action of a series of SRO mineral colloids in the biodegradation of readily degradable organic and more resistant polymer substrates remains to be uncovered.

Jenkinson and Rayner[119] developed a simulation model in which soil organic matter is separated into five compartments: decomposable plant material (DPM, half-life of 0.165 years), resistant plant material (RPM, 2.31 years), soil biomass (B10, 1.69 years), physically stabilized organic matter (POM, 49.5 years), and chemically stabilized organic matter (COM, 1980 years). For a unitary input of plant material (1 t of fresh plant C per hectare per year) under steady conditions, the model predicts that after 10,000 years the soil will contain 0.01 t of C in DPM, 0.47 t in RPM, 0.28 t in B1O, 11.3 t in POM, and 12.2 t in COM. This is considered to be a useful representation of the turnover of organic matter in cropped soils. The protective action appears to be largely due to the adsorption of the organic matter on mineral colloids and/or the formation of organomineral complexes. The extent of contribution of SRO and crystalline mineral colloids to the protective action remains to be established.

Catalysis of SRO mineral colloids of soils also plays a vital role in abiotic transformations of natural organic components, as discussed in Section V. The catalytic power of these mineral colloids in the pathway of turnover of organic C in nature thus merit increasing attention.

B. REPLENISHMENT AND CYCLING OF NUTRIENTS

Microbial by-products and resistant plant residues adsorbed on soil particles have turnover times of years. Fulvic acids have turnover times of hundreds of years, whereas HAs and humins usually approach thousands of years in their turnover time.[120] Although the HAs and humins constitute by far the majority of the organic matter in a system, they contribute only a small proportion to the annual cycling of C within the soil because of their very slow turnover rate. The undecomposed litter, soil biomass, and microbial metabolites along with the plant residues constitute the active fraction of organic matter that has a prominent role in replenishing the majority of nutrients essential for plant growth and human and animal health and in the cycling of nutrient elements within the geocycle on an annual basis.

The effect of a series of SRO mineral colloids of soils, which differ in structural configuration and surface properties, on the biodegradation and turnover of N-, P-, and S-bearing organics and associated nutrients is largely unknown. The ability of a series of SRO mineral colloids to degrade organics by catalyzing the decarboxylation and deamination of organic compounds, the polymerization of polyphenolics, and the polycondensation of the associated organics is even less understood.

C. FORMATION OF HUMIC SUBSTANCES

The formation of humic substances can proceed through biotic and abiotic processes. Research data indicate that oxidative polymerization of polyphenols is important in the formation of humic substances.[22] The oxidative polymerization of polyphenols can be accelerated enzymatically and nonenzymatically.[121,122] Short-range ordered mineral colloids have been shown to be reactive in promoting the abiotic oxidative polymerization of polyphenols, resulting in the formation of humic substances, as discussed in Section V. However, the similarities and differences in the mechanistics of the catalytic effects of SRO mineral colloids and enzymes on the formation of humic substances remain to be investigated. The influence of the structural configuration and surface properties of a wide range of SRO mineral colloids of soils on their catalysis of the formation of humic substances needs to be studied.

D. ENZYMATIC ACTIVITY AND MICROBIAL EVENTS

Allophanic clays adsorb considerable amounts of enzymes, such as proteases and amylase and their substrate, and thereby affect the action of these enzymes.[123] Ross[124] reported that the original activities of α-amylase, β-amylase, and invertase were decreased by the presence of allophane to 57, 12, and 20%, respectively. The research on the interaction of enzymes with soil inorganic constituents has largely dealt with crystalline clay minerals.[125] The role of SRO mineral colloids in affecting enzymatic activities should receive increasing attention.

Stotzky[126] reported that the utilization of organics by microbes seems to be modified when the organics are bound on metal oxide-coated clays, compared with clays of homoionic to nonpolymeric cations. As the surface properties of these metal oxides vary with pH and ionic factors, the release of such bound organics may occur as the pH and ionic environments of natural microhabitats fluctuate. Once the organics have been released from the hydrous metal oxide-mineral complexes, they could be either degraded by soil microbes or complexed by mineral colloids. The role of SRO mineral colloids in the dynamics of the transformations of organic compounds and associated microbial events thus should be studied further.

E. FORMATION OF SHORT-RANGE ORDERED MINERAL COLLOIDS

Organic compounds are a perturbing factor in the crystallization of Al and Fe precipitation products and aluminosilicates.[2,17,18] Interactions of organic ligands with Al, Fe, and Si and subsequent incorporation of these ligands into the structural network promote the formation of SRO mineral colloids.

Ligands that form polydentate complexes generally have a stronger ability to cause structural distortion of the precipitation products of Al than those with fewer functional groups.[18] The implications of these findings are evident in the formation of SRO organomineral complexes in soil.[127,128] Besides hydroxy-Al interlayering of expansible-layer silicates,[61] the formation of crystalline Al hydroxides in soils and sediments can be hampered through the complexation of organic ligands with hydroxy Al, especially in environments where the accumulation of organic components tends to occur. Substantial concentrations of crystalline Al hydroxides are thus absent in soils rich in organic matter in the temperate regions.

The transformations of ferrihydrite to crystalline Fe oxides and oxyhydroxides are effectively retarded by complexation with organic components.[129] Furthermore, the transformations of Fe(III)-FA complexes to hematite/goethite are even more inhibited.

Humic substances are strongly bound to Al released from parent volcanic ash by weathering, which limits the coprecipitation of Si and Al.[114] The accumulation of humus in the Dystrandepts A_1 and Spodosol A_2 and Bh horizons favors the formation of hydroxy-Al-humus complexes, "allophane-like" constituents, and/or opaline silica and retard the formation of allophane and imogolite, which are formed in the Dystrandept B and (B) and Spodosol B_s horizons.[114,130-134]

F. MODIFICATIONS OF SURFACE REACTIVITY OF MINERAL COLLOIDS

The adsorption of organic ligands on surfaces of SRO mineral colloids modifies the electrical charge on colloidal particles. Several studies have shown that the pH_{iep} of alumina suspensions is decreased by the adsorption of natural dissolved organic matter.[52,53,135]

Noncrystalline Al hydroxides precipitated in the absence of interfering ions are very unstable and revert rapidly to the crystalline state,[136] causing the specific surface of Al hydroxides to be substantially reduced during aging. In the presence of complexing organics, the occupation of coordination sites of Al by complexing organic ligands would retard the reversion of precipitation products to the crystalline state, as that would require the removal of organic ligands from within the structure. The maintenance of the noncrystalline state with an accompanying high specific surface by these organic ligands stabilizes the high surface reactivity of SRO precipitation products of Al.[32] Furthermore, the surface charge and surface acidity of these precipitation products are substantially changed, and their ability to bind cations and anions are also significantly modified.[137] Similar studies on the influence of the incorporation of organic ligands into the structural network of SRO iron oxides and aluminosilicates on their surface properties remain to be undertaken.

G. BIOAVAILABILITY OF NUTRIENTS AND POLLUTANTS

The incorporation of N-, P-, and S-bearing organics into the structural network of SRO mineral colloids and the adsorption of these organics on their surfaces may directly affect the bioavailability of these nutrients. Research needs to be conducted in this important area of science.

Modifications of the surface reactivity of SRO mineral colloids by the adsorption of dissolved organics[52,135] may alter the affinity of these colloids toward nutrients and pollutants. Consequently, the dynamics and bioavailability of these nutrients and pollutants could be modified.

H. EFFECT OF AL ON PLANT GROWTH

Aluminum toxicity is a major problem in acid soils (pH <5.5). Toxicities caused by Al have been noted in acid soils in several regions of the eastern U.S., Canada, and the tropics.[138,139] However, acid soils that are rich in native organic matter, or amended with large quantities of organic residues, give low Al^{3+} concentrations in the soil solution and result in good growth of crops under conditions where toxicities would otherwise occur. The influence of organic matter on the amelioration of Al toxicities is evidently due to the reaction of organic ligands with Al, resulting in the formation of less toxic hydroxy Al-organic complexes.

I. FORMATION OF STABLE AGGREGATE

Organic matter is considered to be indispensable for promoting the good structure of a wide range of soils, particularly those representative of Mollisols, Alfisols, Ultisols, and

Inceptisols.[9] Complexation of organic ligands with polyvalent cations (e.g., Al, Fe, and Ca) on clay surfaces is involved in stabilizing soil aggregates. Therefore, abiotic transformation of organic components through complexation with sesquioxidic components and SRO $CaCO_3$ should influence the stability of soil aggregates and associated soil physical properties.

J. ECOLOGICAL AND HEALTH EFFECTS

Several processes are involved in the mobilization and transport of toxic metals in the pedosphere. Complexation of these metals with organic ligands is certainly involved in the processes.[137,140] Toxic metals can be released through weathering of parent rocks and/or desorption from soil colloids by complexing organic ligands. The released metals will contaminate the biosphere through uptake by plants and transport to groundwater and stream, river, and lake waters. Of equal importance is the modification of metal toxicities by abiotic transformations of organics through their complexation with metals in soil solutions and other natural waters.

VII. SUMMARY AND CONCLUSIONS

Abiotic transformations of natural organic components in soils and associated environments are substantially influenced by SRO mineral colloids. Major processes include the formation of metal-organic complexes in soil solutions and other natural waters, incorporation of organics into the structural network of SRO mineral colloids, retention of organics on the surfaces of the mineral colloids, and abiotic degradation of organics by catalysis of the mineral colloids.

The extent of the formation and the stability of soluble metal-organic complexes depends on the affinity of organics toward the metals, namely, the stability constants of the metal-organic complexes and the concentration of organics. The formation of metal-organic complexes in soil solutions and other natural waters would influence further transformations of organics by abiotic and biotic processes. This process should also influence the transformations, transport, and toxicities of metals in terrestrial and aquatic ecosystems.

The extent of the incorporation of organic components into the structural network of SRO Al and Fe oxides and aluminosilicates depends on the strength of organic ligands in promoting the formation of the SRO mineral colloids, which in turn varies with the affinity of the organic ligands for Al. The influence of these reactions on the biodegradation of organics in the pedosphere and associated environments merits attention. The impact of these reactions on the surface reactivity of soil mineral colloids pertaining to the transformations, dynamics, and bioavailability of nutrients and pollutants in the environment should not be overlooked.

The retention of organic components on surfaces of SRO mineral colloids affects abiotic and biotic transformations of organic components, especially the active organic fraction of soils, namely, decomposing plant materials, microbial metabolites, and plant residues. This process would thus influence the replenishment and cycling of nutrients in the environment. Enzymatic activities have been shown to be influenced by the interactions of enzymes with SRO mineral colloids. The role of SRO mineral colloids in influencing enzymatic activities and related microbial events should receive more attention. The retention of organics by SRO mineral colloids also modifies the surface charge and potential of soil particles and contributes to the formation of stable aggregates.

Natural organic components such as polyphenolics and organic acids have been shown to be degraded by the catalysis of SRO Mn, Fe, and Al oxides and/or aluminosilicates. Abiotic oxidative transformations of these organics include polymerization, polycondensation, ring cleavage, decarboxylation, and deamination reactions. These reactions contribute to the formation of humic substances. Furthermore, abiotic oxidative degradation of organics can be a pathway of turnover of organic C, N, P, and S in nature that has not been considered in the past. The catalytic role of a series of SRO mineral colloids, which differ in structural configuration and surface reactivity, in affecting the dynamics of nutrient-bearing organics via abiotic degradation processes in natural environments deserves increasing attention for the years to come.

ACKNOWLEDGMENTS

This study was supported by Grant GP2383-Huang of the Natural Sciences and Engineering Research Council of Canada.

REFERENCES

1. **Hsu, P. H.,** Aluminum oxides and oxyhydroxides, in *Minerals in Soil Environments,* Dixon, J. B. and Weed, S. B., Eds., Soil Science Society of America, Madison, WI, 1989, 331.
2. **Schwertmann, U. and Taylor, R. M.,** Iron oxides, in *Minerals in Soil Environments,* Dixon, J. B. and Weed, S. B., Eds., Soil Science Society of America, Madison, WI, 1989, 379.
3. **Wada, K.,** Allophane and imogolite, in *Minerals in Soil Environments,* Dixon, J. B. and Weed, S. B., Eds., Soil Science Society of America, Madison, WI, 1989, 331.
4. **Wolfe, N. L., Mingelgrin, U., and Miller, G. C.,** Abiotic transformations in water, sediments, and soil, in *Pesticides in the Soil Environment: Processes, Impacts, and Modeling,* Cheng, H. H., Ed., Soil Science Society of America, Madison, WI, 1990, 103.
5. **Demolon, A. and Barbier, G.,** Conditions de formation et constitution du complexe argilo-humique des sols, *C. R. Acad. Sci. Paris,* 188, 654, 1929.
6. **Waksman, S. A. and Iyer, K. R. N.,** Contribution to our knowledge of the chemical nature and origin of humus. IV. Fixation of proteins by lignins and formation of complexes resistant to microbial decomposition, *Soil Sci.,* 36, 69, 1933.
7. **Greenland, D. J.,** Interaction between clays and organic compounds in soils. I. Mechanisms of interaction between clays and defined organic compounds, *Soils Fert.,* 28, 415, 1965.
8. **Schnitzer, M. and Kodama, H.,** Reactions of minerals with soil humic substances, in *Minerals in Soil Environments,* Dixon, J. B. and Weed, S. B., Eds., Soil Science Society of America, Madison, WI, 1977, 741.
9. **Stevenson, F. J. and Vance, G. F.,** Naturally occurring aluminum-organic complexes, in *The Environmental Chemistry of Aluminum,* Sposito, G., CRC Press, Boca Raton, FL, 1989, 117.
10. **Kwong, Ng, Kee, K. F., and Huang, P. M.,** Influence of citric acid on the crystallization of aluminum hydroxides, *Clays Clay Miner.,* 23, 164, 1975.
11. **Kwong, Ng, Kee, K. F., and Huang, P. M.,** Comparison of the influence of tannic acid and selected low-molecular-weight organic acids on precipitation products of aluminum, *Geoderma,* 26, 179, 1981.
12. **Lind, C. J. and Hem, J. D.,** Effects of Organic Solutes on Chemical Reactions of Aluminum, *U.S. Geological Survey Water Supply Paper 1827-G.*
13. **Kodama, H. and Schnitzer, M.,** Effect of fulvic acid on the crystallization of Fe(III)-oxides, *Geoderma,* 19, 279, 1977.
14. **Kodama, H. and Schnitzer, M.,** Effect of fulvic acid on the crystallization of aluminum hydroxides, *Geoderma,* 24, 195, 1980.
15. **Inoue, K. and Huang, P. M.,** Influence of citric acid on the natural formation of imogolite, *Nature (London),* 308, 58, 1984.
16. **Violante, A. and Huang, P. M.,** Influence of inorganic and organic ligands on the formation of aluminum hydroxides and oxyhydroxides, *Clays Clay Miner.,* 33, 181, 1985.
17. **Huang, P. M.,** Ionic factors affecting the formation of short-range ordered aluminosilicates, *Soil Sci. Soc. Am. J.,* 55, 1172, 1991.
18. **Huang, P. M.,** Ionic factors affecting aluminum transformations and the impact on soil and environmental sciences, *Adv. Soil Sci.,* 8, 1, 1988.

19. **Shindo, H. and Huang, P. M.,** Role of Mn(IV) oxide in abiotic formation of humic substances in the environment, *Nature (London),* 298, 363, 1982.
20. **Shindo, H. and Huang, P. M.,** Significance of Mn(IV) oxide in abiotic formation of organic nitrogen complexes in natural environments, *Nature (London),* 308, 57, 1984.
21. **Wang, M. C. and Huang, P. M.,** Significance of Mn(IV) oxide in the abiotic ring cleavage of pyrogallol in natural environments, *Sci. Total Environ.,* 113, 147, 1992.
22. **Huang, P. M.,** Role of soil minerals in transformations of natural organics and xenobiotics in soil, in *Soil Biochemistry, Vol. 6,* Bollag, J.-M. and Stotzky, G., Eds., Marcel Dekker, New York, 1990, 29.
23. **Rovira, A. D.,** Plant root exudates, *Bot. Rev.,* 35, 35, 1969.
24. **Bruckert, S.,** Influence des composes organiques solubles sur la pedogenese en melien acide, *Ann. Agron.,* 21, 421, 1970.
25. **Huang, P. M.,** Organo-alumino polymer associations and their significance in soil and environmental sciences, in *Soil Colloids and Their Associations in Aggregates,* de Boodt, M. F., Hayes, M., and Herbillon, A., Eds., Plenum Press, New York, 1991, 179.
26. **Jackson, T. A.,** Humic matter in natural waters and sediments, *Soil Sci.,* 119, 56, 1975.
27. **Reuter, J. H. and Perdue, E. M.,** Importance of heavy metal-organic matter interactions in natural waters, *Geochim. Cosmochim. Acta,* 41, 325, 1977.
28. **Stevenson, F. J.,** *Humus Chemistry: Genesis, Composition, Reactions,* Wiley-Interscience, New York, 1982.
29. **Vinkler, P., Lakatos, B., and Meisel, J.,** Infrared spectroscopic investigations of humic substances and their metal complexes, *Geoderma,* 15, 231, 1976.
30. **Boyd, S. A., Sommers, L. E., and Nelson, D. W.,** Infrared spectra of sewage sludge fractions: evidence for an amide metal binding site, *Soil Sci. Soc. Am. J.,* 43, 893, 1979.
31. **Piccolo, A. and Stevenson, F. J.,** Infrared spectra of Cu^{2+}, Pb^{2+}, and Ca^{2+} complexes of soil humic substances, *Geoderma,* 27, 195, 1981.
32. **Kwong, Ng, Kee, K. F., and Huang, P. M.,** The relative influence of low-molecular-weight complexing organic acids on the hydrolysis and precipitation of aluminum, *Soil Sci.,* 128, 337, 1979.
33. **Kwong, Ng, Kee, K. F., and Huang, P. M.,** Nature of hydrolytic precipitation products of aluminum as influenced by low-molecular-weight complexing organic acids, in *Proc. 6th Int. Clay Conf., Oxford,* Mortland, M. M. and Farmer, V. C., Eds., Elsevier, Amsterdam, 1979, 527.
34. **Violante, A. and Huang, P. M.,** Influence of oxidation treatments on surface properties and reactivities of short-range ordered precipitation products of aluminum, *Soil Sci. Soc. Am. J.,* 53, 1402, 1989.
35. **Anderson, J. U.,** An improved pretreatment for mineralogical analysis of samples containing organic matter, *Clays Clay Miner.,* 10, 380, 1963.
36. **Lavkulich, L. M. and Wiens, J. H.,** Comparison of organic matter destruction by hydrogen peroxide and sodium hypochlorite and its effects on selected mineral constituents, *Soil Sci. Soc. Am. Proc.,* 34, 755, 1970.
37. **Violante, A. and Huang, P. M.,** Effect of tartaric acid and pH on the nature and physicochemical properties of short-range ordered aluminum precipitation products, *Clays Clay Miner.,* 40, 462, 1992.
38. **Inoue, K. and Huang, P. M.,** Influence of selected organic ligands on the formation of allophane and imogolite, *Soil Sci. Soc. Am. J.,* 50, 1623, 1986.
39. **Inoue, K. and Huang, P. M.,** Effect of humic and fulvic acids on the formation of allophane, in *Proc. 8th Int. Clay Conf.,* Schultz, L. G., van Olphen, H., and Mumpton, F. A., Eds., Clay Minerals Society, Denver, 1987, 221.
40. **Tokashiki, Y. and Wada, K.,** Weathering implications of the mineralogy of clay fractions of two Ando soils, Kyushu, *Geoderma,* 14, 47, 1975.
41. **Tait, J. M., Yoshinaga, N., and Mitchell, B. D.,** The occurrence of imogolite in some Scottish soils, *Soil Sci. Plant Nutr. (Tokyo),* 24, 145, 1978.
42. **Wada, K. and Greenland, D. J.,** Selective dissolution and differential infrared spectroscopy for characterization of "amorphous" constituents in soil clays, *Clay Miner.,* 8, 241, 1970.
43. **Farmer, V. C., Russell, J. D., and Smith, B. F. L.,** Extraction of inorganic forms of translocated Al, Fe, and Si from a Podzol Bs horizon, *J. Soil Sci.,* 34, 571, 1983.

44. **Lou, G. and Huang, P. M.,** Nature and charge properties of X-ray noncrystalline aluminosilicates as affected by citric acids, *Soil Sci. Soc. Am. J.,* 53, 1287, 1989.
45. **Lou, G. and Huang, P. M.,** Impact of structural perturbation of allophanic materials by citric acid on their surface alteration, in *Proc. 9th Int. Clay Conf.,* Vol. 2, Farmer, V. C. and Tardy, Y., Eds., Publié avec le concours de l'Université Louis Pasteur et du Centre National de la Recherche Scientifique, Strasbourg, 1990, 137.
46. **Blume, H. P. and Schwertmann, U.,** Genetic evaluation of profile distribution of aluminum, iron, and manganese oxides, *Soil Sci. Soc. Am. Proc.,* 33, 437, 1969.
47. **Campbell, A. S. and Schwertmann, U.,** Iron oxide mineralogy of placic horizons, *J. Soil Sci.,* 35, 569, 1984.
48. **Krishnamurti, G. S. R. and Huang, P. M.,** Kinetics of Fe(II) oxygenation and the nature of hydrolytic products as influenced by ligands, in *Proc. 9th Int. Clay Conf.,* Vol. 1, Farmer, V. C. and Tardy, Y., Eds., Publié avec le concours de l'Université Louis Pasteur et du Centre National de la Recherche Scientifique, Strasbourg, 1990, 195.
49. **Parfitt, R. L., Fraser, J. D., and Farmer, V. C.,** Adsorption on hydrous oxides. II. Oxalate, benzoate, and phosphate on gibbsite, *J. Soil Sci.,* 28, 40, 1977.
50. **Parfitt, R. L., Fraser, A. R., and Farmer, V. C.,** Adsorption of hydrous oxides. III. Fulvic acid and humic acid on goethite, gibbsite, and imogolite, *J. Soil Sci.,* 28, 289, 1977.
51. **Kummert, R. and Stumm, W.,** The surface complexation of organic acids on hydrous γ-Al_2O_3, *J. Colloid Interface Sci.,* 75, 373, 1980.
52. **Davis, J. A. and Gloor, R.,** Adsorption of dissolved organics in lake water by aluminum oxide. Effect of molecular weight, *Environ. Sci. Technol.,* 15, 1223, 1981.
53. **Davis, J. A.,** Adsorption of natural dissolved organic matter at the oxide/water interface, *Geochim. Cosmochim. Acta,* 46, 2381, 1982.
54. **Bowers, A. R. and Huang, C. P.,** Adsorption characteristics of polyacetic amino acids on hydrous γ-Al_2O_3, *J. Colloid Interface Sci.,* 105, 197, 1985.
55. **McBride, M. B. and Wesselink, B.,** Chemisorption of catechol on gibbsite, boehmite, and amorphous alumina surfaces, *Environ. Sci. Technol.,* 22, 703, 1988.
56. **Huang, P. M., Wang, T. S. C., Wang, M. K., Wu, M. H., and Hsu, N. W.,** Retention of phenolic acids by noncrystalline hydroxy-aluminum and iron compounds and clay minerals of soils, *Soil Sci.,* 123, 213, 1977.
57. **Goh, T. B. and Huang, P. M.,** Formation of hydroxy-Al-montmorillonite complexes as influenced by citric acid, *Can. J. Soil Sci.,* 64, 411, 1984.
58. **Goh, T. B. and Huang, P. M.,** Influence of citric and tannic acids on hydroxy-Al interlayering in montmorillonite, *Clays Clay Miner.,* 34, 37, 1986.
59. **Rich, C. I.,** Hydroxy interlayers in expansible phyllosilicates, *Clays Clay Miner.,* 16, 15, 1968.
60. **Singer, A. and Huang, P. M.,** Effect of humic acid on aluminum interlayering in montmorillonite, *Soil Sci. Soc. Am. J.,* 57, 271, 1993.
61. **Barnhisel, R. I. and Bertsch, P. M.,** Chlorite and hydroxy-interlayered vermiculite and smectite, in *Minerals in Soil Environments,* Dixon, J. B. and Weed, S. B., Eds., Soil Science Society of America, Madison, WI, 1989, 729.
62. **Allen, B. L. and Hajak, B. F.,** Mineral occurrence in soil environments, in *Minerals in Soil Environments,* Dixon, J. B. and Weed, S. B., Eds., Soil Science Society of America, Madison, WI, 1989, 199.
63. **Inoue, T. and Wada, K.,** Adsorption of humified clover extracts by various clays, in *Trans. Int. Congr. Soil Sci. 9th,* Vol. 3, Adelaide, Australia, 1968, 289.
64. **Reeve, R. and Fergus, I. G.,** Black and white waters and their possible relationship to the Podzolization process, *Aust. J. Soil Res.,* 21, 59, 1982.
65. **Evans, L. T. and Russell, E. W.,** The adsorption of humic and fulvic acids by clays, *J. Soil Sci.,* 10, 119, 1959.
66. **Levashkevich, G. A.,** Interaction of humic acids with hydroxides of iron and aluminum, *Pochvovedenie,* 4, 58, 1966.
67. **McKeague, J. A., Brydon, J. E., and Miles, N. M.,** Differentiation of forms of extractable iron and aluminum in soils, *Soil Sci. Soc. Am. Proc.,* 35, 33, 1971.
68. **Tipping, E.,** The adsorption of aquatic humic substances by iron oxides, *Geochim. Cosmochim. Acta,* 45, 191, 1981.

69. **McKenzie, R. M.,** The synthesis of birnessite, cryptomelane, and some other oxides and hydroxides of manganese, *Mineral. Mag.*, 38, 493, 1971.
70. **Schnitzer, M.,** Quo vadis soil organic matter research? Panel discussion papers, Publ. 5, Whither soil research?, in *Int. Congr. Soil Sci. 12th,* New Delhi, 1982, 67.
71. **Chang, H. M. and Allan, G. G.,** Oxidation, in *Lignins,* Sarkanen, K. V. and Ludwig, C. H., Eds., Wiley-Interscience, New York, 1971, 433.
72. **Schnitzer, M.,** Recent findings on the characterization of humic substances extracted from soils from widely differing climatic zones, in *Soil Organic Matter Studies,* Vol. 2, IAEA-SM-211/7, International Atomic Energy Agency, Vienna, 1977, 117.
73. **Schnitzer, M. and Skinner, S. I. M.,** Organo-metallic interactions in soils. I. Reactions between a number of metal ions and the organic matter of a Podzol Bh horizon, *Soil Sci.*, 96, 86, 1963.
74. **Wagner, G. H. and Stevenson, F. J.,** Structural arrangement of functional groups in soil humic acid as revealed by infrared analyses, *Soil Sci. Soc. Am. Proc.*, 29, 43, 1965.
75. **Theng, B. K. G., Wake, J. R. H., and Posner, A. M.,** The infrared spectrum of humic acid, *Soil Sci.*, 102, 70, 1966.
76. **Hatcher, P. G., Schnitzer, M., Dennis, L. W., and Maciel, G. E.,** Aromaticity of humic substances in soils, *Soil Sci. Soc. Am. J.*, 45, 1089, 1981.
77. **Wilson, M. A.,** Applications of nuclear magnetic resonance spectroscopy to the study of the structure of soil organic matter, *J. Soil Sci.*, 32, 167, 1981.
78. **Preston, C. M., Rauthan, B. S., Rodger, C., and Ripmeester, J. A.,** A hydrogen-1, carbon-13, and nitrogen-15 nuclear magnetic resonance study of *p*-benzoquinone polymers incorporating amino nitrogen compounds ("synthetic humic acids"), *Soil Sci.*, 134, 277, 1982.
79. **Wilson, M. A. and Goh, K. M.,** Proton-decoupled pulse Fourier-transform ^{13}C magnetic resonance of soil organic matter, *J. Soil Sci.*, 28, 645, 1977.
80. **Lee, J. S. K. and Huang, P. M.,** Photochemical effects on the abiotic transformations of polyphenolics as catalyzed by Mn(IV) oxide (abstr.), in *First Workshop of Working Group MO of the Int. Soc. Soil Sci.*, Edmonton, Canada, August 11 to 15, 1992.
81. **Shindo, H. and Huang, P. M.,** Comparison of the influence of Mn(IV) oxide and tyrosinase on the formation of humic substances in the environment, *Sci. Total Environ.*, 117/118, 103, 1992.
82. **Wang, M. C. and Huang, P. M.,** Polycondensation of pyrogallol and glycine and the associated reactions as catalyzed by birnessite, *Sci. Total Environ.*, 62, 435, 1987.
83. **Lehmann, R. G. and Cheng, H. H.,** Reactivity of phenolic acids in soil and formation of oxidation products, *Soil Sci. Soc. Am. J.*, 52, 1304, 1988.
84. **Liu, S.-Y., Minard, R. D., and Bollag, J.-M.,** Oligomerization of syringic acid, a lignin derivative, by a phenoloxidase, *Soil Sci. Soc. Am. J.*, 45, 1100, 1981.
85. **Bollag, J.-M., Liu, S. Y., and Minard, R. D.,** Ezymatic oligomerization of vanillic acid, *Soil Biol. Biochem.*, 14, 157, 1982.
86. **Pohlman, A. A. and McColl, J. G.,** Organic oxidation and manganese and aluminum mobilization in forest soils, *Soil Sci. Soc. Am. J.*, 53, 686, 1989.
87. **McBride, M. B.,** Oxidation of dihydroxybenzenes in aerated aqueous suspensions of birnessite, *Clays Clay Miner.*, 37, 341, 1989.
88. **McBride, M. B.,** Oxidation of 1,2- and 1,4-dihydroxybenzene by birnessite in acidic aqueous suspension, *Clays Clay Miner.*, 37, 479, 1989.
89. **Kung, K. H. and McBride, M. B.,** Electron transfer processes between hydroquinone and hausmannite (Mn_3O_4), *Clays Clay Miner.*, 36, 297, 1988.
90. **Wertz, J. E. and Vivo, J. L.,** Electron spin resonance of semiquinone, *J. Chem. Phys.*, 23, 2441, 1955.
91. **Fukuzumi, S., Ono, Y., and Keii, T.,** ESR studies on the formation of *p*-benzosemiquinone anion over manganese dioxide, *Int. J. Chem. Kinet.*, 7, 535, 1975.
92. **Stone, A. T. and Morgan, J. J.,** Reduction and dissolution of manganese (III) and manganese (IV) oxides by organics. I. Reaction with hydroquinone, *Environ. Sci. Technol.*, 18, 450, 1984.
93. **Stone, A. T. and Morgan, J. J.,** Reduction and dissolution of manganese (III) and manganese (IV) oxides by organics. II. Survey of the reactivity of organics, *Environ. Sci. Technol.*, 18, 617, 1984.
94. **Jaurequi, M. A. and Reisenauer, H. M.,** Dissolution of oxides of manganese and iron by root exudate components, *Soil Sci. Soc. Am. J.*, 46, 314, 1982.

95. **Traina, S. J. and Doner, H. E.,** Copper-manganese (II) exchange on a chemically reduced birnessite, *Soil Sci. Soc. Am. J.*, 49, 307, 1985.
96. **Lehmann, R. G., Cheng, H. H., and Harsh, J. B.,** Oxidation of phenolic acids by soil iron and manganese oxides, *Soil Sci. Soc. Am. J.*, 51, 352, 1987.
97. **Stone, A. T.,** Reductive dissolution of manganese (III)/(IV) oxides by substituted phenols, *Environ. Sci. Technol.*, 21, 979, 1987.
98. **Scheffer, F., Meyer, B., and Niederbudde, E. A.,** Huminstoffbildung unter Katalytischer Einwirkung naturlich vorkommender Eisenverbindungen im Modelversuch, *Z. Pflanzenernaehr. Dueng. Bodenkd.*, 87, 26, 1959.
99. **Shindo, H. and Huang, P. M.,** Catalytic effects of manganese (IV), iron (III), aluminum, and Si oxides on the formation of humic polymers, *Soil Sci. Soc. Am. J.*, 48, 927, 1984.
100. **Wang, T. S. C., Wang, M. C., and Huang, P. M.,** Catalytic synthesis of humic substances by using aluminas as catalysts, *Soil Sci.*, 136, 226, 1983.
101. **Eaton, D. G.,** Complexing of metal ions with semiquinones. An electron spin resonance study, *Inorg. Chem.*, 3, 1268, 1964.
102. **Ziechmann, W.,** Die Darstellung von Huminsäuren im heterogenen System mit neutraler Reaktion, *Z. Pflanzenernaehr. Dueng. Bodenkd.*, 84, 155, 1959.
103. **Gordon, R. L. and Harris, G. W.,** Effect of particle-size on the quantitative determination of quartz by X-ray diffraction, *Nature (London)*, 175, 1135, 1955.
104. **Iler, R. K.,** *The Chemistry of Silica*, John Wiley & Sons, New York, 1979, 3.
105. **Ribault, L. L.,** Presence d'une pellicule de silica amorphe a la surface den cristaux de quartz des formations sableuses, *C. R. Seances Sci. Ser. D*, 272, 1933, 1971.
106. **Larson, R. A. and Hufnal, J. M.,** Oxidative polymerization of dissolved phenols by soluble and insoluble inorganic species, *Limnol. Oceanogr.*, 25, 505, 1980.
107. **Wang, M. C.,** Catalytic Role of Selected Soil Minerals in the Abiotic Formation of Humic Substances and the Associated Reactions, Ph.D. thesis, University of Saskatchewan, Saskatoon, 1987.
108. **Kyuma, K. and Kawaguchi, K.,** Oxidative changes of polyphenols as influenced by allophane, *Soil Sci. Soc. Am. Proc.*, 28, 371, 1964.
109. **Kumada, K. and Kato, H.,** Browning of pyrogallol as affected by clay minerals, *Soil Sci. Plant Nutr. (Tokyo)*, 16, 195, 1970.
110. **Thorpe, J. and Smith, G. D.,** Higher categories of soil classification: order, suborder, and great soil group, *Soil Sci.*, 67, 117, 1949.
111. **Aomine, S. and Jackson, M. L.,** Allophane determination in Ando soils by cation exchange capacity delta value, *Soil Sci. Soc. Am. Proc.*, 23, 210, 1959.
112. **Zunino, H., Borie, F., Aguilera, S., Martin, J. P., and Haider, K.,** Decomposition of ^{14}C-labeled glucose, plant and microbial products, and phenols in volcanic ash-derived soils of Chile, *Soil Biol. Biochem.*, 14, 37, 1982.
113. **Wada, K. and Aomine, S.,** Soil development on volcanic materials during the quaternary, *Soil Sci.*, 116, 170, 1973.
114. **Wada, K. and Higashi, T.,** The categories of aluminum- and iron-humus complexes in Ando soils determined by selective dissolution, *J. Soil Sci.*, 27, 357, 1976.
115. **Griffith, S. M. and Schnitzer, M.,** The isolation and characterization of stable metal-organic complexes from tropical volcanic soils, *Soil Sci.*, 120, 126, 1975.
116. **Broadbent, F. E., Jackson, R. N., and McNicoll, J.,** Mineralization of C and N in some New Zealand allophanic soils, *Soil Sci.*, 98, 118, 1964.
117. **Ino, Y. and Monsi, M.,** On the decomposition of soil organic matter in humic allophanic soils at Mount Kirigamine, *Bot. Mag.*, 77, 168, 1964.
118. **Martin, J. P., Zunino, H., Peirano, P., Caiozzi, M., and Haider, K.,** Decomposition of ^{14}C labeled lignins, model humic acid polymers, and fungal mealanins in allophanic soils, *Soil Biol. Biochem.*, 14, 289, 1982.
119. **Jenkinson, D. S. and Rayner, J. H.,** The turnover of soil organic matter in some of the Rothamsted classical experiments, *Soil Sci.*, 123, 298, 1977.
120. **Paul, E. A. and Van Veen, J. A.,** The use of tracers to determine the dynamic nature of organic matter, in *Trans. Int. Congr. Soil Sci. 11th*, Vol. 3, Edmonton, Canada, 1978, 61.
121. **Haider, K., Martin, J. P., and Filip, Z.,** Humus biochemistry, in *Soil Biochemistry*, Vol. 4, Paul, E. A. and McLaren, A. D., Eds., Marcel Dekker, New York, 1975, 175.

122. **Flaig, W., Beutelspacher, H., and Rietz, E.,** Chemical composition and physical properties of humic substances, in *Soil Components, Vol. 1. Organic Components,* Gieseking, J. E., Ed., Springer-Verlag, New York, 1975, 1.
123. **Kobayashi, Y. and Aomine, S.,** Mechanism of inhibitory effect of allophane and montmorillonite on some enzymes, *Soil Sci. Plant Nutr.,* 13, 189, 1967.
124. **Ross, D. J.,** Invertase and amylase activities as influenced by clay minerals, soil clay fractions, and topsoils under grassland, *Soil Biol. Biochem.,* 15, 287, 1983.
125. **Burns, R. G.,** Interaction of enzymes with soil minerals and organic colloids, in *Interactions of Soil Minerals with Natural Organics and Microbes,* Huang, P. M. and Schnitzer, M., Eds., Soil Science Society of America, Madison, WI, 1986, 429.
126. **Stotzky, G.,** Influence of soil mineral colloids on metabolic processes, growth, adhesion, and ecology of microbes and viruses, in *Interactions of Soil Minerals with Natural Organics and Microbes,* Huang, P. M. and Schnitzer, M., Eds., Soil Science Society of America, Madison, WI, 1986, 305.
127. **Mitchell, B. D., Farmer, V. C., and McHardy, W. J.,** Amorphous inorganic materials in soils, *Adv. Agron.,* 16, 327, 1964.
128. **Campbell, A. S., Young, A. W., Livingstone, L. G., Wilson, M. A., and Walker, T. W.,** Characterization of poorly-ordered alumino-silicates in a vitric Andosol from New Zealand, *Soil Sci.,* 123, 362, 1977.
129. **Schwertmann, U., Kodama, H., and Fisher, W. R.,** Mutual interactions between organics and iron oxides, in *Interactions of Soil Minerals with Natural Organics and Microbes,* Huang, P. M. and Schnitzer, M., Eds., Soil Science Society of America, Madison, WI, 1986, 223.
130. **Higashi, T. and Wada, K.,** Size fractionation, dissolution analysis, and infrared spectroscopy of humus complexes in Ando soils, *J. Soil Sci.,* 28, 653, 1977.
131. **Farmer, V. C., Russell, J. D., and Berrow, M. L.,** Imogolite and proto-imogolite allophane in spodic horizons: evidence for a mobile aluminum silicate complex in Podzol formation, *J. Soil Sci.,* 31, 673, 1980.
132. **Farmer, V. C., Russell, J. D., and Smith, B. F. L.,** Extraction of inorganic forms of translocated Al, Fe, and Si from a Podzol Bs horizon, *J. Soil Sci.,* 34, 571, 1983.
133. **Farmer, V. C.,** Possible roles of a mobile hydroxyaluminum orthosilicate complex (proto-imogolite) and other hydroxyaluminum and hydroxy-iron species in Podzolization, in *Migrations Organo-Minérals dans les Sols Tempérés,* Colloq. Int. The Centre National de la Recherche Scientifique No. 303, 1981, 275.
134. **Anderson, H. A., Berrow, M. L., Farmer, V. C., Hepburn, A., Russell, J. D., and Walker, A. D.,** A reassessment of Podzol formation process, *J. Soil Sci.,* 33, 125, 1982.
135. **Bowers, A. R. and Huang, C. P.,** Adsorption characteristics of metal-EDTA complexes onto hydrous oxides, *J. Colloid Interface Sci.,* 110, 575, 1985.
136. **Greenland, D. J.,** Interactions between humic and fulvic acids and clays, *Soil Sci.,* 111, 34, 1971.
137. **Huang, P. M. and Violante, A.,** Influence of organic acids on crystallization and surface properties of precipitation products of aluminum, in *Interactions of Soil Minerals with Natural Organics and Microbes,* Huang, P. M. and Schnitzer, M., Eds., Soil Science Society of America, Madison, WI, 1986, 159.
138. **Foy, C. D.,** Effects of aluminum on plant growth, in *The Plant Root and its Environment,* Carson, E. W., Ed., University Press of Virginia, Charlottesville, 1974, 601.
139. **Hargrove, W. L. and Thomas, G. W.,** Effects of organic matter on exchangeable aluminum and plant growth in acid soils, in *Chemistry of the Soil Environment,* Dowdy, R. H. et al., Eds., Spec. Publ. No. 40, American Society of Agronomy, Madison, WI, 1981, 151.
140. **Robert, M. and Berthelin, J.,** Role of biological and biochemical factors in soil mineral weathering, in *Interactions of Soil Minerals with Natural Organics and Microbes,* Huang, P. M. and Schnitzer, M., Eds., Soil Science Society of America, Madison, WI, 1986, 453.

Chapter 12

Influence of Pyrogallol on the Catalytic Action of Iron and Manganese Oxides in Amino Acid Transformation

M. C. Wang

CONTENTS

I. INTRODUCTION

Two mechanisms have been proposed[1] for the reaction of phenols with amino acids in the presence of phenoloxidase: the addition of amino acid to phenol, and the deamination and decarboxylation of amino acid. Andreux et al.[2] prepared N-containing polymers by reacting catechol with glycine at pH 7.9. The CO_2 released by the biodegradation of the polymers was derived from the carboxyl carbon of the glycine incorporated into the polymers.[2]

Manganese oxides are powerful causal agents of the polymerization of phenolic compounds.[3-5] Hydroquinone was oxidized and polymerized in aqueous solution between pH 3 to 7 in the presence of iron oxides.[6] Birnessite has a catalytic effect on the deamination[5] and decarboxylation[7] of glycine. Manganese oxides are very common in soils,[8-10] and birnessite is one of the most common forms.[10] Short-range ordered iron oxides are commonly present in soils.[11] The amino acids serine, phenylalanine, proline, methionine, and cysteine are also found in soil.[12] Pyrogallol is a common phenol in soils.[1,13]

Structural differences that influence abiotic mineralization remain to be established among serine, phenylalanine, proline, methionine, and cysteine. The present study investigated the role of certain structural features of these amino acids in the reactions which occur during their mineralization. In addition, the present study investigated the effect of pyrogallol on the catalysis of the mineralization of these same amino acids by iron and manganese oxides.

II. MATERIALS AND METHODS

A. MATERIALS

After preparation of birnessite[10] and short-range ordered iron oxide,[14] these compounds were lightly ground, suspended in deionized distilled water at a 1:10 (w/v) ratio, and

0-87371-914-X/95/$0.00+$.50

dispersed by ultrasonification. Particle size fractions of <2 μm were collected by sedimentation[15] and freeze dried. The freeze-dried oxides were examined by X-ray powder diffraction (XRD) and infrared (IR) absorption spectroscopy. Analytical grade L-serine (2-amino-3-hydroxypropionic acid), L-phenylalanine (α-amino-β-phenylpropionic acid), L-proline (2-pyrrolidinecarboxylic acid), DL-methionine (2-amino-4-methylthiobutanoic acid), L-cysteine (α-amino-β-thiopropionic acid), and pyrogallol (1,2,3-trihydroxybenzene) were obtained from Sigma Chemical Company (St. Louis, MO).

B. DETERMINATION OF RELEASED CO_2 AND NH_3

Four hundred milligrams of autoclaved manganese oxide or iron oxide was suspended in about 25 ml of sterilized aqueous solution containing 0.2 g kg^{-1} of the antiseptic thimerosal. The pH of the suspension was adjusted to 6.00 with 0.01 *M* HCl or 0.01 *M* NaOH over several days, until it was stabilized. Amino acid, amino acid-pyrogallol, manganese oxide- or iron oxide-amino acid, and manganese oxide- or iron oxide-amino acid-pyrogallol systems were examined in the experiment. The amounts and distribution of CO_2 and NH_3 were determined as described by Wang.[16]

C. EXAMINATION OF MICROBIAL ACTIVITY

Duplicate suspensions were prepared. A small amount of each reaction suspension was inoculated onto agar plates of GYEN,[17] AIA,[18] and Czapek-Dox[18] at both the beginning and the end of the reaction period. All plates were incubated at 25°C for 40 d and then examined for any microbial growth.

D. DETERMINATION OF MINERALIZED NH_4^+ AND SO_4^{-2} AND OF DISSOLVED MN AND FE

At the end of the reaction period, the pH of the suspensions in the 500-ml Erlenmeyer flasks[16] was measured. Amounts of mineralized NH_4^+ and SO_4^{-2}, and of dissolved Mn and Fe in the suspensions, were determined as described by Wang.[16]

III. RESULTS AND DISCUSSION

There was no growth of microorganisms in any of the study systems; all the reactions observed in this study were, thus, abiotic processes. Manganese oxide catalyzed the decarboxylation of amino acids much more strongly than did iron oxide (Figure 1). It seems that the ability of manganese and iron oxides to enhance the decarboxylation of amino acids is related to the reducing power of amino acids, especially the electron-withdrawing power of –COOH and –SH groups in the amino acids.[19] Due to the high reducing power of cysteine,[20] the amount of CO_2 released from the manganese oxide- and iron oxide-L-cysteine systems was greater than that in the corresponding systems for the other amino acids (Figure 1).

At the end of the reaction period, the pH of the suspensions of the manganese oxide-amino acid-pyrogallol systems increased substantially above their initial value of 6.00 (Figure 2). This increase in pH was likely due to the consumption of protons and electrons in the reduction of Mn(IV) of manganese oxide to Mn(II).[21] It appears that the consumption of protons in this reduction process exceeded the generation of protons caused by the oxidative decarboxylation of amino acids or the ring cleavage of pyrogallol, or both, leading to the net increase of pH (Figure 2). The pH of the suspensions of the iron oxide-amino acid-pyrogallol systems was much lower than 6.00 (Figure 2), which can be attributed to the generation of protons in these reaction systems. The final pH of the iron oxide-L-cysteine system was much higher than its initial value of 6.00, while those of the other iron oxide-amino acid systems were almost the same as, or lower than 6.00 (Figure 2). This is attributable to the properties of the transformed products of the amino acids

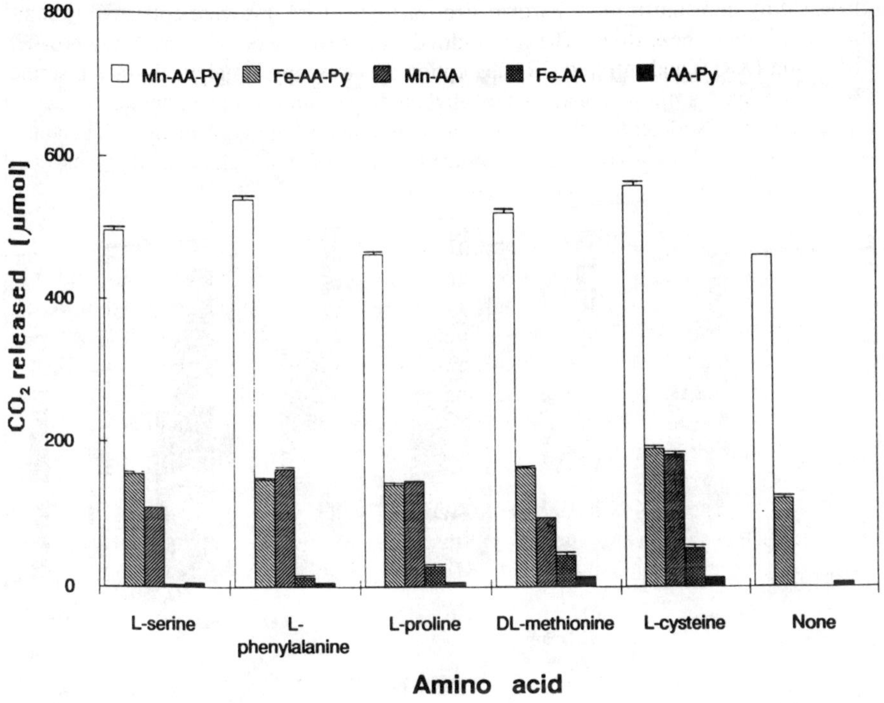

Figure 1 Quantities of CO_2 released from the manganese oxide-amino acid-pyrogallol (Mn-AA-Py), iron oxide-amino acid-pyrogallol (Fe-AA-Py), manganese oxide-amino acid (Mn-AA), iron oxide-amino acid (Fe-AA), and amino acid-pyrogallol (AA-Py) systems at the end of a 90-h reaction period.

as catalyzed by the iron oxide. In the L-serine-, L-phenylalanine-, L-proline-, and L-cysteine-pyrogallol systems, as well as in the pure pyrogallol system, the final pH of the solutions decreased substantially below the initial value of 6.00 (Figure 2). This decrease in pH was attributed to the generation of protons during the polycondensation of amino acids and pyrogallol[16] and during the polymerization of pyrogallol.[22]

The ability of manganese oxide to catalyze the deamination of amino acids seems to be related to both the position of $-NH_2$ groups in the amino acids and their reducing power.[20] The N in the –NH group of L-proline is in the ring structure; thus, the activation energy for the ring cleavage of the structure to release NH_3 may be higher than in cases where the N is not in the ring. This would explain why the amount of NH_3 released from the L-proline system was lower than for the other amino acids (Figure 3). In the iron oxide-amino acid systems, the deamination of amino acids did not follow the same trend as that seen in the manganese oxide-amino acid systems (Figure 3). This may be attributable to the different catalytic effect of iron oxide from that of manganese oxide to convert amino acid-N to NH_3, NH_4^+, and polymer-N. The deamination of amino acids as catalyzed by manganese oxide was much reduced in the presence of pyrogallol (Figure 3). This reduction in deamination may be due to incorporation of the NH_2 groups into pyrogallol-amino acid-derived N-bearing macromolecules. No deamination of proline in the manganese oxide-proline-pyrogallol system could be detected, indicating that the NH group in the ring structure of proline was not cleaved. In the iron oxide-amino acid-pyrogallol systems, deamination did not follow the same trend as that seen in the manganese oxide-amino acid-pyrogallol systems (Figure 3). This also may be attributable

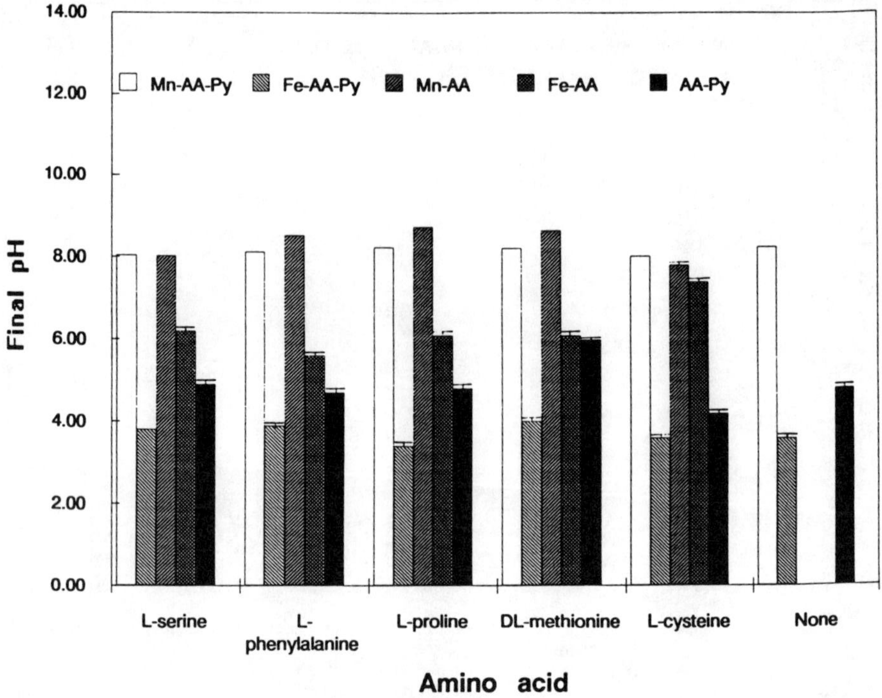

Figure 2 Final pH of the suspensions of manganese oxide-amino acid-pyrogallol, iron oxide-amino acid-pyrogallol, manganese oxide-amino acid, iron oxide-amino acid, and amino acid-pyrogallol systems.

to the different catalytic effect of iron oxide from that of manganese oxide to convert amino acid-N to NH_3, NH_4^+, and polymer-N in the presence of pyrogallol.

The amount of SO_4^{-2} produced in the manganese oxide-L-cysteine system was 140 times higher than that produced in the manganese oxide-DL methionine system (Figure 4). This was attributed to the higher reducing power of L-cysteine as compared to DL-methionine.[20] However, the amounts of SO_4^{-2} produced in the L-cysteine- and DL-methionine-iron oxide systems were negligible (Figure 4). In the absence of manganese oxide or iron oxide, the amounts of SO_4^{-2} produced in the DL-methionine and L-cysteine systems in the presence or absence of pyrogallol were negligible (data not shown). There was little difference in the amount of SO_4^{-2} produced by the manganese oxide and iron oxide systems containing DL-methionine and L-cysteine, in either the presence or absence of pyrogallol. Pyrogallol therefore appears to have little influence on the stimulation by manganese oxide of the SO_4^{-2}-producing mineralization of DL-methionine and L-cysteine.

Dissolved Mn adsorbed on the sediments (Figure 5) in the reaction systems were mainly in the form of Mn^{2+}, as established by high-performance ion-exchange chromatography. The quantity of dissolved Mn was much higher in the manganese oxide-amino acid-pyrogallol systems than in the corresponding systems without pyrogallol (Figure 5). This is attributed to the high reducing power of pyrogallol toward the Mn of Mn oxides. Pyrogallol, however, had only a slight effect on the dissolution of iron oxide (Figure 5). The production of Mn^{2+} in the reaction systems confirmed that the increase in the final pH of the manganese oxide-amino acid systems, both with or without pyrogallol (Figure 2), was due to the consumption of protons and electrons in the reduction of Mn(IV) to Mn(II).[21]

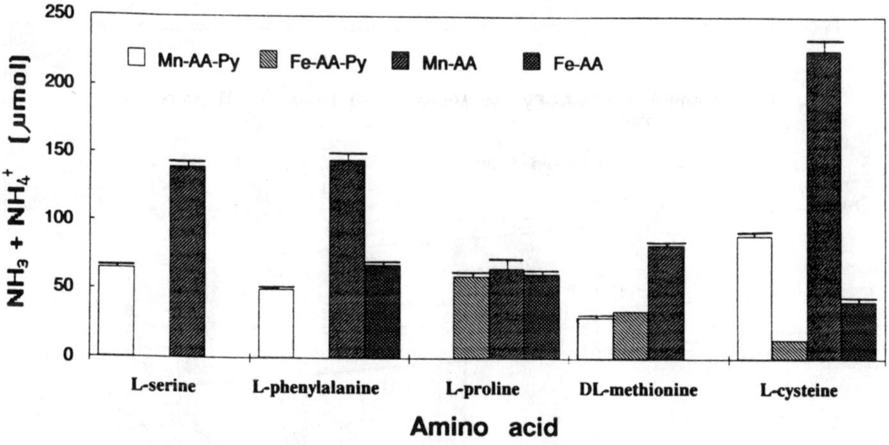

Figure 3 Mineralization of amino acid-N to NH_3 and HN_4^+ in manganese oxide-amino acid-pyrogallol, iron oxide-amino acid-pyrogallol, manganese oxide-amino acid, and iron oxide-amino acid systems.

IV. SUMMARY AND CONCLUSIONS

The catalytic effect of iron and manganese oxides (<2 μm) on the mineralization of five amino acids, and the effect of pyrogallol on this mineralization was studied in systems free of microbial activity. The amino acids studied were L-serine, L-phenylalanine, L-proline, DL-methionine, and L-cysteine. In the absence of both iron and manganese oxides, no released CO_2 or NH_3 could be detected from any of the amino acids at the end of a 90-h reaction period. The quantities of CO_2 released from the iron oxide- and manganese oxide-amino acid systems were 4 ± 2, 13 ± 3, 28 ± 4, 43 ± 4, and 53 ± 4 μmol per 400 mg of iron oxide, and 109 ± 0, 162 ± 3, 145 ± 2, 95 ± 2, and 182 ± 3 μmol per 400 mg of manganese oxide, for the amino acids in the order as listed above, respectively. The corresponding amounts of NH_3 released were: ND (not detectable), 68 ± 2, 62 ± 2, ND, and 42 ± 6 μmol from the iron oxide-amino acid systems, and 140 ± 4, 145 ± 5, 66 ± 6,

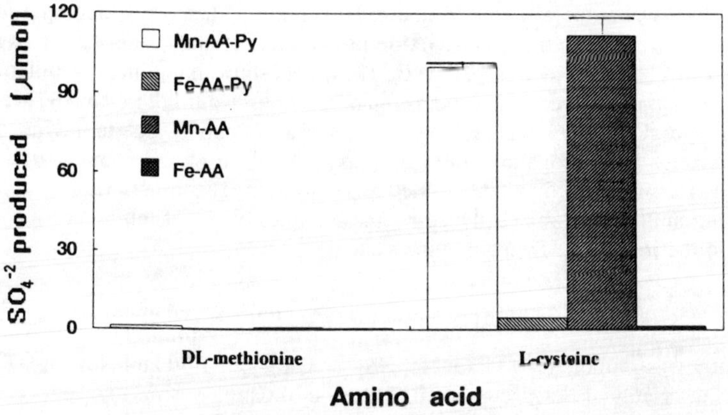

Figure 4 Mineralization of DL-methionine and L-cysteine to SO_4^{-2} in manganese oxide-amino acid-pyrogallol, iron oxide-amino acid-pyrogallol, manganese oxide-amino acid, and iron oxide-amino acid systems.

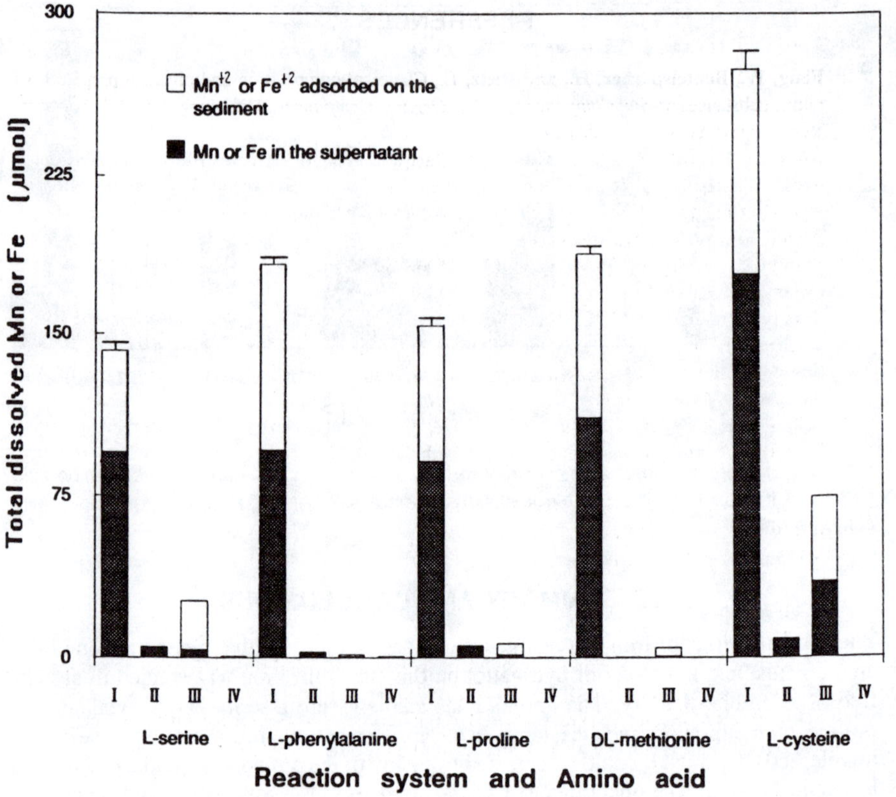

Figure 5 Dissolution of Mn and Fe in manganese oxide-amino acid-pyrogallol (I), iron oxide-amino acid-pyrogallol (II), manganese oxide-amino acid (III), and iron oxide-amino acid (IV) systems.

83 ± 2, and 226 ± 8 μmol from the manganese oxide-amino acid system. Only negligible amounts of released SO_4^{-2} could be detected from the DL-methionine and L-cysteine systems in the absence of iron and manganese oxides. The corresponding amounts of SO_4^{-2} released from the manganese oxide-DL-methionine and -L-cysteine systems were 0.8 ± 0.0 and 112 ± 7 μmol, respectively. These data show that mineralization of N from amino acids, as catalyzed by iron and manganese oxides, was inhibited by pyrogallol. The effect of pyrogallol on N mineralization of amino acids may be due to the incorporation of NH_2 groups into pyrogallol-amino acid-derived N-bearing macromolecules. One exception to this was the iron oxide-DL-methionine system. The mineralization of S from S-containing amino acids, as catalyzed by these oxides, was not inhibited by pyrogallol, except in the iron oxide-DL-methionine system.

ACKNOWLEDGMENTS

This study was supported by Grant No. NSC77-0409-B005-50 and NSC78-0409-B005-49 from the National Science Council, Republic of China.

REFERENCES

1. **Flaig, W., Beutelspacher, H., and Rietz, E.,** Chemical composition and physical properties of humic substances, in *Soil Components. Vol. 1. Organic Components,* Gieseking, J. E., Ed., Springer-Verlag, New York, 1975, chap. 1.
2. **Andreux, F., Golebiowska, D., Chone, T., Jacquin, F., and Methche, M.,** Characterisation et transformations en milieu mull d'un modele humique issu de l'autoxydation du systeme catechol-glycine et marque selectivement au carbon-14, in *Soil Organic Matter Studies II,* International Atomic Energy Agency, Vienna, 1977, 43.
3. **Shindo, H. and Huang, P. M.,** Role of Mn(IV) oxide in abiotic formation of humic substances in the environment, *Nature (London),* 298, 363, 1982.
4. **Shindo, H. and Huang, P. M.,** Catalytic effects of manganese(IV), iron(III), aluminum and silicon oxides on the formation of phenolic polymers, *Soil Sci. Soc. Am. J.,* 48, 927, 1984.
5. **Shindo, H. and Huang, P. M.,** Significance of Mn(IV) oxide in abiotic formation of organic nitrogen complexes in natural environments, *Nature (London),* 308, 57, 1984.
6. **Scheffer, F., Meyer, B., and Niederbudde, E. A.,** Himinstoffbildung under Katalytishcher Einwirkung naturlich vorkommender Eisenverbindungen im Modelversuch, *Z. Pflanzenernaehr. Bodenkd.,* 87, 26, 1959.
7. **Wang, M. C. and Huang, P. M.,** Polycondensation of pyrogallol and glycine and the associated reactions as catalyzed by birnessite, *Sci. Total Environ.,* 62, 435, 1987.
8. **McKeague, J. A., Damman, A. W. H., and Heringa, P. K.,** Iron-manganese and other pans in some soils of Newfoundland, *Can. J. Soil Sci.,* 48, 243, 1968.
9. **Taylor, R. M.,** The association of manganese and cobalt in soils—further observations, *J. Soil Sci.,* 19, 77, 1968.
10. **McKenzie, R. M.,** The synthesis of birnessite, cryptomelane, and some other oxides and hydroxides of manganese, *Mineral. Mag.,* 38, 493, 1971.
11. **Schwertmann, U. and Taylor, R. M.,** The significance of oxides for the surface properties and the usefulness of synthetic oxides as models for their study, *Int. Soc. Soil Sci. Bull.,* 60, 62, 1981.
12. **Bremner, J. M.,** Nitrogenous compounds, in *Soil Biochemistry.* Vol. 1, McLaren, A. D. and Peterson, G. H., Eds., Marcel Dekker, New York, 1967, 19.
13. **Martin, J. P. and Haider, K.,** A comparison of the use of phenolase and peroxidase for synthesis of model humic acid-type polymers, *Soil Sci. Soc. Am. J.,* 44, 983, 1980.
14. **Huang, P. M., Wang, T. S. C., Wang, M. K., Wu, M. H., and Hsu, N. W.,** Retention of phenolic acids by noncrystalline hydroxy-aluminum and -iron compounds and clay minerals of soils, *Soil Sci.,* 123, 213, 1977.
15. **Jackson, M. L.,** *Soil Chemical Analysis—Advanced Course, 2nd Ed.,* Jackson, M. L., Madison, WI, 1979, chap. 3.
16. **Wang, M. C.,** Catalysis of nontronite in phenols and glycine transformations, *Clays Clay Miner.,* 39, 202, 1991.
17. **Germida, J. J. and Casida, L. E., Jr.,** Myceloid growth of *Arthrobacter globiformis* and other *Arthrobacter* species, *J. Bacteriol.,* 144, 1152, 1980.
18. **Olson, B. M., McKercher, R. B., and Germida, J. J.,** Microbial populations in trifluralin-treated soil, *Plant Soil,* 76, 379, 1984.
19. **Morrison, R. T. and Boyd, R. N.,** *Organic Chemistry,* 3rd ed., Allyn & Bacon, Boston, 1974, chap. 20, 37.
20. **Cornell, R. M., Giovanoli, R., and Schneider, W.,** Effect of cysteine and manganese on the crystallization of noncrystalline iron(III) hydroxide at pH 8, *Clays Clay Miner.,* 38, 21, 1990.
21. **Weast, R. C., Ed.,** *Handbook of Chemistry and Physics,* 59th ed., CRC Press, Boca Raton, FL, 1978.
22. **Wang, M. C. and Huang, P. M.,** Pyrogallol transformations as catalyzed by nontronite, bentonite, and kaolinite, *Clays Clay Miner.,* 37, 525, 1989.

Chapter 13

Photochemical Effect on the Abiotic Transformations of Polyphenolics as Catalyzed by Mn(IV) Oxide

J. S. K. Lee and P. M. Huang

CONTENTS

I. INTRODUCTION

Polyphenols occur widely in soils as decomposition products of plant and animal materials as well as microbial metabolites.[1-7] They are also present in aquatic environments such as lakes and rivers,[8] surface waters and groundwaters,[9] sediments,[10] sediment interstitial water,[11] and air-water interfaces.[12] These terrestrial and aquatic polyphenols are regarded as important precursors in the formation of humic substances in soils and sediments.[6,13-15] Mn(IV) oxides are commonly found in soils and sediments,[16-20] and they are considered to be ubiquitous above pH 5 as mineral coatings and nodules on or near well-aerated surfaces.[21] Birnessite is one of the most common Mn(IV) oxides in terrestrial and aquatic surroundings.[17] Shindo and Huang[22,23] showed that birnessite is very powerful in promoting the abiotic degradation of diphenols (hydroquinone, resorcinol, and catechol), leading to the subsequent formation of humic polymers. Recently, Wang and Huang[24] reported the unprecedented effect of birnessite in promoting the abiotic release of CO_2 through its ability to cleave the ring structure of pyrogallol under ambient temperature and pressure. However, the role of light in influencing the promoting effect of Mn(IV) oxide in the abiotic transformations of polyphenols remains unknown.

The objective of this study was to investigate the photochemical and associated thermal effects on the role of birnessite in catalyzing the abiotic degradation of four polyphenols: hydroquinone, resorcinol, catechol, and pyrogallol. Pyrogallol is a common polyphenol in soils.[5,13] Catechol, resorcinol, hydroquinone, and their derivatives are present in the chemical degradation products of humic substances[12,25-27] and are widely distributed in nature.[28] These polyphenolics differ not only in the number, but also the position of their hydroxyl groups. Hence, they may serve as functional models in facilitating our understanding of the transformations of natural and xenobiotic phenolic compounds in natural environments.

0-87371-914-X/95/$0.00+$.50

II. MATERIALS AND METHODS

A. MATERIALS

Birnessite was synthesized according to the method recommended by McKenzie.[17] The synthetic sample was washed with deionized distilled water, collected by ultrafiltration (0.01 μm pore size filter), and subsequently freeze dried. The freeze-dried birnessite was examined by X-ray diffraction (XRD) and infrared (IR) spectroscopic analyses. The specific surface of this product was estimated by the ethylene glycol monoethyl ether (EGME) retention method.[29] Pyrogallol (1,2,3-trihydroxybenzene) was obtained from BDH Chemicals. Three diphenols, hydroquinone (1,4-dihydroxybenzene), resorcinol (1,3-dihydroxybenzene), and catechol (1,2-dihydroxybenzene), were also used. Catechol was obtained from BDH Chemicals, while hydroquinone and resorcinol were obtained from Aldrich Chemical Company, Inc.

B. METHODS

1. Determination of Released CO_2

The apparatus used to determine the amount and distribution of CO_2 released was adopted from Wang and Huang.[30] One hundred milligrams of birnessite was suspended in 20 ml of autoclaved deionized distilled water containing 0.2 g kg^{-1} thimerosal (an antiseptic). The suspension was adjusted to pH 7.00 with 0.01 *M* HCl. The birnessite suspension was then transferred with a pipette to an autoclaved 500-ml Erlenmeyer flask. The flask containing the suspension was then weighed and supplemented with sterilized deionized distilled water (pH 7.00) to adjust the volume of the suspension to 25 ml. An autoclaved 50-ml polyethylene centrifuge tube containing 5 ml of 500 g kg^{-1} KOH was placed in the flask, its upper end leaning against the wall of the flask. In addition, a 20-ml autoclaved polyethylene vial containing 5 ml of 0.1 *M* polyphenol solution (adjusted to pH 7.00) was placed in the Erlenmeyer flask. The flask was capped tightly with a rubber stopper, which was connected by Tygon tubing to a manometer containing Brodie solution,[31] shaken slowly to mix the polyphenol with the birnessite suspension, and gently agitated under different light intensities, atmospheres, and temperatures for a reaction period of 90 h.

To investigate the influence of light, samples were studied under four different light intensities, ranging from 0 (dark) to 500 μE m^{-2} s^{-1}. To safeguard against exposure to light for the samples in the dark, all procedures (ranging from preparation of suspensions and solutions to placement of all components in the Erlenmeyer flask) were performed in a dark room with the aid of a red lamp, used in a photographic dark room. The light meter showed a reading of 0 μE m^{-2} s^{-1} even when the phytotron was as close as 2 cm to the red lamp. In addition, Erlenmeyer flasks used for the samples in the dark were painted black and covered with black tape. When the flask was capped with a rubber stopper, the light intensity within the flask was also 0 μE m^{-2} s^{-1}. Preliminary studies were carried out under laboratory lighting, and the intensity was found to have a value of 10 ± 2 μE m^{-2} s^{-1}. More stringent control of illumination was accomplished in a light chamber in which light intensity was controlled by adjusting the distance of the lamp rack from the samples. Two light intensities, 300 and 500 μE m^{-2} s^{-1}, were used. In nature, a level of 500 μE m^{-2} s^{-1} is the lower limit on a typical cloudy day, on which the intensity is 500 to 2000 μE m^{-2} s^{-1}.[32] Thus, the levels of illumination used in this experiment should be sufficient to demonstrate the effect of light as encountered in a natural environment. All light intensities were measured with a LI-COR (model LI-189SA) Quantum Sensor.

To elucidate the role of O_2 in the systems studied, samples were conducted under O_2, N_2, and air atmospheres. Gaseous O_2 and N_2 with 99.995 and 99.99% purity, respectively, were used in the 15-min flushing procedures. In addition, studies were conducted under five different temperatures (5, 15, 25, 35, and 45°C) in a constant-temperature oscillating water bath.

2. Detection of Microbial Activity

A small sample of the reaction suspension was inoculated, with a platinum loop, onto agar plates of GYEN,[33] AIA, and Czapek[34] at both the beginning and end of the reaction period. All inoculated plates were incubated at 25°C for 15 d to examine the existence of any microbial growth in the systems studied.

3. Determination of pH, Absorbances, and Dissolved O_2

At the end of a reaction period, the pH values of the suspensions were measured. The suspensions were then centrifuged at 7800 g for 10 min, and the absorbances of the supernatants at 472 and 664 nm were determined. The dissolved O_2 concentrations in the systems studied were determined throughout the entire reaction period with an Orion 820 dissolved-oxygen probe (a two-electrode system with an oxygen-permeable membrane).

All experiments were replicated three times.

III. RESULTS AND DISCUSSION

Growth of microorganisms was not evident in the systems studied (Table 1), indicating that all reactions observed in this study were abiotic.

As light intensity was increased from 0 (dark) to 500 $\mu E\ m^{-2}\ s^{-1}$, the amounts of CO_2 released from the birnessite-polyphenol and polyphenol systems under an air atmosphere at the end of a 90-h reaction period increased (Table 2). The increase in CO_2 released as a result of an increase in light intensity from 0 to 10 $\mu E\ m^{-2}\ s^{-1}$ was undoubtedly the consequence of a photochemical effect, since the temperature under these two light intensities was 25°C. However, as light intensity increased from 10 to 300 and then to 500 $\mu E\ m^{-2}\ s^{-1}$, there was a concomitant increase in temperature from 25.0 to 27.5 and then to 29.5°C, respectively. The increase in CO_2 released as a result of this temperature increase was 12.4, 9.3, 9.5, and 18.7 $\mu mol\ °C^{-1}$ in the birnessite-hydroquinone, birnessite-resorcinol, birnessite-catechol, and birnessite-pyrogallol systems, respectively. When the temperature was increased from 5 to 45°C in the dark (0 $\mu E\ m^{-2}\ s^{-1}$), the CO_2 released also increased (Figure 1). The increase in CO_2 released in this instance was 12.7, 13.3, 15.9, and 19.4 $\mu mol\ °C^{-1}$ in the birnessite-hydroquinone, birnessite-resorcinol, birnessite-catechol, and birnessite-pyrogallol systems, respectively. Evidently, the increase in CO_2 released, as light intensity increased from 10 to 500 $\mu E\ m^{-2}\ s^{-1}$, was largely due to the increase in temperature, imparting an associated thermal effect on the ring cleavage of the polyphenolics as catalyzed by birnessite. These data indicate that both photochemical and thermal energies contributed to the abiotic generation of CO_2 from the birnessite-polyphenol and polyphenol systems.

The release of CO_2 from the birnessite-polyphenol systems in the dark under an atmosphere of air at the end of a 90-h reaction period was 16 to 30 times higher than those from the polyphenol systems (Table 2). Similarly, the release of CO_2 from the birnessite-polyphenol systems in the dark under an atmosphere of O_2 was 13 to 18 times higher than those from the polyphenol systems (Table 3). Wang and Huang[24] have shown that birnessite is effective in promoting the abiotic generation of CO_2 through its ability to cleave the ring structure of pyrogallol under ambient temperature and pressure. Our results show that birnessite is effective in promoting the abiotic release of CO_2 from the polyphenols even in the dark. The amounts of CO_2 released from the birnessite-polyphenol under an N_2 atmosphere, however, were drastically reduced (Table 3), indicating the importance of the synergistic effect of birnessite and oxygen in promoting the release of CO_2 from the polyphenols. The birnessite used in this study was poorly crystalline to X-ray and has a specific surface of 246 ± 4 $m^2\ g^{-1}$. Thus, this oxide should have extensive exposed edge surfaces. These edges contain manganese of high oxidation numbers, such

Table 1 Examination of Growth of Microorganisms in the Birnessite-Polyphenol and Polyphenol Systems During a 15-d Incubation at 25°C

Birnessite	Polyphenol	Thimerosal[a]	Inoculum[b]	Growth of microorganisms in various media: GYEN[c]	AIA[d]	Czapek-Dox[e]
+[f]	Hydroquinone	+	–[g]	NG[h]	NG	NG
–	Hydroquinone	+	–	NG	NG	NG
+	Resorcinol	+	–	NG	NG	NG
–	Resorcinol	+	–	NG	NG	NG
+	Catechol	+	–	NG	NG	NG
–	Catechol	+	–	NG	NG	NG
+	Pyrogallol	+	–	NG	NG	NG
–	Pyrogallol	+	–	NG	NG	NG
–	—	–	+	G[i]	G	G

[a] Used as an antiseptic. It does not affect the oxidation process of phenolic compounds.[35]
[b] Centrifugate from the soil-water suspension (soil:water = 1:10) centrifuged at 1000 g.
[c] Agar for incubation of bacteria, medium-contained glucose, yeast extract, nutrient broth, and agar.
[d] Actinomycete isolation agar for incubation of actinomycetes.
[e] Czapek-Dox agar for incubation of fungi.
[f] In the presence.
[g] In the absence.
[h] No growth.
[i] Growth detected.

as Mn(III) and Mn(IV),[35] and hence are able to adsorb and polarize oxygen molecules, leading to the cleavage of the ring structure of polyphenols and the subsequent release of CO_2.

The data on the dissolved O_2 concentration of the systems provide additional evidence on the role of birnessite and O_2 in the ring cleavage of polyphenols (Figure 2).

Table 2 Release of CO_2 from the Birnessite-Polyphenol and Polyphenol Systems as Influenced by Different Light Intensities Under an Air Atmosphere at the End of a 90-h Reaction Period

Systems	Light intensity ($\mu E\ m^{-2}\ s^{-1}$): 0[a]	10[b]	300[c]	500[c]
	μmol[d]			
Birnessite + hydroquinone	299.8 (±1.7)	317.0 (±2.1)	351.3 (±1.3)	373.4 (±1.0)
Hydroquinone	10.0 (±1.7)	12.5 (±0.3)	13.3 (±0.4)	14.7 (±0.2)
Birnessite + resorcinol	320.7 (±2.4)	352.1 (±1.5)	364.7 (±1.1)	391.8 (±0.8)
Resorcinol	12.5 (±0.4)	15.6 (±1.1)	16.5 (±0.3)	17.4 (±0.6)
Birnessite + catechol	386.2 (±0.9)	417.9 (±2.0)	438.3 (±1.5)	460.1 (±0.8)
Catechol	17.9 (±0.4)	20.7 (±0.1)	21.8 (±0.2)	23.7 (±0.0)
Birnessite + pyrogallol	469.1 (±0.5)	481.5 (±0.9)	525.6 (±1.1)	565.0 (±1.4)
Pyrogallol	28.7 (±0.1)	30.5 (±0.3)	34.4 (±0.3)	36.8 (±0.5)

[a] In the dark.
[b] Under laboratory lighting; fluctuation = ±2 $\mu E\ m^{-2}\ s^{-1}$.
[c] In light chamber; fluctuations = ±25 $\mu E\ m^{-2}\ s^{-1}$.
[d] CO_2 released from systems (initial pH = 7.00) with and without 100 mg birnessite suspended in 30 ml sterilized distilled water which contained 0.2 g kg^{-1} thimerosal (antiseptic) and 0.5 mmol polyphenol.

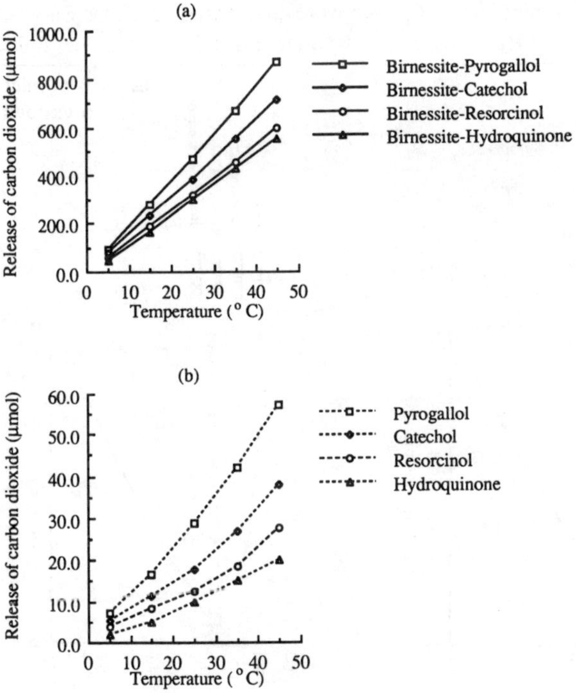

Figure 1 Release of CO_2 from (a) birnessite-polyphenol and (b) polyphenol systems as influenced by temperature under an air atmosphere in the dark at the end of a 90-h reaction period.

Table 3 Release of CO_2 from the Birnessite-Polyphenol and Polyphenol Systems Under N_2 and O_2 Atmospheres in the Dark and Ambient Conditions at 25°C at the End of a 90-h Reaction Period

System	Dark[a]		Ambient[b]	
	O_2[c]	N_2[d]	O_2	N_2
	μmol[e]			
Birnessite + hydroquinone	337.2 (±1.2)	7.0 (±0.3)	350.4 (±1.1)	8.9 (±1.0)
Hydroquinone	18.6 (±0.3)	0.0 (±0.0)	20.3 (±0.1)	0.0 (±0.0)
Birnessite + resorcinol	368.5 (±1.0)	8.3 (±0.3)	387.9 (±1.0)	11.7 (±0.1)
Resorcinol	21.7 (±0.3)	0.0 (±0.0)	24.4 (±0.4)	0.0 (±0.0)
Birnessite + catechol	413.4 (±1.5)	11.6 (±0.4)	445.0 (±0.8)	13.3 (±0.4)
Catechol	26.6 (±0.2)	0.0 (±0.0)	29.2 (±0.3)	0.0 (±0.0)
Birnessite + pyrogallol	497.9 (±1.0)	17.5 (±0.3)	520.4 (±1.7)	21.5 (±0.1)
Pyrogallol	37.0 (±0.4)	0.0 (±0.0)	40.3 (±0.5)	0.0 (±0.0)

[a] Light intensity = 0 $\mu E\ m^{-2}\ s^{-1}$.

[b] Under laboratory lighting, light intensity = 10±2 $\mu E\ m^{-2}\ s^{-1}$.

[c] Systems flushed with 99.998% (purity) O_2 gas.

[d] Systems flushed with 99.99% (purity) N_2 gas.

[e] CO_2 released from systems (initial pH = 7.00) with and without 100 mg birnessite suspended in 30 ml sterilized distilled water which contained 0.2 g kg^{-1} thimerosal (antiseptic) and 0.5 mmol polyphenol.

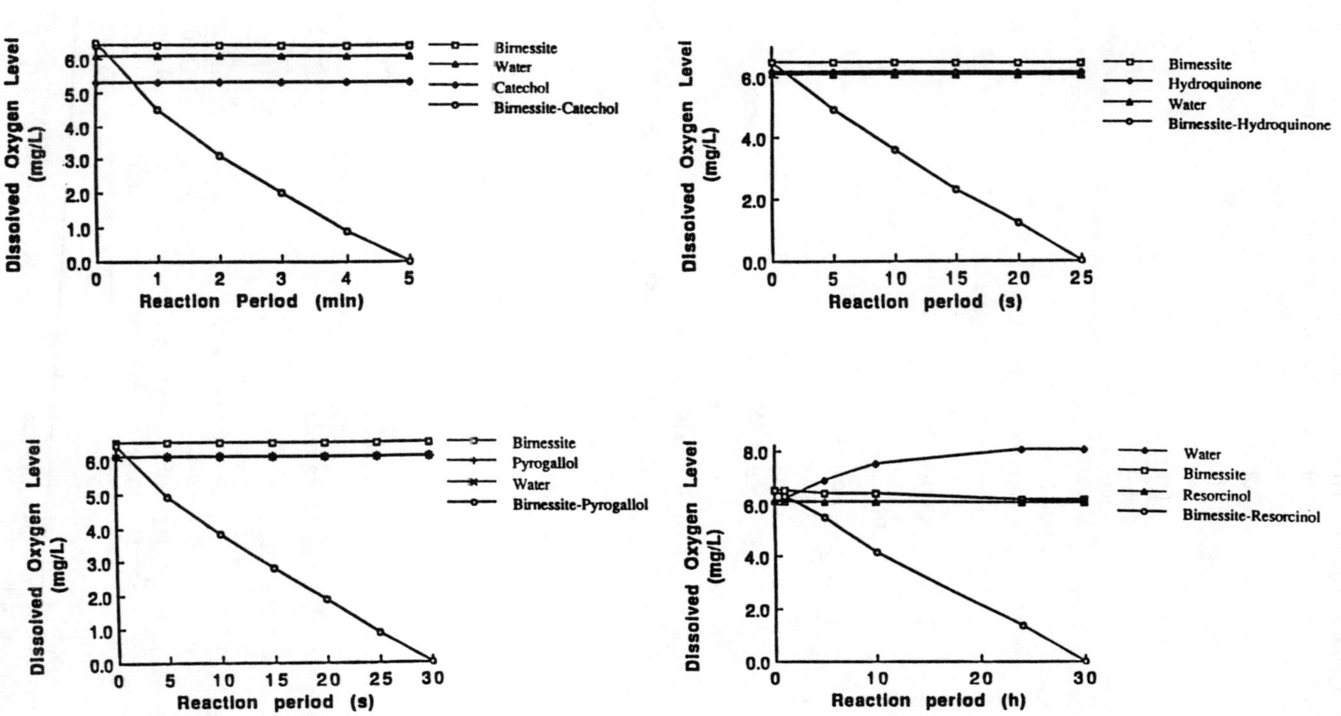

Figure 2 Dissolved O_2 level in the birnessite, birnessite-polyphenol, polyphenol, and sterilized deionized distilled water systems under an air atmosphere at a temperature of 25°C in the dark as a function of time.

The dissolved O_2 concentration of the systems after 15 min of N_2 flushing was found to be 0.0 to 0.1 mg l^{-1}, which indicated that only a very trace amount of dissolved O_2 remained in the systems. The dissolved O_2 content of sterilized deionized distilled water (containing thimerosal) increased from 6.1 to 8.2 mg l^{-1} in 24 h and remained at this level for the rest of the 90-h reaction period (Figure 2d), indicating dissolution of O_2 as time elapsed. The polyphenol solution and birnessite suspension both showed a decline in dissolved O_2 level at the end of the reaction (Figures 2a to d). The slight decrease in dissolved O_2 level in the polyphenol systems indicates some consumption of O_2 as the autooxidative reaction of the polyphenols occurred. The greater decline in dissolved O_2 level in the birnessite system was apparently due to adsorption of O_2 by the edges of birnessite. The complete depletion of dissolved O_2 in the birnessite-polyphenol systems in the reaction period ranging from 25 s to 30 h (Figures 2a to d) was reached before the end of the 90-h reaction period. In addition, the release of CO_2 from the systems continued to proceed even after the dissolved O_2 was depleted (Figure 3). These two situations appear contradictory to the hypothesis that the synergistic effect of birnessite and O_2 is crucial in the ring cleavage of polyphenols. However, an estimate of the amount of O_2 in the air-space above the suspension showed that there was ample O_2 available in the system. The initial amounts of dissolved O_2 present in the birnessite-polyphenol systems under O_2 and air atmospheres in the dark were only 0.15 and 0.04%, respectively, of the amount of O_2 in the airspace above the suspension. Hence, the additional amounts of O_2, enabling continual release of CO_2 even after the dissolved O_2 was depleted, must have derived from the airspace above the suspension. These gaseous O_2 molecules apparently dissolved continuously in the suspension, and were adsorbed and polarized by the edges of birnessite so rapidly that the O_2 probe was not able to detect the trace level of O_2 molecules in the solution phase. McBride[37] measures dissolved O_2, during the oxidation of polyphenols by Mn oxides, to calculate the rate of O_2 consumption. Our studies indicate that measurement of dissolved O_2 in such systems may not provide an accurate scenario of O_2 consumption. Atmospheric oxygen may continuously diffuse into the aqueous phase and may be rapidly adsorbed on the surfaces of Mn oxides without being detected as dissolved O_2.

The sequence of CO_2 released from the birnessite-polyphenol systems at the end of a 90-h reaction period followed the order birnessite-pyrogallol > birnessite-catechol > birnessite-resorcinol > birnessite-hydroquinone (Tables 2 and 3). This is apparently due to the number and position of the hydroxy groups of the polyphenols. Pyrogallol (1,2,3-trihydroxybenzene) has three consecutive hydroxyls and hence should be most conducive to ring cleavage. Catechol (1,2-dihydroxybenzene), which has two hydroxyls in the *ortho* position, should be more easily cleaved than resorcinol (1,3-dihydroxybenzene), with two *meta* hydroxyls, which is in turn more easily cleaved than hydroquinone (1,4-dihydroxybenzene), with *para* hydroxyls.

When light intensity was increased from 0 (dark) to 500 µE m^{-2} s^{-1}, there was a consistent decrease in the absorbances at 472 and 664 nm of the supernatants in the birnessite-polyphenol systems under an atmosphere of air at the end of a 90-h reaction period (Tables 4 and 5). However, the absorbances increased when the temperature was raised from 5 to 45°C (Figure 4), indicating the effect of thermal energy in the oxidative degradation of polyphenols as catalyzed by birnessite. Thus, thermal degradation of polyphenols in the presence of birnessite gave rise to higher absorbances, which are opposite to the photochemical degradation effect. Despite an increase in temperature associated with increasing light intensity and an increase in absorbance as temperature increased from 5 to 45°C, the decrease in absorbance as light intensity increased from 0 to 500 µE m^{-2} s^{-1} must be attributed to the effect of photofragmentation of the polyphenolics and their degradation products.

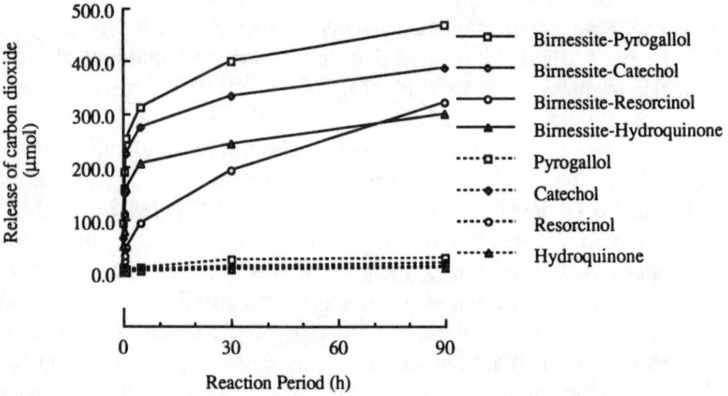

Figure 3 Release of CO_2 in birnessite-polyphenol and polyphenol systems under an air atmosphere at a temperature of 25°C in the dark as a function of reaction period.

Table 4 Absorbances at 472 nm of the Supernatants in the Birnessite-Polyphenol and Polyphenol Systems as Influenced by Different Light Intensities Under an Air Atmosphere at the End of a 90-h Reaction Period

	Light intensity (µE m^{-2} s^{-1})			
Systems	**0**[a]	**10**[b]	**300**[c]	**500**[c]
Birnessite + hydroquinone	19.25 (±0.12)	6.94 (±0.08)	3.10 (±0.03)	2.22 (±0.05)
Hydroquinone	0.04 (±0.02)	0.07 (±0.00)	0.09 (±0.01)	0.11 (±0.00)
Birnessite + resorcinol	16.19 (±0.05)	5.58 (±0.02)	2.49 (±0.08)	1.70 (±0.04)
Resorcinol	0.04 (±0.00)	0.08 (±0.02)	0.10 (±0.00)	0.13 (±0.00)
Birnessite + catechol	15.21 (±0.04)	4.87 (±0.01)	2.26 (±0.03)	1.53 (±0.06)
Catechol	0.09 (±0.00)	0.11 (±0.02)	0.14 (±0.01)	0.17 (±0.01)
Birnessite + pyrogallol	12.14 (±0.06)	3.10 (±0.02)	1.09 (±0.04)	0.42 (±0.04)
Pyrogallol	0.12 (±0.00)	0.18 (±0.02)	0.21 (±0.03)	0.27 (±0.05)

[a] In the dark.
[b] Under laboratory lighting; fluctuation = ±2 µE m^{-2} s^{-1}.
[c] In light chamber; fluctuations = ±25 µE m^{-2} s^{-1}.

Table 5 Absorbances at 664 nm of the Supernatants in the Birnessite-Polyphenol and Polyphenol Systems as Influenced by Different Light Intensities Under an Air Atmosphere at the End of a 90-h Reaction Period

	Light intensity (µE m^{-2} s^{-1})			
Systems	**0**[a]	**10**[b]	**300**[c]	**500**[c]
Birnessite + hydroquinone	5.99 (±0.07)	2.01 (±0.06)	1.22 (±0.04)	0.79 (±0.03)
Hydroquinone	0.01 (±0.00)	0.01 (±0.00)	0.03 (±0.01)	0.06 (±0.01)
Birnessite + resorcinol	4.82 (±0.02)	1.24 (±0.02)	0.89 (±0.08)	0.51 (±0.07)
Resorcinol	0.01 (±0.00)	0.03 (±0.00)	0.05 (±0.00)	0.08 (±0.01)
Birnessite + catechol	4.07 (±0.03)	0.86 (±0.01)	0.55 (±0.05)	0.31 (±0.03)
Catechol	0.02 (±0.00)	0.04 (±0.01)	0.06 (±0.00)	0.10 (±0.01)
Birnessite + pyrogallol	3.08 (±0.03)	0.32 (±0.03)	0.25 (±0.01)	0.19 (±0.02)
Pyrogallol	0.02 (±0.00)	0.06 (±0.01)	0.08 (±0.00)	0.12 (±0.00)

[a] In the dark.
[b] Under laboratory lighting; fluctuation = ±2 µE m^{-2} s^{-1}.
[c] In light chamber; fluctuations = ±25 µE m^{-2} s^{-1}.

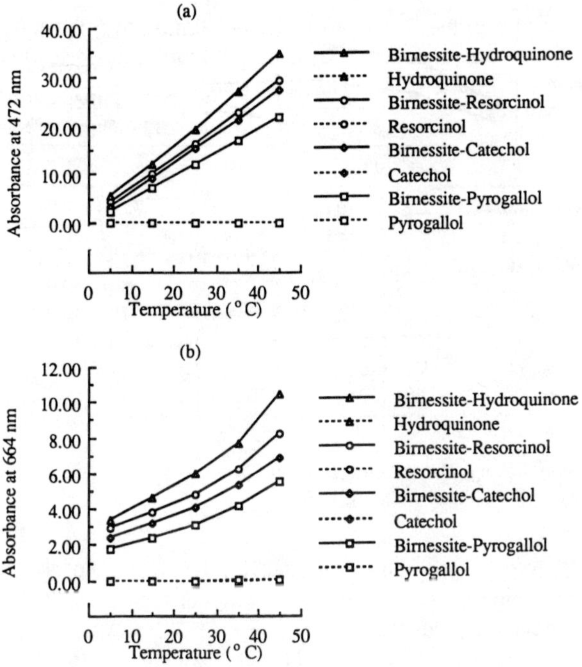

Figure 4 Absorbances at (a) 472 and (b) 664 nm of the supernatants in the birnessite-polyphenol and polyphenol systems as influenced by temperature under an air atmosphere in the dark.

The final pH values of the suspensions in the birnessite-polyphenol systems were higher than their initial pH value of 7.00 (Table 6). This observation, which is in agreement with the findings by Shindo and Huang,[22,23] is attributed to the consumption of protons, as birnessite was being reduced, which overrode the generation of protons during the oxidation of the polyphenols. As the light intensity increased from 0 to 500 μE m^{-2} s^{-1}, the final pH values decreased. This is in agreement with the absorbance data (Table 4). In the absence of birnessite, autooxidation of the polyphenols generated protons, which were not consumed and gave rise to a drop in final pH (Table 6). The final pH of the polyphenol systems in the dark, however, remained virtually the same as their initial pH of 7.00. This was attributed to the relatively little autooxidation, as indicated by absorbance (Table 4), of the polyphenols and hence less production of protons. Thus, pH changes were not evident. Nevertheless, as light intensity increased from 0 to 500 μE m^{-2} s^{-1}, autooxidation became substantial and so also the production of protons, thus resulting in a persistent drop in final pH.

Our data indicated that at least two simultaneous but different processes were responsible for the transformations of the polyphenolics as catalyzed by Mn(IV) oxide in both the presence and absence of light. The abiotic ring cleavage of the polyphenolics, resulting in the evolution of CO_2, was enhanced thermally and photochemically. However, thermal and photochemical degradation of the polyphenolics also contributed to the opposite effect, as seen in the absorbances. The thermal effect enhanced the polymerization of the polyphenolics and polycondensation of their degradation products, whereas the photochemical effect promoted fragmentation of the polyphenolics and their transformation products. The data showed that the photochemical effect did promote the release of CO_2, but, more significantly, it contributed to the fragmentation of the polyphenolics.

Table 6 Final pHs of the Suspensions in the Birnessite-Polyphenol and Polyphenol Systems as Influenced by Different Light Intensities Under an Air Atmosphere at the End of a 90-h Reaction Period

Systems	Light intensity ($\mu E\ m^{-2}\ s^{-1}$)			
	0[a]	10[b]	300[c]	500[c]
Birnessite + hydroquinone	8.31 (±0.01)	7.65 (±0.03)	7.54 (±0.04)	7.41 (±0.02)
Hydroquinone	7.05 (±0.06)	5.97 (±0.02)	5.66 (±0.07)	5.37 (±0.03)
Birnessite + resorcinol	8.56 (±0.01)	7.82 (±0.01)	7.61 (±0.04)	7.50 (±0.01)
Resorcinol	6.97 (±0.02)	5.46 (±0.03)	5.14 (±0.02)	4.89 (±0.05)
Birnessite + catechol	8.61 (±0.02)	7.99 (±0.04)	7.79 (±0.04)	7.64 (±0.03)
Catechol	7.00 (±0.03)	5.01 (±0.01)	4.72 (±0.06)	4.48 (±0.03)
Birnessite + pyrogallol	8.97 (±0.02)	8.16 (±0.03)	8.03 (±0.02)	7.87 (±0.04)
Pyrogallol	7.01 (±0.03)	4.53 (±0.01)	4.35 (±0.02)	4.11 (±0.03)

[a] In the dark.
[b] Under laboratory lighting; fluctuation = ±2 $\mu E\ m^{-2}\ s^{-1}$.
[c] In light chamber; fluctuations = ±25 $\mu E\ m^{-2}\ s^{-1}$.

This was attributed to the photochemical effect on the cleavage of the ring structure of the polyphenolics by the adsorbed O_2 on birnessite, which gave rise to some CO_2 release but primarily fragmented products.

The effects of light and temperature on the metal oxide may differ in terms of the ease with which the electrons deposited on the oxide surface can be transferred from the exterior to the interior of the oxide. This is important because the ability of the surface metal centers to accept electrons, and hence the ability to oxidize organics, will be dictated by the rate at which these metals are reduced and solubilized. In addition, the effects of light and temperature may influence the ease with which electrons can be transferred from the organics to the oxide, which is governed by the ease with which the electrons can be dislodged from the parent organics. A mathematical model which describes the competing photochemical and thermal effects may provide further elucidation of the processes and may shed light on the transformations of other organics, such as polynuclear aromatic compounds. The fate and transformations of the organics may depend on the competing effects of ring cleavage and polymerization. On the one hand, the edges of birnessite have the ability to adsorb and polarize O_2 molecules, which can in turn catalyze the ring cleavage of the organics. On the other hand, birnessite can catalyze the formation of polyphenol free radicals, which can undergo polymerization leading to the formation of humic polymers. Moreover, fragments produced from the ring cleavage reaction may subsequently undergo polycondensation and polymerization. The stoichiometry of the reactions involved in the systems warrants in-depth studies. Use of specifically labeled ring carbon compounds could provide insights into the mechanism of the ring cleavage reaction and hence the relative extent of ring cleavage and polymerization.

These findings indicate that the photochemical and thermal effects on the abiotic degradation of polyphenolics via the inorganic components in soils and sediments merit attention. The findings also imply that the pathway of C turnover through the abiotic degradation of polyphenolics in the photic zones of soils and aquatic environments, where sunlight and oxygen abound, may differ from those in the subsurface and submerged layers of soils and aquatic sediments, where sunlight does not penetrate and oxygen is depleted. Global and geographical variations in climatic patterns, ranging from those of the tropics to those of the temperate regions, and seasonal temperature fluctuations, ranging from summer to winter, may also impart a substantial influence on the humification process via the abiotic ring cleavage and associated degradative transformations of natural and xenobiotic polyphenolics as promoted by the inorganic components, such as Mn(IV) oxides, in terrestrial and aquatic environments. Furthermore, the photochemical effect on abiotic-biotic interactions in the transformations of polyphenolics certainly merits attention.

IV. SUMMARY AND CONCLUSIONS

Mn(IV) oxides have been shown to have unprecedented abilities in promoting the abiotic ring cleavage of pyrogallol, resulting in the release of CO_2, and degradation of polyphenolics (hydroquinone, resorcinol, catechol, and pyrogallol), leading to the subsequent formation of humic polymers under ambient temperature and pressure. However, the role of light in the abiotic transformations of polyphenolics as catalyzed by Mn(IV) oxides remains unknown. The objective of our study was to investigate the significance of photochemical and associated thermal effects on the abiotic degradation of four polyphenols — hydroquinone, resorcinol, catechol, and pyrogallol — in the presence of birnessite (δ-MnO_2), one of the most common Mn(IV) oxides in soils and sediments. Our data showed that the abiotic release of CO_2 in the birnessite-polyphenol and polyphenol systems increased as light intensity was increased from 0 to 500 $\mu E\ m^{-2}\ s^{-1}$ and as temperature was increased from 5 to 45°C at a light intensity of 0 $\mu E\ m^{-2}\ s^{-1}$ (dark). The synergistic effect of birnessite and O_2 in the abiotic ring cleavage of polyphenols was implicitly evident in the CO_2 release and dissolved O_2 data in the systems under N_2, air, and O_2 atmospheres. Contrary to the CO_2 release, the absorbances of the supernatants in the birnessite-polyphenol and polyphenol systems decreased as the light intensity was increased from 0 to 500 $\mu E\ m^{-2}\ s^{-1}$, despite an increase in absorbance as the temperature was increased from 5 to 45°C in the dark. This was probably a consequence of photofragmentation of the polyphenolics and the nature of their degradation products. These findings indicate that the photochemical and thermal effects on the abiotic degradation of polyphenolics via the inorganic components in soils and sediments merit attention. The findings also imply that the pathway of C turnover through the abiotic degradation of polyphenolics in the photic zones of soils and aquatic environments, where sunlight and oxygen abound, may differ from those in the subsurface and submerged layers of soils and aquatic sediments, where sunlight does not penetrate and oxygen is depleted. Furthermore, global and geographical variations in climatic patterns, ranging from those of the tropics to those of the temperate regions, and seasonal temperature fluctuations, ranging from summer to winter, may impart a substantial influence on the humification process via the abiotic ring cleavage and associated degradative transformations of natural and xenobiotic polyphenolics as promoted by the inorganic components, such as Mn(IV) oxides, in terrestrial and aquatic environments.

REFERENCES

1. **Coulson, C. B., Davies, R. I., and Lewis, D. A.,** Polyphenols in plant, humus and soil. I. Polyphenols of leaves, litter, and superficial humus from mull and mor sites, *J. Soil Sci.,* 11, 20, 1960.
2. **Coulson, C. B., Davies, R. I., and Lewis, D. A.,** Polyphenols in plant, humus and soil. II. Reduction and transport by polyphenols of iron in model soil columns, *J. Soil Sci.,* 11, 30, 1960.
3. **Henderson, M. E. K.,** Release of aromatic compounds from birch and spruce sawdusts during decomposition by white-rot fungi, *Nature (London),* 175, 634, 1955.
4. **Jensen, V.,** Decomposition of angiosperm tree leaf litter, in *Biology of Plant Litter Decomposition,* Vol. 1, Dickinson, C. H. and Pugh, G. J. G., Eds., Academic Press, New York, 1974, chap. 3.
5. **Martin, J. P. and Haider, K.,** in *Lignin Biodegradation: Microbiology, Chemistry, and Potential Applications,* Kirk, T. K., Higuchi, T., and Chang, H., Eds., CRC Press, Boca Raton, FL, 1980, chap. 4.

6. **Wang, T. S. C., Huang, P. M., Chou, C.-H., and Chen, J.-H.,** The role of soil minerals in the abiotic polymerization of phenolic compounds and formation of humic substances, in *Interactions of Soil Minerals with Natural Organics and Microbes,* Huang, P. M. and Schnitzer, M., Eds., SSSA Spec. Publ. No. 17, Madison, WI., 1986, chap. 8.
7. **Whitehead, D. C., Dibb, H., and Hentley, R. D.,** Bound phenolic compounds in water extracts of soil, plant roots and leaf litter, *Soil Biol. Biochem.,* 15, 133, 1983.
8. **Cranwell, P. A.,** Environmental organic chemistry of rivers and lakes, both water and sediment, in *Environmental Chemistry,* Vol. 1, Eglinton, G., Ed., Chemical Society, London, 1975, chap. 2.
9. **Larson, R. A. and Hufnal, J. M., Jr.,** Oxidative polymerization of dissolved phenols by soluble and insoluble inorganic species, *Limnol. Oceanogr.,* 25, 505, 1980.
10. **Degens, E. T.,** Diagenesis of organic matter, in *Diagenesis in Sediments,* Larsen, G. and Chilingar, G. V., Eds., Elsevier, Amsterdam, 1967, chap. 7.
11. **Brown, F. S., Baedecker, M. J., Nissenbaum, A., and Kaplan, I. R.,** Early diagenesis in a reducing fjord, Saanich Inlet, British Columbia. III. Changes in organic constituents of sediment, *Geochim. Cosmochim. Acta,* 36, 1185, 1972.
12. **Wangersky, P. J.,** The surface film as a physical environment, *Annu. Rev. Ecol. Syst.,* 7, 161, 1976.
13. **Flaig, W., Beutelspacher, H., and Rietz, E.,** Chemical composition and physical properties of humic substances, in *Soil Components. Vol. 1, Organic Components,* Gieselking, J. E., Ed., Springer-Verlag, New York, 1975, chap. 1.
14. **Hayes, M. H. B. and Swift, R. S.,** The chemistry of soil organic colloids, in *The Chemistry of Soil Constituents,* Greenland, D. J. and Hayes, M. H. B., Eds., John Wiley & Sons, New York, 1978, chap. 3.
15. **Kononova, M. M.,** *Soil Organic Matter,* Pergamon Press, New York, 1966.
16. **McKeague, J. A., Damman, A. W. H., and Heringa, P. K.,** Iron, manganese and other pans in some soils of Newfoundland, *Can. J. Soil Sci.,* 48, 243, 1968.
17. **McKenzie, R. M.,** The synthesis of birnessite, cryptomelane, and some other oxides and hydroxides of manganese, *Mineral. Mag.,* 38, 493, 1971.
18. **McKenzie, R. M.,** Manganese oxides and hydroxides, in *Minerals in Soil Environments,* Dixon, J. B. and Weeds, S. B., Eds., Soil Science Society of America, Madison, WI, 1989, chap. 9.
19. **Santos, M. C. D., Mermut, A. R., Anderson, D. W., and St. Arnold, R. J.,** Micromorphology of three Gray Luvisols in east-central Saskatchewan, *Can. J. Soil Sci.,* 65, 717, 1985.
20. **Taylor, R. M.,** The association of manganese and cobalt in soils—further observations, *J. Soil Sci.,* 19, 77, 1968.
21. **Bartlett, J. R.,** Soil redox behavior, in *Soil Physical Chemistry,* Sparks, D. L., Ed., CRC Press, Boca Raton, FL, 1986, chap. 7.
22. **Shindo, H. and Huang, P. M.,** Role of Mn (IV) oxide in abiotic formation of humic substances in the environment, *Nature (London),* 298, 363, 1982.
23. **Shindo, H. and Huang, P. M.,** Catalytic effects of manganese (IV), iron (III), aluminum, and silicon oxides on the formation of phenolic polymers, *Soil Sci. Soc. Am. J.,* 48, 927, 1984.
24. **Wang, M. C. and Huang, P. M.,** Significance of Mn (IV) oxide in the abiotic ring cleavage of polyphenol in natural environments, *Sci. Total Environ.,* 113, 147, 1992.
25. **Burges, N. A., Hurst, H. M., and Walkden, B.,** The phenolic constituents of humic acid and their relation to the lignin of the plant cover, *Geochim. Cosmochim. Acta,* 28, 1547, 1964.
26. **Haider, K., Martin, J. P., and Filip, Z.,** Humus biochemistry, in *Soil Biochemistry,* Vol. 4, Paul, E. A. and McLaren, A. D., Eds., Marcel Dekker, New York, 1975, chap. 6.
27. **Schnitzer, M. and Khan, S. U.,** *Humic Substances in the Environment,* Marcel Dekker, New York, 1972.
28. **Harborne, J. B. and Simmonds, N. W.,** The natural distribution of the phenolic aglycones, in *Biochemistry of Phenolic Compounds,* Harborne, J. B., Ed., Academic Press, London, 1964, chap. 2.
29. **Eltantawy, I. W. and Arnold, P. W.,** Reappraisal of ethylene glycol monoethyl ether (EGME) method for surface area estimation of clays, *J. Soil Sci.,* 24, 232, 1973.
30. **Wang, M. C. and Huang, P. M.,** Abiotic ring cleavage of pyrogallol and the associated reactions as catalyzed by a natural soil, *Sci. Total Environ.,* 81, 501, 1989.
31. **Umbreit, W. W., Burris, R. H., and Stauffer, J. F.,** *Manometric Techniques,* Burgess Publishing, Minneapolis, MN, 1964, 305.

32. **Sunda, W. G., Huntsman, S. A., and Harvey, G. R.,** Photoreduction of manganese oxides in seawater and its geochemical and biological implications, *Nature (London),* 301, 234, 1983.
33. **Germida, J. J. and Casida, L. E., Jr.,** Myceloid growth of *Arthrobacter globiformis* and other *Arthrobacter* species, *J. Bacteriol.,* 144, 1152, 1980.
34. **Olson, B. M., McKercher, R. B., and Germida, J. J.,** Microbial population in trifluralin-treated soil, *Plant Soil,* 76, 379, 1984.
35. **Wang, T. S. C., Wang, M. C., Ferng, Y. L., and Huang, P. M.,** Catalytic synthesis of humic substances by natural clays, silts, and soils, *Soil Sci.,* 135, 350, 1983.
36. **Giovanoli, R., Staehli, E., and Feitknecht, W.,** Uber oxihydroxide des vierwertigen mangans mit schichtengitter 2. Mangan (III)-manganat (IV), *Helv. Chim. Acta,* 53, 453, 1970.
37. **McBride, M. B.,** Adsorption and oxidation of phenolic compounds by iron and manganese oxides, *Soil Sci. Soc. Am. J.,* 51, 1466, 1987.

Chapter 14

Potential of the Supercritical Fluid Extraction Technique for Characterizing Organic-Inorganic Interactions in Soils

H. H. Cheng, J. Gan, W. C. Koskinen, and L. J. Jarvis

CONTENTS

I. INTRODUCTION

Soils have different capacities to affect the processes governing the fate of a chemical in the environment, and thus are capable of modifying the potential impact of the chemical as an environment-degrading pollutant. The major processes involved can be termed retention, transformation, and transport.[1] Depending upon the properties of the soils and chemicals involved and the environmental conditions, each process could interact with the other processes and affect the process kinetics. For instance, the retention process can limit the ability of soil microorganisms to biotransform a chemical and retard its transport by water in the soil.

The retention process is governed by the mechanisms that bind the chemical to the soil matrix. Therefore, an understanding of the interactions between these organic chemicals and the inorganic soil particle surfaces is of particular importance in assessing the fate and impact of the organic chemicals on environmental quality. However, little is known about the nature of the specific bonds involved in the binding of any organic chemical on the soil mineral surfaces. The main obstacle to elucidating binding mechanisms appears to be a lack of appropriate tools to characterize the bonds.

The extractability of a chemical from soil is related to the ability of an extracting agent, e.g., organic or aqueous solvents or mechanical force, to break the bonds. Therefore, it should be possible to elucidate the forces involved in organic-inorganic interactions from a determination of the effectiveness of the extraction method. For instance, Cheng[2] has shown that the effectiveness of certain solvent extraction methods can be associated with the ability of these solvents to break ionic, hydrogen, or hydrolytic bonds.

Supercritical fluid extraction (SFE) was initially developed for large-scale chemical processing applications to extract a particular component from a solid matrix, such as the extraction of caffeine from coffee. SFE is increasingly used in environmental analysis for the extraction of organic pollutants from environmental solids. It has advantages over the

0-87371-914-X/95/$0.00+$.50

commonly used extraction by organic solvents. Supercritical fluids (SFs) have much better mass transfer characteristics than liquid solvents, and they have solute diffusivities an order of magnitude higher and viscosities an order of magnitude lower than the liquid solvents. Solvent strength of the SF can be controlled; it depends on the pressure and temperature used for extraction. An SF's solvent strength increases with increasing density. SFs normally are gases under ambient temperature and pressure; thus, they can also be easily evaporated to facilitate the analyte concentration steps after the extraction. Moreover, SFs are relatively inert and nontoxic, and have little cost associated with the disposal of waste solvents.

SFE is basically a nonselective extraction method, similar to the use of organic solvents. However, extraction selectivity can be obtained by modifying SFE parameters. Commonly used SFs have a range of Hildebrand solubility parameters that can be used for selective solubilization. Differences in characteristics between SFs with similar Hildebrand solubility parameters can also be exploited for selectivity. For instance, SF-CO_2 and SF-N_2O have similar Hildebrand solubility parameters, but the increased extractability of 2,3,7,8-tetrachlorodibenzo-*p*-dioxin (2,3,7,8-TCDD) by N_2O[3] is possibly related to its small permanent dipole moment, making it more effective than SF-CO_2 at displacing 2,3,7,8-TCDD from sorptive sites.

The solubility of a chemical in a nonpolar SF has been shown to increase markedly by the addition of a polar cosolvent. For instance, the 620% increase in the solubility of 2-aminobenzoic acid in SF-CO_2 by the addition of 3.5 mol% methanol to SF-CO_2 was attributed to the methanol forming strong hydrogen bonds with the solute.[4] Modifiers can also provide extraction selectivity. Ethanol was more effective than methanol added to SF-CO_2 in extracting linuron from soil, whereas methanol was a more effective modifier for diuron extraction from soil.[5] This could be due to either differences in solute solubility in SF or occupation of the soil matrix sorption sites by the modifier.

The potential of the SFE technique in breaking more recalcitrant bonds has not yet been explored fully. For instance, Lehmann et al.[6,7] have demonstrated that phenolic acids are readily oxidized in the presence of manganese dioxide (MnO_2) or soils, and the reaction products are strongly held on the matrix and are extremely difficult to extract. To test the potential of the SFE technique for elucidating organic-inorganic binding mechanisms, we determined the extraction efficiency of SFE in recovering either ferulic acid (FerA) or its reaction products from MnO_2, a highly reactive soil mineral. The objective of this study was to test the effectiveness of various operating modes of SFE in extracting FerA and its reaction products from a FerA-MnO_2 system.

II. MATERIALS AND METHODS

A. CHEMICALS

FerA (4-hydroxy-3-methoxycinnamic acid) and MnO_2 were purchased from Sigma Chemical Co. All the organic solvents used were either high-performance liquid chromatography (HPLC) or pesticide grade.

B. REACTION KINETICS OF FERA AND MnO_2

MnO_2 (0.1 or 1.0 g) was incubated with 3 ml of 100 mg l^{-1} aqueous FerA solution (molar ratio FerA/MnO_2 = 1.344×10^{-3} and 1.344×10^{-4}, respectively) in 7-ml aluminum foil-capped glass vials in the dark at 25.0 ± 0.2°C. At selected time intervals, the solution phase was analyzed for FerA by HPLC. The FerA solution (0.10 ml) was diluted to 1.0 ml with HPLC mobile phase and injected into the HPLC. A Hewlett Packard HP 1090® with a photodiode array detector was used for the analysis. The HPLC conditions were: Alltech Adsorbosphere HS C18 5-μm column (250 × 4.6 mm); methanol:1% acetic acid

(45:55, v:v) mobile phase; 1.0 ml min^{-1} flow rate; detection wavelength of 317 to 329 nm and injection volume of 100 μl.

C. EXTRACTION AND ANALYSIS OF FERA-MnO_2 REACTION PRODUCTS

MnO_2 (1.0 g) was incubated with 3 ml of 100 mg l^{-1} aqueous FerA solution (molar ratio FerA/MnO_2 = 1.344 × 10^{-4}) in aluminum foil-capped 7-ml glass vials in the dark at 25.0 ± 0.2°C for 5 d. HPLC analysis of the solution showed that less than 1% of the original ferulic acid remained unreacted. MnO_2 and solution were filtered with Whatman No. 1 filter paper, and MnO_2 residues were further rinsed with 2.0 ml of deionized water. The oxide residues were air dried and transferred to aluminum foil-capped 7-ml glass vials and stored at –20°C until extracted.

Nine different solvent systems, listed in Figures 2 and 3 (see Section III.B.), were used for extraction. The oxide complexes (1 g) were shaken mechanically overnight with 3.0 ml of each solvent system. Organic solvent extracts were filtered through 0.4-μm nylon syringe filters, evaporated to near dryness, and redissolved in 1.0 ml HPLC mobile phase. The NaOH extract was acidified with 1.0 *M* HCl to a pH of 2 to 3. HCl and acidified NaOH extracts were then saturated with NaCl and extracted with equal volumes of diethyl ether (× 2) and ethyl acetate (× 2) consecutively. The organic extracts were evaporated to near dryness and redissolved in 1.0 ml HPLC mobile phase.

The air-dried oxide complexes (1.0 g) were also extracted using a Hewlett Packard 7680A SFE Module® with supercritical CO_2 as the extracting fluid. Additional air-dried oxide complexes were rewet with 1 ml deionized water or methanol prior to extraction. The SFE pressure was 34.5 MPa, and the SFE temperature was varied from 45 to 120°C, with the corresponding SF CO_2 density ranging from 0.6 to 0.9 g ml^{-1}. The extracted chemicals were trapped on stainless steel balls or C18 ODS powder at 67°C, then redissolved in methanol. The methanol solution was evaporated to a small volume and then diluted to 1.0 ml with mobile phase for HPLC analysis.

The HPLC conditions were: mobile phase, 40% methanol-60% H_2O for 10 min, increasing to 90% methanol-10% H_2O for 20 min, and then decreasing to 40% methanol-60% H_2O for 25 min; detection wavelength of 279 to 291 nm and other conditions as described previously.

III. RESULTS AND DISCUSSION

A. REACTION KINETICS OF FERA AND MnO_2

The rate of loss of FerA from the solution increased with decreasing FerA/MnO_2 molar ratio (Figure 1). At the molar ratio FerA/MnO_2 = 1.334 × 10^{-4} (Figure 1B), the reaction could be depicted as following pseudo first-order kinetics; almost no FerA was detectable in the solution after 72 h. In contrast, at the higher ratio system FerA/MnO_2 = 1.334 × 10^{-3} (Figure 1A), the reaction slowed considerably after 72 h. During the first 72 h, 59.6% of FerA reacted, and only an additional 2.3% of FerA had reacted from 72 to 144 h. This evidence is in accordance with the postulation by Lehmann and Cheng[7] that oxidation of FerA takes place on the exposed surface of the MnO_2 mineral, and the oxidized products are sorbed onto the oxide surface, thus covering up the available sites for oxidation by MnO_2. Complexes formed on the oxide surface were found to be undesorbable in water. Thus, the reaction kinetics is not governed solely by the quantity of MnO_2 present, but also by the availability of exposed mineral surface for reaction.

B. SOLVENT EXTRACTABILITY OF FERA-MnO_2 REACTION PRODUCTS

HPLC analyses of the organic solvent extracts of FerA-MnO_2 reaction products are shown in Figure 2. Numerous peaks were detected in all extracts; some of the peaks had

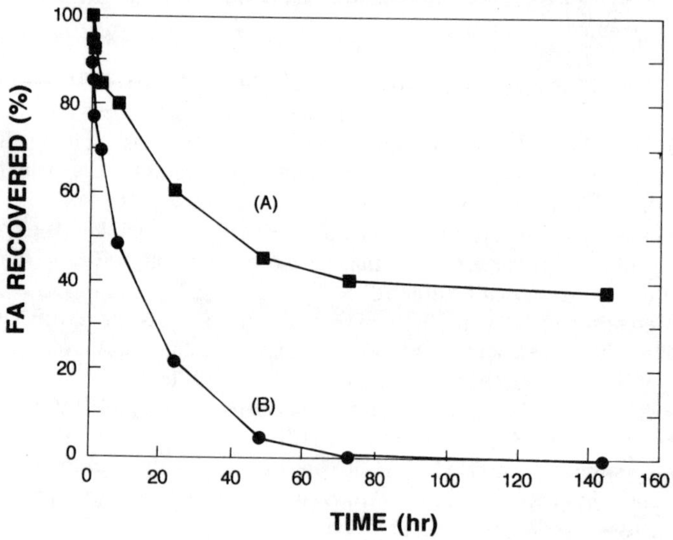

Figure 1 Reaction kinetics of FerA and MnO_2 at FerA/MnO_2 molar ratios. (A) 1.344×10^{-3}; (B) 1.344×10^{-4}.

shorter retention times, and others had longer retention times than that for FerA (3.8 min), suggesting that both oxidation of FerA and polymerization of the reaction products may have occurred. However, all of the organic solvent systems failed to extract any significant amount of the products from the samples, indicating that very strong bonds were involved. HCl and NaOH extracted significantly more chemical than the organic solvents (Figure 3) (note the difference in the axis scales of Figures 2 and 3). One large peak eluted before FerA with HCl extraction, and a group of large peaks eluted before FerA with NaOH extraction, indicating increased polarity of the extracted compounds. The UV spectra of the two peaks in the NaOH extracts were similar, but different from that of the main peak in the HCl extract (Figure 4).

It appears that extraction of the FerA-MnO_2 reaction products from FerA-MnO_2 samples with NaOH and HCl was more effective than with organic solvents even though no FerA was found in the NaOH and HCl extracts, while trace amount of FerA were always found in all of the organic solvent extracts. The extracted compounds bore little structural similarity to FerA, as evidenced by the UV spectra.

C. SFE EXTRACTABILITY OF FERA-MnO_2 REACTION PRODUCTS

Various temperatures and corresponding solvent densities were tested for SF extraction of FerA-MnO_2 reaction products with and without prewetting the sample with water or methanol. SFE was more effective in extraction of the reaction products that the organic solvents, but was comparable to NaOH and HCl. Generally, HPLC chromatograms of SFE extracts were similar to those (shown in Figure 3) of organic solvent extracts (i.e., SFE extracts also contained trace amounts of FerA). However, the SF extract of the FerA-MnO_2 reaction products using SF-CO_2 at 34.5 MPa and 120°C showed two large peaks with retention times shorter than FerA (Figure 3) but longer than those in NaOH and HCl extracts. UV spectra of the two large peaks showed absorption maxima at higher wavelengths than the peaks in NaOH and HCl extracts (Figure 4), but were similar to the FerA

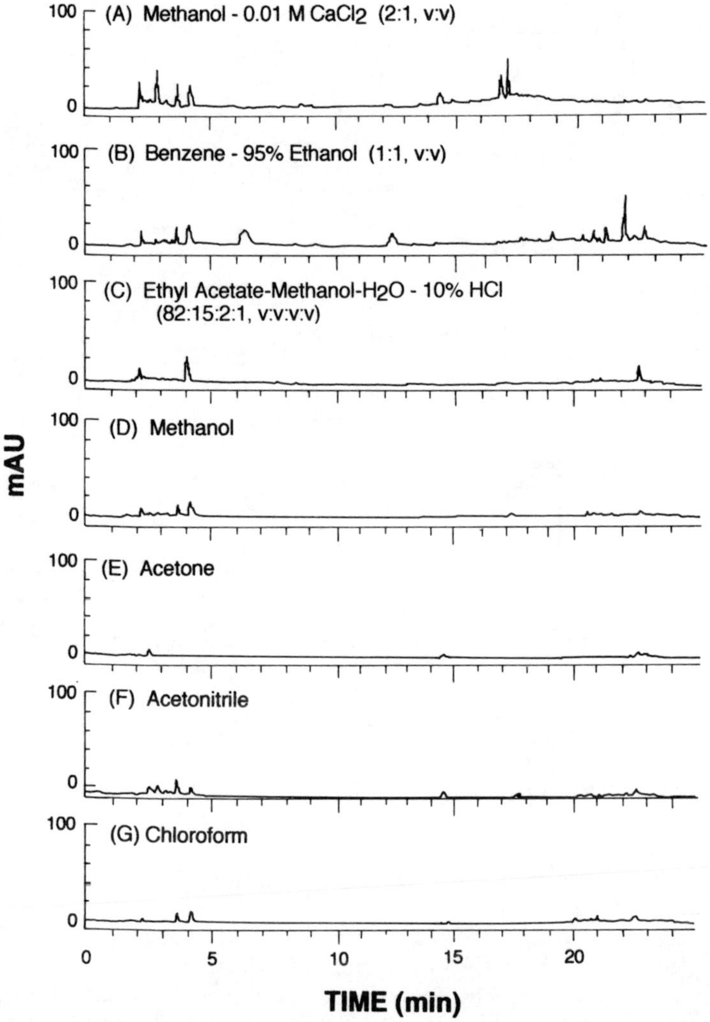

Figure 2 HPLC chromatograms of organic solvent extracts of FerA-MnO_2 reaction products. FerA retention time = 3.8 min.

UV spectra. It appears that either SF extracted different reactions products from the FerA-MnO_2 or was less destructive to the FerA-MnO_2 system than NaOH or HCl. The increased extractability at 120°C compared to lower temperatures can be attributed to increased solubilization of precipitated reaction products, an increased rate of desorption of the sorbed reaction products, or an increased rate of decomposition of a reaction product. Additional research will be necessary to elucidate the temperature effects further.

IV. SUMMARY AND CONCLUSIONS

The purpose of this study was to explore the potential of the SFE technique for extracting organic chemicals from the soil matrix. This study represents a preliminary effort toward the elucidation of organic-inorganic interactions in soils. Even though the

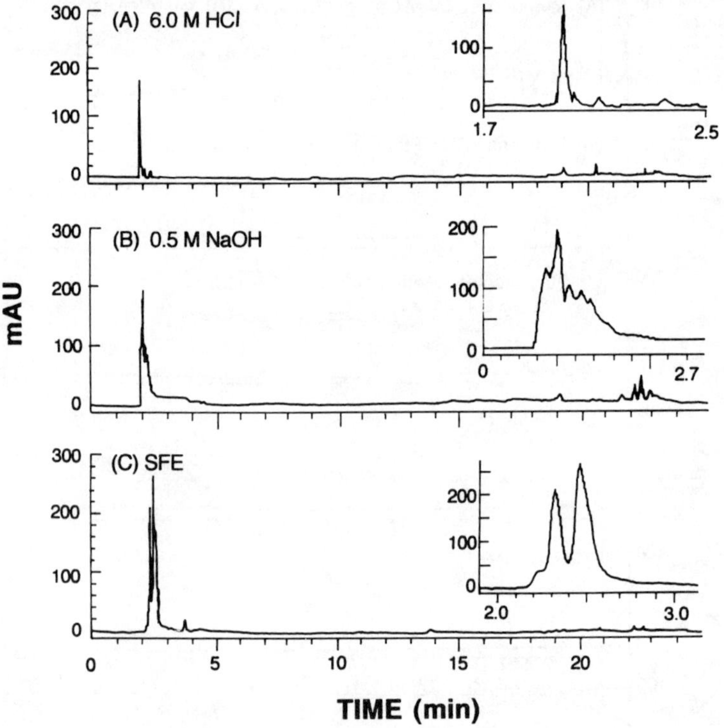

Figure 3 HPLC chromatograms of HCl, NaOH, and SFE extracts of FerA-MnO_2 reaction products. FerA retention time = 3.8 min.

ferulic acid-manganese oxide system is complex because of the rapid oxidation of FerA at the mineral surface and subsequent polymerization of the reaction products, SFE was shown to be more effective than organic solvent extraction and equal to acid or base extraction in recovering FerA and its reaction products. Based on retention times, number of peaks, and UV spectra, SFE extracts different products than NaOH or is less destructive during the extraction process. Further studies will be aimed at characterizing the nature of the chemical species recovered and elucidating the possible mechanisms of binding in this as well as other organic-inorganic systems.

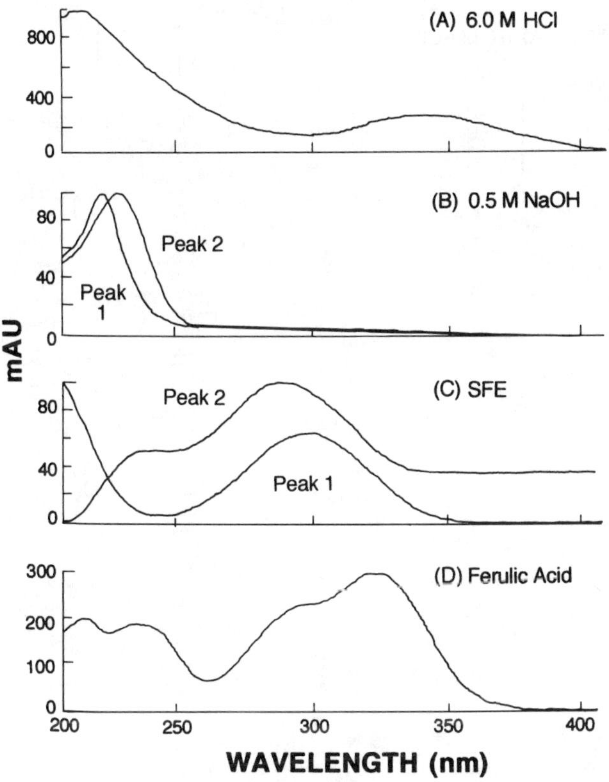

Figure 4 UV spectra of FerA and major peaks recovered from HCl, NaOH, and SFE extracts.

REFERENCES

1. **Cheng, H. H.,** Pesticides in the soil environment: an overview, in *Pesticides in the Soil Environment: Processes, Impacts, and Modeling,* Cheng, H. H., Ed., Soil Science Society of America, Madison, WI, 1990, 1.
2. **Cheng, H. H.,** Organic residues in soils: mechanisms of retention and extractability, *Int. J. Environ. Anal. Chem.,* 39, 165, 1990.
3. **Onsuka, F. I. and Terry, K. A.,** Supercritical fluid extraction of 2,3,7,8-tetrachlorodibenzo-*p*-dioxin from sediment samples, *J. High Resolut. Chromatogr.,* 12, 357, 1989.
4. **Dobbs, J. M., Wong, J. M., Lahiere, R. J., and Johnston, K. P.,** Modification of supercritical fluid phase behavior using polar cosolvents, *Ind. Eng. Chem. Res.,* 26, 56, 1987.
5. **Wheeler, J. R. and McNally, M. E. P.,** Supercritical fluid extraction and chromatography of representative agricultural products with capillary and microbore columns, *J. Chromatogr. Sci.,* 27, 534, 1989.
6. **Lehmann, R. G., Cheng, H. H., and Harsh, J. B.,** Oxidation of phenolic acids by soil iron in manganese oxides, *Soil Sci. Soc. Am. J.,* 51, 352, 1987.
7. **Lehmann, R. G., and Cheng, H. H.,** Reactivity of phenolic acids in soil and formation of oxidation products, *Soil Sci. Soc. Am. J.,* 52, 1304, 1988.

Chapter 15

Dissolution and Fractionation of Calcium-Bound and Iron- and Aluminum-Bound Humus in Soils

J.-M. Xu and K.-N. Yuan

CONTENTS

I. INTRODUCTION

Metallic ions play an important role in the formation and stabilization of soil organomineral complexes, mainly as bonding bridges between the humic substances and the clay minerals. The Ca-bound and sesquioxide-bound humus have been considered the most basic types of soil-bound humus. Tyulin[1] developed a fractional extraction method for separating them, which involved extractions of Ca-bound humus with 0.004 mol l^{-1} NaOH after a preliminary decalcification with 0.05 mol l^{-1} HCl, and then extractions of sesquioxide-bound humus with 0.01 mol l^{-1} NaOH following a treatment of the residue with 0.025 mol l^{-1} H_2SO_4. Kellerman[2] recommended the use of Tamm's reagent to remove active sesquioxides, and then extraction of the sesquioxide-bound humus with dilute NaOH. Kononova and Bel'Chikova[3] estimated the content of Ca-bound humus according to the difference between the humus extracted with 0.1 mol l^{-1} NaOH and that with 0.1 mol l^{-1} NaOH + 0.1 mol l^{-1} $Na_4P_2O_7$ mixture from two portions of the soil sample, respectively. Fu[4] suggested that soil-bound humus might be divided into three fractions: the loosely combined humus (H1) first extracted with 0.1 mol l^{-1} NaOH, the stably combined humus (H2) sequentially extracted with 0.1 mol l^{-1} NaOH + 0.1 mol l^{-1} $Na_4P_2O_7$ mixture from soil, and the tightly combined humus (H3) remaining in the soil residue. Recently, Xu[5] reported that the loosely combined humus was mainly associated with Fe and Al, and the stably combined humus mainly with Ca. Although the methods given above are really helpful in determining the stability of combination between the soil humus and mineral constituents, they are almost incapable of separating Ca-bound or Fe- and Al-bound humus from soils due to the complexity of the soil composition and associated humus forms. Therefore, we developed a method for separating Ca-bound and Fe- and Al-bound humus from soils on the basis of the dissolution of humus and metallic ions by various extractants from synthetic organomineral complexes and soils.

0-87371-914-X/95/$0.00+$.50

II. MATERIALS AND METHODS

A. SOIL SAMPLES

The nine soil samples used in this experiment were collected from surface horizons (0 to 20 cm) of the typical zonal soils in East China. Generally speaking, sesquioxides and 1:1 layer clay minerals such as kaolinite in soils showed a decreasing trend, whereas exchangeable bases and 2:1 layer clay minerals such as montmorillonite showed an increasing trend as one proceeds from south to north. Table 1 presents some basic characteristics of these soil samples.

B. SYNTHETIC ORGANOMINERAL COMPLEXES

The clay minerals used were montmorillonite (Mont) from Tonglu, Zhejiang Province, China, and kaolinite (Kaol) from Suzhou, Jiangsu Province, China. Samples were ground to pass a 200-mesh sieve. The organic material used was fresh milk vetch. It was washed with distilled water, oven dried at 105°C, and then passed through an 18-mesh sieve. In order to imitate the actual composition of the soils used, Ca-saturated Mont and Fe-saturated (mostly in the form of iron oxides) and Al-saturated Kaol were first prepared by the method outlined by Varadachari et al.[6] Then the wet Ca-Mont, Fe-Kaol, and Al-Kaol were mixed with organic material, separately, at a ratio of 2:1. Distilled water was added to obtain a paste which was incubated at a constant temperature of 25°C for about half a year. All the samples were dried at room temperature and powdered to pass through a 60-mesh sieve. The light fraction of organic matter was removed from these samples by an ethanol-bromoform mixture having a density of 1.8 g cm^{-3}, according to the procedure described by Fu,[4] leaving the synthetic organomineral complexes in the samples.

Table 1 Some Basic Characteristics of the Soils Used[a]

Sample no.	Soil name	Locality	pH	O.M. (g kg^{-1})	CEC (cmol(+) kg^{-1})	Exch Ca (cmol(+) kg^{-1})	Free Fe (g kg^{-1})	Free Al (g kg^{-1})
1	Chernozem	Heilongjiang (northeast)	6.98	34.3	37.15	31.24	9.27	1.95
2	Black soil	Heilongjiang (northeast)	6.62	48.0	35.35	28.88	10.92	2.40
3	Brown soil	Shandong (east)	7.70	30.4	10.83	11.87	6.43	1.32
4	Yellow-brown soil	Jiangsu (east)	6.55	16.7	20.05	16.79	21.29	4.00
5	Red soil	Zhejiang (southeast)	5.51	8.5	12.42	6.06	29.92	4.73
6	Red soil	Jiangxi (southeast)	5.16	27.2	9.80	3.52	17.79	6.00
7	Red soil	Jiangxi (southeast)	4.63	9.3	10.76	0.13	31.19	3.57
8	Lateritic red soil	Guangdong (south)	4.01	37.6	13.21	0.08	38.89	12.25
9	Latosol	Hainan (south)	4.82	36.9	8.48	0.67	49.52	8.53

[a] Free Fe and Al were determined by ICP, exchangeable Ca by the 1 mol l^{-1} neutral NH_4OAc method, and other items by the routine methods.

The organomineral complexes originated from Ca-Mont, Fe-Kaol, and Al-Kaol are referred to as Ca complex, Fe complex, and Al complex, respectively, in this chapter.

C. METHODS

The synthetic organomineral complexes were used as test materials for the extraction of Ca, Fe, Al, and organic C by different reagents. The extractants employed were: 0.1 mol l^{-1} $Na_4P_2O_7$, 0.1 mol l^{-1} NaOH + 0.1 mol l^{-1} $Na_4P_2O_7$ mixture, 0.1 mol l^{-1} NaOH, 0.5 mol l^{-1} $(NaPO_3)_6$, and 0.5 mol l^{-1} neutral Na_2SO_4. Two-gram portions of the prepared synthetic samples were added to the same number of 100-ml centrifuge tubes, each containing 20 ml of different extracting solutions. The tubes were shaken for 1 h, allowed to stand overnight, and then centrifuged. The residues were again shaken with another 20 ml of the corresponding extracting solution.

The procedure was repeated until the extracts were almost free of Ca and organic C. All extracts were collected separately in volumetric flasks. Then the samples were washed with distilled water until almost no color of organic C could be observed in the washings. All the washings and extracts were combined. The content of organic C in the extracts was analyzed by dichromate oxidation with external heating, and the contents of Fe, Al, and Ca were determined using an inductively coupled plasma (ICP) 16500 XR made by Perkin-Elmer (U.S.).

The soil samples were sequentially extracted first by 0.5 mol l^{-1} Na_2SO_4 and then by 0.1 mol l^{-1} NaOH + 0.1 mol l^{-1} $Na_4P_2O_7$ mixture. The extraction with each reagent followed the procedure described above.

III. RESULTS AND DISCUSSION

A. DISSOLUTION OF SYNTHETIC ORGANOMINERAL COMPLEXES WITH DIFFERENT EXTRACTANTS

All the extractants employed were able to dissolve humus from different organomineral complexes, indicating that these extractants could destroy the metallic bridges between the humus and the clay minerals to varying degrees (Table 2). Among the five extractants, the 0.1 mol l^{-1} NaOH + 0.1 mol l^{-1} $Na_4P_2O_7$ mixture was the most efficient in the extraction of C due to the strong complexation of $P_2O_7^{-4}$ with multivalent metallic ions such as Ca, Fe, and Al and the great dissolubility of humus in the mixed solution. In the Al- and Fe-complexes, the content of Al was much higher than that of Fe and Ca in the mixed solution and was the highest compared with that in the other extractants (Table 2). The high pH of the mixture was the direct cause of the low Fe and Ca content in the mixture, while the formation of soluble aluminate resulted in high Al content in alkaline solution.[7] The 0.1 mol l^{-1} $Na_4P_2O_7$ extractant was quite efficient in extracting humus from the Ca- and Al-complexes, as was 0.1 mol l^{-1} NaOH from Fe- and Al-complexes. Both were close to the 0.1 mol l^{-1} NaOH + 0.1 mol l^{-1} $Na_4P_2O_7$ mixture in efficiency, but the humus extracted by 0.1 mol l^{-1} NaOH from the Ca complex was much lower than that by the mixture, which was in agreement with the conclusion that 0.1 mol l^{-1} NaOH mainly extracted humus combined with Fe and Al without dissolving humus combined with Ca.[5] The 0.5 mol l^{-1} $(NaPO_3)_6$ extractant was moderately efficient in extracting humus. The PO_3^- could strongly complex with Ca, and hence the content of Ca was very high in this extracting solution. However, the results listed in Table 2 indicate that PO_3^- also dissolved a certain amount of Fe and Al. In the 0.5 mol l^{-1} Na_2SO_4 extracting solution, the contents of both humus and Ca were a close approximation to that in 0.5 mol l^{-1} $(NaPO_3)_6$, but no Fe or Al could be detected, which was different from the case in 0.5 mol l^{-1} $(NaPO_3)_6$. This might indicate that the 0.5 mol l^{-1} neutral Na_2SO_4 solution was specific to the extraction of humus combined with Ca.

Table 2 Dissolution of Synthetic Organomineral Complexes with Various Extractants (g kg^{-1} complex)

	Ca complex				Fe complex				Al complex			
Extractant	**Ca**	**Fe**	**Al**	**C**	**Ca**	**Fe**	**Al**	**C**	**Ca**	**Fe**	**Al**	**C**
0.1 mol l^{-1} NaOH + 0.1 mol l^{-1} $Na_4P_2O_7$	0.35	0.02	0.46	3.14	0.17	0.02	0.35	2.65	0.13	0.01	1.86	2.31
0.1 mol l^{-1} $Na_4P_2O_7$	1.29	0.33	1.10	3.06	0.40	3.20	0.24	1.65	0.65	0.07	1.37	2.27
0.1 mol l^{-1} NaOH	0.05	0.00	0.21	1.42	0.00	0.01	0.33	2.42	0.00	0.01	1.81	2.30
0.5 mol l^{-1} $(NaPO_3)_6$	8.57	0.43	1.02	1.80	0.36	1.71	0.15	1.91	0.37	0.02	0.78	1.50
0.5 mol l^{-1} Na_2SO_4	8.66	0.01	0.00	1.86	Tr.[a]	Tr.	Tr.	Tr.	Tr.	Tr.	Tr.	Tr.

[a] Tr. = trace.

In brief, the efficiency of extractants in extracting both humus and metallic ions varied with the types of organomineral complexes. The 0.5 mol l^{-1} Na_2SO_4 extractant was the most efficient in extracting Ca from Ca complexes, as was 0.1 mol l^{-1} $Na_4P_2O_7$ in extracting Fe from Fe complexes and 0.1 mol l^{-1} NaOH in extracting Al from Al complexes. The 0.1 mol l^{-1} NaOH + 0.1 mol l^{-1} $Na_4P_2O_7$ mixture was the best extractant in the extraction of C from different organomineral complexes.

B. SEQUENTIAL DISSOLUTION OF SOIL HUMUS WITH SELECTED EXTRACTANTS

The above results indicate that among all the extractants used in this study, 0.5 mol l^{-1} neutral Na_2SO_4 solution was the most effective in extracting Ca-bound humus without destroying Fe- and Al-bound organomineral complexes. Therefore, 0.5 mol l^{-1} Na_2SO_4 was selected for extracting the humus bound with Ca from soils with different base saturation percentages. The results obtained are listed in Table 3. It can be observed that Ca dominated in the 0.5 mol l^{-1} Na_2SO_4 extracts of all the soils with little Fe and Al, which was consistent with the data of the synthetic organomineral complexes in Table 2. In acid soils, the much lower content of Ca in the extractant was directly related to the low content of exchangeable and total Ca, especially in soil samples 7 and 8. The content of organic carbon was significantly correlated with that of Ca in the extracting solution ($r = 0.889$, $n = 7$). In spite of some Al in the extracting solution of samples 7 and 8, the percentage of extracted humus was below 4%, which also indicated that 0.5 mol l^{-1} Na_2SO_4 solution did not dissolve Fe- and Al-bound humus.

Following the first extraction with 0.5 mol l^{-1} Na_2SO_4, the mixed solution of 0.1 mol l^{-1} NaOH + 0.1 mol l^{-1} $Na_4P_2O_7$ was used to dissolve sequentially the residual soils because of its higher efficiency in extracting Ca-, Fe-, and Al-bound humus. These results are also presented in Table 3. It was found that little or no Ca existed in the mixed extracting solution, demonstrating that 0.5 mol l^{-1} Na_2SO_4 solution extracted Ca-bound humus to a maximum level, but did not dissolve the Fe- and Al-bound humus. On the other hand, it is clear that the mixed extracting solution only extracted the Fe- and Al-bound humus. The percentage of organic C in the extract to the total organic C of the soil was significantly related to the content of extractable Al content ($r = 0.781$, $n = 7$).

It is evident that sequential extraction first with 0.5 mol l^{-1} Na_2SO_4 and then with 0.1 mol l^{-1} NaOH + 0.1 mol l^{-1} $Na_4P_2O_7$ mixture can efficiently separate or isolate the two different binding forms of humus, Ca-bound humus and Fe- and Al-bound humus. The metallic ions, mainly acting as a coordination center, can bind the functional groups of humus with the surface groups of clay minerals. In fact, the organomineral complexes of soil are composed of organic matter, minerals, and metallic ions. The stability of the complexes depends on the type of metallic ions as well as the type and composition of clay mineral and organic matter. It is generally accepted that Ca-bound humus belongs to the outer complex, and Fe- and Al-bound humus to the inner complex. In spite of a weak ligand, SO_4^{2-} can destroy Ca-bound organomineral complexes and extract Ca-bound humus. However, SO_4^{2-} is too weak to destroy the Fe- and Al-bound inner complex.[4] Only strong ligands such as OH and P_2O_7 can complex with Fe and Al, and release the combined humus.

C. A PROPOSED METHOD FOR SEPARATING CA-BOUND AND FE- AND AL-BOUND HUMUS FROM SOILS

On the basis of the above described dissolution tests on synthetic organomineral complexes and soils, a method was suggested to separate Ca-bound and Fe- and Al-bound humus by 0.5 mol l^{-1} neutral Na_2SO_4 and 0.1 mol l^{-1} NaOH + 0.1 mol l^{-1} $Na_4P_2O_7$ mixture from various soils. The procedure is given as follows.

Table 3 Successive Dissolution of Soils with Selected Extractants (g kg^{-1})[a]

Sample no.	0.5 mol l^{-1} Na_2SO_4					0.1 mol l^{-1} NaOH + 0.1 mol l^{-1} $Na_4P_2O_7$				
	Al	Fe	Ca	C	C/Ct (%)	Al	Fe	Ca	C	C/Ct (%)
1	0.00	0.00	4.81	5.79	29.10	0.85	0.02	0.06	4.13	20.76
2	0.00	0.00	4.61	8.56	30.78	0.80	0.01	0.05	6.69	24.03
3	0.00	0.01	1.80	3.82	21.69	1.11	0.03	0.01	2.98	16.96
4	0.00	0.01	2.73	1.28	13.20	0.78	0.02	0.00	2.37	24.42
5	0.00	0.01	1.02	0.41	8.28	3.18	0.01	0.00	1.89	38.61
6	0.00	0.00	0.60	0.76	4.81	2.80	0.02	0.00	5.29	33.48
7	0.18	0.01	0.04	0.22	4.07	2.78	0.02	0.00	1.48	27.41
8	0.25	0.03	0.03	0.51	2.36	5.48	0.21	0.00	8.10	37.17
9	0.01	0.01	0.12	0.52	2.43	2.46	0.18	0.00	8.00	37.38

[a] C was the organic C in 0.5 mol l^{-1} Na_2SO_4 extracting solution or that in 0.1 mol l^{-1} NaOH + 0.1 mol l^{-1} $Na_4P_2O_7$ mixture after treatment with 0.5 mol l^{-1} Na_2SO_4, and Ct was the total organic C in soil.

1. Suspend 5.00 g of soil sample in 25 ml of ethanol-bromoform mixture having a density of 1.80 g cm^{-3} in a 100-ml polyethylene centrifuge tube and sonicate 5 min on an Ultrasonic Generator-CSF-3A (made in China). Centrifuge the suspension for 5 min at 4000 rpm. Decant the supernatant, redisperse the sample by stirring in another 25 ml of the heavy liquid, and centrifuge immediately. Repeat the procedure several times until no light fraction is found in the supernatant. Wash the residue (heavy fraction) in the centrifuge tube three times with acetone. Air dry the heavy-fraction samples, crush gently, and store in a bottle.
2. Put 2.00 g of the heavy fraction into a 100-ml centrifuge tube containing 20 ml of 0.5 mol l^{-1} neutral Na_2SO_4 solution, shake for 1 h, allow to stand overnight, and centrifuge. Shake the residue in another 20 ml of the solution and centrifuge again. Repeat this procedure several times until no calcium can be detected in the Na_2SO_4 extracting solution. Collect all the extracts in a 100- or 250-ml volumetric flask. Wash the residue with dilute Na_2SO_4 solution and distilled water until no color of organic C can be observed in the washings. Combine all the washings with the extracts. The extracted portion is Ca-bound humus.
3. Following the removal of Ca-bound humus, disperse the residue in 20 ml of 0.1 mol l^{-1} NaOH + 0.1 mol l^{-1} $Na_4P_2O_7$ mixture, shake for 1 h, allow to stand overnight, and centrifuge. Shake the residue with another 20 ml of the mixture and centrifuge again. Repeat this procedure several times until the extracting solution is almost colorless. Collect all the extracts in the same flask. The organic matter in these extracts is Fe- and Al-bound humus.

IV. SUMMARY AND CONCLUSIONS

Based on the data on selective dissolution of synthetic organomineral complexes and soils, a fractionation scheme was developed to separate Ca-bound and Fe- and Al-bound humus by extraction with 0.5 mol l^{-1} neutral Na_2SO_4 and then with 0.1 mol l^{-1} NaOH + 0.1 mol l^{-1} $Na_4P_2O_7$ mixture. The fractionation method developed is useful for better understanding the nature of organomineral complexes of soils and sediments.

REFERENCES

1. **Tyulin, A. F.,** Peptizing methods of determining soil physico-chemical properties, *Sov. Soil Sci.,* 4-5, 3, 1943 (in Russian).
2. **Kellerman, V. V.,** Physico-chemical characteristics of water stable aggregates of various soils, in *Soil Physico-Chemical Problems and Research Methods,* Antipov-Karataev, I. N., Ed., AH Press, CCCP, 1959, 3 (in Russian).
3. **Kononova, M. M. and Bel'Chikova, N. P.,** Rapid methods of determining the composition of humus in mineral soils, *Sov. Soil Sci.,* 10, 75, 1961 (in Russian).
4. **Fu, J. P.,** Fractionation of soil organo-mineral complexes, in *Soil Colloid Part II: Methods for Soil Colloid Research,* Hseung, Y., Ed., Science Press, Beijing, 1985, 40 (in Chinese).
5. **Xu, J. M.,** Studies on the Calcium-Bound and Iron- and Aluminum-Bound Organo-Mineral Complexes in Soils, Ph.D. thesis, Zhejiang Agricultural University, 1990 (in Chinese).
6. **Varadachari, C., Mondal, A. H., and Ghosh, K.,** Some aspects of clay-humus complexation: effect of exchangeable cations and lattice charge, *Soil Sci.,* 151, 220, 1991.
7. **Carballas, M., Cabaneiro, A., Guitian-Ribera, F., and Carballas, T.,** Organo-metallic complexes in Atlantic humiferous soils, *An. Edafol. Agrobiol.,* 38, 1033, 1979.

Chapter 16

The Quality of Soil Organic Matter as Characterized by Solid CPMAS ^{13}C-NMR and Py-FIMS

C. M. Monreal, E. G. Gregorich, M. Schnitzer, and D. W. Anderson

CONTENTS

I. INTRODUCTION

For many years, chemical characterization of soil organic matter (SOM) involved the use of alkaline and acid solutions to extract OM tightly bound to soil inorganic colloids. Only 20 to 60% of the SOM was extracted, and the extreme pH values during extraction may produce artifacts.[1] During the last decade, advances made in cross-polymerization and magic-angle spinning (CPMAS) solid ^{13}C-nuclear magnetic resonance (NMR) and pyrolysis-field ionization mass spectrometry (Py-FIMS) have assisted in the study of soil organic matter structure. CPMAS solid ^{13}C-NMR is a nondestructive method that has helped to characterize OM in mineral and organic soils.[2-5] In comparison, Py-FIMS is a destructive method that has enabled the identification of single macromolecules in humic substances and whole soils with low organic C concentration.[6,7] Both methods are useful tools to study the effects of farming practices on the structure and transformations of SOM. The aims of this study were to (1) characterize OM in whole and particle-size fractions of two Gleysols by CPMAS solid ^{13}C-NMR and Py-FIMS and (2) investigate potential soil components and processes involved during OM transformation *in vivo* soil systems.

II. MATERIALS AND METHODS

A. SOILS AND PARTICLE SIZE FRACTIONS

Samples from the 0- to 5-cm layer of Ah and Ap horizons were taken from adjacent replicated plots (45 × 200 m) in uncultivated and cultivated Gleysols in 1991. Four composite soil samples were taken from each of the four replicates in the cultivated and

0-87371-914-X/95/$0.00+$.50

forested plots. Soil samples were mixed thoroughly to obtain a representative soil sample. Both soils were silty clay loam in texture. The cultivated soil had been cropped for approximately 90 years, the last 25 years under corn. The native soil supported a mixed hardwood forest. Plant residues returned to the forest floor consisted mainly of death branches and leaves of poplar *(Populus alba)*, alder *(Alnus* spp.), oak *(Quercus rubra)*, and maple *(Acer rubrum)*. Residues returned to the cultivated soil included plant debris from barley *(Hordeum vulgare)*, alfalfa *(Medicago sativa)*, and corn *(Zea mays)*. The bulk density of the forest and cultivated soils were 0.68 and 1.18 mg m^{-3}, respectively. The concentration of organic C in the forest samples were 10.03% (whole), 17.54% (sand), 9.47% (silt), and 7.50% (clay). The corresponding values in the cultivated soil were 1.92, 1.07, 1.61, and 2.76%, respectively. Fractions of sand (>50 μm), silt (2 to 50 μm), and clay (<2 μm) were obtained by ultrasonically dispersing soil samples at 300 W for 8 min in an ice bath.[8] Sand was separated by wet sieving, and silt and clay fractions by centrifugation and decantation.[9] Samples were subsequently freeze dried and ground to pass a 100-mesh sieve before analysis.

B. CPMAS ^{13}C-NMR

Soil samples were analyzed on a Bruker CXP-200 solid-state CPMAS ^{13}C-NMR equipped with a 7-mm Doty Scientific magic angle solid probe. Saphire rotors with Kel-F turbines and caps holding up to 500 mg of sample were used. All spectra were collected at room temperature and at 50.13 MHz under the following conditions: sweep width 20 kHz; data size 2 K, zero filled to 4 K; line broadening factor of 75 Hz applied prior to Fourier transform; cross-polarization contact time 2 ms; recycle delay 3 to 5 s, and number of scans 5000 to 25,000. Spinning speeds were optimized at 3 to 3.5 kHz through observation of the ^{79}Br resonance. Chemical shifts were referenced externally to tetramethylsilane (TMS) via adamantane as a secondary reference.

Spin-counting experiments were conducted to determine the extent to which all carbons contributed to the final spectra of soil samples. A known amount of soil sample was mixed thoroughly with KBr (10% w/w) using a mortar and pestle and delivered to the sample holder. For quantitative purposes, each soil and particle-size fraction was thoroughly mixed with an exact weight of glycine (NH_2CH_2COOH, 10% w/w). The entire CPMAS ^{13}C-NMR spectrum for each spiked and unspiked sample (–50 to 300 ppm) was integrated and then baseline fitted to remove the soil NMR spectra. The intensities of spectra signals from the soil and glycine were subsequently obtained. The $\%C_{observed}$ was calculated from

$$\%C_{observed} = \frac{M_g}{Mx} \cdot \frac{C_g}{Cx} \cdot \frac{Ix}{I_g} \cdot 100$$

where M_g and Mx are the weights of the glycine and soil sample, respectively, C_g and Cx are the percent C in glycine and soil, respectively, and Ix and I_g are the integrated intensities of the soil and glycine, respectively. A similar approach was used by Botto et al.[10] in a quantitative study of coals.

The chemical shifts in the NMR spectra were grouped into areas representing C in aliphatics (0 to 107 ppm), aromatics: aromatic + phenolic (110 to 160 ppm), carboxyl (160 to 200 ppm), and ketone (200 to 220 ppm). The mass of C type (C_t) per volume of soil was calculated as follows:

$$C_t\left(\text{kg ha}^{-1}\,5\text{ cm}^{-1}\right) = A\left(\text{m}^2\text{ ha}^{-1}\right) \cdot D(\text{m}) \cdot BD\left(\text{kg m}^{-3}\right) \cdot C(\%) \cdot SI(\%) \qquad (1)$$

where A = soil area, D = soil depth, BD = soil bulk density, C = percent organic carbon, and SI = percent signal intensity for each C type. Equation 1 assumes that nonobserved carbon is equally distributed among all types of C, and was multiplied by $\%C_{observed}$ to provide minimum and maximum mass values for each type of C. Soil aromaticity was defined as the ratio of aromatic C to aliphatic and aromatic C.

C. PYROLYSIS FIELD IONIZATION MASS SPECTROMETRY AND SOIL C

About 100 to 200 μg of soil were analyzed by Py-FIMS as described by Schulten and Schnitzer.[11] Total C was measured by dry combustion and inorganic C by acid digestion.[12] Soil organic C (Co) was determined by the difference between total and inorganic C.

III. RESULTS AND DISCUSSION

A. ^{13}C-NMR CHARACTERIZATION

In our study, some NMR spectra showed a low signal-to-noise ratio. Figures 1A and B are shown as examples of the best and worst spectra obtained for whole and particle size fractions. Spin-counting experiments allowed one to define a quantitative range with upper and lower limits. These limits correspond to assumed and measured $\%C_{observed}$ values (Table 1). This quantitative approach takes into consideration any potential errors caused by the integration of spectra with a low signal-to-noise ratio. Noteworthy, a mass balance showed that the sum of the mass of C types in particle size fractions corresponded to the mass of C types measured separately in whole samples. These results suggest that the low signal-to-noise ratio in spectra did not affect quantification. The recovery of aromatic C in forest and cultivated samples was overestimated by 13 and 9%, respectively (Table 1). This may be due to effects caused by the presence of paramagnetic and ferromagnetic species found in natural samples.[13] A separate parallel study involving the pretreatment of soil samples with $SnCl_2$ and $Na_2S_2O_4$ decreased the magnetic susceptibility of soil samples and improved the quality of NMR spectra, but resulted in a loss of up to 20% of the soil organic carbon (data not shown). Therefore, we consider that the values of Table 1 represent quantitative information on the different C types measured in their own geophysical and chemical environment. Based on the latter, the NMR analysis showed that all different types of C were observed in the spectra of all samples (Figures 1A and B, Table 1). Spin-counting experiments showed that most (81 to 106%) of the C in whole and particle size fractions was observed by solid ^{13}C-NMR (Table 1). These values are higher than the range reported for five fractions of humic/fulvic-glycine mixture (30 to 62%).[14]

1. Forest Soil

Results are discussed on the basis that "unseen" C is equally distributed among all C types. The recovery of Co and types of C in particle size fractions varied between 96 and 113% of that found in whole soil samples (Table 1). Aliphatic C comprised 62% of the total Co mass, followed by aromatic (18%), carboxylic (14%), and ketonic carbon (6%). A predominance of aliphatic rather than aromatic structures has been reported for soils of Australia and Canada.[3-5] In our study, the absolute amount of aliphatics was distributed in similar proportions among particle size fractions. The silt fraction stabilized OM and contained more than twice the mass of aromatic, carboxylic, and ketonic carbon than the sand or clay particles.

C aromaticity was 23% in whole soils, highest in the silt (32%), and lowest in the clay (19%), and it was lower than the 35% aromaticity typically reported for chemically extracted soil humic acids.[7] The accumulation of aromatic C in the silt of the forest soil may be due to abiotic and biotic processes:

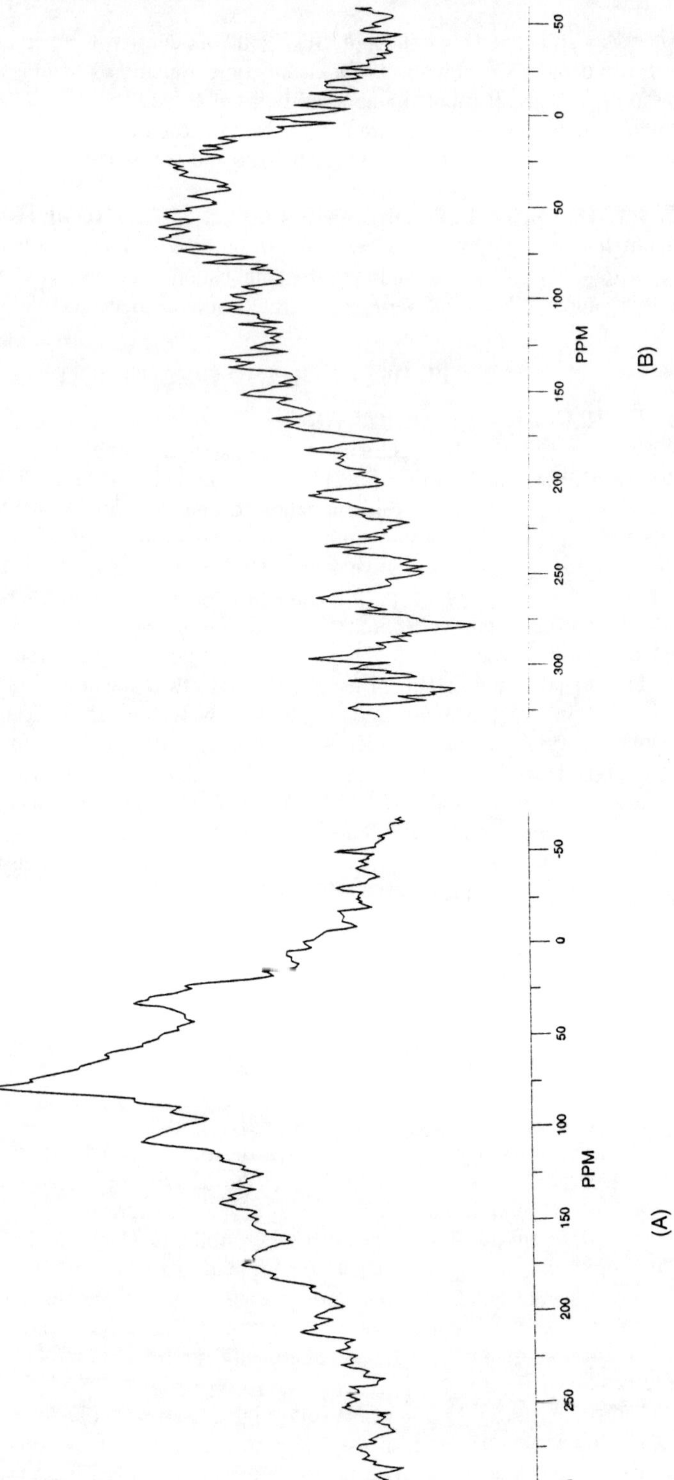
-50
0
50
100
150
200
250
300
PPM
(B)
-50
0
50
100
150
200
250
300
PPM
(A)

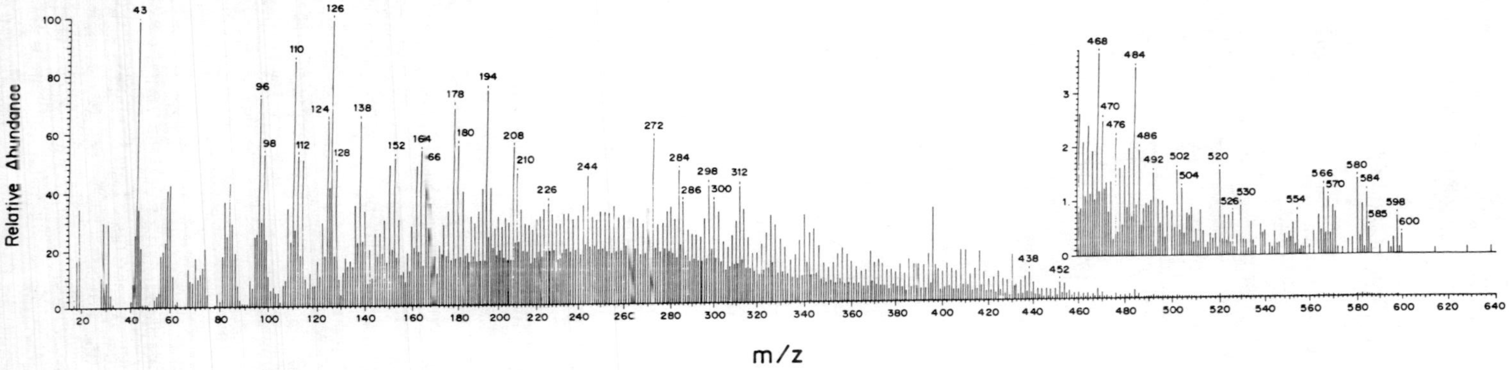

(C)

(D)

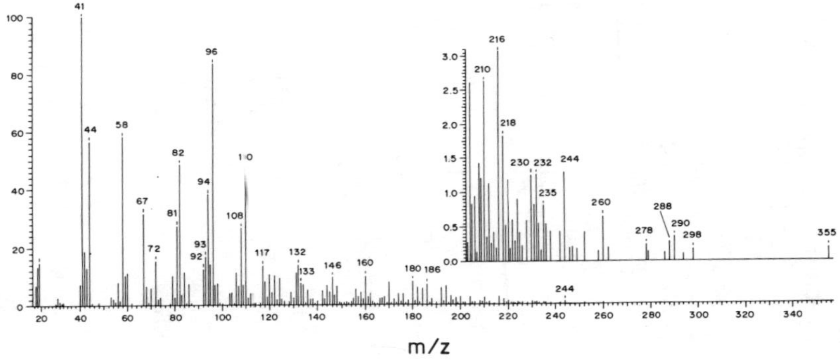

Figure 1 Solid CPMAS ^{13}C-NMR spectra obtained for particle size fractions in the 0- to 5-cm depth. (A) Sand of the forest; (B) clay of the cultivated soil. Pyrolysis-field ionization mass spectra for whole-soil samples; (C) of the forest; and (D) cultivated soil.

Table 1 $\%C_{observed}$ and Distribution of Different Types of C in the 0- to 5-cm Depth of Two Gleysols as Characterized by CPMAS ^{13}C-NMR

Soil fraction	Org. C (t ha⁻¹)	$\%C_{obs}$	Type of C (t ha⁻¹) Aliphatic	Aromatic	Carboxyl	Ketonic
			Forest			
Whole		100a[a]	20.5	6.0	5.2	2.4
	34.1	85m	17.4	5.1	4.4	2.1
Sand		100a	6.0	1.7	1.0	0.3
	9.0	102m	6.1	1.8	1.0	0.3
Silt		100a	7.4	3.6	3.5	1.6
	16.1	94m	7.0	3.3	3.3	1.5
Clay		100a	6.2	1.5	0.8	0.5
	8.9	103m	6.3	1.6	0.8	0.5
% recovery[b]	99.7		95.6	113.3	101.9	100.0
			Cultivated			
Whole		100a	7.1	2.2	1.4	0.6
	11.3	82m	5.9	1.8	1.1	0.5
Sand		100a	0.5	0.2	0.2	0.07
	1.0	81m	0.4	0.2	0.1	0.1
Silt		100a	2.7	1.0	0.6	0.25
	4.6	97m	2.6	1.0	0.6	0.2
Clay		100a	3.8	1.2	0.7	0.3
	6.0	85m	3.2	1.1	0.6	0.2
% recovery	102.0		98.6	109.0	107.1	103.3

[a] a = assumed, m = measured.

[b] $$\% \text{ recovery} = \frac{C_{(t)sand} + C_{(t)silt} + C_{(t)clay}}{C_{(t)whole}} \cdot 100$$

It is assumed that "unseen" carbon was equally distributed among all C types.

1. Abiotic stabilization of products derived from partially and totally decomposed plant lignins[15] and biosynthetic processes of soil biota.[16] In a different study, the age for Co stabilized in coarse and fine silt fractions of a Black cultivated Chernozem varied between 800 and 965 years.[17]
2. Reduced microbial and enzymatic activities. Highly polymerized humic materials form intricate networks of aliphatic and aromatic moieties,[18] making organic substrates inaccessible for decomposition.
3. Aromatic oxidases within aggregates high in silt may be repressed or inhibited by cations or metals adsorbed to clay minerals.

Further studies are warranted on the chemical and physical structure of OM and its interaction with inorganic matrices, biochemical processes, and the accumulation of different types of C in the silt.

2. Cultivated Soil

The amount of Co and types of C recovered in particle size fractions varied between 99 and 109% of that found in whole soils. Soil cultivation reduced the Co mass by 67%, and the loss of aliphatic, aromatic, carboxylic, and ketonic C was 66, 65, 75, and 76%, respectively. Co in the sand, silt, and clay was reduced by 89, 72, and 32%, respectively (Table 1). The distribution of the different C types was similar to that in the forest samples: aliphatic > aromatic > carboxylic > ketonic. Clay contained more than 52% of all the organic structures present in whole samples, and less than 12% was associated with the sand.

Aromaticity in the cultivated whole soil was 23%, highest in the sand (30%), and slightly lower in the silt (27%) and clay fractions (25%). A parallel study conducted on the same soil, and that was based on the $\delta^{13}C$ of the C4 corn residue, showed that the average turnover times for corn residue C associated with sand, silt, and clay were 56, 128, and 98 years, respectively. The slower turnover of corn residue C in the finer particle size fractions helps in part to explain the accumulation of organic structures in the clay and silt.

In whole samples, aromatic and aliphatic C decreased at similar average rates (0.72 vs. 0.70% $year^{-1}$, respectively), suggesting similar average turnover rates for both types of structures. Carboxylic and ketonic C decreased more rapidly (0.84% $year^{-1}$, on average). These results support the concept of OM stability based on physicochemical adsorption and physical protection of OM by soil aggregates,[19] rather than the concept of chemical recalcitrance of aromatics proposed earlier in studies of soil humic acids and in soil organic matter models.[20-23]

Decomposition of aromatic structures may be associated with the biotic and abiotic processes of organic matter transformation. Aromatic ring cleavage requires displacement of π-electrons from the molecular orbitals. Microbial oxygenation, enzymatic oxidation of aromatics,[24,25] and physicochemical reactions between Fe^{3+} and Al^{3+} of inorganic colloids and aromatic compounds can catalyze the displacement of π-electrons *in vivo* soil systems and pure clay-substrate studies.[26,27] The splitting of aromatic rings by abiotic or biotic processes may result in an increase of aliphatic C. This hypothesis is supported by changes in aliphatic/Co and aromatic/aliphatic C ratios observed in the forest and cultivated samples. In whole samples, the aliphatic/Co ratio increased slightly (0.60 to 0.63) after 90 years of cultivation. This slight increase was paralleled by a larger increase of the aliphatic/Co ratio (0.46 to 0.59) and a decrease of the aromatic/aliphatic ratio (0.48 to 0.36) in the silt fraction. The latter shows a relative proportional increase in aliphatic C at the expense of aromatic C in cultivated samples. Further research is needed to define the relative contributions of biotic and abiotic processes transforming soil organic matter *in vivo* .

B. PYROLYSIS FIELD IONIZATION MASS SPECTROMETRY

1. Forest and Cultivated Soils

Mass signals from the Py-FIMS spectra show striking differences in the pyrolysis products of both soils. The forest samples showed a great variety and high relative abundance of polysaccharides corresponding to pentose units at m/z 114 and 132, hexose subunits at m/z 126, 144, and 162,[28] and aldohexose subunits of cellulose at m/z 126 and 144 (Figures 1C and D).[29]

Mass signals for n-fatty acids were abundant in the forest and almost nonexistent in the cultivated sample. A great proportion of the mass signals for n-fatty acids in the forest sample (m/z 396 to 564) is associated with pyrolysis products of fatty acids synthesized by plants and/or insects (n-C26 to n-C38).[30] In contrast, the cultivated soil shows a mass signal at m/z 298 (n-C19), which suggests the presence of a microbially synthesized n-fatty acid.[31] The CPMAS ^{13}C-NMR and the Py-FIMS data support the hypothesis of rapid turnover for soil fatty acids and that losses of soil lipids from the cultivated soil may be associated with microbial metabolism.

The forest sample showed a higher relative abundance of aromatics than the cultivated sample (Figure 1D). Structures such as guaiacol, methyl guaiacol, ethenyl guaiacol, vanillin, syringol, coniferyl aldehyde, coniferyl alcohol, ferulic acid, sinapyl aldehyde, and sinapyl alcohol were found in the forest sample. These biomarkers have been previously identified with signals at m/z 124, 138, 150, 152, 154, 178, 180, 194, 208, and 210, respectively.[32] These results are consistent with the decrease in aromatic C with cultivation as observed by CPMAS ^{13}C-NMR.

Mass signals for N-acetyl amino acids are found at m/z 137 and 167. The latter structures are found in peptidoglycans and chitin of soil organisms and were found in higher abundance (36%) in the forest relative to that in the cultivated samples (<3%). The forest sample showed signals at m/z 386 and 396, which are fingerprints for cholesterol and ergosterol.[11] None of these signals were evident in the cultivated sample. Cholesterol and ergosterol are common to eukaryotic organisms such as fungi.[30,31] Differences in the type and abundance of fatty acids, N-acetyl amino acids, cholesterol, and ergosterol between the forest and cultivated soil may reflect differences in total soil microbial biomass and in bacterial and fungal populations between both soils. This hypothesis needs further testing.

IV. SUMMARY AND CONCLUSIONS

The chemical characterization of OM in whole soils and particle size fractions by CPMAS ^{13}C-NMR and Py-FIMS permit the following conclusions:

1. Between 46 and 72% of the soil organic C mass consisted of aliphatic structures. Aromatic C comprised between 17 and 23% of all organic C in whole and particle size fractions of the forest and cultivated samples.
2. Cultivation decreased the mass of C and all classes of organic compounds. Most organic structures were lost from the sand and silt fractions. Aromatic and aliphatic C were lost at similar rates (0.7% year^{-1}).
3. This study, in contrast to work based on alkali extraction of humic materials, indicates that decreases in OM and increased biological recalcitrance are not a consequence of an increased content of resistant aromatic structures. Cleavage of aromatic structures by biotic and abiotic processes may account for fewer aromatic structures in the cultivated Gleysol.

ACKNOWLEDGMENTS

This research was undertaken as part of the Soil Quality Evaluation Program through funding provided from the National Soil Conservation Program to the Canada-Saskatchewan Agreement on Soil Conservation.

REFERENCES

1. **Worobey, B. L. and Webster, G. R. B.,** Indigenous 13C-NMR structural features of soil humic substances, *Nature (London),* 292, 256, 1981.
2. **Wilson, M. A., Pugmire, R. J., Zilm, K. W., Goh, K. M., Heng, S., and Grant, D.,** Cross-polarization ^{13}C NMR spectroscopy with 'magic angle' spinning characterizes organic matter in whole soils, *Nature (London),* 294, 648, 1981.
3. **Oades, J. M., Vassallo, A. M., Waters, A. G., and Wilson, M. A.,** Characterization of organic matter in particle size and density fractions from a red-brown earth, *Aust. J. Soil Res.,* 25, 71, 1987.
4. **Arshad, M. A., Schnitzer, M., Angers, D. A., and Ripmeester, J. P.,** Effects of till vs. no-till on the quality of soil organic matter, *Soil Biol. Biochem.,* 22, 595, 1990.
5. **Preston, C. M., Axelson, D. E., Levesque, M., Mathur, S. P., Dinel, H. D., and Dudley, R. L.,** Carbon-13 NMR and chemical characterization of particle-size separates of peats differing in degree of decomposition, *Org. Geochem.,* 14, 393, 1989.
6. **Schulten, H.-R.,** Pyrolysis and soft ionization mass spectrometry of aquatic/terrestrial humic substances and soils, *J. Anal. Appl. Pyrol.,* 12, 149, 1987.

7. **Schnitzer, M.,** Soil organic matter. The next 75 years, *Soil Sci.,* 151, 41, 1991.
8. **Gregorich, E. G., Kachanoski, R. G., and Voroney, R. P.,** Ultrasonic dispersion of aggregates: distribution of organic matter in size fractions, *Can. J. Soil Sci.,* 68, 395, 1988.
9. **Anderson, D. W., Saggar, S., Bettany, J. R., and Stewart, J. W. B.,** Particle size fractions and their use in studies of soil organic matter. I. The nature and distribution of forms of carbon, nitrogen and sulfur, *Soil Sci. Soc. Am. J.,* 45, 767, 1981.
10. **Botto, R. E., Wilson, R., and Winans, R. E.,** Evaluation of the reliability of solid ^{13}C NMR spectroscopy for the quantitative analysis of coals: study of whole coals and maceral concentrates, *Energy Fuels,* 1, 173, 1987.
11. **Schulten, H.-R. and Schnitzer, M.,** Aliphatics in soil organic matter in fine-clay fractions, *Soil Sci. Soc. Am. J.,* 54, 98, 1990.
12. **Tiessen, H., Bettany, J. R., and Stewart, J. W. B.,** An improved method for the determination of carbon in soils and soil extracts by dry combustion, *Commun. Soil Sci. Plant Anal.,* 12, 211, 1981.
13. **Snape, C. E., Axelson, D. E., Botto, R. E., Delpuech, J. J., Tekely, P., Gerstein, B. C., Pruski, M., Maciel, G. E., and Wilson, M. A.,** Quantitative reliability of aromaticity and related measurements on coals by 13C NMR. A debate, *Fuel,* 68, 547, 1989.
14. **Wilson, M. A. and Vassallo, A. M.,** Compositional and solid-state nuclear magnetic resonance study of humic and fulvic acid fractions of soil organic matter, *Anal. Chem.,* 59, 551, 1987.
15. **Flaig, W.,** Effects of microorganisms in the transformation of lignin to humic substances, *Geochim. Cosmochim. Acta,* 28, 1523, 1964.
16. **Martin, J. P. and Haider, K.,** Phenolic polymers of *Stachybotrys chartarum,* and *Epicoccum nigrum* in relation to humic acid formation, *Soil Sci.,* 107, 260, 1969.
17. **Anderson, D. W. and Paul, E. A.,** Organo-mineral complexes and their study by radiocarbon dating, *Soil Sci. Soc. Am. J.,* 48, 298, 1984.
18. **Schulten, H.-R., Plage, B., and Schnitzer, M.,** A chemical structure for humic substances, *Naturwissenschaften,* 78, 311, 1991.
19. **Tisdall, J. M. and Oades, J. M.,** Organic matter and water-stable aggregates in soils, *J. Soil Sci.,* 33, 141, 1982.
20. **Anderson, D. W.,** Processes of humus formation and transformation in soils of the Canadian Great Plains, *J. Soil Sci.,* 30, 77, 1979.
21. **Jenkinson, D. S. and Rayner, J. H.,** The turnover of soil organic matter in some of the Rothamsted classical experiments, *Soil Sci.,* 123, 298, 1977.
22. **McGill, W. B., Hunt, H. W., Woodmanse, R. G., and Reuss, J.,** Phoenix, a model of the dynamics of carbon and nitrogen in grassland soil, *Ecol. Bull.,* 33, 49, 1981.
23. **Parton, W. J., Schimel, D. S., Cole, C. V., and Ojima, D. S.,** Analysis of factors controlling soil organic matter levels in great plains grasslands, *Soil Sci. Soc. Am. J.,* 51, 1173, 1987.
24. **Haider, K. and Martin, J. P.,** Humic acid-type phenolic polymers from *Asperguillus sydowi* culture medium, *Strachybotrys* spp. cells and autoxidized phenol mixture, *Soil Biol. Biochem.,* 2, 145, 1970.
25. **Kirkpatrick, D. S., Biggs, S. R., Conway, B., Finn, C. M., Hawkins, D. R., Honda, T., Ishida, M., and Powell, G. P.,** Metabolism of N-(2,3- dichlorophenyl)-3,4,5,6-tetrachlorophthalamic acid (Techlofthalam) in paddy soil and rice, *J. Agric. Food Chem.,* 29, 1149, 1981.
26. **Wang, C. W. and Huang, P. M.,** Pyrogallol transformations as catalyzed by nontronite, bentonite and kaolinite, *Clays Clay Miner.,* 37, 525, 1989.
27. **Wang, M. C. and Huang, P. M.,** Significance of Mn(IV) oxide in the abiotic ring cleavage of pyrogallol in natural environments, *Sci. Total Environ.,* 113, 147, 1992.
28. **Hempfling, R., Simmleit, N., and Schulten, H.-R.,** Characterization and chemodynamics of plant constituents during maturation, senescence and humus genesis in spruce ecosystems, *Biogeochemistry,* 13, 27, 1991.
29. **Schnitzer, M., Schulten, H.-R., Schuppli, P., and Angers, D. A.,** Organic matter extraction from soils with water at high pressure and temperatures, *Soil Sci. Soc. Am. J.,* 55, 102, 1991.
30. **Salisbury, F. B. and Ross, C. W.,** *Plant Physiology,* Wadsworth, Belmont, CA, 1978, chap. 14.
31. **Hardwood, J. L. and Russell, N. J.,** *Lipids in Plants and Microbes,* George Allen and Unwin Ltd., London, 1984, chap. 3.
32. **Schulten, H.-R. and Simmleit, N.,** Impact of ozone on high-molecular constituents of beech leaves, *Naturwissenschaften,* 73, 618, 1986.

Chapter 17

Extracellular Polysaccharides: An Interface Between Microorganisms and Soil Constituents

C. Chenu

CONTENTS

I. INTRODUCTION

The production of extracellular polysaccharides (EPS) is widespread among microorganisms such as bacteria, fungi, or algae, and occurs in many natural environments, such as water, soils, or plants. The polymers occur as a capsule or as a slime and constitute the outermost envelope component of the cell.

In soils, polysaccharides as a whole have received much attention because of a prominent role in soil aggregation,[1,2] but there is very little specific knowledge about the production and nature of soil microbial EPS.

The occurrence and specific functions of EPS of soil microorganisms remain the subject of speculation and debate. Extracellular polysaccharides, located at the outermost surface of microorganisms, are an interface between the cells and their immediate surroundings, i.e., the soil solution or atmosphere, soil mineral or organic constituents. Flows or exchanges of water, gases, and nutrients occur through this layer and are thus likely to be affected by it. Extracellular polysaccharides are often assumed to provide protection to microorganisms,[3,4] but, again, there is little experimental evidence of it. In addition, EPS may interact with the surrounding soil constituents by adsorbing onto them and are thus involved in the formation of organomineral associations in soils. Therefore, EPS make a significant contribution to soil aggregate stability.

The present chapter focuses on an organic matter component of soils that may not be quantitatively very important but that has a key position in the interations between microorganisms and soil mineral constituents. This review discusses the occurrence, characteristics, and implications of an EPS interface in soils.

0-87371-914-X/95/$0.00+$.50

II. EPS PRODUCTION IN SOILS

Many soil-inhabiting microorganisms are able to produce EPS in pure culture,[3-5] although it is difficult to assess whether microbial species that are known for producing EPS *in vitro* do so in soils. Direct demonstration of EPS production in soils was provided by observing thin sections of soils with transmission electron microscopy. Many bacteria and some fungi are surrounded by a layer of organic material that is cytochemically identified[6] as polysaccharide (Figure 1).[7-11] Extracellular polysaccharides usually appear as a dense amorphous granular matrix or as strongly entangled fibers (Figure 1). Capsules as well as slimes can be observed, and the EPS layers are 0.05 to 0.5 μm thick. Extracellular polysaccharide matrices are widely and easily observed in thin soil sections, although the proportion of soil microorganisms that are actually surrounded by EPS have not been determined, and obviously it is not possible by this method to quantify the secretion.

It is clearly very difficult to give any quantitative estimation of EPS in soil. The main reason for such a lack of knowledge is that EPS are a fraction of soil organic matter that cannot be specifically isolated or measured since there is no major difference in the composition of microbial EPS from that of plant EPS, and from that of other microbial polysaccharides located in the cell wall or cytoplasm.

Microorganisms produce polysaccharides which contain very little arabinose or xylose, whereas plant polysaccharides contain little mannose or galactose.[12-13] Hence, the relative amounts of (galactose + mannose)/(xylose + arabinose) or of xylose/mannose have been used as an index of the origin of soil polysaccharides.[13-16] Several attempts have been made to separate soil organic fractions in which the polysaccharides would be

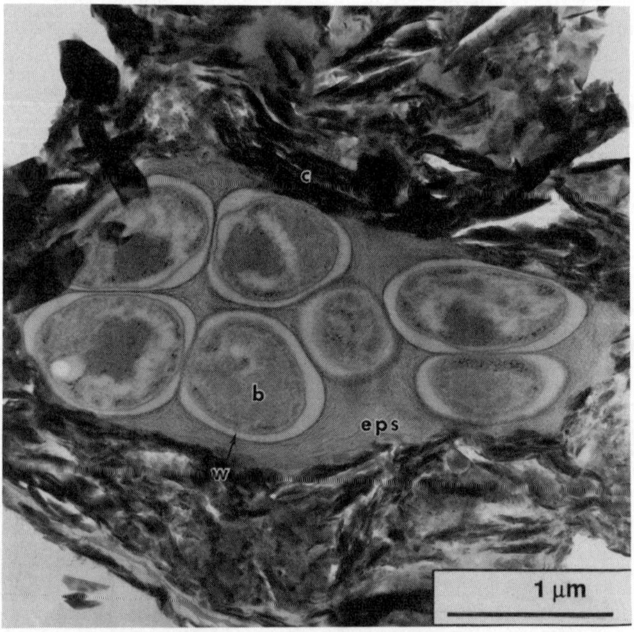

Figure 1 Thin section of soil observed by transmission electron microscopy. The polysaccharides were stained with silver according to Thierry[6] and are visualized by the silver grains. (Photograph by G. Villemin, Centre de Pédologie Biologique, Vandoeuvre, France.) b = bacteria, c = clay, w = wall, EPS = extracellular polysaccharides.

mostly of microbial or plant origin. In some studies, densimetrically separated heavy fractions of soil (d >2 or >1.74) were found enriched in microbial polysaccharides[16-18] whereas in others the light and heavy fractions of soil could not be distinguished on this basis.[19-21] Particle size separation was also unable to produce fractions of only one origin, although clay and silt fractions were always enriched in microbial polysaccharides.[21,22]

Information on the production of microbial EPS come from pure culture studies and demonstrates that both the amounts and nature of EPS may vary with environmental conditions. In liquid culture, EPS production is essentially associated with the stationary phase.[23,24] Polysaccharide synthesis is, on a per cell basis, stimulated when carbon is available in excess, but when nitrogen,[17,23,25-29] phosphorus,[23-24,30] or sulfur[24,31] are limiting. A number of bacterial cells can, however, synthesize EPS in high yields even under carbon limitations.[31,32] The presence of adsorbing surfaces, such as sand surfaces, markedly increased the EPS synthesis of several soil bacteria.[17,30,33] In the only study performed in conditions close to those prevailing in soils, i.e., presence of solid surfaces and unsaturated conditions, EPS production per unit biomass was higher than that in liquid culture.[17] As in liquid culture, N limitation enhanced EPS production.[17]

The amounts of EPS produced by edaphic algae or bacteria, originating or calculated from a wide range of data, are reported in Table 1.[22-40] Synthesis of EPS ranged from 0.02 to six times the dry weight of the cell. In many cases, the microorganisms allocated more carbon to EPS production than to protoplasm synthesis (ratios >1). Based on these data, a rough estimate of the levels of EPS that may be expected in soils is presented in Table 2.

Table 1 Quantitative Estimation of Extracellular Polysaccharide Production by Edaphic Algae or Bacteria

		EPS per cell[a]	EPS per 10^9 cells[b]	
Organism	**Conditions**	**(mg mg^{-1})**	**(mg per 10^9 cells)**	**Refs.**
algae				
Chlamydomonas	Liquid culture[c]	0.45		34
Chlamydomonas	Liquid culture	0.17–1.3		35
Chlamydomonas	Soil		1–43 (total polys.)	36
Porphyridium	Liquid culture	0.32		37
Porphyridium	Liquid culture	0.8–1.7		27
bacteria				
Klebsiella	Liquid culture	2.2–6		24
Azotobacter	Liquid culture	3		31
Azotobacter	Liquid culture	1–2.5		23
Rhizobium	Liquid culture	1		32
Rhizobium	Liquid culture	0.03–0.23	0.03–0.21	38
Xanthomonas	Liquid culture	6		31
Pseudomonas	Liquid culture	0.06–0.6		39
Pseudomonas	Liquid culture	0.3–1		17
Pseudomonas	Unsaturated sand	2.2–4.7		17
Nostoc	Liquid culture	0.3–0.6		34
Nostoc	Soil	0.15–0.25		40
Soil bacterium	Liquid culture	0.02–0.03	0.02–0.03	30
Soil bacterium	Saturated sand	0.11		30

[a] For calculations, the following cell compositions were assumed[41]: carbon, 47%; protein, 65%; RNA, 7%. The carbon content of EPS is 40%. When EPS occurred both as capsules and slime, the given data correspond to the soluble slime fraction only.[17,27,37]

[b] When original data were expressed as EPS per number of cells, 10^9 cells of bacteria were assumed to weigh 1.1 mg.

[c] All liquid cultures are in the stationary phase.

Table 2 An Estimation of the Amounts of Extracellular Polysaccharides Associated with Microorganisms in Soils

Soil[a]	Biomass[a] (mg C g^{-1} soil)	C content[a] (mg C g^{-1} soil)	EPS[b] (mg g^{-1} soil)		EPS (% of soil (organic matter)		EPS[c] (% of soil polysaccharides)	
			0.1	0.5	0.1	0.5	0.1	0.5
Arable	0.22	10.0	0.05	0.23	0.2	1.1	2.1	11.7
Deciduous forest	1.23	36.2	0.26	1.31	0.3	1.6	3.1	18.1
Permanent grassland	1.26	39.4	0.27	1.34	0.3	1.5	3.1	17.0

[a] From Reference 42. The carbon content of the biomass is taken as 47%.
[b] Two levels of EPS per unit biomass were considered: 0.1 or 0.5 EPS/dry weight of cells.
[c] An average of 11% of the soil organic matter as polysaccharides is considered.[13]

The ratios of EPS per unit biomass of 0.05 or 0.5 were chosen, which appear reasonable considering that (1) not all soil microorganisms produce EPS and (2) nutritional conditions in soils may not be favorable for EPS synthesis. It appears that EPS associated with living microorganisms may represent 0.02 to 1.4 mg g^{-1} of soil, that is, 0.1 to 1.5% of soil organic matter and 1 to 16% of soil polysaccharides. Hence, it represents a minor but not negligible proportion of soil components. Since these estimates concern only the EPS associated with living microorganisms, the total amount of EPS present in soils will be higher. Microbial polysaccharides will persist for some time in soils after the death of the microorganism; they have short residence times in soils, comparable to those of the biomass itself.[43]

III. STRUCTURE AND PROPERTIES OF EPS OF SOIL MICROORGANISMS

In contrast with the scant information available on soil EPS *in situ,* a great deal of knowledge has been obtained *in vitro* on the composition, structure, and properties of microbial EPS because they have wide industrial applications. Several of the commercially used EPS are produced by soil-inhabiting microorganisms, such as dextrans from *Leuconostoc* bacteria, xanthan from *Xanthomonas campestris,* curdlan from *Agrobacter tumefaciens* or *radiobacter,* scleroglucan from the fungi *Sclerotium,* the EPS from *Porphyridium* unicellular algae, etc.

Extracellular polysaccharides may be produced in general as discrete capsules or as loose slimes, more or less attached to the cells.[44] Usually, one strain produces one structurally defined polysaccharide, but no general relationship exists between the taxonomy and composition of EPS: EPS are usually characteristic of a given species, if not of a strain.[23] Unfortunately, the main trait of microbial EPS as a whole is their diversity.

Glucose is the most common monomer, followed by galactose, mannose, and their acidic or amino derivatives. Extracellular polysaccharides most frequently carry a net negative charge due to glucuronic, galacturonic, or mannuronic acids or to nonsugar units (pyruvic acid, succinic acid). Monomers usually define a repeating sequence such as that of xanthan (Figure 2). The composition or even the complete structure of EPS is known for a wide range of soil microorganisms, in particular for rhizobia.[23,45-48] Most EPS have ordered conformations of a buckled ribbon or helical type.[45] These are stabilized by the alignment and cooperative weak interactions between long ordered sequences. A general feature with polysaccharides is that different chains are involved, and that the ordered sequences are interrupted, for example, by a different monomer. Such "kinks" lead one chain to bind to several others through ordered zones, the result being a three-dimensional network.[49] Chains with ordered conformations being more or less interconnected give particular solution properties: EPS form very viscous solutions or even gels. EPS are therefore widely used as thickening, gelling, or suspending agents in the food industry.[50] The physical structure of many EPS consists of fibers, a few nanometers wide and >100 nm in length, that may be extensively interconnected.[19] Another striking property of EPS is their ability to retain water: they may retain as much as 100 times their weight in water (Figure 3).

Hence, from such data, EPS-producing soil microorganisms may be viewed as surrounded by a colloid or gel phase that is porous, contains mostly water, and is able to retain cations. This matrix is deformable, and its architecture may respond to changes in the ambient environment.

IV. ROLES OF EPS FOR SOIL MICROORGANISMS

What is the role of the EPS matrix for soil microorganisms? An intriguing point is that bacteria or fungi, with no extracellular polysaccharide, do occur in soils and that, quite often, mutation of bacteria toward a non-EPS-producing form does not affect its viability.

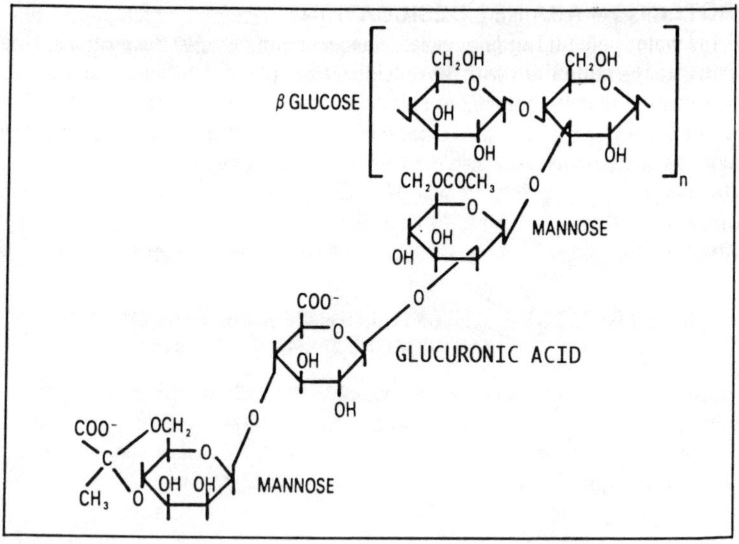

Figure 2 Structure of xanthan, an extracellular polysaccharide produced by the soil bacterium *Xanthomonas campestris*.

Hence, EPS are not essential for the viability of bacteria, although it seems unlikely that evolution would have preserved species producing functionless substances that require carbon and energy for their synthesis. Hence, EPS may have some competitive advantages. Some of their presumed roles are presented and discussed here. The role of EPS in interactions between microorganisms and plants is not considered here, although there is very good evidence that extracellular polysaccharides are involved in surface recognition processes in plant-microorganism or prey-predator relationships.

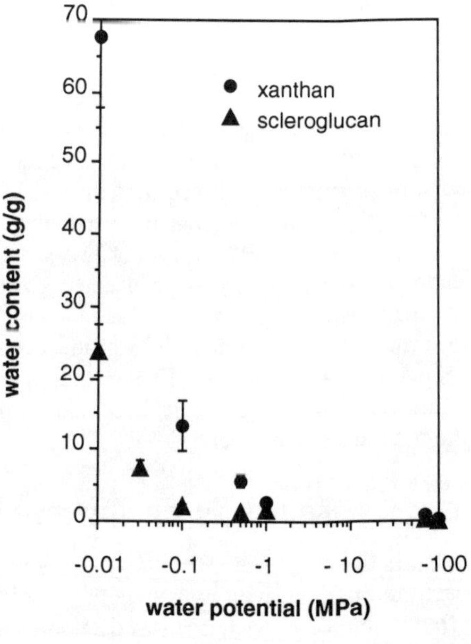

Figure 3 Water-release curves of microbial extracellular polysaccharides as a function of water potential.[67]

A. PROTECTION AGAINST DESICCATION

In soil, the water potential undergoes large and sometimes rapid fluctuations. Unicellular organisms are in equilibrium with the outside water potential because water moves freely across biological membranes. When soil dries, microorganisms loose water, experience plasmolysis, and eventually die. The main function that has been suggested for EPS is to protect bacteria and unicellular algae against desiccation.[3,4,10,30]

1. Survival to Desiccation and EPS

Several studies investigated the survival of EPS-producing vs. -nonproducing strains of bacteria when exposed to desiccation (e.g., *Klebsiella aerogenes,*[51] *Pseudomonas solanacearum,*[52] *Pseudomonas* sp,[17] and various rhizobia[53-57]). Actually, no clear relationship has been found between EPS production and survival to drying. On the other hand, the addition of purified EPS to bacteria enhanced their resistance to desiccation.[53] It may be noted that polysaccharides are frequently used as inoculant carriers for microbes.[8-60] Severe bacteria mortality occurs when drying the inoculum, but it is reduced with polymer.[61]

Recently, the physiological responses of soil bacteria during a desiccation event have been studied.[62] Soil pseudomonad were grown on sands and subjected to a desiccation event. The bacteria increased their EPS production during desiccation: carbon was allocated to polysaccharide synthesis rather than to protein synthesis (Figure 4), suggesting a role of EPS as protectants.

2. Water-Related Properties of EPS

Another approach to this question is to investigate whether the properties of the extracellular polysaccharides are consistent with a protective role.

First, let us consider the water retention curves of EPS. The very high amounts of water held by many EPS of microorganisms, especially at high water potentials, are consistent with the idea of an external storage of water. At low water potential, the amounts of water retained by EPS are still large (Figure 3), and greater than those retained by bacterial cells.[63-65] Water may be preferentially adsorbed by EPS rather than by the cell itself: rather than a prevention from drying, a more thorough drying of the cells would then be achieved with EPS. This concept is consistent with the findings of Bushby and Marshall[63] and Al Rashidi et al.[64] that the less desiccation-sensitive strains of rhizobia were those retaining less water at low water potentials, and with the fact that for many strains, a better viability during storage is ensured at very low water potentials rather than at intermediate ones.[61]

Extracellular polysaccharides of soil microorganisms examined to date have a high capacitance at high water potentials, as have plant extracellular polysaccharides[65]: small changes in water potential (Ψ) are associated with substantial losses of water (Figure 3). The high capacitance provides buffering against short-term water fluctuations; when dry conditions are met EPS can loose large amounts of water by evaporation with only small changes in Ψ: the bacteria will remain in a relatively high Ψ environment. However, such a mechanism cannot be at play at low water potentials, given the low capacitance of EPS in that range. The same traits can be extended to the immediate soil surroundings of bacteria when impregnated with EPS. Extracellular polysaccharides were shown to enhance the water retention of clay and sand materials.[62,66,67]

Microbial EPS were also shown to decrease the drying rates of sands and clays. Both the rates of water loss and water potential decrease were lowered.[62,67] It takes more time for a sand or clay amended with EPS to reach low water potentials than for a pure clay or sand. It means more time available for metabolic adjustment of the bacteria. The buffering effect of EPS against water potential fluctuations may operate during both desiccation and rehydration of soils.[67] The sudden increase in water potential during rain events is drastic for microorganisms.[68]

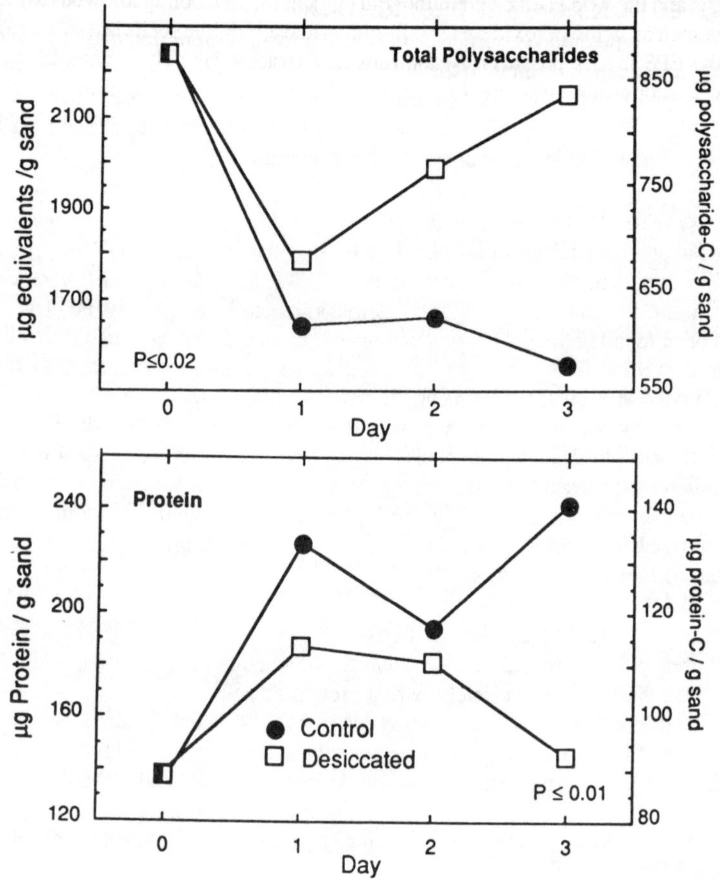

Figure 4 Polysaccharide and protein production by a soil *Pseudomonas* during desiccation.[62]

Another limiting factor for microorganisms when a soil dries is that less water is available for diffusion and water paths become discontinuous as the soil fabric desaturates. The diffusion of nutrients may then be drastically reduced. The EPS matrices have small-sized pores and shrink upon desiccation; they are likely to remain water saturated in a wide range of water potentials.[67,69] It can be hypothesized, then, that EPS can increase the range of water potential at which the diffusion of nutrients can take place.

Given the rather conflicting set of results found in the literature, it is difficult to draw conclusions on this point. Although the direct measurements of bacterial survival to desiccation in relation to EPS production are not convincing, we emphasize that the properties of EPS are compatible with a protective role. The protective role of EPS was usually assumed to consist of a storage of water for the bacteria or algae.[3] It is more likely that EPS act as buffers, reducing the rates and amplitudes of water potential fluctuations. In addition, EPS may allow the diffusion of nutrients to persist in the immediate environment of microbes, as water potential decreases.

B. METABOLISM

Most microorganisms appear unable to use their EPS as a reserve of carbon and energy,[4] but polysaccharides may have an indirect role in nutrition as suitable media for enzyme activities. The EPS of white rot fungi were shown to establish a physical junction between the fungus and the wood cells.[70] Immunogold cytochemical labeling allowed extracellular enzymes such as lignin peroxidases,[70,71] β–glucosidase,[72] or hemicellulases[73] to be located within the EPS matrix produced by the fungus. Extracellular polysaccharides may then serve as a supporting structure for extracellular enzymes. Furthermore, an important retention of water by the EPS matrix could help to maintain an environment suitable for enzyme activities in a wide range of soil water potentials.

C. IONIC BARRIERS

Bacteria rely entirely on diffusion for their livelihood. They cannot reach out and ingest a food substrate, nor can they disperse away a toxic substance. This applies in particular to metal ions that can turn, over a narrow concentration range, from a growth-promoting element to a toxic one. In soil environments, ion and metal concentrations naturally fluctuate widely. It can be predicted that acidic polysaccharides located as external layers around the cells must exert a profound influence on the traffic of charged molecules into and out of the cells. Extracellular polysaccharides have been suggested to be modulators of metal ion concentration, and consequently of pH, in the vicinity of microbes.[74] Several studies demonstrated that the secretion of EPS increased the tolerance of bacteria to metals,[74,75] e.g., a *Bradyrhizobium* was found to be more sensitive to aluminum when grown on media which depleted its EPS production,[75] or that growth on metal-enriched media resulted in increased EPS production.[76] The amount of EPS produced by 20 strains of rhizobia was well correlated with their tolerance to Al and acidity.[77] In another study, however, a nonmucoid mutant of rhizobia and the parent strain had the same tolerance to aluminum.[78] However, all these studies were conducted *in vitro,* and there is no assessment of such a function of EPS under soil conditions.

The mechanism of protection is a decrease in the free ion concentration at the cell surface, as a result of the metal complexation by the EPS. Complexation involves organics, e.g., pyruvates or succinates that are present in most microbial extracellular polysaccharides (Section II). Anionic sites such as carboxylates[79] are the most abundant and most effective electron-donating group of EPS.[74] In the study of Cunningham and Munns with rhizobia,[77] however, the chelating ability of extracted EPS did not correlate with the sensitivity of the strain to aluminum.[80] A feature unique to polysaccharides is that two or more electron-donating groups, on separate polymer chains, may bind with a polyvalent metal.[74] Steric fit then becomes very important. A well-known example is the egg-box model that describes the complexation of calcium by two chains of alginate.[45] In many cases such as the latter, the cation induces conformational ordering of EPS and promotes gel formation.[74,81]

Considering the anionic groups and the physical structure of EPS, these polymers can be presumed to affect more generally the diffusion of ions, substrates, or metabolites in or out of the cells, acting as adsorbents or as molecular sieves.

D. ADHESION OF MICROORGANISMS TO SOIL PARTICLES

Another suggested role of microbial EPS is to facilitate or enhance cell adhesion to soil particles.[82] Extracellular polysaccharides can be involved in the adhesion of microorganisms to mineral particles, being the outermost surface of microorganisms. The attachment of bacterial cells to solid surfaces has been demonstrated to enhance EPS production,[30] and as the production of EPS continues after an initial attachment of the cells, EPS have been suggested to contribute to a firmer adhesion as well.[83] The role of EPS in adhesion to solids is quite well documented in freshwater or marine environments,[82] and prediction of what occurs in soil is often extrapolated from these environments. In soils, evidence for the involvement of EPS in adhesion to solid surfaces is mostly indirect, i.e., many EPS are able to adsorb on clay minerals or sands (see following). Some direct, but qualitative, evidence also comes from the observation with SEM of the attachment of EPS-producing microorganisms to sand grains[84-87] or to clay particles.[52] An amorphous film coats the bacteria and forms a meniscus between the solid and the cell. With low-temperature scanning electron microscopy (SEM) the fibrillar nature of this coat can be observed as desiccation proceeds slowly.[87] Similar evidence comes from transmission electron microscopy (TEM) observations of the contact between clay and microorganisms, with specific staining of polysaccharides (Figure 5). Considering the relative size of most clay particles (0.1 to 2 μm) and bacteria, it is mainly the adhesion of clay to EPS surfaces that is observed.

V. INTERACTIONS OF EPS WITH SOIL MINERALS

A. ADSORPTION OF EPS ON CLAY MINERALS

The adsorption of polysaccharides to clay minerals has been studied for many years and is not reviewed here. The adsorption of microbial polysaccharides follows main traits common to other polymers. Neutral or anionic polysaccharides adsorb to clay surfaces, often in high amounts and with high affinity and low reversibility. Adsorption of anionic EPS is pH dependent, and involves polyvalent cations.[88-91] As frequently stated, the individual bonds are weak ones (hydrogen bondings, Van der Waals forces), but the result over a large macromolecule is a very strong binding.

The adsorption leads to an interparticle bridging of the clay by EPS strands.[66,69] The ability of polysaccharides to form intermolecular linkages is outstanding. The combined effects of adsorption and cross-links between different EPS strands set up a very effective bridging, leading to an organomineral network.[92] The size of EPS molecules can attain that of soil clay particles[69] (also cf. Section IV), and along with their fibrillar shape, it allows EPS to entangle clay particles. A very probable consequence of large size and strong interaction with clay surroundings is that EPS do not diffuse much in soil fabrics, but remain close to where they have been excreted.

B. EPS-MEDIATED AGGREGATION IN CLAY MATERIALS

In several laboratory experiments, microorganisms or roots were grown in artificial soils made of pure clays.[93-96] By SEM and TEM, the contact between the clay particles and fungal or bacterial cells was observed. The clay particles tended to align tangentially to the cell wall of fungi or to the surface of EPS-embedded bacterial colonies. Fibrillar material, assumed to be polysaccharides in SEM observations, were produced by the fungi and impregnated the surrounding clay platelets.[94] Similar enmeshment of clay particles by polysaccharide fibers was observed in the immediate vicinity of bacteria.[96] "Nascent aggregates"[96] were thus formed with sizes of 1 to 20 μm, depending on the size and shape of the cell or colony.[95]

C. EPS-MEDIATED AGGREGATION IN SOILS

Are these structures formed in soils as well? Bacteria embedded within an EPS matrix and coated with clay particles are quite a common feature in TEM observations of thin sections of soils (Figure 1).[7-11] However the abundance of clay-coated bacteria compared to that of "free" ones is undetermined.

A very common fact when dispersing a soil by mechanical means is that some microaggregates, resistant to dispersion, remain in the silt or clay fractions and may account for a significant proportion of the soil.[14,97,98] By TEM observation,[97,99,100] many of these microaggregates are identified as bacteria coated with EPS and clay minerals, as shown in Figure 5.

Such aggregates of microbial origin are outstanding in regard to soil aggregation, quantitatively, inasmuch as the microaggregates are observed to persist after death of the bacteria (Figures 5b and c) and because they represent a specific level of aggregation (1

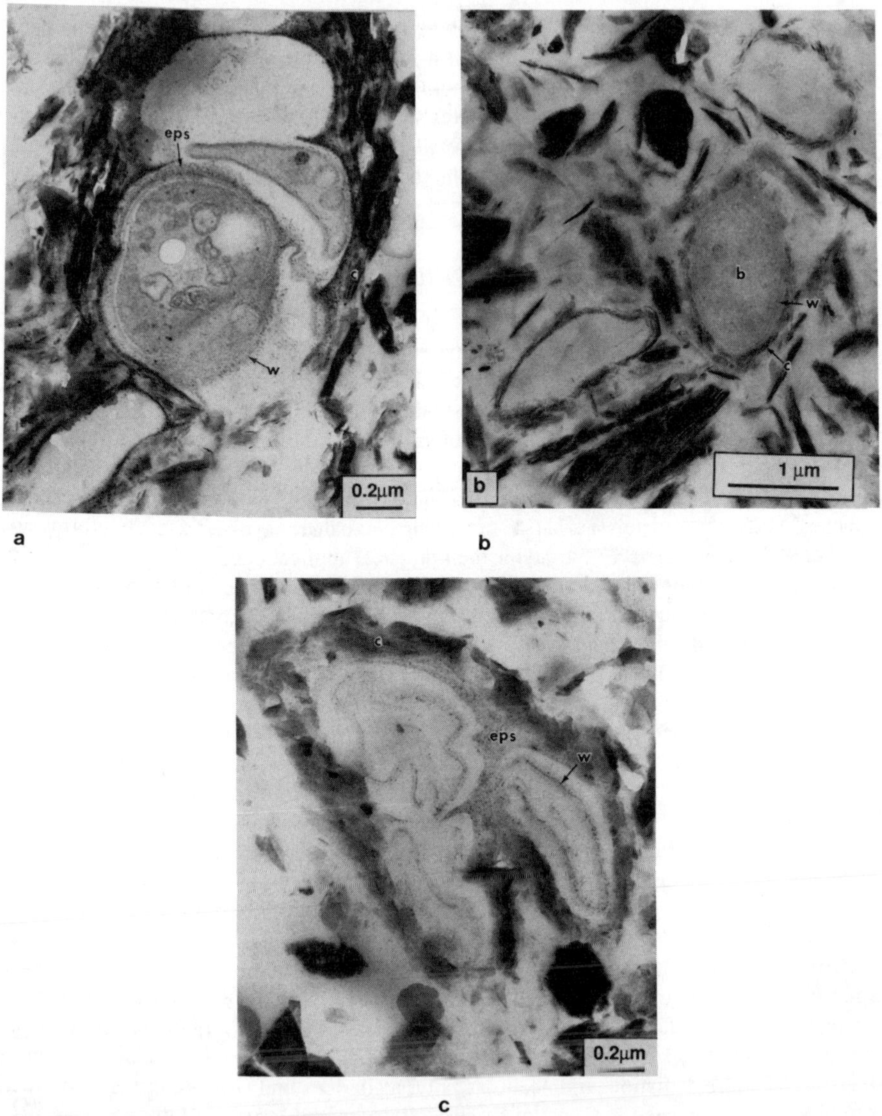

Figure 5 Ultrasonic-resistant micro-aggregates in the coarse clay fraction (0.2 to 2 µm) of a silty soil. (a) Microorganism surrounded by EPS and clay particles; (b) live and dead (empty) bacteria (U/Pb staining); (c) remnants of dead microorganisms. Only the wall and EPS remain. In a and c, the polysaccharides were stained with silver[6] and are visualized by the silver grains. b = bacteria, w = wall, EPS = extracellular polysaccharide, c = clay. Photograph by C. Chenu.

to 20 µm). The enmeshment of such microaggregates by hyphae, roots, or their binding by other agents leads to increases in water-stable aggregates in greater size classes. The participation of microbial polysaccharides in the stability of soil aggregates has been demonstrated in numerous studies.[101] Bacterial cells and EPS located in microaggregates may in turn be physically protected from biodegradation.[14]

The interaction of EPS with its surroundings leads at least some bacteria and fungi to live in a microenvironment constituted by an inner layer of polysaccharide and an outer layer of clays which is more or less impregnated with polysaccharide. Considering the effects of EPS coatings alone, we may wonder how bacteria function in the soil microenvironment where the diffusion of nutrients, metabolites, and gases is likely to be restricted. The main question is probably whether bacteria embedded in such structures are physiologically active or whether they represent some form of resistance of non-sporulating bacteria to drastic conditions in soils as suggested by Kilbertus et al.[10] To date, these questions remain unanswered but are central to an understanding of microbial ecology in soils.

VI. CONCLUSION

In this chapter, we have emphasized the function of polysaccharides in forming a frequent interface between microorganisms and the surrounding soil constituents. Studies are, however, needed to quantify the secretion of EPS *in situ* and to identify the nutritional and environmental factors of their production under soil conditions.

The adhesive and gel-like properties of EPS allow them to bind soil particles into stable aggregates and attribute to polysaccharides a major role in soil aggregation. A potential exists for improving the physical quality of soils with microbial polysaccharides.[102] Mass-cultured polysaccharide-producing microalgae have been successfully evaluated as soil conditioners,[103,104] although their practical applications remain limited, mainly because of an incomplete understanding of the factors that govern colonization of soil and polysaccharide production and because of cost limitations.[105] Carbon and N availability can be controlled with cover crop and fertilizer in order to promote microbial EPS production.[18] Management of EPS production *in situ* appears to be a potentially effective method to improve aggregate stability in cultivated soils.

We have discussed some of the possible functions of EPS for soil microorganisms. Extracellular polysaccharides should confer some advantages on its producer, but these are not clearly elucidated at present. Extracellular polysaccharides undoubtedly have a major role in specific recognition processes in plant-microorganism associations. These were not discussed here. Extracellular polysaccharides may also have adaptive advantages for soil microorganisms by (1) establishing a physical junction between the solid substratum or food source and the microorganism and (2) buffering environmental fluctuations in water potential as well as in pH or free ion concentration. For these functions, the macromolecular properties of polysaccharides are essential. Hence, EPS may have an integral function and help survival in a variety of nonideal conditions. Knowledge of the factors which control the survival of microorganisms in soils is necessary, as specialized microorganisms are now being introduced to soils for agriculture or soil bioremediation. Extracellular polysaccharides and the microorganisms that produce them also have a potential in agriculture as soil structure-improving agents, inoculant carriers for introduced microorganisms, or even as agents for metal complexation in the bioremediation of soils.

ACKNOWLEDGMENTS

The author wishes to thank E. Roberson and M. Robert for thoughtful discussions, and G. Villemin for the photograph of Figure 1.

REFERENCES

1. **Martin, J. P.,** Decomposition and binding action of polysaccharides in soils, *Soil Biol. Biochem.*, 3, 33, 1971.
2. **Lynch, J. M. and Bragg, E.,** Microorganisms and soil aggregate stability, *Adv. Soil Sci.*, 30, 133, 1985.

3. **Hepper, C.,** Extracellular polysaccharides of soil bacteria, in *Soil Microbiology,* Walker, N., Ed., John Wiley & Sons, New York , 1975, 93.
4. **Dudman W. F.,** The role of surface polysaccharides in natural environments, in *Surface Carbohydrates of the Procaryotic Cell,* Sutherland, I., Ed., Academic Press, New York, 1977, 357.
5. **Hebbar, K. P., Guéniot, B., Heyraud, A., Colin Morel, P., Heulin, T., Balandreau, J., and Rinaudo, M.,** Characterization of exopolysaccharides produced by rhizobacteria, *Appl. Microbiol. Biotechnol.,* in press.
6. **Thierry J. P.,** Mise en évidence de polysaccharides sur coupes fines en microscopie électronique, *J. Microsc.,* 6, 987, 1967.
7. **Foster, R. C.,** Localization of organic materials in situ in ultrathin sections of natural soil fabrics using cytochemical techniques, in *Submicroscopic Techniques as Applied to Thin Sections of Undisturbed Soil Materials,* Bisdom, E. B. A., Ed., International Working Group on Submicroscopy, Wageningen, The Netherlands, 1981, 309.
8. **Foster R. C.,** Microenvironments of soil microorganisms, *Biol. Fertil. Soils,* 6, 189, 1988.
9. **Proth, J.,** *Evolution de la Microflore d'une Rendzine Forestière Récemment Privée de ces Apports Naturels en Litière de Charme. Etude Microbiologique et Ultrastructurale,* Ph.D. thesis, Université de Nançy, 1978.
10. **Kilbertus, G., Proth, J., and Verdier, R. B.,** Effets de la dessiccation sur les bactéries gram-négatives d'un sol, *Soil Biol. Biochem.,* 11, 109, 1979.
11. **Villemin, G. and Toutain, F.,** Contribution à la mise au point d'une méthode d'observation des microagrégats du sol en microscopie électronique à transmission, in *Soil Micromorphology,* Proc. VIIth Int. Work. Meet. Soil Micromorphology, Fedoroff, N., Bresson, L. M., and Courty, M. A., Eds., Association Francaise pour l'Etude des Sols, Plaisir, France, 1987, 43.
12. **Oades, J. M. and Wagner, G. H.,** Biosynthesis of sugars in soils incubated with ^{14}C glucose and ^{14}C dextran, *Soil Sci. Soc. Am. Proc.,* 35, 914, 1971.
13. **Cheshire, M. V.,** *Nature and Origin of Carbohydrates in Soils,* Academic Press, London, 1979.
14. **Oades, J. M.,** Soil organic matter and structural stability mechanisms and implications for management, *Plant Soil,* 76, 319, 1984.
15. **Murayama, S.,** Changes in polysaccharide composition during the decomposition of straws under field conditions, *Soil Sci. Plant Nutr.,* 30, 367, 1984.
16. **Oades, J. M.,** Studies on polysaccharides. III. Composition of polysaccharides in some Australian soils, *Aust. J. Soil Res.,* 5, 103, 1972.
17. **Roberson, E. B.,** Extracellular Polysaccharide Production by Soil Bacteria: Environmental Control and Significance in Agricultural Soils, Ph.D. thesis 9228833, University of California, Berkeley, 1991.
18. **Roberson, E. B., Sarig, S., and Firestone, M. K.,** Cover crop management of polysaccharide-mediated aggregation in an orchard soil, *Soil Sci. Soc. Am. J.,* 55, 734, 1991.
19. **Murayama, S., Cheshire, M. V., Mundie, C. M., Sparling, G. P., and Shepherd, H.,** Comparison of the contributions to soil organic matter fractions, particularly carbohydrates, made by plant residues and microbial products, *J. Sci. Food Agric.,* 3, 125, 1979.
20. **Dalal, R. C. and Henry, R. J.,** Cultivation effects on carbohydrate contents of soil and soil fractions, *Soil Sci. Sci. Am. J.,* 52, 1361, 1988.
21. **Cheshire, M. V. and Mundie, C. M.,** The distribution of labelled sugars in soil particle size fractions as means of distinguishing plant and microbial residues, *J. Soil Sci.,* 32, 65, 1981.
22. **Jarman, T. R., Deavin, L., Slocombe, S., and Righelato, R. C.,** Investigation of the effect of environmental conditions on the rate of exopolysaccharide synthesis in *Azotobacter vinelandii, J. Gen. Microbiol.,* 107, 59, 1978.
23. **Sutherland, I. W.,** Biosynthesis and composition of gram-negative bacterial extracellular and wall polysaccharides, *Annu. Rev. Microbiol.,* 39, 243, 1985.
24. **Elliot, L. F. and Lynch, J. M.,** The effects of available carbon and nitrogen in straw on soil and ash aggregation and acetic acid production, *Plant Soil,* 78, 335, 1984.
25. **Kroen, W. K. and Rayburn, W. R.,** Influence of growth status and nutrients on extracellular polysaccharide synthesis by the soil algae *Chlamydomonas mexicana* (Chlorophycae), *J. Phycol.,* 20, 253, 1984.
26. **Linton, J. D., Jones, D. S., and Woodward, S.,** Factors that control the rate of exopolysaccharide production by *Agrobacterium radiobacter* NICB 11883, *J. Gen. Microbiol.,* 133, 2979, 1987.

27. **Arad, S. M., Friedfman, O. D., and Rotem, A.,** Effect of nitrogen on polysaccharide production in a *Porphyridium* sp., *Appl. Environ. Microbiol.,* 54, 2411, 1988.
28. **Mody, B. R., Bindra, M. O., and Modi, V. V.,** Capsule development and structural characterization of acidic extracellular polysaccharides secreted by cowpea rhizobia, *Curr. Microbiol.,* 20, 145, 1990.
29. **Auer, D. P. F. and Seviour, R. J.,** Influence of varying N sources on polysaccharide production by *Aerobasidium pullulans* in batch culture, *Appl. Microbiol. Biotechnol.,* 32, 637, 1990.
30. **Bengtson, G.,** Bacterial exopolymer and PHB production in fluctuating ground water habitats, *FEMS Microbiol. Ecol.,* 86, 15, 1991.
31. **Jarman, T. R. and Pace, G. W.,** Energy requirements for microbial exopolysaccharide synthesis. *Arch. Microbiol.,* 137, 231, 1984.
32. **Hollander, J. A., Battenhaussen, C. W., and Stouthamer, A. H.,** Growth yields, polysaccharide production and energy conservation in continuous cultures of *Rhizobium trifolii, Antonie van Leuvenhoek J. Microbiol. Serol.,* 45, 401, 1979.
33. **Uhlinger, D. J. and White, D. C.,** Relationship between physiological status and formation of extracellular polysaccharide glycocalyx in *Pseudomonas atlantica, Appl. Environ. Microbiol.,* 45, 64, 1983.
34. **Flaibani, A., Olsen, Y., and Painter, T. J.,** Polysaccharides in desert reclamation: composition of exocellular proteoglycan complexes produced by filamentous blue green and unicellular edaphic algae, *Carbohydr. Res.,* 190, 235, 1989.
35. **Barclay, W. R. and Lewin, R. A.,** Microalgal polysaccharide production for the conditioning of agricultural soils, *Plant Soil,* 88, 159, 1985.
36. **Metting, B.,** Population dynamics of *Chlamydomonas sajao* and its influence on soil aggregate stabilisation in the field, *Appl. Environ. Microbiol.,* 51, 1161, 1986.
37. **Arad, S. M., Adda, M., and Cohen, E.,** The potential of production of sulfated polysaccharides from *Porphyridium, Plant Soil,* 89, 117, 1985.
38. **Cunningham, S. D. and Munns, J. E.,** The correlation between extracellular polysaccharide production and acid tolerance in *Rhizobium, Soil Sci. Soc. Am. J.,* 48, 1273, 1984.
39. **Chao, W. L. and Chen, C. L.,** Role of exopolymer and acid-tolerance in the growth of bacteria in solutions with high copper ion concentration, *J. Gen. Appl. Microbiol.,* 37, 363, 1991.
40. **Ernst, A., Chen, T. W., and Böger, P.,** Carbohydrate formation in rewetted terrestrial cyanobacteria, *Oecologia,* 72, 574, 1987.
41. **Herbert, D., Philips, P. J., and Strange, R. E.,** Chemical analysis of microbial cells, in *Methods in Microbiology,* Norris, R. J. and Ribbons, D. W., Eds., Academic Press, London, 1971, 209.
42. **Jenkinson, D. S. and Ladd, J. N.,** Microbial biomass in soil: measurement and turnover, in *Soil Biochemistry,* Vol. 5, Paul, E. A. and Ladd, J. N., Eds., Marcel Dekker, New York, 1981, 415.
43. **Ladd, J. N., Jocteur Monrozier, L., and Amato, M.,** Carbon turnover and nitrogen transformations in and alfisol and vertisol amended with [U-^{14}C] glucose and [^{15}N] ammonium sulfate, *Soil Biol. Biochem.,* 4, 359, 1992.
44. **Beveridge, T. J. and Graham, L. J.,** Surface layers of bacteria, *Microbiol. Rev.,* 55, 684, 1991.
45. **Powell, D. A.,** Structure, solution properties and biological interactions of some microbial exocellular polysaccharides, in *Microbial Polysaccharides and Polysaccharases,* Berkeley, R. C. W., Gooday, G. W., and Ellwood, D. C., Ed., Academic Press, New York, 1979.
46. **Fleet, G. H. and Phaff, H. J.,** Fungal glucans, structure and metabolisms, in *Plant Carbohydrates II. Extracellular Carbohydrates,* Tanner, W. and Loewus, F. A., Eds., Springer-Verlag, New York, 1981, 416.
47. **Mort, A. J. and Bauer, W. D.,** Composition of the capsular and extracellular polysaccharides of Rhizobium japonicum, *Plant Physiol.,* 66, 158, 1980.
48. **Hollingsworth, R. I., Dazzo, F. B., Hallenga, K., and Muselman, B.,** The complete structure of the trifoliin A lectin binding capsular polysaccharide of *Rhizobium trifolii* 843, *Carbohydr. Res.,* 172, 97, 1988.
49. **Morris, E. and Norton, I. T.,** Polysaccharide aggregation in solutions and gels, in *Aggregation Processes in Solution. Studies in Physical and Theoretical Chemistry,* Elsevier, Amsterdam, 1983, 17.
50. **Cottrell, I. W.,** Industrial potential of fungal and bacterial polysaccharides, in *Fungal Polysaccharides,* Sanford, P. A. and Matzua, K., Eds., ACS Symp. Ser. 126, American Chemical Society, Washington, D.C., 1980, 251.

51. **Bitton, G., Henis, Y., and Lahav, N.,** Influence of clay minerals, humic acid and bacterial capsular polysaccharide on the survival of *Klebsiella aerogenes* exposed to drying and heating in soils, *Plant Soil,* 45, 65, 1976.
52. **Schmit, J. and Robert, M.,** Action des argiles sur la survie d'une bactérie phytopathogène *Pseudomonas solanacearum* E.F.S., *C. R. Acad. Sci. Paris,* 229, II, 733, 1984.
53. **Peña-Cabriales, J. J. and Alexander, M.,** Survival of *Rhizobium* in soils undergoing drying, *Soil Sci. Soc. Am. J.,* 43, 962, 1979.
54. **Bushby, H. V. A. and Marshall, K. C.,** Some factors affecting the survival of root-nodule bacteria on desiccation, *Soil Biol. Biochem.,* 9, 143, 1977.
55. **Osa-Afiana, L. O. and Alexander, M.,** Clays and the survival of *Rhizobium* in soil during dessiccation, *Soil Sci. Soc. Am. J.,* 46, 285, 1982.
56. **Hartel, P. G. and Alexander, M.,** Role of extracellular polysaccharide production and clays in the desiccation tolerance of Cowpea *Bradyrhizobia, Soil Sci. Soc. Am. J.,* 50, 1193, 1986.
57. **Mary, P., Ochin, D., and Tailliez, R.,** Growth status of rhizobia in relation to their tolerance to low water activities and desiccation stresses, *Soil Biol. Biochem.,* 18, 179, 1986.
58. **Bashan, Y.,** Alginate beads as synthetic inoculant carrier for slow release of bacteria that affect plant growth, *Appl. Environ. Microbiol.,* 51, 1089, 1986.
59. **Lewis, J. A. and Papavizas, G. C.,** Application of *Trichoderma* and *Gliocladium* in alginate pellets for control of *Rhizoctonia* damping-off, *Plant Pathol.,* 36, 438, 1987.
60. **Jung, G., Mugnier, J., Diem, H. G., and Dommergues, Y. R.,** Polymer entrapped *Rhizobium* as an inoculant for legumes, *Plant Soil,* 65, 219, 1982.
61. **Mugnier, J. and Jung, G.,** Survival of bacteria and fungi in relation to water activity and the solvent properties of water in biopolymer gels, *Appl. Environ. Microbiol.,* 50, 108, 1985.
62. **Roberson, E. B. and Firestone, M. K.,** The relationship between desiccation and exopolysaccharide production by a soil *Pseudomonas, Appl. Environ. Microbiol.,* 58, 1284, 1992.
63. **Bushby, H. V. A. and Marshall, K. C.,** Water status of rhizobia in relation to their susceptibility to desiccation and to their protection by montmorillonite, *J. Gen. Microbiol.,* 99, 19, 1977.
64. **Al-Rashidi, R. K., Loynachan L. R., and Frederick, L. R.,** Desiccation tolerance of four strains of *Rhizobium japonicum, Soil Biol. Biochem.,* 14, 489, 1982.
65. **Morse, S. R.,** Water balance in *Hemizonia luzulifolia:* the role of extracellular polysaccharides, *Plant, Cell Environ.,* 13, 39, 1990.
66. **Chenu, C.,** Influence of a fungal polysaccharide, scleroglucan, on clay microstructures, *Soil Biol. Biochem.,* 21, 299, 1989.
67. **Chenu, C.,** Clay- or sand-polysaccharide associations as models for the interface between micro-organisms and soil: water-related properties and microstructure, *Geoderma,* 56, 143, 1993.
68. **Kieft, T. L., Soroker, E., and Firestone, M. K.,** Microbial biomass response to a rapid increase in water potential when dry soil is wetted, *Soil Biol. Biochem.,* 19, 119, 1987.
69. **Chenu, C. and Jaunet, A. M.,** Cryoscanning electron microscopy of microbial extracellular polysaccharides and their association with minerals, *Scanning,* 14, 360, 1992.
70. **Ruel, K. and Joseleau, J. P.,** Involvement of an extracellular glucan sheath during the degradation of populus wood by *Phanerochaete chrysosporium, Appl. Environ. Microbiol.,* 57, 374, 1991.
71. **Daniel, G., Nilson, T., and Perterson, B.,** Intra- and extracellular localization of lignin peroxydase during the degradation of solid wood and wood fragments by *Phanerochaete chrysosporium* by using transmission electron microscopy and immunogold labelling, *Appl. Environ. Microbiol.,* 55, 871, 1989.
72. **Gallagher, I. M. and Evans, C. S.,** Immunogold cytochemical labelling of β glucosidase in the white rot fungi Coriolus versicolor, *Appl. Microbiol. Biotechnol.,* 32, 588, 1990.
73. **Green, F., II, Clausen, C. A., Larsen, M. J., and Hihhley, T. J.,** Immunoscanning electron microscopic localization of extracellular wood degrading enzymes within the fibrillar sheath of the brown-rot fungus *Postia placenta, Can. J. Microbiol.,* 38, 898, 1992.
74. **Geesey, G. G. and Jang, L.,** Interactions between metal ions and capsular polymers, in *Metal Ions and Bacteria,* Beveridge, T. J. and Doyle, R. J., Eds., John Wiley and Sons, New York, 1991, 325.
75. **Leon-Barrios, M., Gutiérrez-Navarro, A. M., Pérez-Galdona, R., Díaz-Siverio, J., Trujillo, J., and Corzo, J.,** Acidic exopolysaccharide from *Bradyrhizobium (Chamaecytisus proliferus), J. Appl. Bacteriol.,* 72, 91, 1992.

76. **Cunningham, S. D. and Munns, J. E.,** Effects of rhizobial extracellular polysaccharide on pH and aluminum activity, *Soil Sci. Soc. Am. J.,* 48, 1276, 1984.
77. **Chao, W. L. and Chen, C. L. F.,** Role of exopolymer and acid tolerance in the growth of bacteria in solutions with high copper ion concentration, *J. Gen. Appl. Microbiol.,* 37, 363, 1991.
78. **Kingsley, M. T. and Bohlool, B. B.,** Extracellular polysaccharide is not responsible for aluminium tolerance of *Rhizobium leguminosarum bv. Phaseoli* CIAT899, *Appl. Environ. Microbiol.,* 58, 1095, 1992.
79. **Tan, E. L. and Loutit, M. W.,** Concentration of molybdenum by extracellular material produced by rhizosphere bacteria, *Soil Biol Biochem.,* 8, 461, 1976.
80. **Cunningham, S. D. and Munns, J. E.,** The correlation between extracellular polysaccharide production and acid tolerance in *Rhizobium, Soil Sci. Soc. Am. J.,* 48, 1273, 1984.
81. **Ferala, N. F., Champlin, A. K., and Fekete, F. A.,** Morphological differences in the capsular polysaccharide of nitrogen-fixing *Azotobacter chroococcum* as a function of iron and molybdenum starvations, *FEMS Microbiol. Lett.,* 33, 137, 1986.
82. **Sutherland, I. W.,** Microbial exopolysaccharides — their role in microbial adhesion in aqueous systems, *Crit. Rev. Microbiol.,* 40, 173, 1984.
83. **Allison, D. G. and Sutherland, I. W.,** The role of exopolysaccharides in adhesion of freshwater bacteria, *J. Gen. Microbiol.,* 133, 1319, 1987.
84. **Campbell, R. and Porter, R.,** Low temperature scanning electron microscopy of microorganisms in soil, *Soil Biol. Biochem.,* 14, 241, 1982.
85. **Robbins, R. J., Hall, D. O., Turner, R. J., and Rhodes, M. J. C.,** Mucilage acts to adhere cyanobacteria and cultured plant cells to biological and inert surfaces, *FEMS Microbiol. Lett.,* 34, 155, 1986.
86. **Chaubal, R., Wilmot, V. A., and Wynn, K. W.,** Visualization, adhesiveness and cytochemistry of the extracellular matrix produced by uredinospore germ tubes of *Puccinia sorghi, Can. J. Bot.,* 69, 2044, 1991.
87. **Roberson, E. B., Chenu, C., and Firestone, M. K.,** Microstructural changes in bacterial exopolysaccharides during desiccation, *Soil Biol. Biochem.,* 25, 1299, 1993.
88. **Theng, B. K. G.,** *Formation and Properties of Clay-Polymer Complexes,* Elsevier, Amsterdam, 1979.
89. **Chenu, C., Pons, C. H., and Robert, M.,** Interaction of kaolinite and montmorillonite with neutral polysaccharides, in *Proc. International Clay Conference, Denver, 1985,* Schultz, L. G., Van Olphen, H., and Mumpton, F. A., Eds., Clay Minerals Society, Bloomington, IL, 1987, 375.
90. **Aly, S. M. and Letey, J.,** Polymer and water quality effects on flocculation of montmorillonite, *Soil Sci. Soc. Am. J.,* 52, 1453, 1988.
91. **Khandall, R. K., Chenu, C., Lamy, I., and Tercé, M.,** Adsorption of different polymers on kaolinite and their effect on flumequine adsorption, *Appl. Clay Sci.,* 6, 343, 1992.
92. **Chenu, C. and Guérif, J.,** Mechanical strength of clay minerals as influenced by an adsorbed polysaccharide, *Soil Sci. Soc. Am. J.,* 55, 1076, 1991.
93. **Guckert, A., Breisch, H., and Reisinger, O.,** Interface sol — racine. I. Etude au microscope électronique des relations mucigel — argiles — microorganismes, *Soil Biol. Biochem.,* 7, 241, 1975.
94. **Dorioz, J. M. and Robert, M.,** Etude expérimentale de l'interaction entre champignons et argiles: conséquences sur la microstructure des sols, *C. R. Acad. Sci. Paris,* 295(II), 511, 1982.
95. **Dorioz, J. M. and Robert, M.,** Aspects microscopiques des relations entre les microorganismes ou végétaux et les argiles. conséquences sur les microorganisations et la microstructuration des sols, in *Soil Micromorphology. Proc. VIIth Int. Work. Meet. Soil Micromorphology,* Fedoroff, N., Bresson, L. M., and Courty, M. A., Eds., AFES, Plaisir, France, 1987, 353.
96. **Dormaar, J. F. and Foster, R. C.,** Nascent aggregates in the rhizosphere of perenial ryegrass *(Lolium perenne* L.), *Can. J. Soil Sci.,* 71, 465, 1991.
97. **Jocteur Montrozier, L., Ladd, J. N., Fitzpatrick, R. W., Foster, R. C., and Raupach, M.,** Physical properties, mineral and organic components and microbial biomass content of size fraction in soils of contrasting aggregation, *Geoderma,* 49, 37, 1991.
98. **Balesdent, J., Pétraud, J. P., and Feller, C.,** Effet des ultrasons sur la distribution granulométrique des matières organiques des sols, *Sci. Sol,* 29, 95, 1991.
99. **Tiessen, H. and Stewart, J. W. B.,** Light and electron microscopy of stained microaggregates: the role of organic matter in soil aggregation, *Biogeochemistry,* 5, 312, 1988.

100. **Feller, C., François, C., Villemin, G., Portal, J. M., Toutain, F., and Morel, J. L.,** Nature des matières organiques associées aux fractions argileuses d'un sol ferrallitique, *C. R. Acad. Sci. Paris,* 312(II), 1491, 1991.
101. **Robert, M. and Chenu, C.,** Interactions between microorganisms and soil minerals, in *Soil Biochemistry,* Vol. 7, Stotzky, G. and Bollag, J. M., Eds., Marcel Dekker, New York, 1991, 30.
102. **Gouzou, L., Burtin, G., Philippy, R., Bartoli, F., and Heulin, T.,** Effect of inoculation with *Bacillus polymixa* on soil aggregation in the wheat rhizosphere: preliminary examination, *Geoderma,* 56, 479, 1993.
103. **Metting, B.,** Dynamics of wet and dry aggregate stability from a three year microalgal soil conditioning experiment in the field, *Soil Sci.,* 143, 139, 1987.
104. **Rao, D. L. N. and Burns, R. G.,** The effect of surface growth of blue green algae and bryophytes on some microbiological, biochemical and physical properties, *Biol. Fertil. Soils,* 9, 239, 1990.
105. **Zimmerman, W. J.,** Microalgal biotechnology and applications in agriculture, in *Soil Microbial Ecology,* Metting, F. B., Ed., Marcel Dekker, New York, 1992, 457.

Chapter 18

Low-Molecular-Weight Aliphatic Carboxylic Acids in Some Andisols of Japan

M. Tani, T. Higashi, and S. Nagatsuka

CONTENTS

I. INTRODUCTION

Among low-molecular-weight aliphatic carboxylic acids (LACAs) in soils, oxalic acid and citric acid both have relatively low dissociation constants, and have attracted much attention with respect to their roles in soil formation processes through their complexation reactions with metals and capacity of proton release.[1-3] In Japan, much research has been conducted on the effects of volatile organic acids, mainly acetic and propionic acids, on the growth of rice plants as a phytotoxin.[4] Consequently, the actual amounts and compositions of LACAs in the forest soils of Japan have not been investigated.

Although there were some studies on LACAs in the forest soil solution of Spodosols[5] because LACAs were thought to be involved in Podzolization, and in Ultisols and Alfisols[6,7] related to the reaction with aluminum, few studies on LACAs in other soil orders have been conducted. Andisols especially should have received more attention, because they have more distinctive properties, described as "Andic soil properties,"[8] that influence the adsorption-desorption of LACAs in soils. This characteristic is due to short-range ordered minerals such as allophane and imogolite that have the ability to develop positive charges. The purposes of this study were (1) to investigate the distribution of LACAs in some Andisol profiles of Japan in terms of their amounts and compositions and (2) to establish the influential effects of soil solution pH and the contents of amorphous Al and Fe on the adsorption of LACAs in the soils.

II. MATERIALS AND METHODS

A. SOIL SAMPLES

Soil sampling sites are located at Tsukuba city, Ibaraki, Japan (Table 1). Two types of Andisols, Wet Andosols[9] (Aquic Melanudands) and light-colored Andosols (Typic Hapludands) were selected for this study. To some degree, the sites differ in the soil color of the topsoil and in the moisture condition of the subsoil. The soil samples of wet Andosols were taken from six horizons on November 21, 1990, and those of light-colored

0-87371-914-X/95/$0.00+$.50

Table 1 Brief Description of the Two Andisol Profiles

Sample number	Horizon	Depth (cm)	Soil color[a]
	Wet Andosols (Aquic Melanudands)		
SKA1	A1	0–15	7.5YR2/2
SKA2	A2	15–31	7.5YR2/2
SKA3	2A3	31–51	7.5YR2/1
SKA4	2A4	51–69	7.5YR2/3
SKA5	3A5	69–91	7.5YR2/1.5
SKB1	3Bw	91–133	7.5YR4/6
	Light-colored Andosols (Typic Hapludands)		
TKA1	A1	0–11	7.5YR3/3
TKA2	A2	11–41	7.5YR3.5/4
TKB1	Bw21	41–75	7.5YR4/6
TKB2	Bw2/2wAb	75–100	7.5YR4/4
			7.5YR4/6

[a] By Munsell color system.

Andosols from four horizons on May 11, 1991. The vegetation on both Andosols is lowland forest of Japanese red pine *(Pinus densiflora* Sieb. et Zucc.), with an understory of bamboo species *(Pleioblastus Chino* Makino). Field-moist soil samples were passed through a 7-mm sieve after removal of roots or litter as possible, and stored at 4°C until the extraction of LACAs.

B. ANALYTICAL METHODS

Soil samples were analyzed for soil pH by glass electrode, the contents of amorphous Al and Fe by the method of Blakemore et al.,[10] and total carbon and nitrogen contents by a CN analyzer (Yanaco MT-500). The amounts and composition of LACAs in soil samples were determined by the method of Tani et al.[11] Ten kinds of LACAs — formic, acetic, propionic, butyric, lactic, oxalic, fumaric, succinic, malic, and citric acids — were identified and determined simultaneously by high-performance liquid chromatography (HPLC). The method includes the following three steps.

1. Crude extract was extracted with ultrapure water and NH_4-phosphate buffer (0.1 *M* $NH_4H_2PO_4$-H_3PO_4, pH 2) from field-moist soil samples, and separated (shaken for 30 min, centrifuged at 10,000 g for 10 min.), followed by physical purification (removal of fine suspended particles and hydroxides in crude extract by filtration through a 0.025-μm Millipore filter).
2. Chemical purification and concentration. The filtrates were acidified to pH 1.6 to 1.7 with diluted H_2SO_4 and treated with PVP (polyvinyl pyrrolidone) resin to remove phenolic substances. This was followed by continuous extraction with diethyl ether (4 h) and evaporation to dryness under reduced pressure. The samples were redissolved with 2 ml of ultrapure water and treated with a cation-exchange cartridge.
3. Analysis with HPLC. Two types of columns were used: TSK gel ODS805M and TSK gel SCX (TOSOH Corp.). The LACAs were detected by UV (210 nm) and EC analyses, respectively. The details of analytical conditions of HPLC and the validities of the HPLC analysis are described in our previous paper.[11]

III. RESULTS AND DISCUSSION

A. LOW-MOLECULAR-WEIGHT ALIPHATIC CARBOXYLIC ACIDS (LACAS)

The amounts of ten LACAs in soil samples obtained from extractions by ultrapure water and NH_4-phosphate buffer solution are given in Table 2. These extractants were used to differentiate the forms of LACAs in the soils. Water-soluble forms were extracted by ultrapure water, while those adsorbed on soil surfaces were evaluated by the difference between extractions by phosphate buffer and ultrapure water.

The total amounts of LACAs, the sum of water-soluble and adsorbed forms, extracted by phosphate buffer ranged from 131.4 to 404.3 μmol kg^{-1} oven-dried soil. This range of values is comparable with those in the upper A and B horizons in some forest soils in Japan.[11] As for the total amounts of LACAs in Wet Andosols (SK) and Light-colored Andosols (TK), the TK samples always showed higher amounts in any horizons; this is probably due to the difference in sampling season. Low-molecular-weight organic acids are in a dynamic state in soils, so they are likely affected by climatic conditions and biological activities depending on the season. Five kinds of LACAs (formic, acetic, oxalic, succinic, and citric acids) accounted for about 90% of the total amounts of LACAs in all samples (Table 2). However, the amounts of formic and acetic acids, relative to the total amounts of LACAs, in subsoils of Andisols were higher compared with those of podozolic soil (Haplorthods) and brown forest soil (Dystrochreps) in Japan.[11]

As for the profile distribution of LACAs, the water-soluble LACAs were highest in SKA4 and TKB1, while those adsorbed in SKB1 and TKB1 (Figure 1), the observed distribution of water-soluble LACAs is due to the increase in volatile LACAs (formic acid and especially acetic acid) in subsoils at about 60 cm in depth in both Andisols (Table 2). It was generally recognized that in Andisols of Japan, due to low bulk density and high

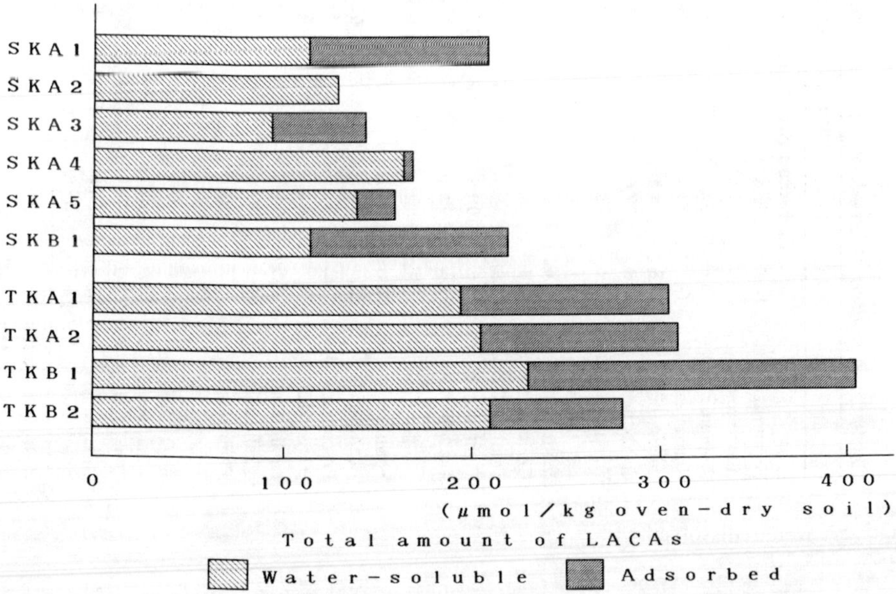

Figure 1 Total amounts of low-molecular-weight aliphatic carboxylic acids (LACAs) in soil samples (SK and TK).

Table 2 Amounts of Low-Molecular-Weight Aliphatic Carboxylic Acids (LACAs) in Soil Samples

	SKA1	SKA2	SKA3	SKA4	SKA5	SKB1	TKA1	TKA2	TKB1	TKB2
Extracted with Ultrapure Water (μmol kg^{-1} oven-dried soil)										
Formic acid	14.8	38.8	26.5	37.6	35.1	14.4	29.7	34.5	50.1	40.7
Acetic acid	15.0	41.5	17.9	43.7	49.3	50.8	22.8	31.0	60.0	46.0
Propionic acid	Tr.[a]	Tr.	Tr.	Tr.	Tr.	Tr.	4.9	2.6	4.8	4.1
Butyric acid	Tr.	Tr.	Tr.	Tr.	Tr.	Tr.	Tr.	Tr.	Tr.	Tr.
Lactic acid	1.8	18.5	6.5	32.3	13.6	1.4	4.1	0.7	5.3	4.1
Oxalic acid	58.8	27.6	27.8	31.4	26.6	27.9	43.7	53.1	32.6	45.3
Fumaric acid	0.1	Tr.	Tr.	Tr.	Tr.	Tr.	0.1	0.2	0.1	0.1
Succinic acid	17.1	9.4	8.0	9.3	9.6	14.9	39.1	36.1	29.3	27.6
Malic acid	ND[b]	Tr.	ND	ND	ND	ND	4.7	Tr.	3.4	3.7
Citric acid	8.3	8.4	7.6	8.9	8.2	6.8	44.9	48.1	43.3	36.5
Total LACAs	115.9	144.2	94.4	166.5	142.3	116.1	194.0	206.3	228.8	208.1
Total VCAs[c]	31.6	98.8	50.9	113.7	97.9	66.5	61.5	68.8	120.1	94.9
Total NVCAs[d]	94.3	45.4	43.4	52.8	44.4	49.6	132.5	137.5	108.6	113.1
Extracted with NH_4-Phosphate Buffer (μmol kg^{-1} oven-dried soil)										
Formic acid	24.6	32.7	32.9	48.0	38.3	54.2	50.1	59.0	60.5	39.8
Acetic acid	25.9	22.3	41.9	42.3	47.1	75.1	49.4	64.7	55.0	55.9
Propionic acid	5.2	Tr.	Tr.	Tr.	Tr.	Tr.	14.0	4.9	6.0	4.7
Butyric acid	Tr.	Tr.	Tr.	Tr.	Tr.	Tr.	Tr.	Tr.	Tr.	Tr.
Lactic acid	17.6	11.1	4.3	6.5	6.3	7.7	7.4	12.0	8.7	7.0
Oxalic acid	89.6	40.8	39.4	46.8	44.2	62.1	95.5	87.3	185.0	92.9
Fumaric acid	0.4	0.1	0.2	0.1	0.1	0.1	0.6	0.2	0.1	0.1
Succinic acid	15.1	10.3	9.6	11.1	10.6	11.5	36.8	31.9	31.2	29.5
Malic acid	1.6	0.4	1.2	3.2	2.8	0.6	3.1	3.2	5.7	5.0
Citric acid	27.7	13.7	14.0	12.5	10.6	11.5	49.4	44.4	51.7	47.3
Total LACAs	207.7	131.4	143.4	170.6	159.9	222.7	306.3	307.6	404.3	282.2
Total VCAs	73.3	66.0	79.1	96.8	91.7	137.0	120.9	140.7	130.2	107.4
Total NVCAs	134.4	65.3	64.3	73.8	68.2	85.7	185.5	167.0	274.1	174.8

[a] Trace amounts.
[b] No peak detected.
[c] Total amounts of volatile LACAs (VCAs: formic, acetic, propionic, butyric, and lactic acids).
[d] Total amounts of nonvolatile LACAs (NVCAs: oxalic, fumaric, succinic, malic, and citric acids).

water retention ability,[12] especially in subsoils, the moisture contents tend to be higher in field-moist conditions than other soils (Table 3). Hence, high moisture content would be easily maintained throughout the year, resulting in an increase of anaerobic conditions and enhancing the production of volatile LACAs in the subsoils of Andisols. This is similar to hydromorphic soils such as paddy soils.[1] In a previous paper,[11] we also found that the high moisture contents of field-moist soil samples in the B horizon of volcanic ash soils could be responsible for the production of volatile organic acids by bacteria under relatively anaerobic conditions. This is probably due to the relatively large number of micropores instead of macropores. In a recent study, some researchers ignored the detection of acetic acid in soil solution.[7] However, an increase of acetic and formic acids would be expected in subsoils of Ultisols with poor water permeability in argillic horizons. The total amounts of adsorbed LACAs also increased in the subsoils of the Andisols studied; this was mainly caused by the increase of the amounts of adsorbed oxalic acid (Table 2).

B. EFFECTS OF SOIL SOLUTION PH AND CONTENTS OF AMORPHOUS AL AND FE ON THE AMOUNTS OF ADSORBED LACAS

The amounts of adsorbed volatile LACAs (VCAs) were higher in SKA1, SKA3, SKB1, TKA1, and TKA2. The pHs of crude extracts from these five horizons, except TKA1, were also higher than other horizons (Table 3). Here, the pH of crude extracts by ultrapure water was supposed to be related to soil solution pH. Linear regression analysis between the total amounts of adsorbed VCAs and soil solution pH showed a positive correlation coefficient ($r = 0.962$) when a sample of TKA1 was excluded, where the amounts of adsorbed formic acid relative to the total amounts of adsorbed VCAs were higher than others (Figure 2a). Since the dissociation constants of formic and acetic acids are relatively higher than other LACAs, it was expected that soil solution pH strongly affected the adsorption of these VCAs. No distinct positive correlation was found between the total amounts of adsorbed nonvolatile LACAs (NVCAs) and soil solution pH (Figure 2b).

Table 3 Some General Properties of Soil Samples

Sample	pH[a] DW	pH[a] CE	Alox (g kg^{-1})	Feox[b] (g kg^{-1})	T-C (g kg^{-1})	T-N (g kg^{-1})	*MC*[c] (kg kg^{-1})
SKA1	4.61	6.06	39.3	16.7	81.9	5.0	0.75
SKA2	5.60	5.50	42.3	18.5	40.4	2.1	0.68
SKA3	5.54	5.64	52.1	21.4	64.2	3.3	1.00
SKA4	5.44	5.23	58.9	22.9	56.7	2.9	1.12
SKA5	5.61	5.42	56.4	20.9	55.0	2.8	1.17
SKB1	5.63	6.42	74.8	13.3	18.2	1.0	1.12
TKA1	5.06	4.68	44.7	22.9	68.8	4.5	1.03
TKA2	5.21	6.39	64.3	28.7	35.6	2.3	0.97
TKB1	5.52	5.27	77.6	35.8	23.8	1.6	1.24
TKB2	5.65	5.37	73.8	30.4	18.7	1.2	1.40

[a] pH (DW): pH of air-dried soil samples (soil:water = 1:2.5), pH (CE): pH of crude extracts by ultrapure water.

[b] Amorphous aluminum and iron extracted with acid ammonium oxalate (pH 3.0).

[c] Moisture contents of field-moist soil samples determined by drying for 16 h at 105°C.

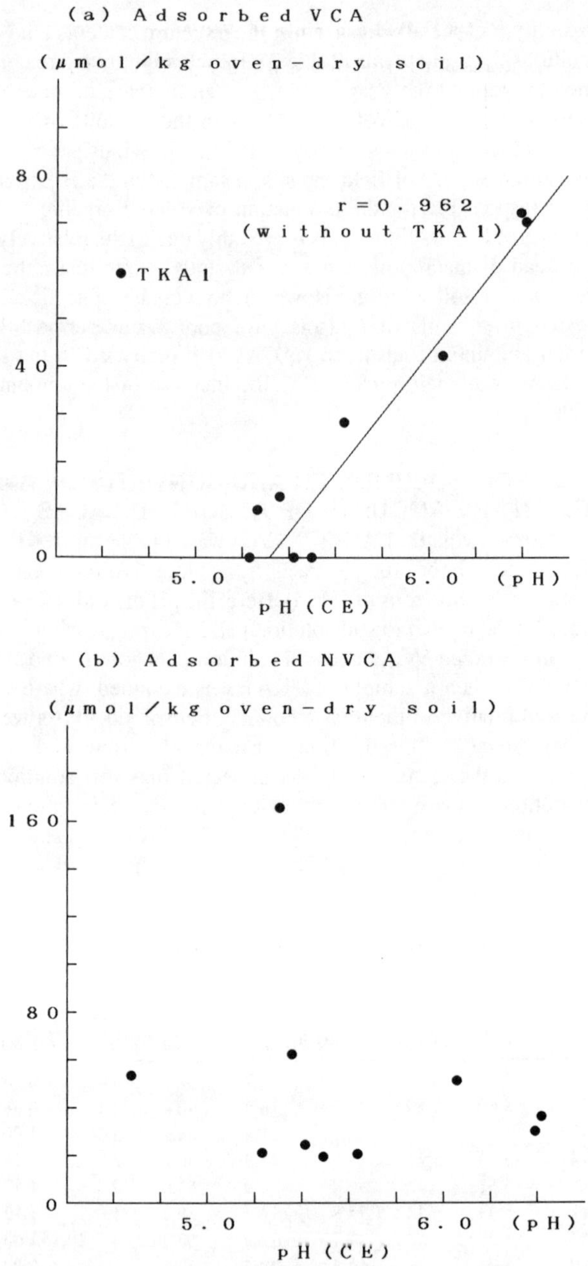

Figure 2 Correlation of soil solution pH with total amounts of adsorbed (a) VCAs and (b) NVCAs.

As for the amounts of adsorbed NVCAs, an increase of adsorbed oxalic acid in subsoils was evident. There was also a lot of adsorbed citric acid in the subsoils (Table 2). It is known that short-range ordered minerals are abundant in Andisols; they are, in fact, one of the most diagnostic inorganic components for this order. [As shown in Table 3, the contents of amorphous Al and Fe, including Al and Fe binding to humus to some extent, extracted by

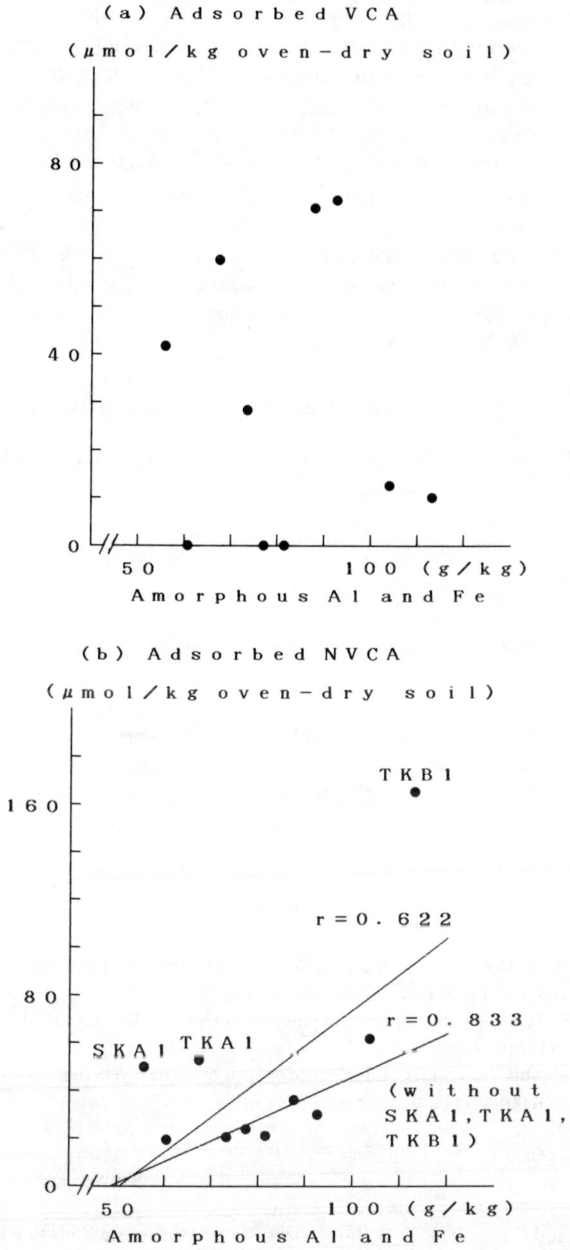

Figure 3 Correlation of contents of amorphous Al and Fe with total amounts of adsorbed (a) VCAs and (b) NVCAs.

acid ammonium oxalate tended to increase more in subsoils than topsoils.] It was assumed that the increase in the total amounts of adsorbed NVCAs in the subsoils of Andisols was influenced by the increase in the contents of amorphous Al and Fe. As expected, linear regression analysis between the total amounts of adsorbed NVCAs and the contents of amorphous Al and Fe showed a relatively high correlation coefficient ($r = 0.622$). Moreover, the value of the correlation coefficient increased to r = 0.833 when we analyzed for only seven soil samples, excluding SKA1, TKA1, and TKB1, where extremely higher amounts of adsorbed oxalic acid were observed compared with other NVCAs (Figure 3b). However, no apparent positive correlation was found between the total amounts of adsorbed VCAs and the contents of amorphous Al and Fe (Figure 3a).

From the results of the distribution of LACAs in some Andisol profiles in Japan, it was obvious that the adsorption processes of LACAs were highly correlated with soil solution pH in the case of VCAs, mainly formic and acetic acids, and with the contents of amorphous Al and Fe in the case of NVCAs, mainly oxalic and citric acids. These phenomena could be attributed to the dissociation of LACAs caused by a rise of soil solution pH and the change in the capacity of adsorption-positive sites on the surface of amorphous Al and Fe of Andisols.

IV. SUMMARY AND CONCLUSIONS

We examined the profile distribution of ten low-molecular-weight aliphatic carboxylic acids (LACAs) in some Andisols of Japan with respect to their amounts and compositions. The total amounts of LACAs ranged from 131.4 to 404.3 μmol kg^{-1} oven-dried soil. Five LACAs (formic, acetic, oxalic, succinic, and citric acids) accounted for about 90% of the composition of all samples from different horizons. The following conclusions about the distribution of LACAs were reached:

1. The volatile LACAs, mainly consisted of formic and acetic acids, were higher in subsoils than topsoils, and tended to increase with the capacity of water retention related to the actual moisture content of field-moist soil samples.
2. Two soil factors, soil solution pH and contents of amorphous Al and Fe, affected the amounts and the compositions of adsorbed LACAs. Specifically, it was shown that soil solution pH had an effect on the adsorption processes of volatile LACAs (VCAs), and the contents of amorphous Al and Fe on that of nonvolatile LACAs (NVCAs).

REFERENCES

1. **Stevenson, F. J.,** Organic acids in soil, in *Soil Biochemistry,* Vol. 1, McLaren, A. D. and Peterson, G. H., Eds., Marcel Dekker, New York, 1967, chap. 5.
2. **Bruckert, S.,** Influence des composés organiques solubles sur la pédogénese en milieu acide, *Ann. Agron.,* 21, 421, 1970.
3. **Huang, P. M. and Violante, A.,** Influence of organic acids on crystallization and surface properties of precipitation products of aluminum, in *Interaction of Soil Minerals with Natural Organics and Microbes,* Huang, P. M. and Schnitzer, M., Eds., SSSA Spec. Publ. No. 17, Soil Science Society of America, Madison, WI, 1986, chap 6.
4. **Takijima, Y.,** Determination of organic acids in paddy soils, *Soil Sci. Plant Nutr.,* 28, 7, 1960 (in Japanese).
5. **Fox, T. R. and Comerford, N. B.,** Low-molecular-weight organic acids in selected forest soils of the southeastern USA, *Soil Sci. Soc. Am. J.,* 54, 1136, 1990.
6. **Hue, N. V., Craddock, G. R., and Adams, F.,** Effect of organic acids on aluminum toxicity in subsoils, *Soil Sci. Soc. Am. J.,* 50, 28, 1986.
7. **Tam, S. C. and McColl, J. G.,** Aluminum-binding ability of soluble organics in Douglas fir litter and soil, *Soil Sci. Soc. Am. J.,* 55, 1421, 1991.
8. **Soil Survey Staff,** *Keys to Soil Taxonomy,* 4th ed., Soil Management Support Services, Virginia, 1990, 23.
9. **Amano, Y.,** Classification of cultivated soils in Japan, *JARQ,* 18, 275, 1985.
10. **Blakemore, L. C., Searle, P. L., and Daly, B. K.,** Extractable iron, aluminum and silicon, in *Methods for Chemical Analysis of Soils,* Blakemore, L. C. et al., Eds., N Z Bur. Sci. Rep. 80, DSIR, Lower Hutt, New Zealand, 1987, 71.

11. **Tani, M., Higashi, T., and Nagatsuka, S.,** Dynamics of low-molecular-weight aliphatic carboxylic acids (LACAs) in forest soils. I. Amount and composition of LACAs in different types of forest soils in Japan, *Soil Sci. Plant Nutr.,* 39, 485, 1993.
12. **Shoji, S. and Ono, T.,** Physical and chemical properties and clay mineralogy of Andosols from Kitakami Japan, *Soil Sci.,* 126, 297, 1978.

Chapter 19

Relationship Between Organic Acids and Microorganisms on a Long-Term Cropping Site in Southeastern Australia

W. J. Slattery, P. M. Mele, and G. R. Steed

CONTENTS

I. INTRODUCTION

Traditional cropping systems in temperate regions of Australia involve cultivation and stubble burning. These practices increase the potential for soil degradation,[1] decrease total organic C,[2] and are associated with decreases in microbial activity and microbial biomass,[3,4] compared with conservation tillage systems that use direct-drilling methods and stubble retention.

In the same region of southeastern Australia, about 12 million ha of agriculturally productive land is affected by soil acidity. It has been recognized that a major limiting factor associated with plant and root growth on strongly acid soils [pH <4.5 (0.01 mol $CaCl_2$ L^{-1})] is the presence of toxic Al.[5-9] Aluminum in soils has been shown to exist in both monomeric and polymeric forms.[10-12] The role of low-molecular-weight organic acid ligands in complexing Al and Fe has been demonstrated.[13-16] Natural organic acids in soil solutions are derived from plant and animal residues as well as microbial metabolism.[17] Simple aliphatic acids in soil solutions have been shown to be continuously produced through the activity of microorganisms in the soil.[18,19]

Under cropping systems, the type of rotation and hence vegetative residues would be important in influencing the concentration of specific organic acids, which in turn will dictate to some extent the diversity of the microbial population that colonize the residues.[17] Alternatively, the diversity of the population may influence the kind of organic acids present in the soil.

Accurate quantification of soil microbial populations is well recognized as being problematic. Viable count procedures using artificial growth media severely underestimate populations.[20,21] The choice of this procedure in this study was based upon resource

0-87371-914-X/95/$0.00+$.50

availability and validated by the selection of nutrient-poor media, selective media, and/or soil pretreatment for improved accuracy. Interpretation of results is mindful of these shortcomings.

In this study, we attempted to identify the dominant low-molecular-weight organic acids that have the potential to complex with Al on a moderately acidic, sandy clay loam soil with long-term (11 years) tillage treatments and correlate these changes with a previous microbial study. In order to understand the origin of the low-molecular-weight organic acids, we attempted to relate the incidence of aliphatic and aromatic acids with soil microbial populations.

II. MATERIALS AND METHODS

A. SITE DESCRIPTION

The tillage study was situated at the Rutherglen Research Institute (146° 25'E., 36° 06'S.) in northeastern Victoria, Australia, on a duplex soil (Dy 3.33,[22] with a sandy clay loam surface horizon, which ranges over 100 to 200 mm in depth, as described in Carter and Mele.[23] The tillage experiment was initiated in 1981. Continuous wheat *(Triticum aestivum* L.) was grown in the experiments from 1981 to 1986, followed by peas *(Pisum sativum* L.) (1987), wheat (1988), lupin *(Lupinus angustifolius* L.) (1989), and wheat (1990). Two tillage treatments were used for the present study: DDR — direct drilling, stubble retained (4 to 6 t ha^{-1} per annum) and DDB — direct drilling, stubble burnt (in late March, early April).

B. SOIL SAMPLING

Soil samples were obtained from the two tillage treatments in 1988, 1991, and 1992. In 1988 and 1991, the samples were obtained in the months of April and January, respectively, at three depths: 0 to 25, 25 to 50, and 50 to 100 mm. Viable microbial counts were performed on the 1988 sampling, whereas biomass C and organic C (0 to 25 mm) was performed on the 1991 sampling. In 1992, the samples were obtained from the two tillage treatments in January, at three depths (0 to 25, 25 to 50, and 50 to 100 mm) for, soil pH, Al, Mn, organic C, and organic acids.

C. MICROBIAL ANALYSIS

Microbial biomass determinations were made using a modified procedure of the Jenkinson and Powlson[24] chloroform fumigation technique. After the 10-d incubation, the samples were extracted with 2 *M* KCl (1:5, soil:extractant ratio) and the extract analyzed for ninhydrin-reactive N according to the method of Amato and Ladd.[25] The ninhydrin-reactive N (fumigated minus control) multiplied by a constant, 21, provides an estimate of microbial biomass C.[25] Viable counts of the various groups of soil microorganisms were determined by the plate dilution technique using selective media,[26] on field-moist sieved (<2 mm) soil.

D. SOLID-PHASE EXTRACTIONS

All soils were dried immediately after sampling from the field in a fan-forced oven at 40°C for 24 h, rumbled to pass through a 2-mm sieve, and stored in sealed plastic bags at room temperature. Aluminum, manganese, and pH were measured in a 1:5, soil to 0.01 mol $CaCl_2$ l^{-1} extractant suspension. The Al content of the extract was determined using a modified colorimetric procedure of Wilson and Sergeant,[26] where pyrocatechol violet was used as the color-developing reagent. Manganese was determined by atomic absorption spectroscopy. Organic carbon was measured by the method of Heanes,[28] which requires a chromic acid digestion followed by adsorption at 600 nm.

E. SOLUTION PHASE

Soil solutions were prepared by placing 200 g of oven-dried (40°C) soil (air-dried soil was not available) into polyethylene bags and moistened to 10 kPa water content and incubated at room temperature for 24 h.[29] The soil solution was extracted using the method of Gillman.[30] Immediately following extraction, the soil solutions were filtered (0.22 μ*M* membrane filter) for electrical conductivity, pH, and monomeric Al. Monomeric Al in the presence of organic acids was determined using the short-term colorimetric procedure developed by Kerven et al.[31] Total soluble Al was determined by the lanthanum and iron method.[32] Short-chain aliphatic and aromatic organic acids were determined by high-performance liquid chromotography (HPLC) after sample concentration and clean-up.[29]

III. RESULTS

A. SOIL CHEMISTRY

Organic C, pH, Al, and Mn were not significantly different ($p < 0.05$) between the two stubble treatments after a 10-year period (Table 1).

Soil solution pH was significantly ($p < 0.05$) lower in the 25- to 50-mm soil depth, for both treatments, corresponding to a high level of monomeric Al. Total soil solution Al decreased significantly ($p < 0.05$) with soil depth for the stubble-retained treatment but showed no significant ($p < 0.05$) change for the stubble-burnt treatment.

Monomeric Al (Al-mono) for the three soil depths (0 to 25, 25 to 50, and 50 to 100 mm) was not significantly ($p < 0.05$) different between the treatments. The concentration of complexed Al (Al-comp) was significantly ($p < 0.05$) higher in the 0- to 25-mm soil depth where stubble was retained compared to stubble burning. The distribution of complexed Al in the stubble-retained treatment was predominantly (94%) in the 0- to 25-mm soil depth compared to the stubble-burnt treatment, where complexed Al was more evenly distributed throughout the three (0 to 25, 25 to 50, and 50 to 100 mm) soil depths (Table 2).

Concentrations of low-molecular-weight aliphatic acids were significantly ($p < 0.05$) higher in all three soil depths for stubble-burnt treatments, and the concentration of low-molecular-weight aromatic acids was significantly ($p < 0.05$) higher in the 0- to 25-mm soil depth for the stubble-retained treatment (Table 3).

B. MICROBIAL BIOMASS AND VIABLE COUNTS

Microbial biomass C measured in 1988 showed a significant ($p < 0.05$) increase for the stubble-retained treatment at all depths (Table 3). The values for viable counts given in Table 4 are ratios of the numbers in the stubble-retained treatment divided by the numbers

Table 1 Some soil Properties of a Sandy Clay Loam Under Two Residue Management Systems for 10 Years

Tillage system	Depth (mm)	Bulk density[a] (mg m^{-3})	Organic C[a] (g kg^{-1})	pH[b] (H_2O)	pH[b] ($CaCl_2$)	A1[b] (mg kg^{-1})	Mn[b] (mg kg^{-1})
DDR	0–25	1.23	11.8	6.51	5.70	—	—
	0–100	1.30	9.2	5.98	4.75	5.9	12.0
DDB	0–25	1.33	11.9	6.56	5.75	—	—
	0–100	1.30	9.3	6.06	4.84	6.1	12.0
SE		0.05	0.3	0.07	0.06	0.7	8.0

[a] 1991 sampling.
[b] 1992 sampling.

Table 2 Soil Solution Properties (1992 Sampling) of a Sandy Clay Loam Under Two Residue Management Systems for 11 Years

Depth (mm)	Al total (μ*M*)	Al mono (μ*M*)	Al org (μ*M*)	% Al comp	pH	I
			DDR			
0–25	9.4	0.5	8.9	95	5.92	0.003
25–50	5.8	5.2	0.6	9	4.62	0.006
50–100	1.8	1.8	—	—	5.15	0.002
			DDB			
0–25	3.9	0.5	3.4	87	5.51	0.003
25–50	4.2	3.0	1.2	28	4.61	0.008
50–100	2.9	0.5	2.4	83	5.60	0.002
LSD ($p = 0.05$)	1.9	2.2	4.0		0.42	

in the stubble-burnt treatment. Viable counts for fluorescent pseudomonads, facultative anaerobes, and actinomycetes were significantly higher in the stubble-burnt treatments for all soil depths. Gram-negative bacteria were significantly higher in the stubble-retained treatment only at the 50- to 100-mm soil depth (Table 4).

IV. DISCUSSION

Burning of wheat stubble had little influence on the organic C levels in these direct-drilled systems. This finding is not consistent with other long-term minimum tillage studies, which suggest that the burning of stubble will decrease organic C.[2,33]

Under minimum or no-tillage systems, it has been reported that there is a higher proportion of aliphatic C (70 to 90%) compared with aromatic C.[34] Since both treatments used in this study represent no-till systems, a change in aliphatic and aromatic proportions between the two treatments may not be expected. Since microorganisms produce aliphatic acids through the decomposition of organic C, a greater concentration of aliphatic acids found in the stubble-burnt treatment could be related to several factors: (1) an increase in the biochemical activity within the soil associated with a larger pool of active microorganisms, (2) a greater accumulation of organic acids due to a lower active microbial population, or (3) a changed microbial population that results in an excess of succinic acid due to the selective microbial decomposition of aromatic compounds within lignin.[35,36] Increased microbial activity in the stubble-burnt treatments is postulated as resulting from an increase in the rate of N mineralization.[26]

A major failing with this study is that soil samplings for microbial biomass and specific populations were taken 1 and 3 years before the sampling for soil solution organic acids. Seasonal changes in microbial biomass have, however, been observed,[37] and at this site,[23] biomass C was always higher in the stubble-retained treatment compared with the stubble-burnt treatment. This finding is further verified by an earlier study[38] on this same site using the same treatments, in which biomass C was higher in the stubble-retained treatments. It is reasonable, then, to assume that the biomass C levels are of a similar difference between the two treatments in 1992 as they were in 1991.

In future studies, concurrent analysis of microbial populations and soil solution chemistry will be performed on the same samples. Although microbial biomass C, highest in DDR treatments, is a relatively small proportion of the total organic C pool (2%),[24] the rapid turnover rates ensure that new substrate C will be incorporated into the humus. It

Table 3 Short-Chain Organic Acid Concentrations (μM) in Soil Solutions as Measured by HPLC After Filtration Through Sep-Pak C-18 Cartridges

Depth (mm)	Biomass C[a] (μg C g^{-1} soil)	Aliphatic acids[b]				Aromatic acids[b]		Total	
		Citric	Tartaric	Malic	Succinic	Phthalic	*p*-OH Benzoic	Aliphatics	Aromatics
					DDR				
0–25	420	—	—	—	3.80	0.72	0.44	3.8	1.16
25–50	309	—	—	—	1.75	0.37	—	1.8	0.37
50–100	306	3.75	—	1.45	3.92	0.04	—	9.1	0.04
					DDB				
0–25	133	0.64	—	—	10.50	0.49	—	11.1	0.49
25–50	124	1.01	—	—	11.90	0.21	—	12.9	0.21
50–100	106	5.14	0.77	2.07	14.60	0.06	—	22.6	0.06
LSD ($p = 0.05$)	34	1.72			3.09	0.21			

[a] 1991 sampling.
[b] 1992 sampling.

Table 4 Microbial Populations (1988 Sampling) in a Sandy Clay Loam Under Different Stubble Management Systems for 8 years

	Depth (mm)		
	0–25	25–50	50–100
Spore-forming aerobes	0.71	1.27	1.02
Fluorescent pseudomonads	0.12[b]	0.09[c]	0.10[b]
Gram-negative bacteria	1.28	0.80	1.73[a]
Facultative anaerobes	0.18[b]	0.19[b]	0.39[a]
Actinomycetes	0.87	0.41[a]	0.23[b]
Fungi	0.43[a]	0.76	0.81

Note: Values are expressed as the ratio of viable counts in retained treatment (DDR) divided by the burnt treatment (DDB).

[a] $p < 0.05$.
[b] $p < 0.01$.
[c] $p < 0.001$.

is this intermediary process of decomposition and transformation that leads to the formation of short-chain aliphatic and aromatic acids important in the complexing of toxic Al. Stubble-burnt treatments contained a higher concentration of aliphatic acids, consistent with the view that simple aliphatic acids are continuously produced by the action of microorganisms.[19] A higher concentration of succinate in the stubble-burnt treatment is consistent with the higher number of some (facultatively anaerobic bacteria, fluorescent pseudomonads, actinomycetes, and fungi) culturable microorganisms, since the microbial breakdown of aromatic compounds to produce succinate by ring cleavage is performed by several kinds of microorganisms.[35,36] Since organic acids extracted from soil solutions represent the balance between production, sorption, and decomposition, the accumulation of any one organic acid may be representative of a change in the microbial population that prevents or accelerates the processes of production or decomposition. The presence of *p*-OH benzoic acid in the stubble-retained treatments may be related to the presence of this acid in wheat straw.[17]

For the three depths combined, the stubble-burnt treatment contained a higher concentration of the strongly Al-complexing organic ligands (citrate, tartrate, and malate). In contrast to this, the complexed Al was higher in the stubble-retained treatment, suggesting that other anions not detected are complexing Al, or that the aromatic acid ligands play a more significant role. If we consider the concentration of organic acids for each individual depth, then, in the stubble-retained treatments, the 0- to 25-mm depth is the most significant Al-complexing depth, whereas for stubble-burnt treatments, Al-complexing occurred throughout, down to 100 mm. Where stubbles were retained, virtually all of the Al was complexed in the 0- to 25-mm soil depth, compared with stubble-burnt treatments where Al was leached down the soil profile. The higher soil pH for this depth in the stubble-retained treatment is consistent with strong Al-complexing preventing Al from participating in hydrolysis reactions and a subsequent lowering of soil pH.

V. SUMMARY AND CONCLUSIONS

The specific relationship between microbial populations and organic acids in soil solutions are not clear in this preliminary study. Relationships between organically complexed Al and Al toxicity are more clearly defined. However, the importance of different stubble-managed systems needs further investigation, especially on acid soils. This study identifies the need to investigate further the origins of important Al-complexing organic ligands. The need to identify farming systems that ameliorate the effects of soil acidity by complexing toxic Al will be beneficial to maintaining crop yields on acid soils.

ACKNOWLEDGMENTS

The authors gratefully acknowledge Dr. G. R. Steed for the use of a long-term tillage trial at the Rutherglen Research Institute, and the Grains Research and Development Corporation of Australia and the National Soil Conservation Program of Australia for financial support of this work at the Rutherglen Research Institute.

REFERENCES

1. **Carter, M. R., and Steed, G. R.,** The effects of direct-drilling and stubble retention on hydraulic properties at the surface of duplex soils in north-eastern Victoria, *Aust. J. Soil Res.*, 30, 505, 1992.
2. **Chan, K. Y., Roberts, W. P., and Heenan, D. P.,** Organic carbon and associated soil properties of a red earth after 10 years of rotation under different stubble and tillage practices, *Aust. J. Soil Res.*, 30, 71, 1992.
3. **Carter, M. R.,** Microbial biomass as an index for tillage induced changes in soil biological properties, *Soil Till. Res.*, 7, 29, 1986.
4. **Hart, P. B. S., August, J. A., Ross, C. W., and Julian, J. F.,** Some biochemical and physical properties of Tokomaru silt loam under pasture and after 10 years of cereal cropping, *N. Z. J. Agric. Res.*, 31, 77, 1988.
5. **Kamprath, E. J.,** Exchangeable aluminum as a criterion for liming leached mineral soils, *Soil Sci. Soc. Am. Proc.*, 34, 252, 1970.
6. **Reeve, N. G., and Sumner, M. E.,** Lime requirements of natal Oxisols based on exchangeable aluminum, *Soil Sci. Soc. Am. Proc.*, 34, 595, 1970.
7. **Webber, M. D., Hoyt, P. B., Nyborg, M., and Corneau, D.,** A comparison of lime requirement methods for acid Canadian soils, *Can. J. Soil Sci.*, 57, 361, 1977.
8. **Farina, M. P. W., Sumner, M. E., Plank, C. O., and Letzsch, W. S.,** Exchangeable aluminum and pH as indicators of lime requirement of corn, *Soil Sci. Soc. Am. J.*, 44, 1036, 1980.
9. **Hoyt, P. B. and Nyborg, M.,** Field calibration of liming responses of four crops using soil pH, Al and Mn, *Plant Soil*, 102, 21, 1987.
10. **Coleman, N. T. and Thomas, G. W.,** The basic chemistry of soil acidity, in *Soil Acidity and Liming*, Pearson, R. W. and Adams, F., Eds., American Society of Agronomy, 1970, chap. 1.
11. **Hsu, P. H.,** Aluminum hydroxides and oxyhydroxides, in *Minerals in Soil Environments*, Dixon, J. B. and Weed, S. B., Eds., Soil Science Society of America, Madison, WI, 1977, 99.
12. **Huang, P. M.,** Aluminum and the fate of nutrients and toxic substances in terrestrial and freshwater environments, in *Encyclopedia of Systems and Controls*, Singh, M., Ed., Pergamon Press, Oxford, U.K., 1987, 262.
13. **Kwong, Ng Kee K. F. and Huang, P. M.,** Surface reactivity of aluminum hydroxides precipitated in the presence of low molecular weight organic acids, *Soil Sci. Soc. Am. J.*, 43, 1107, 1979.
14. **Violante, A. and Violante, P.,** Influence of pH, concentration and chelating power of organic anions on the synthesis of aluminum hydroxides and oxyhydroxides, *Clays Clay Mineral.*, 28, 425, 1980.
15. **Stevenson, F. J.,** *Humus Chemistry*, John Wiley & Sons, New York, 1982.
16. **Inoue, K. and Huang, P. M.,** Influence of selected organic ligands on the formation of allophane and imogolite, *Soil Sci. Soc. Am. J.*, 50, 1623, 1986.
17. **Huang, P. M. and Violante, A.,** Influence of organic acids on crystallisation and surface properties of precipitation products of aluminum, in *Interactions of Soil Minerals with Natural Organics and Microbes*, Huang, P.M. and Schnitzer, M., Eds., Soil Science Society of America, Madison, WI, 1986, chap. 6.
18. **Lewis, J. A. and Starkey, R. L.,** Decomposition of plant tannins by some soil microorganisms, *Soil Sci.*, 107, 235, 1969.

19. **Flaig, W.**, Organic compounds in soil, *Soil Sci.*, 111, 19, 1971.
20. **Atlas, R. M. and Bartha, R.**, Microbial Ecology, Fundamentals and Applications, 2nd ed., Benjamin Cummings, Menlo Park, CA, 1987, 203.
21. **Pickup, R. W.**, Development of molecular methods for the detection of specific bacteria in the environment, *J. Gen. Microbiol.*, 137, 1009, 1991.
22. **Northcote, K. J.**, *A Factual Key for the Recognition of Australian Soils*, Rellim Technical Publications, Glenside, S. Australia, 1979
23. **Carter, M. R. and Mele, P. M.**, Changes in microbial biomass and structural stability at the surface of a duplex soil under direct drilling and stubble retention in north-eastern Victoria, *Aust. J. Soil Res.*, 30, 493, 1992.
24. **Jenkinson, D. S. and Powlson, D. S.**, The effects of biocidal treatments on metabolism in soil. V.A. method for measuring soil biomass, *Soil Biol. Biochem.*, 8, 209, 1976.
25. **Amato, M. and Ladd, J. N.**, Assay for microbial biomass based on ninhydrin-reactive nitrogen in extracts of fumigated soil, *Soil Biol. Biochem.*, 20, 107, 1988.
26. **Wilson, A. D. and Sergeant, G. A.**, The colorimetric determination of aluminium in minerals by pyrocatechol violet, *Analyst*, 88, 109, 1963.
27. **Mele, P. M.**, Characterisation of Microbial Populations in Conservation Farming Practices in North-Eastern Victoria, Ph.D. thesis, LaTrobe University, Melbourne,1993.
28. **Heanes, D. L.**, Determination of total organic-C in soils by an improved chromic acid digestion and spectrophotometric procedure, *Commun. Soil Sci. Plant Anal.*, 15, 1191, 1984.
29. **Kerven, G. L.**, unpublished data, 1992.
30. **Gillman, G. P.**, *A Centrifuge Method for Obtaining Soil Solution*, Reprint No. 16, CSIRO Australia, Division of Soils, 1976.
31. **Kerven, G. L., Edwards, D. G., Asher, C. J., Hallman, P. S., and Kokot, S.**, Aluminium determination in soil solution. II. Short-term colorimetric procedures for the measurement of inorganic monomeric aluminium in the presence of organic acid ligands, *Aust. J. Soil Res.*, 27, 91, 1989.
32. **Menzies, N. W., Kerven, G. L., Bell, L. C., and Edwards, D. G.**, Determination of total soluble aluminium in soil solution using pyrocatechol violet, lanthanum and iron to discriminate against micro-particulates and organic ligands, *Commun. Soil Sci. Plant Anal.*, 23, 2525, 1992.
33. **Pikul, J. L., Jr. and Allmaras, R. R.**, Physical and chemical properties of a haploxeroll after 50 years of residue management, *Soil Sci. Soc. Am. J.*, 50, 214, 1986.
34. **Arshad, M. A., Schnitzer, M., Angers, D. A., and Ripmeester, J. A.**, Effects of till vs no-till on the quality of soil organic matter, *Soil Biol. Biochem.*, 22, 595, 1990.
35. **Martin, J. P. and Haider, K.**, Decomposition of specifically carbon-14-labeled ferulic acid: free and linked into model humic acid-type polymers, *Soil Sci. Soc. Am. J.*, 40, 377, 1976.
36. **Evans, W. C.**, The microbiological degradation of aromatic compounds, *J. Gen. Microbiol.*, 32, 177, 1963.
37. **Carter, M. R. and Rennie, D. A.**, Dynamics of soil microbial biomass N under zero and shallow tillage for spring wheat, using ^{15}N urea, *Plant Soil*, 76, 157, 1984.
38. **Haines, P. H. and Uren, N. C.**, Effects of conservation tillage farming on soil microbial biomass, organic matter and earthworm populations, in north-eastern Victoria, *Austr. J. Exp. Agric.*, 30, 365, 1990.

Chapter 20

Effect of the Addition of Plant Residues on the Mineralization of Sulfur in Costa Rican Soils

L. Deleón, R. Salas-Camacho, and J. A. Briceño-Salazar

CONTENTS

I. INTRODUCTION

Coffee (*Coffea arabica* L.) is grown in Costa Rica under both shade and full sunlight. Trees such as *Erythrina* sp. and *Inga* sp., utilized for shade, produce dry foliage that becomes a significant contribution to soil organic matter and available soil nutrients. In addition, these processes benefit physical and chemical properties of the soil. In Costa Rica, there are not enough studies dealing with the effect of the incorporation of *Erythrina, Inga,* and *Coffea* foliage residues on the biochemical process in the soil, and the contribution of them to the pool of nutrients. Information is even more scarce in the case of sulfur status and transformation in Costa Rican soils.[1-4]

On the other hand, the coffee plant requirement for sulfur is sizeable and can even be higher than its requirements for phosphorus. Severe deficiencies of sulfur can lead to the coffee plant's death.[3]

Because of such findings, it seems important to study the effect of foliage residues of *Coffea, Erythrina,* and *Inga* added to five different soils of Costa Rica, used to grow coffee, on the mineralization of organic sulfur. For this purpose, soils taken from plots of coffee cultivated under shade and full sunlight were incubated with and without foliage residues under room temperature (average 22°C).

II. MATERIALS AND METHODS

Soil samples were collected in the field at a depth of 0 to 0.15 m from plots of coffee plantations from five locations cultivated under shade and full sunlight. Samples were air dried under greenhouse conditions and sifted through a 2-mm sieve. The analysis reported in Table 1 was performed as follows: pH was measured in water using a soil:water ratio of 1:2.5, organic carbon was determined using wet digestion according to Mebius,[5] total nitrogen was determined by Kjeldahl,[6] and available S was extracted as described by Dick and Tabatabai and measured using an automated ion analyzer.[7] Available P was extracted using a modified Olsen solution[8] readings were performed using the methodology of Murphy and Riley[9] and an automated ion analyzer (FIA),[7] and textural analysis was done by the hydrometer method.

The leaf tissues of *Erythrina, Coffea,* and *Inga* were washed with distilled water, dried at 70°C, and ground to pass a 0.25-mm sieve. Their chemical composition is shown in Table 2. Total nitrogen was determined by Kjeldahl,[6] digestion of Ca, Mg, K, and P was

0-87371-914-X/95/$0.00+$.50

Table 1 Properties of Soils Used

Locality/ Condition	Subgroup	ph H20	Organic Carbon	Total Nitrogen	Micro g element*/g Soil		Clay	Silt	Sand
			g kg-1		S	P	% distribution		
GRECIA	Oxic								
full sunlight	Haplustands	5.7	25	2.0	12	8	43	36	21
shade		5.8	27	2.0	14	9	44	35	21
TURRIALBA	Fluventic								
full sunlight	Eutropepts	5.6	28	2.0	15	105	24	55	21
shade		5.2	35	2.0	17	74	29	51	20
shade									
SALITRAL	Tropohuaults								
full sunlight		5.2	32	2.6	18	31	37	38	25
shade		5.6	32	2.6	19	46	35	38	27
PEREZ ZELEDON	Ustoxic								
full sunlight	Paleustults	5.1	36	2.7	21	8	41	45	14
shade		5.5	42	2.7	22	9	35	47	18
DOTA	Fluventic								
full sunlight	Eutropepts	5.7	52	4.6	23	135	27	48	25
shade		5.0	66	5.2	25	232	22	56	22

* Extractable S
Extractable P

Table 2 Elemental Content of Foliar Tissues Used

	g kg^{-1}					
Foliar tissue	**N**	**Ca**	**K**	**Mg**	**P**	**S**
Erythrina sp.	32.0	8	21	7	2	2
Inga sp.	31.0	5	15	2	2	2
Coffea arabica	24.0	8	24	4	1	2

performed with a nitroperchloric mixture of 1:5, the bases were analyzed using atomic absorption spectrometry, P readings were done using Murphy and Riley methodology,[9] samples for total S were prepared using dry digestion, and analysis of sulfate using FIA.[7,10]

Sulfur mineralization was measured using an incubation method at room temperature during a period of 3 months (0, 0.5, 1.0, 1.5, 2.0, 2.5, and 3.0 months). Soils for the five localities under both sunlight and shade were mixed, in triplicate, with the three dried residues. The totals of the replicate plus the controls (without addition of residue) were multiplied to use a destructive sampling, i.e., the sample was discarded once extracted. For this purpose, a 20-g sample of soil, placed in a 125-ml plastic vial, was mixed with 0.448 g of residue, and water was added according to its water-holding capacity (corrections were made for humidity to prevent analysis error). The vial was covered with aluminum foil, permitting gaseous interchange. After each period of incubation, the sample was extracted for 1 h with 50 ml of 100 μg ml^{-1} $Ca(H_2PO_4)_2$ solution.

The SO_4^{2-} determination was performed with an automated ion analyzer.[7] For this purpose, sulfur extracted with $Ca(H_2PO_4)_2$ solution was compared with standards from 1 to 50 mg SiL (prepared from Na_2SO_4) through a turbidimetric method.

III. RESULTS AND DISCUSSION

The effect of the different treatments can be observed in Figures 1 to 3, which show the extractable sulfur accumulated at each time. There was a marked variation in the levels of S-SO_4^{2-} found in the five locations during the first two measurements of incubation (first month) that can be considered as indexes of net sulfur mineralization.

In the remaining period, there was a decrease and stabilization of the extractable sulfur. Similar behavior on organic sulfur mineralization has been observed by Janzen and Kucey,[11] and by Kowalenko and Lowe,[12] who noted a significant influence of the addition of crop residues on the process of mineralization. It is documented that organic sulfur mineralization is affected by organic matter amendments and plant growth,[10,13] although, data obtained under laboratory conditions must be interpreted carefully because the method of incubation affects the amount of S-SO_4^{2-} released.[14] As a matter of fact, the pattern of mineralization shown in Figures 1 to 3 is different from those obtained by different authors[15,16] who utilized leaching tubes for incubation and concluded that the amount of organic sulfur mineralized is proportional to the elapsed time of incubation, and also from data reported by Maynard et al.[14] and Saggar et al;[17] who utilized a closed incubation method and found little mineralization or net immobilization of sulfur.

The data obtained can be explained as follows: the decreased immobilization can be interpreted as an immobilization reaction, that is, the microorganisms in the soil take away S-SO_4^{2-} for their metabolism from that produced in the first stages. Another explanation is that the S-SO_4^{2-} produced could be held by soil colloids.

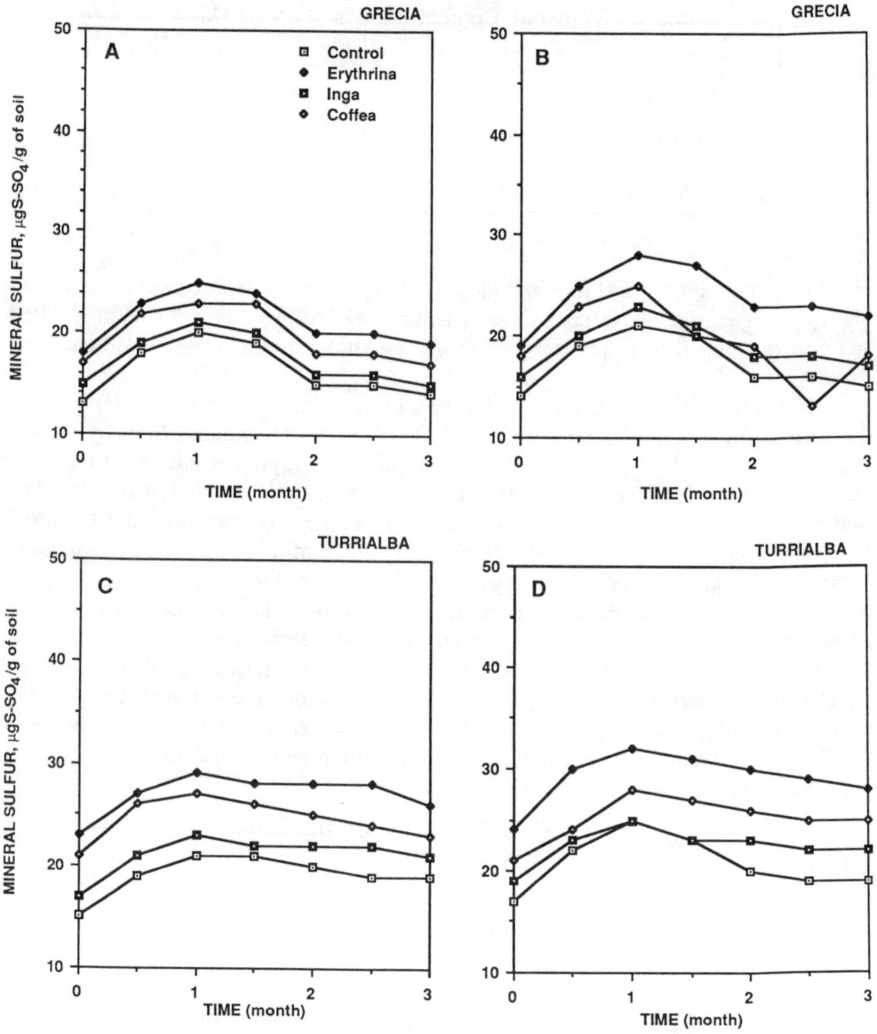

Figure 1 Extractable mineral sulfur in soils incubated at room temperature with and without addition of air-dried foliar tissue. Grecia (Oxic Haplustands) under (a) full sunlight and (b) shading; Turrialba (Fluventic Eutropepts) under (c) full sunlight and (d) shading.

There was no significant difference between soils under shade compared to those under full sunlight, but there was between plant residue treatments and the control. The highest mineral sulfur in the five locations studied was obtained by *Erythrina* (with the highest content of N), followed by *C. arabica.* Moreover, there was a significant correlation between soil organic matter and mineral sulfur ($r = 0.93$), as can be observed in Table 3. For example, the soil of Dota (g kg^{-1} C = 52 and 66 under full sunlight and shade, respectively) presented the highest rate of sulfur mineralization, and the soil of Grecia (g kg^{-1} C = 25 and 27 under full sunlight and shade, respectively) showed the lowest sulfur mineralization.

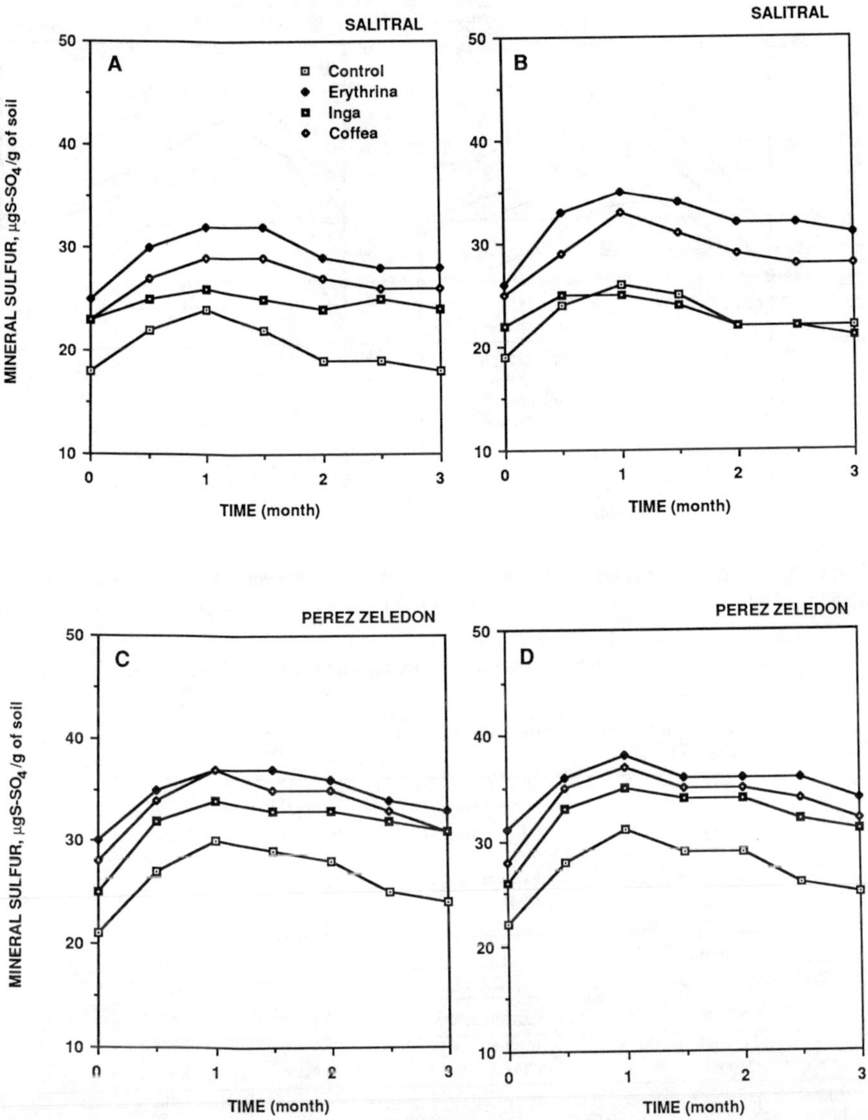

Figure 2 Extractable mineral sulfur in soils incubated at room temperature with and without addition of air-dried foliar tissue. Salitral (Tropohumults) under (a) full sunlight and (b) shading; Pérez Zeledón (Ustoxic Paleustults) under (c) full sunlight and (d) shading.

IV. SUMMARY AND CONCLUSIONS

Addition of foliar residues produced an increase of the mineral sulfur levels of the studied soils. The effect followed the order *Erythrina* sp. > *C. arabica* > *Inga* sp.; only in the first two cases were there significant differences with respect to the control.

There was not a significant difference in the amount of mineral sulfur in any of the localities studied when samples from sunlight were compared with those under the shade condition.

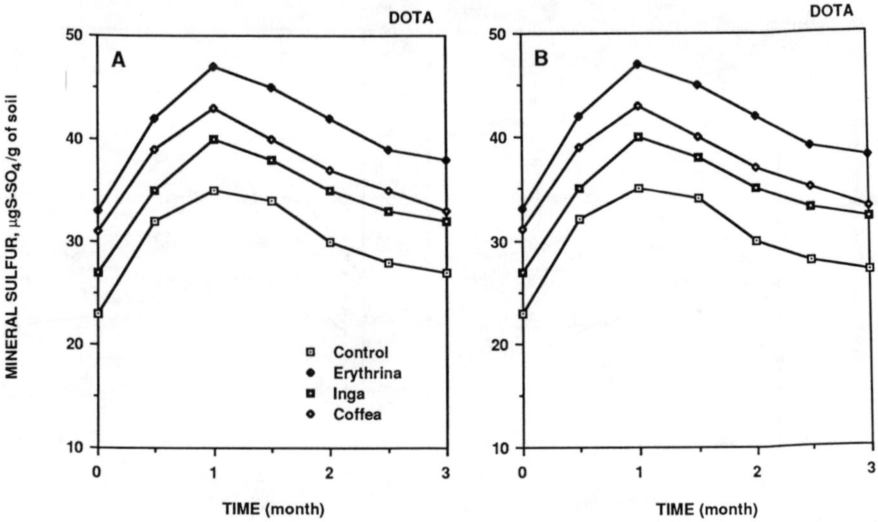

Figure 3 Extractable mineral sulfur in soils incubated at room temperature with and without addition of air-dried foliar tissue. Dota (Fluventic Eutropepts) under (a) sunlight and (b) shading.

Table 3 Treatment Effect on the Mineralization of Sulfur in Samples from Soils Under Full Sunlight and Shading During the Period of Incubation (Average) With and Without Dried Foliar Tissue.

		micro g S/g soil	
Locality	**Foliar tissue**	**Full sunlight**	**Shading**
Finca San	Erythrina sp.	22.0 a+	24.6 a++
Antonio	Coffea arabica	20.4 b	21.6 a
GRECIA,	Inga sp.	18.6 c	19.7 b
Alajuela	Control	17.0 c	18.1 b
Hacienda	Erythrina sp.	27.8 a	29.0 a
La Isabel	Coffea arabica	24.7 b	25.7 b
TURRIALBA,	Inga sp.	21.4 c	22.4 c
Cartago	Control	19.4 c	20.9 c
Hacienda	Erythrina sp.	29.6 a	32.0 a
SALITRAL,	Coffea arabica	27.4 a	29.2 a
Desamparados	Inga sp.	24.3 b	23.7 b
San José	Control	20.9 b	23.3 b
Cafetalera	Erythrina sp.	34.7 a	35.4 a
San Pedrito,	Coffea arabica	33.2 ab	33.4 ab
PEREZ ZELEDON,	Inga sp.	31.7 b	32.3 b
San José	Control	27.3 a	27.2 c
Finca	Erythrina sp.	41.7 a	43.2 a
Jeremias Ureña	Coffea arabica	35.6 b	38.9 b
DOTA,	Inga sp.	34.9 b	35.7 c
San José	Control	30.4 b	33.5 c

(+, ++) Similar letters in the same column, for the same locality, correspond to values not different at $P < 0.05$ according to Duncan multiple test.

Mineral sulfur had values proportional to the organic matter content of the soils, and was higher in the first month of incubation.

ACKNOWLEDGMENTS

This study was supported by University of Costa Rica VI-733-89-007 funds. The authors are indebted to Ing. Alvaro Segura from ICAFE (Costa Rica) and Ing. Victor Quiroga for the sampling and data analysis, respectively.

REFERENCES

1. **Balcázar Ch., A.,** Estudio del Contenido de S en Algunos Suelos del Pacífico Sur de Costa Rica, Tesis de grado, Facultad de Agronomía, Universidad de Costa Rica, 1976, 65.
2. **Bornemisza, E.,** Categorías del Azufre en Suelos de la Meseta Central de Costa Rica, Ministerio de Agricultura e Industria, laboratorios de química, 1958, 7.
3. **Bornemisza, E.,** *Problemas del Azufre en Suelos y Cultivos de Mesoamérica,* Editorial de la Universidad de Costa Rica, Primera edición, 1990, 101.
4. **Bornemisza, E., Valverde, E., and Alvarado, A.,** Disponibilidad de azufre en algunos suelos del Atlántico Norte de Costa Rica, *Agron. Costarric.,* 2, 147, 1978.
5. **Mebius, L. J.,** A rapid method for the determination of organic carbon in soil, *Anal. Chim. Acta,* 22, 120, 1960.
6. **Bremner, J. M. and Mulvaney, C. S.,** Nitrogen-total, in *Methods of Soil Analysis, Part 2. Chemical and Microbiological Properties,* Page, A. L., Ed., Agron. Monogr. No. 9, 2nd ed., American Society of Agronomy and Soil Science Society of America, Madison, WI, 1982, 595.
7. **Lachat Instruments,** *Methods Manual for the QuickChem Automated Ion Analizer,* Quick Chem Method No. 12-116-10-1-B, 1988.
8. **Olsen, S. R., Cole, C. V., Watanabe, F. S., and Dean, L. A.,** Estimation of available phosphorus in soils by extraction with sodium bicarbonate. U.S. Department of Agriculture Circular, 939, 1954.
9. **Murphy, J., and Riley, J. P.,** A modified single solution method for determination of phosphate in natural waters, *Anal. Chim. Acta,* 27, 31, 1962.
10. **Tabatabai, M. A.,** Sulfur, in *Methods of Soil Analysis, Part 2, Chemical and Microbiological Properties,* Page, A. L., Ed., Agron. Monogr. No. 9, 2nd ed., American Society of Agronomy and Soil Science Society of America, Madison, WI, 1982, 501.
11. **Janzen, H. H. and Kucey, R. M.,** C, N, and S mineralization of crop residues, *Plant Soil,* 106, 35, 1988.
12. **Kowalenko, C. G. and Lowe, L. E.,** Mineralization of sulfur from four soils and its relationship to soil carbon, nitrogen and phosphorus, *Can. J. Soil Sci.,* 55, 9, 1975.
13. **Tabatabai, M. A.,** *Sulfur in Agriculture,* 1st ed., Agron. Monogr. No. 27, American Society of Agronomy and Soil Science Society of America, Madison, WI, 1985, 405.
14. **Maynard, D. G., Stewart, J. W., and Bettany, J. R.,** Sulfur and nitrogen mineralization in soils compared using two incubation techniques, *Soil Biol. Biochem.,* 15, 251, 1983.
15. **Pirela, H. J. and Tabatabai, M. A.,** Sulfur mineralization rates and potentials of soils, *Biol. Fertil. Soils,* 6, 26, 1988.
16. **Ellert, B. H. and Bettany, J. R.,** Comparison of kinetic models for describing net sulfur and nitrogen mineralization, *Soil Sci. Soc. Am. J.,* 52, 1692, 1988.
17. **Tabatabai, M. A. and Al-Khafaji, A. A.,** Comparison of nitrogen and sulfur mineralization in soils, *Soil Sci. Soc. Am. J.,* 44, 1000, 1980.
18. **Barrow, N. J.,** A comparison of the mineralization of nitrogen and of sulfur from decomposing organic materials, *Aust. J. Agric. Res.,* 11, 960, 1960.
19. **Barrow, N. J.,** Studies on mineralization of sulfur from soil organic matter, *Aust. J. Agric. Res.,* 11, 306, 1960.
20. **Saggar, S., Bettany, J. R., and Stewart, W. B.,** Sulfur transformations in relation to carbon and nitrogen in incubated soils, *Soil Biol. Biochem.,* 13, 499, 1981.

Part II: Transformations of Natural and Anthropogenic Organic Compounds as Affected by Soil Minerals and Microorganisms

B. Anthropogenic Organics

Chapter 21

Sorption and Biodegradation of Organic Contaminants in Soils: Conceptual Representations of Process Coupling

P. S. C. Rao, C. A. Bellin, and L. S. Lee

CONTENTS

I. INTRODUCTION

Sorption plays a major role in regulating the rates and magnitudes of several environmental processes that govern the fate and transport of organic contaminants in soils, sediments, and aquifers. The concentration (or activity) of an organic chemical in the solution phase is determined to a large extent by the solute distribution between the solution and the organic/inorganic solid phases. Sorption leads to a retardation of solute transport in both liquid and vapor phases, thus influencing the residence time in a given zone (e.g., crop root zone), the time of arrival at a specified depth (e.g., water table), and the rates of mass loss (e.g., volatilization, leaching). Sorption can also influence the bioavailability of organic compounds, thereby controlling directly or indirectly the rate of biodegradation and uptake by plants and other biota. Thus, it is essential to understand both qualitative and quantitative aspects of the explicit nature of sorption and biodegradation, and the coupling between these two processes. Here, we briefly review several examples of such coupled processes in soils.

II. SORPTION OF CONTAMINANTS

A. EQUILIBRIUM APPROACHES

Considerable progress has been made over the past 2 decades in understanding the sorption of organic contaminants from aqueous solutions by natural sorbents, including soils, sediments, and aquifer solids. The term "sorption" is most often used to represent sorbate uptake from solution by a sorbent, but the specific mechanisms responsible for such uptake are still points of much debate. The prevailing paradigm for sorption of

0-87371-914-X/95/$0.00+$.50

nonpolar organic chemicals, that the process can be equated to "partitioning," is based on two principal assumptions: (1) soil organic matter, viewed as a predominantly hydrophobic polymeric phase, is the primary sorbent component and (2) solute uptake results primarily from a hydrophobic exclusion of a nonpolar solute from a polar solvent (water). Chiou[1] suggested that the following features common to sorption data for most nonpolar solutes support the partitioning model: (1) linearity of sorption isotherms over a considerable concentration range, even as the aqueous solubility (S_w) limit is approached, (2) a linear, inverse relationship between S_w and the equilibrium sorption coefficient (K_p), (3) generally low, exothermic heats of sorption (about 15 to 20 kJ mol^{-1}), and (4) absence of competitive sorption when solutes are present in mixtures.

The phenomenological model of partitioning has permitted the synthesis of sorption data for a large number of organic chemicals, as evident in compilations of K_{oc} values (sorption coefficient, K_p, normalized to organic carbon content, OC).[2] However, numerous exceptions to this general model continue to be discussed. This is particularly the case for (1) sorbents that have low organic carbon contents, as in subsoils and many aquifers solids (cf., Ball and Roberts[3]) and (2) organic solutes (e.g., pesticides) that have functional moieties which can result in site-specific interactions with the sorbent constituents (cf., Mingelgrin and Gerstl[4]). Thus, increasing deviations from the partitioning model are likely as the reactivity and number of functional groups on the sorbent or the sorbate increase, or as the organic matter content decreases (organic carbon content < 0.01%). Variations in reported K_{oc} values may be attributed to the following factors[5]: (1) variations in the nature of organic matter, including differences in functional groups, aromaticity, origin, and degree of humification, etc., (2) the contribution from other sorbent surfaces (e.g., clay type and content), (3) the effects of other soil factors (e.g., soil pH, surface area, etc.), and (4) differences in experimental techniques used to measure isotherms.

An alternate to the partitioning model is that sorption should be described as an adsorption process, i.e., as solute accumulation at a two-dimensional solid-liquid interface. Implicit in the partitioning model for sorption is the assumption that the sorbent can be thought of as a three-dimensional phase. Thus, discrimination between these two approaches may depend upon the ability to ascertain whether the sorbent should be viewed as a two- or three-dimensional object. In sorbents with low organic carbon contents (as in most aquifer solids and subsoils), the thickness of the organic matter coatings are calculated to be only several hundreds of angstroms, if it is assumed that uniform organic films cover the available mineral surface area. Such thin organic films may be conceptualized more as hydrophobic interfaces where adsorption occurs, rather than as a bulk hydrophobic phase into which solutes partition. Such sorbent phases may then be treated as being analogous to the stationary phases of column-packing materials employed in reverse-phase liquid chromatography (RPLC). Much thicker organic coatings and discrete organic particles are more likely in most natural sorbents in which the organic matter is nonuniformly distributed. Thus, resolution of this issue is particularly important for describing organic contaminant sorption in such media.

B. NONEQUILIBRIUM APPROACHES

With mounting evidence that equilibrium approaches fail to describe sorption of organic solutes, especially during transport through soils and aquifers, efforts during the past decade have focused on experimental characterization and on development of conceptual models for predicting sorption nonequilibrium. It might be expected that a broad range of rates of approaches to sorption equilibrium (in fact, a continuous distribution of rate constants) should exist for heterogeneous media, such as soils and aquifer solids. However, much of the evidence indicates[6] that nonequilibrium sorption may be approximated as a two-step process: a fraction of the sorption is completed rapidly (essentially instantaneous), while the approach to equilibrium of the remaining fraction is much slower.

The consensus view is that the rate-limiting step for the slow-sorption component in a majority of the cases, particularly for nonpolar organic solutes, is intrasorbent diffusion. Thus, the approach to sorption equilibrium is controlled by the accessibility of sorption domains, and not by chemical reaction kinetics. Current debate centers on the choice between two conceptual models: the retarded intraparticle diffusion (RIPD) model and the intraorganic matter diffusion (IOMD) model. Both modeling approaches are based on intrasorbent solute diffusion.

In the first approach (RIPD model), sorption is assumed to occur within the intraparticle microporous regions of sorbent particles, where solute diffusion is severely restricted or hindered, and is retarded by sorption on pore walls. Data in support of this model have been presented by Wu and Gschwend[7] and Ball and Roberts,[8] among others. Note that in the RIPD model, the sorbent is viewed as a rigid, nondeformable microporous sorbent. Since the diffusion path lengths are small, as approximated by the soil particle radii, the restrictions to solute diffusion arise from the extremely tortuous and sinuous nature of the interior micropore structure. Furthermore, it is assumed that the interior pore walls are completely coated with organic matter, and that all of the sorption occurs instantaneously on these surfaces.

In the alternate conceptual model (IOMD), nonequilibrium arises not due to restricted and retarded diffusion in the aqueous phase filling the micropores, but rather due to diffusion within the polymer-like matrix of organic matter.[6,9] The conformation of such a flexible, deformable, organic elastomer is thought to change when permeated by either solutes or solvents. For example, modifications to the solvent (e.g., electrolyte strength, pH, cosolvent addition, etc.) may cause observable swelling or shrinking of soil organic matter. Much like the changes brought about in interlamellar regions of phyllosilicates upon cation exchange or hydration, solute uptake into the interior regions of organic matter can also result in local-scale alterations in conformation, even though the macroscale changes may not be discernable. These conformational changes, in turn, can moderate the rate of solute uptake or desorption.

While both RIPD and IOMD models have been shown to provide a satisfactory description of sorption nonequilibrium, the choice of one model over the other is limited by the following factors: (1) the inability to provide independent estimates of the required model parameters, (2) an inadequate understanding of either the internal microscale physical structure of the sorbent of the microporous sorbent particles or the chemical structure of organic matter, and (3) the intricate association (e.g., coatings) of organic and inorganic sorbent constituents. Simply put, the geometry of the domains where organic solutes are sorbed is not well understood.

Figure 1 reproduces the schematic representation proposed[10,11] for the heterogeneous nature of the interfaces that must be considered in geometric modeling of contaminant sorption on natural sorbents. A computer simulation[12,13] of the three-dimensional chemical structure for kerogen and similar organic matter is presented in Figure 2. Notice that the three-dimensional structure of kerogen and soil organic matter is conceptualized as a heterogeneous distribution of "islands" of hydrophobic regions (comprising aromatic ring structures) connected by "bridges" of aliphatic compounds, and "polar regions" where specific substituents are located. The relative size and arrangement of these islands, bridges, and sites is likely to vary with the type of organic matter and the degree of humification. The hydrophobic islands may provide favorable regions for nonpolar sorbates (such as polycyclic aromatic hydrocarbons), while the polar regions can offer site-specific sorptive interactions for solutes containing reactive moieties (e.g., pesticides). The quantitative description and experimental observation of diffusion-controlled permeation of such a heterogeneous organic matrix is truly a challenging task.

It is evident that while the macroscale perspective of soil organic matter as an organic elastomer suffices in postulating the "partitioning" model for sorption, much more

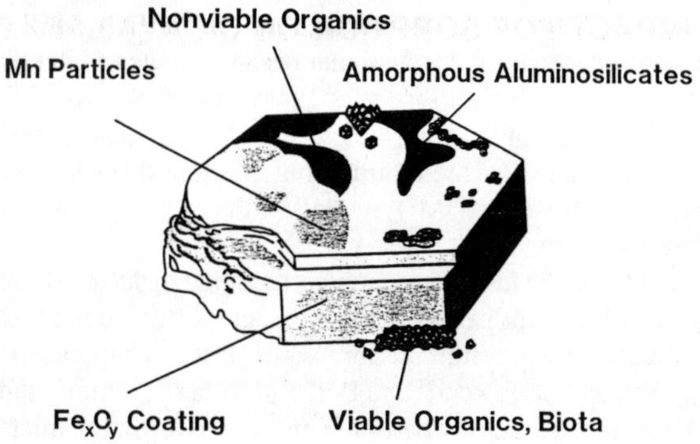

Figure 1 Schematic representation of the association between mineral and organic constituents of soils and sediments.[10,11]

detailed elaborations of the internal chemical and physical structure of organic matter are required for explaining the variations in K_{oc}, or the diffusion-controlled sorption nonequilibrium processes. The practical utilization of information on sorbent structure (Figure 2) in developing mathematical models for contaminant sorption has, to date, not been attempted. Furthermore, current experimental techniques only provide an indirect means of investigating the sources of sorption nonequilibrium. Direct, microscopic investigations using noninvasive techniques (e.g., NMR and IR spectroscopic techniques) are needed to provide a better conceptual view of rate-limiting steps to the sorption of organic solutes.

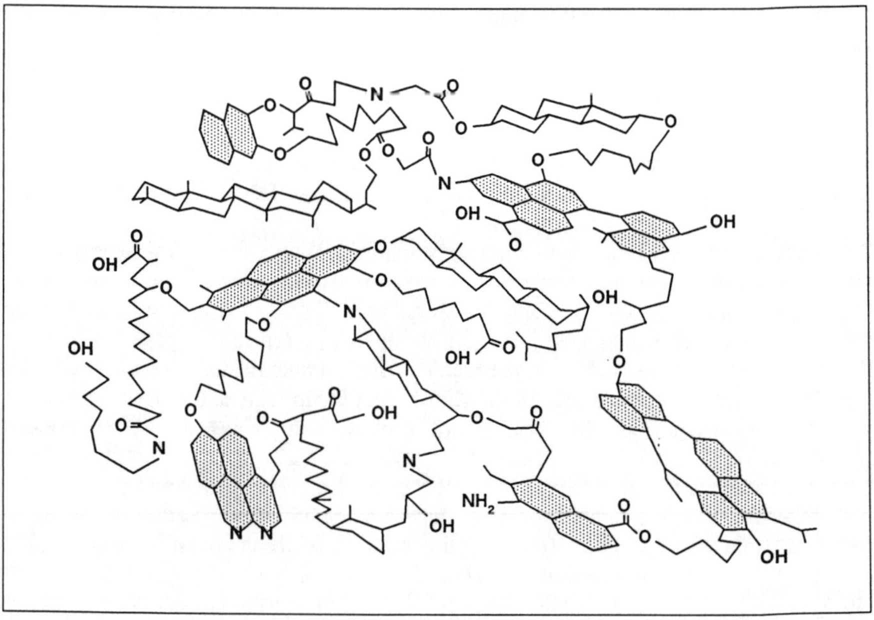

Figure 2 Structural model of kerogen for type II organic matter.[12,13]

III. IMPACTS OF SORPTION ON BIOAVAILABILITY

Biodegradation is a dominant mechanism for attenuation of organic contaminants in soils and aquifers. Despite the variety of microorganisms present in geologic systems, a majority of the literature on biotransformation processes focuses on bacterial isolates. Other microbial groups, including fungi and actinomycetes, may be important in biodegradation, and could alter contaminant sorption and transport. In this chapter, their potential significance is noted; however, the following discussion presents specific cases relating to bacterial biomass.

The recalcitrance of different organic contaminants in a specific soil, or the variations in degradation rates of a specific contaminant in several soils, may be explained to a large extent by understanding the following factors that are known to limit biodegradation kinetics. Physical-chemical processes such as solubility, sorption, and hydrodynamic dispersion lower contaminant concentrations in the solution phase. Furthermore, temperature, soil-water content, and other soil or environmental factors may limit the activity of the microbial consortia. Biotransformation may also be limited by any microbial factor that limits intracellular uptake. Finally, intracellular genetic or biochemical factors (e.g., presence of appropriate enzyme systems, presence and expression of genes) also limit biodegradation rates.

There has been a resurgence of interest in recent years in the issue of bioavailability of sorbed organic contaminants as it relates to bioremediation and environmental toxicity of contaminated soils and sediments. It is generally assumed that organic contaminant in the sorbed phase is not available for direct attack by microorganisms (biodegradation) and uptake by biota (bioavailability) if the following conditions are met: (1) sequestering of the organic solutes in sorptive domains that are inaccessible to microorganisms, (2) the time constant for mass-transfer-limited desorption is larger than the time constant for microbial uptake and biochemical transformation, and (3) microbial degradation of most small organic compounds (molecular mass <600) is believed to be an intracellular process, and not the result of extracellular enzymatic activity. Given these conditions, an inverse relationship between sorption and biodegradation or bioavailability should be expected. This expectation is consistent with experimental observations reported in the literature.[14-27]

Biodegradation rates may be limited by mass transfer [intraparticle diffusion (IPD) and intraorganic matter diffusion (IOMD)] from the interior of the sorbent to the exterior solution. Sorbed contaminants are accessible only after desorption since intraaggregate pores and interior regions of organic matter are generally too small to be accessible to bacteria.[28,29] For several hydrophobic organic compounds, mass transfer has been shown to limit sorption/desorption rates[6,7,9] and biodegradation.[23,24,30-32]

Guerin and Boyd[33] have argued that the influence of contaminant sorption on bioavailability varies with the degradative organism in question, and suggested that organism-specific properties must also be considered. They concluded that the total amount and the rates of naphthalene degradation in soil suspensions by two bacterial isolates were determined by whether the organisms had the ability to directly access sorbed-phase naphthalene. One bacterial isolate (NP-Alk) was judged to be effective in degrading naphthalene only in the solution phase subsequent to desorption, while the other bacterial isolate (*P. putida* 17484) was thought to directly degrade sorbed-phase naphthalene. They concluded that naphthalene biodegradation by the NP-Alk isolate was solely controlled by rate-limited desorption. Based on biodegradation rates that appeared to exceed those estimated by assuming only a solution phase, they concluded that the *P. putida* isolate had the ability to scavenge the sorbed-phase naphthalene.

The experimental observations reported by Guerin and Boyd[33] can be reinterpreted and an alternate explanation offered.[53] By simply invoking rate-limited desorption, naphthalene biodegradation by the *P. putida* isolate could be described by the bicontinuum sorption model coupled with pseudo first-order biodegradation kinetics. For the NP-Alk isolate, naphthalene biodegradation may be controlled by the inaccessibility of the contaminant to the microorganism (i.e., irreversible adsorption, IOMD) or more likely specific microorganism properties (i.e., critical threshold concentrations). Our analysis is based on the following arguments:

1. It should be recognized that if contaminant sorption is reversible, the naphthalene biodegradation rate may be controlled by the rate of desorption, but the total amount degraded is not. This is the case regardless of whether equilibrium or nonequilibrium conditions prevail for sorption-desorption. The total amount of naphthalene degraded decreases with sorption only when sorption is partially or completely irreversible.
2. The organic carbon content of the soils they studied is high enough (0.76 to 5.36%) that hydrophobic partitioning in intraparticle or intraorganic matter regions can be assumed to be the dominant mechanism for naphthalene sorption. As pointed out in earlier discussion, and as evident from scanning electron micrographs of soil-bacteria associations, these types of interior sorptive regions are not likely to be directly accessible to bacteria.
3. The biodegradation kinetics data for the *P. putida* isolate (see Figure 4 in Guerin and Boyd[33]) can be adequately explained for both soils by invoking a bicontinuum sorption model with first-order degradation kinetics, without assigning unique physiologic attributes to the microorganisms.
4. The kinetic data for naphthalene biodegradation by the NP-Alk isolate appear to suggest that either sorption is practically irreversible (untenable given the weight of evidence of published data) or that the sorbed-phase naphthalene is in fact unavailable to this isolate contrary to the conclusion reached by Guerin and Boyd.[33] It is also possible that the total amount of naphthalene degraded may be limited by low contaminant concentrations as a consequence of slow desorption such that the necessary enzymes for biodegradation are not sustained.

In summary, Boyd and Guerin[33] invoked desorption-limited biodegradation as the explanation for naphthalene mineralization by the NP-Alk isolate, but attributed unique physiologic capabilities to the *P. Putida* isolate which allowed it to overcome the desorption constraints. Our arguments suggest that desorption limitations are entirely adequate to describe the data trends for naphthalene mineralization by the *P. putida* isolate, while sorption-based arguments cannot explain the data for the NP-Alk isolate. The foregoing arguments should not be taken, however, to imply that organism-specific factors are unimportant. Even though considerable physiologic diversity of bacteria and other microorganisms is to be expected, direct surface scavenging of sorbed contaminants is yet to be unequivocally demonstrated.

IV. BIOMASS IMPACTS ON CONTAMINANT SORPTION/TRANSPORT

Growth or addition of bacteria may drastically alter the chemical, physical, and microbiological environment of soil surfaces.[34] Therefore, the impact of bacterial biomass on contaminant behavior in porous media near hazardous waste sites is of interest. Marshall[35] summarizes the significance of surfaces on microbial activity as follows: "It is my belief that many microbiologists fail to appreciate the effects of interfaces on microbial populations, despite the widespread occurrence of solid-liquid, gas-liquid, and liquid-liquid interfaces in natural microbial habitats. ...importance must be given to the nature, distribution, and unique physicochemical properties of interfaces, the interaction between microorganisms and interfaces, and the modifying effects of interfaces on the ecology of microorganisms."

In addition to contaminant biodegradation, bacterial addition in porous media may result in (1) contaminant sorption by bacteria (biosorption), (2) bacterial growth or transport through the porous media, leading to hydrologic changes (e.g., pore clogging as a result of physical straining) and biofacilitated contaminant transport, and (3) bacterial adhesion onto soil surfaces, altering the contaminant sorption capacity. In the following discussion, we briefly examine each of these factors.

A. BACTERIAL ADHESION

Factors controlling bacterial sorption on surfaces have been investigated extensively.[34,35] Bacterial adhesion to surfaces may be facilitated by Brownian motion, advective transport, and active movement by motile bacteria. Electrostatic, hydrophobic, and specific chemical interactions enhance bacterial adhesion near surfaces. In soils, most gram-negative bacteria possess a net negative charge. Therefore, electrostatic interactions generally result in bacterial repulsion from negatively charged soil particles. However, bacterial adhesion may be increased by increasing the electrolyte concentration, which leads to a compression of the diffuse double layer, or by decreasing pH, which reduces the overall negative charge of the surfaces. Hydrogen bonding and cation bridging are examples of site-specific interactions between bacteria and soils, whereas hydrophobic partitioning and van der Waals forces are examples of nonspecific mechanisms.

The mechanism of attachment stipulates the distribution of bacterial biomass and reversibility of bacterial adhesion. Adhesion and growth of bacteria on surfaces may result in development of microcolonies, monolayers, and/or biofilms. Initial adhesion processes involve contact of the cell with the surface, freely exchanging between the attached and solution ("free") phase. Bacterial cells that individually and irreversibly attach via cation bridging or extracellular polymers promote monolayer coverage of surfaces.[16] Irreversible adhesion of bacteria to surfaces and other cells coinciding with growth and the production of extracellular polysaccharides may result in biofilm formation.[34] Multilayer formation of microbial biomass limits penetration of oxygen into the biofilm, thus changing the microbial population in response to variations in electron acceptors and donors. Characterization of the distribution of bacteria in soils suggests that microcolonies are predominantly formed.[36] Biofilm formation is commonly assumed in wastewater treatment plants, indicative of high nutrient concentrations and high bacterial populations. Near contaminated sites where high substrate concentrations are likely, high bacterial populations may promote biofilm formation.

B. BACTERIAL TRANSPORT

Physical, chemical, and microbial factors controlling bacterial transport in porous media have recently been summarized.[37-39] Bacterial transport may be limited by physical constraints imposed by the porous media, such as soil structure and pore size distribution.[39] Straining or filtration occurs in soils and aquifers when bacteria that are too large to pass between soil particles become clogged, and are thus restricted from further penetrating into the porous medium.[38,40] Once bacteria become clogged in the soil pores, water flow is restricted, and the path of water flow may be altered.[36] Chemical constraints, such as adhesion of bacteria, may also limit bacterial migration through soils and aquifers.[37,38,41,42]

C. BIOSORPTION

Sorption of hydrophobic organic chemicals,[43,44] metals,[45] and organic amines[46] by various microorganisms has been shown in aquatic systems. A consensus on biosorption mechanisms has not been reached, however, and usually no distinction is made between sorption onto extracellular regions and absorption into the cells. Properties such as aqueous solubility and log K_{ow} (K_{ow} = octanol water partition coefficient) for the contaminant,[47] and bacterial lipid content[48] have been correlated to biosorption of nonpolar organic contaminants. Biosorption of trace metals (e.g., cadmium) has been shown to occur via adsorption onto extracellular bacterial polymers with minimal intracellular uptake.[45] Sorption of organic amines by algae has been described by mechanisms including ion exchange and hydrophobic bonding.[46]

D. BIOMASS IMPACTS ON CONTAMINANT SORPTION

Bacterial adsorption by soils[49] and clay minerals[14] may alter the contaminant sorption capacity of porous media. Under nutrient- and substrate-rich conditions, as is the case near wastes sites, biofilm formation may create diffusional barriers leading to nonequilibrium sorption of contaminants. In some cases, bacterial biomass may physically block sorption sites, thereby reducing contaminant sorption. Physical blockage of sorption sites and biomass-induced changes in the speciation of quinoline, an ionizable organic base, reduced sorption and enhanced transport in laboratory columns.[50] In contrast, bacterial biomass may act as an additional sorbent, thereby increasing contaminant sorption.

E. BIOFACILITATED TRANSPORT

The occurrence of biosorption and bacterial migration, regardless of the underlying mechanisms, suggests the potential for biofacilitated transport of contaminants. Biosorption technology has been commercialized to mobilize metals in the mining industry.[51] Biofacilitated transport of two hydrophobic organic contaminants, DDT and hexachlorobenzene, in sand columns has been recently demonstrated[52] (see Figure 3). The extent of such biofacilitated transport is dependent on the hydrophobicity of the organic contaminant (represented by log K_{ow}) and the bacterial concentration in the aqueous phase. Model calculations,[52] shown in Figure 4, suggest that for very hydrophobic compounds (log K_{ow}>6), enhanced mobility can be significant at bacterial concentrations exceeding 10^6 cells per ml, whereas much higher bacterial densities (>10^8 cells per ml) would be required for organic contaminants with lower hydrophobicity (log K_{ow} >4). Such high bacterial concentrations may be encountered when indigenous bacteria are stimulated (by nutrient additions) or when specific bacteria are introduced during bioremediation of contaminated sites.

IV. SUMMARY AND CONCLUSIONS

Sorption and biodegradation are the two dominant processes that determine the environmental fate of organic contaminants in soils and aquifers. Quantitative aspects of the coupling between these processes are poorly understood, but available literature reviewed here suggests that at least a qualitative analysis is possible. Sorption can sequester the organic contaminants in regions inaccessible to microorganisms and other biota, and thereby can directly or indirectly limit both the rates and extent of biodegradation and bioavailability. Considerable information exists on the equilibrium and kinetic aspects of organic contaminant sorption by soils, sediments, and aquifer solids. Organic contaminants can also be sorbed by the microorganisms (biosorption), but the effects of microorganisms on contaminant sorption and transport have begun to be examined only recently.

The prevailing paradigm for sorption is that the process can be equated to "partitioning" of nonpolar organic solutes from water into the soil organic matter phase. Available data generally support this three-dimensional view of sorption, rather than a two-dimensional representation of sorption as solute accumulation at a hydrophobic interface. Recent investigations have dealt with various factors controlling the rate at which

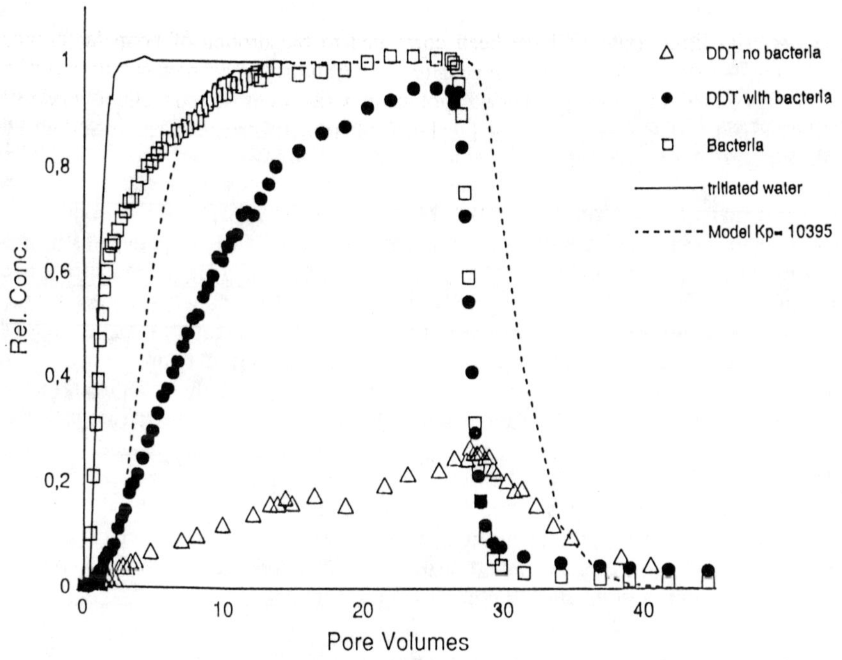

Figure 3 A demonstration of biofacilitated transport of DDT transport through a sand column in the absence or presence of free-living cells (1.2×10^9 cells per ml).[52]

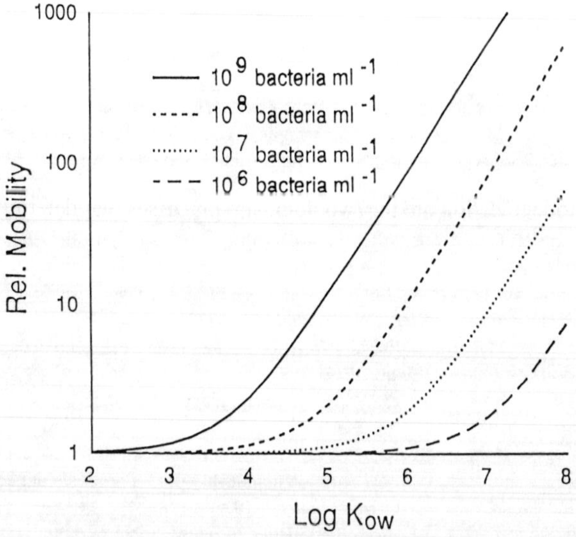

Figure 4 Relative mobility (ratio of apparent contaminant velocities with and without biomass) of an organic contaminant as a function of contaminant hydrophobicity (log K_{ow}) and bacterial density.[52]

sorption equilibrium is attained. Solute diffusion within the intraparticle and intraorganic matter domains has been proposed as the primary rate-limiting step in achieving sorption equilibrium. While the conceptual view of sorption as a three-dimensional bulk hydrophobic phase has served as an adequate conceptual model to interpret equilibrium sorption data, it is insufficient to discriminate between the two major classes of models proposed to describe sorption nonequilibrium. For such a purpose, information is needed on the internal physical structure of the sorbent, principally that of the soil organic matter matrix, as well as the nature of the association of mineral and organic matter constituents.

Appropriate conceptualization of the sorption processes also has important implications in interpreting data for biodegradation kinetics. For example, sequestering of organic solutes within the sorbent matrix (intraparticle and/or intraorganic matter regions) has been suggested to limit bioavailability, which can be a major constraint to bioremediation of contaminated soils. Microorganisms can directly or indirectly influence sorption and transport of organic solutes. The impact of biosorption, biofacilitated transport, and biomass-induced alteration of the sorbent surfaces on contaminant transport are presented. Attachment and growth of microbial biomass on sorbent surfaces can influence the extent and rate of contaminant sorption as a result of either blockage of sorption sites or alteration of the physical-chemical properties of the interfacial regions. Biosorption of the contaminants and transport of the microorganisms through soils and aquifers can result in biofacilitated contaminant transport.

ACKNOWLEDGMENTS

This work was supported in part by USDA-CSRS Hatch funds provided for Regional Research Project W-82. Discussions with several colleagues (Drs. Denie Augustijn, Itaru Okuda, Cliff Johnston, Mark Brusseau, and Zeina Hinedi) were influential in formulating our opinions about the structure of soil organic matter. We are thankful for their insightful comments.

REFERENCES

1. **Chiou, C. T.,** Theoretical considerations of the partition uptake of nonionic organic compounds by soil organic matter, in *Reaction and Movement of Organic Chemicals in Soils,* Sawhney, B. L. and Brown, K., Eds., SSSA Spec. Publ. No. 22, Soil Science Society of America, Madison, WI, 1989, 1.
2. **Wauchope, R. D., Buttler, T. M., Hornsby, A. G., Augustijn-Beckers, P. W. M., and Burt, J. P.,** The SCS/ARC/CES pesticide properties database for environmental decision making. *Rev. Environ. Chem. Toxicol.,* 123, 1, 1991.
3. **Ball, W. P. and Roberts, P. V.,** Long-term sorption of halogenated organic chemicals by aquifer material. I. Equilibrium, *Environ. Sci. Technol.,* 25, 1223, 1991.
4. **Mingelgrin, U. and Gerstl, Z.,** Reevaluation of partitioning as a mechanism of nonionic chemical sorption in soils, *J. Environ. Qual.,* 12, 1, 1983.
5. **Green, R. E. and Karickhoff, S. W.,** Sorption estimates for modeling, in *Pesticides in the Soil Environment: Processes, Impacts, and Modeling,* Cheng, H. H., Ed., Soil Science Society of America, Madison, WI, 1990, 79.
6. **Brusseau, M. L. and Rao, P. S. C.,** Sorption nonideality during organic contaminant transport in porous media, *CRC Crit. Rev. Environ. Control,* 19, 33, 1989.
7. **Wu, S. and Gschwend, P. M.,** Sorption kinetics of hydrophobic organic compounds to natural sediments and soil, *Environ. Sci. Technol.,* 20, 717, 1986.
8. **Ball, W. P. and Roberts, P. V.,** Long-term sorption of halogenated organic chemicals by aquifer material. II. Intra-particle diffusion, *Environ. Sci. Technol.,* 25, 1237, 1991.
9. **Brusseau, M. L., Jessup, R. E., and Rao, P. S. C.,** Nonequilibrium sorption of organic chemicals: elucidation of rate-limiting processes, *Environ. Sci. Technol.,* 25, 134, 1991.

10. **Jenne, E. A.,** Trace element sorption by sediments and soil sites and processes, in *Proc. Symp. "Molybdenum in the Environment,"* Cheppel, W. A. and Peterson, D. K., Eds., Marcel Dekker, New York, 1977, 425.
11. **Leenheer, J. A.,** Organic substance structures that facilitate contaminant transport and transformations in aquatic sediments, in *Organic Substances and Sediments in Water,* Vol. 1, Baker, R. A., Ed., Lewis, Chelsea, MI, 1991, 3.
12. **Behar, F. and Vandenbroucke, M.,** Chemical modelling of kerogens, *Org. Geochem.,* 11, 15, 1987.
13. **Rullkotter, J. and Michaelis, W.,** The structure of kerogen and related materials: a review of recent progress and future trends. *Org. Geochem.,* 16, 829, 1990.
14. **Stotzky, G. and Rem, L. T.,** Influence of clay minerals on microorganisms. I. Montmorillonite and kaolinite on bacteria, *Can. J. Microbiol.,* 12, 547, 1966.
15. **Madsen., E. L.,** Effects of Chemical Speciation on Microbial Mineralization of Organic Compounds, Ph.D. thesis, Cornell University, Ithaca, NY, 1985.
16. **van Loosdrecht, M. C. M., Lyklema, J., Norde, W., and Zehnder, A. J. B.,** Influences of interfaces on microbial activity, *Microbiol. Rev.,* 54, 75, 1990.
17. **Weber, J. B., and Coble, H. D.,** Microbial decomposition of diquat adsorbed on montmorillonite and kaolinite on clays, *J. Agric. Food Chem.,* 16, 475, 1968.
18. **Subba-Rao, R. V. and Alexander, M.,** Effect of sorption on mineralization of low concentrations of aromatic compounds in lake water sediments, *Appl. Environ. Microbiol.,* 44, 659, 1982.
19. **Miller, M. E. and Alexander, M.,** Kinetics of bacterial degradation of benzylamine in a montmorillonite suspension, *Environ. Sci. Technol.,* 25, 240, 1991.
20. **Wszolek, P. C. and Alexander, M.,** Effect of desorption rate on the biodegradation of *n*-alkylamines bound to clay, *J. Agric. Food Chem.,* 27, 410, 1979.
21. **Gordon, A. S. and Millero, F. J.,** Adsorption mediated decrease in the biodegradation rate of organic compounds, *Microb. Ecol.,* 11, 289, 1985.
22. **Dashman, T. and Stotzky, G.,** Microbial utilization of amino acids and a peptide bound on homoionic montmorillonite and kaolinite, *Soil Biol. Biochem.,* 18, 5, 1986.
23. **Robinson, K. G., Farmer, W. S., and Novak, J. T.,** Availability of sorbed toluene in soils for biodegradation by acclimated bacteria, *Water Res.,* 24, 345, 1990.
24. **Smith, S. C., Ainsworth, C. C., Traina, S., and Hicks, R. J.,** The effect of sorption in the biodegradation of quinoline, *Soil Sci. Soc. Am. J.,* 56, 737, 1992.
25. **Lehman, R. G., Miller, J. R., Fontaine, D. D., Laskowski, D. A., Hunter, J. H., and Cordes, R. C.,** Degradation of a sulfonamide herbicide as a function of soil sorption, *Weed Res.,* 32, 197, 1992.
26. **Steen, W. C., Paris, D. F., and Baughman, G. L.,** Effects of sediment sorption on microbial degradation of toxic substances, in *Contaminants and Sediments, Vol. 1, Fate and Transport, Case Studies, Modeling, Toxicity,* Baker, R. A., Ed., Ann Arbor Science Publishers, Ann Arbor, MI, 1980.
27. **Ogram, A. V., Jessup, R. E., Ou, L. T., and Rao, P. S. C.,** Effects of sorption on biological degradation rates of 2,4-dichlorophenoxyacetic acid in soils, *Appl. Environ. Microbiol.,* 49, 582, 1985.
28. **Steinberg, S. M., Pignatello, J. J., and Sawhney, B. L.,** Persistence of 1,2-dibromomethane in soils: entrapment in intra-particle micropores, *Environ. Sci. Technol.,* 21, 1201, 1987.
29. **Scow, K. M. and Alexander, M.,** Effects of diffusion on the kinetics of biodegradation: experimental results with synthetic aggregates, *Soil Sci Soc. Am. J.,* 56, 128, 1992.
30. **Rijnaarts, H. H. M., Bachmann, A., Jumelet, J. C., and Zehnder, A. J. B.,** Effect of desorption and intra-particle mass transfer on the aerobic biomineralization of a-hexachlorocyclohexane in contaminated calcareous soil, *Environ. Sci. Technol.,* 24, 1349, 1990.
31. **Mihelcic, J. R. and Luthy, R. G.,** Degradation of polycyclic aromatic hydrocarbon compounds under various redox conditions in soil-water systems, *Appl. Environ. Microbiol.,* 54, 1182, 1988.
32. **Mihelcic, J. R. and Luthy, R. G.,** Microbial degradation of acenapthalene and naphthalene under denitrification conditions in soil-water systems, *Appl. Environ. Microbiol.,* 54, 1188, 1988.
33. **Guerin, W. F. and Boyd, S. A.,** Differential bioavailability of soil-sorbed naphthalene to two bacterial species, *Appl. Environ. Microbiol.,* 58, 1142, 1992.
34. **Fletcher, M.,** The physiological activity of bacteria attached to solid surfaces, in *Advances in Microbial Physiology,* Vol. 2, Rose, A. H. and Tempest, D. W., Eds., 1991, 54.

35. **Marshall, K. C.,** *Interfaces in Microbial Ecology,* Harvard University Press, Cambridge, MA, 1976, i.
36. **Vandevivere, P. and Baveye, P.,** Saturated hydraulic conductivity reduction caused by aerobic bacteria in sand columns, *Soil Sci. Soc. Am. J.,* 56, 1, 1992.
37. **Tan, Y.,** Transport of Bacteria in Porous Media, Ph.D. thesis, Australian National University, Canberra, ACT, Australia, 1989.
38. **Harvey, R. W.,** Parameters involved in modeling movement of bacteria in groundwater, in *Modeling the Environmental Fate of Microorganisms,* Hurst, C. J., Ed., American Society of Microbiology, Washington, D.C., 1991, 89.
39. **Lindqvist, R. and Enfield, C. G.,** Cell density and non-equilibrium sorption effects on bacterial dispersal in groundwater microcosms, *Microb. Ecol.,* 24, 25, 1992.
40. **Herzig, J. P., Leclerc, D. M., LeGoff, P.,** Flow suspensions through porous media — application to deep infiltration, *Ind. Eng. Chem.,* 62, 8, 1970.
41. **Bales, R. C., Hinkle, S. R, Kroeger, T. W., and Stocking, K.,** Bacteriophage adsorption during transport through porous media: chemical perturbations and reversibility, *Environ. Sci. Technol.,* 25, 2088, 1991.
42. **Harvey, R. W., George, L. H., Smith, R. L., and LeBlanc, D. R.,** Transport of microspheres and indigenous bacteria through a sandy aquifer: results of natural- and forced-gradient tracer experiments, *Environ. Sci. Technol.,* 23, 51, 1989.
43. **Baughman, G. L. and Paris, D. F.,** Microbial bioconcentration of organic pollutants from aquatic systems: a critical review, *CRC Crit. Rev. Microbiol.,* 8, 205, 1981.
44. **Tsezos, M. and Bell, J. P.,** Comparison of the biosorption and desorption of hazardous organic pollutants by live and dead biomass, *Water Res.,* 23, 561, 1989.
45. **Scott, J. A. and Palmer, S. J.,** Sites of cadmium uptake in bacteria used for biosorption, *Appl. Microbiol. Biotechnol.,* 33, 221, 1990.
46. **Crist, R. H., Oberholser, K., and Wong, B.,** Amine-algae interactions: cation exchange and possible hydrogen bonding, *Environ. Sci. Technol.,* 26, 1523, 1992.
47. **Selvakumar, A. and Hsieh, H.-N.,** Correlation of compound properties with biosorption of organic compounds, *J. Environ. Sci. Health,* A23, 543, 1988.
48. **Bitton, G., Dutton, R. J., and Koopman, B.,** Cell permeability to toxicants: an important parameter in toxicity tests using bacteria, *CRC Crit. Rev. Environ. Control,* 18, 177, 1988.
49. **Daniels, S. L.,** The adsorption of microorganisms onto surfaces: a review, *Dev. Ind. Microbiol.,* 13, 211, 1972.
50. **Bellin, C. A. and Rao, P. S. C.,** The impact of bacterial biomass on contaminant sorption and transport in subsurface soils, *Appl. Environ. Microbiol.,* 59, 1813, 1993.
51. **Ehrlich, H. L. and Brierley, C. L.,** *Microbial Mineral Recovery,* McGraw-Hill, New York, 1990.
52. **Lindqvist, R. and Enfield, C. G.,** Biosorption of dichlorodiphenyltrichloroethane and hexachlorobenzene in groundwater and its implications for facilitated transport, *Appl. Environ. Microbiol.,* 58, 2211, 1992.

Chapter 22

The Role of Dissolved Metals and Metal-Containing Surfaces in Catalyzing the Hydrolysis of Organic Pollutants

A. T. Stone and A. Torrents

CONTENTS

I. INTRODUCTION

Synthetic organic compounds subject to hydrolysis are introduced into the environment intentionally (as herbicides and pesticides) and unintentionally (as plasticizers, oil additives, and numerous other products). Predictions concerning pathways and rates of hydrolytic transformations based upon existing chemical information do not always agree with field observations; the complex chemical milieu of natural aquatic environments must be adequately accounted for. Heterogeneous systems (soils, sediments, and aquifers) contain a variety of dissolved constituents and solid surfaces that can affect hydrolysis reactions. These species can alter hydrolysis pathways and rates through phase partitioning, general acid and base catalysis, and metal catalysis.

In the present work, our objective is to examine the role of metal catalysis in bringing about the hydrolysis of organic pollutants in the environment. We begin by reviewing fundamental studies in the laboratory which provide information concerning mechanisms of metal catalysis and the comparative reactivity of various metal species. Next, laboratory experiments on the susceptibility of pesticides and other pollutants to metal catalysis are examined. We end with an assessment of the role of metal catalysis within aquatic environments. We will find that the natural abundance and speciation of metals impose important constraints on catalytic activity.

0-87371-914-X/95/$0.00+$.50

II. PROPERTIES OF METAL IONS

The catalytic effectiveness of a metal ion depends upon its ability to complex reactant molecules (e.g., ester linkage, leaving group, and attacking nucleophile) and shift electron density and conformation in ways favorable for reaction.[1] Variations in complex formation constants among metals in the periodic table reflect differences in metal-ligand bond strengths and solvation forces. General features can be explained by the following rules: (1) complex formation constants generally increase as the charge-to-radius ratio of the metal ion is increased, because the electrostatic contribution to bond formation has increased, (2) polarizable metals and ligands gain additional complex stability through covalent bond formation, and (3) as complex formation constants increase, competition for the metal among available ligands (including OH^- and other inorganic constituents) becomes more pronounced. Using Pearson's[2] approach, tetravalent metals such as Ti^{IV} and trivalent metals such as Fe^{III} and Al^{III} are classic "hard" metals which form strong complexes with classic "hard" ligands. Organic compounds possessing oxygen- and nitrogen-donor atoms are classified as "hard" ligands, while those possessing sulfur-donor atoms are classified as "soft."[2] Although oxygen-donor organic ligands form strong complexes with metals such as Fe^{III}, Al^{III}, and Ti^{IV}, they must compete with hydroxo (OH^-) and oxo (O^{2-}) species that limit metal solubility near neutral pH. The alkali-earth metals Ca^{2+} and Mg^{2+} are also classified as "hard" metals; their s- and p-bonding orbitals form bonds that are primarily ionic in nature.

Polarizability is a measure of the ability of a metal or ligand to participate in covalent bond formation. In the case of transition metals, covalent bond formation is favored by d-bonding orbitals, which project electron density outward toward coordinated ligands. As a consequence, polarizable ligands preferentially bind transitional metal ions. Nitrogen-donor atoms are an important example of this. Although still classified as "hard" donors,[2] nitrogen-donor atoms are more polarizable than oxygen-donor atoms. As a consequence, they show preference for the transition metal Fe^{III} relative to the nontransition metal (and less polarizable) Al^{III}.[3]

Complex formation constants involving divalent first-row transition metals generally obey the Irving-Williams series:

$$Ca^{2+} < Mg^{2+} < Mn^{2+} < Fe^{2+} < Co^{2+} < Ni^{2+} < Cu^{2+} > Zn^{2+}$$

(Nontransition metals Ca and Mg have been included for comparison.) This trend arises primarily from differences in ionic radii, although increases in polarizability with increasing atomic number are also important.[4] The greatest difference in complex formation constants among these metals is observed with "soft" sulfur-donor ligands, followed by nitrogen-donor ligands. Differences are least pronounced for oxygen-donor ligands.[5]

Even when differences in metal-substrate complex formation constants are taken into account, differences in electronic structure cause some metals to emerge as particularly effective catalysts. The often-cited catalytic ability of Cu^{2+} arises from both its high complex formation constants and its increased polarizability relative to lighter first-row transition metals.

In the same way that metal ions can influence the reactivity of organic substrates, they can also influence the speciation and reactivity of nucleophiles. Table 1 lists formation constants for a number of metal-hydroxo complexes. Although metal-hydroxo complexes are less reactive nucleophiles than hydroxide ions, their concentrations can be substantially higher, particularly in the neutral to acidic pH range. Values of $p{*}K_1$ represent the pH where the concentration of the first hydroxo species equals the concentration of the free metal ion. Because $p{*}K_1$ values for strong Lewis acid metals are low, they can

Table 1 Average Crustal Abundance for Selected Metals and Complex Formation Constants ($^*K_1$) for the Following Hydrolysis Reaction: $M^{n+}(aq) + H_2O = MOH^{(n-1)+} + H^+$

Element	Ave. crustal Abundance[6] (%)	Formation of hydroxo complex		$-\log{}^*K_1$
Magnesium	2.76	Mg^{II}:	$Mg^{2+} + H_2O = MgOH^+ + H^+$	11.4[7]
Calcium	4.66	Ca^{II}:	$Ca^{2+} + H_2O = CaOH^+ + H^+$	12.7[7]
Aluminum	8.30	Al^{III}:	$Al^{3+} + H_2O = AlOH^{2+} + H^+$	5.0[8]
Titanium	0.632			
Manganese	0.106	Mn^{II}:	$Mn^{2+} + H_2O = MnOH^+ + H^+$	10.6[8]
		Mn^{III}:	$Mn^{3+} + H_2O = MnOH^{2+} + H^+$	0.40[7]
Iron	6.20	Fe^{II}:	$Fe^{2+} + H_2O = FeOH^+ + H^+$	9.5[7]
		Fe^{III}:	$Fe^{3+} + H_2O = FeOH^{2+} + H^+$	2.2[8]
Cobalt	0.0029	Co^{II}:	$Co^{2+} + H_2O = CoOH^+ + H^+$	9.7[8]
		Co^{III}:	$Co^{3+} + H_2O = CoOH^{2+} + H^+$	−0.48[7]
Nickel	0.0099	Ni^{II}:	$Ni^{2+} + H_2O = NiOH^+ + H^+$	9.9[8]
Copper	0.0068	Cu^{II}:	$Cu^{2+} + H_2O = CuOH^+ + H^+$	8.0[8]
Zinc	0.0072	Zn^{II}:	$Zn^{2+} + H_2O = ZnOH^+ + H^+$	9.0[7]

generate hydroxo species across a wider range of pH. The nucleophilities of various hydroxo complexes (e.g., $FeOH^+$ and $CuOH^+$) are a function of p^*K_1 and the polarizability of the metal.

III. MECHANISMS OF CATALYSIS

The reactivity of hydrolyzable organic pollutants arises from the presence of electrophilic (electron-deficient) sites within the molecule. The S_N2 mechanism (nucleophilic substitution-bimolecular) involves attack of the electrophilic site by OH^- or H_2O, generating a higher coordination number intermediate. Subsequent elimination of the leaving group forms the final hydrolysis product:

S_N2

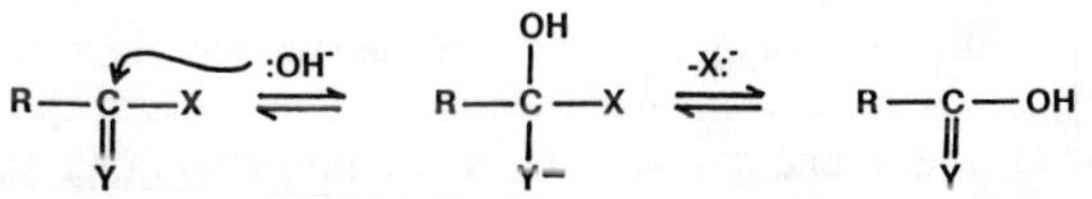

(1)

The S_N1 mechanism (nucleophilic substitution-monomolecular), in contrast, begins with loss of the leaving group to generate a lower coordination number intermediate, followed by generation of the final product by nucleophilic addition:

S_N1

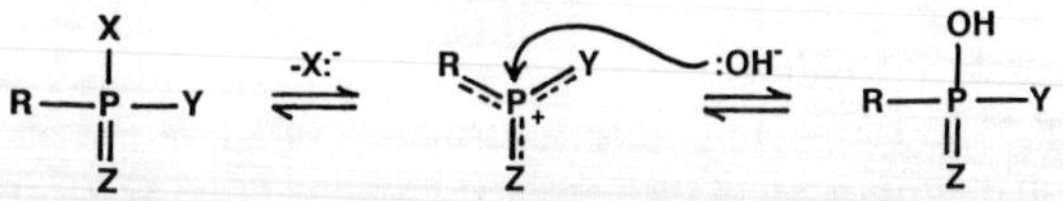

(2)

For each hydrolyzable compound, it is important to establish whether S_N1 or S_N2 hydrolysis is predominant, and whether the rate-limiting step involves generation of the reaction intermediate or its conversion into final products.

It is widely understood that metal ions can catalyze hydrolysis in a manner analogous to acid catalysis; metal ions (and protons) coordinate to the pollutant in such a way that electron density is shifted away from the site of nucleophilic attack, facilitating reaction. Because protons have an extremely high charge density and great polarizing power, metal catalysis is seldom seen in acidic solutions.[9] Metal ions have the advantage, however, in that they can readily coordinate two or more ligand donor sites on a pollutant molecule, and can vastly outnumber protons in neutral and alkaline solutions.[1,10]

In recent years, there has been a growing recognition that metal catalysis can arise from several distinct mechanisms. A comprehensive list (adapted from Suh and Chun[11]) is given below:

Mechanisms of Metal Catalysis

Type 1	The metal coordinates the electrophile, shifting the electronic distribution in the molecule in a way that enhances its reactivity.
Type 2	The metal coordinates to the leaving group, increasing its leaving ability.
Type 3	The metal acts as a center for simultaneous attachment of both the electrophile and attacking nucleophile (template effect).
Type 4	The metal coordinates the nucleophile and induces deprotonation (which increases the reactivity of the nucleophile).
Type 5	Coordination of the substrate with the metal induces conformational changes that facilitate reaction.
Type 6	Coordination of the substrate with the metal makes the molecule more positive, lessening unfavorable electrostatic interaction with the nucleophile.
Type 7	Coordination with the metal blocks inhibitory reverse reaction paths, such as (1) loss of OH^- from a tetrahedral intermediate instead of loss of the leaving group X^-, or (2) nucleophilic attack by X^-.

In the case of S_N2 processes, the hydrolysis of compounds possessing good leaving groups is limited by the rate of nucleophilic attack. As a consequence, these compounds are susceptible to the following categories of metal ion catalysis: 1,3,4,6, and possibly 5. The hydrolysis of compounds possessing poor leaving groups are limited by the breakdown of the tetragonal intermediate; categories 2, 7, and possibly 5 are therefore important. A number of pesticides and other pollutants lie in an intermediate region, where metal catalysis may shift the rate limiting step from nucleophilic attack to breakdown of the tetrahedral intermediate (or vice versa).

A few basic findings concerning reaction mechanisms will now be reviewed. Most studies have been performed on compounds possessing geometric or chemical constraints on reactivity, which allow one pathway to be examined without interference from competitive pathways.

A. CARBOXYLIC ACID ESTERS

In the case of carboxylic acid esters, neither the carbonyl oxygen nor the leaving group oxygen binds metals strong enough to make metal catalysis possible; suitably placed auxiliary ligand donor groups are necessary to enable metal catalysis to take place via chelate formation.

Ni^{2+}- and Cu^{2+}-catalyzed hydrolysis of a series of picolinate esters has been examined by Fife and Przystas.[12] As shown in Figure 1, formation of a five-membered chelate ring can take place. Metal coordination of the carbonyl group withdraws electron density from the electrophilic carbon, promoting nucleophilic attack. Because the ligand-donor ability of the carbonyl oxygen is poor, the auxiliary pyridyl nitrogen is required in order for appreciable complex formation to take place. In order for formation of this five-membered chelate to have an effect on the overall hydrolysis rate, the reaction must be limited by the rate of nucleophilic attack. Type 1 metal catalysis of this kind is observed with esters bearing good leaving groups (with pK_a values below 12.2) such as the nitrophenolate

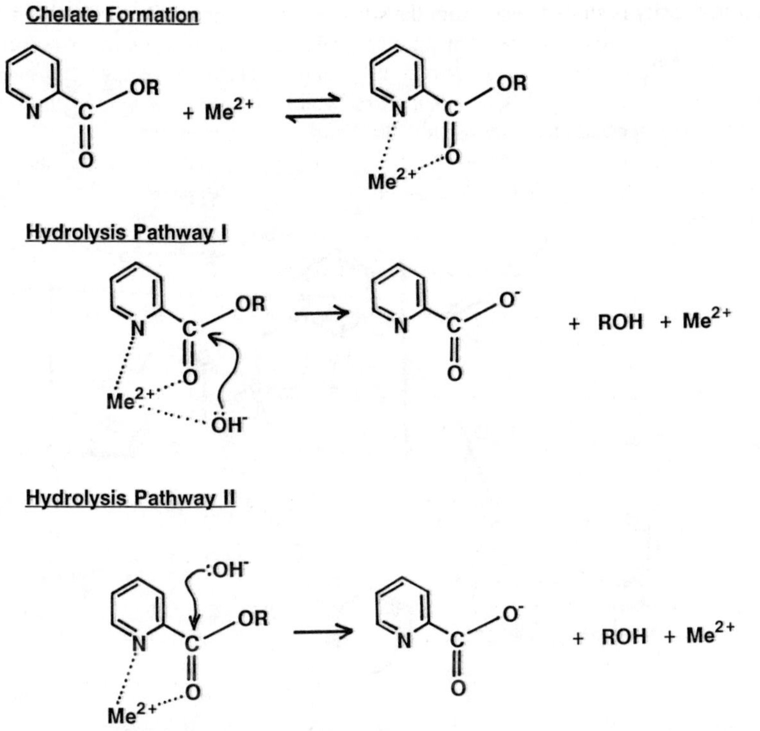

Figure 1 Chelation and subsequent hydrolysis of picolinate esters by Ni^{2+} and Cu^{2+}.[12] ROH: 2,4-dinitrophenol ($pK_a = 4.1$), 4-nitrophenol ($pK_a = 7.1$), phenol ($pK_a = 10.0$), 2,2,2-trifluoroethanol ($pK_a = 12.4$), ethanol ($pK_a = 16.0$).

or phenolate anion.[12] The alcohol moiety in ethyl picolinate is not a good leaving group. For this reason, ethyl picolinate hydrolysis is limited by the breakdown of the tetragonal intermediate, and chelate formation with metal ions does not yield a catalytic effect.[12]

Benzyl esters possess poor leaving groups, and therefore represent an opportunity to explore metal catalysis reactions according to a type 2 mechanism. As shown in Figure 2, 2-pyridylmethyl hydrogen phthalate possesses a pyridyl nitrogen situated in such a way that Ni^{2+}, Cu^{2+}, and Zn^{2+} can chelate the leaving group oxygen. In this case, intramolecular attack by a free carboxylate group takes place, generating an anhydride possessing a tetragonal carbon; chelation facilitates bond breaking with the leaving group, and accelerates the overall hydrolysis reaction.[13]

An example of type 7 catalysis is presented in Figure 3. 3-Carboxyaspirin decomposition involves an intramolecular nucleophilic step followed by external OH^- attack.[11] The free phenolate group generated by the first hydrolytic step is a reactive nucleophile, capable of back reaction. Coordination with a metal ion lowers its nucleophilic reactivity, preventing back reaction.

B. AMIDES AND ANILIDES

Uncatalyzed hydrolysis of amides and anilides are typically 10^4 to 10^6 times slower than the hydrolysis of corresponding esters.[14] RNH^- and R_2N^- are extremely poor leaving groups, which must become protonated or complexed in some manner during the breakdown of the tetragonal intermediate in order for hydrolysis to occur at appreciable rates.[15]

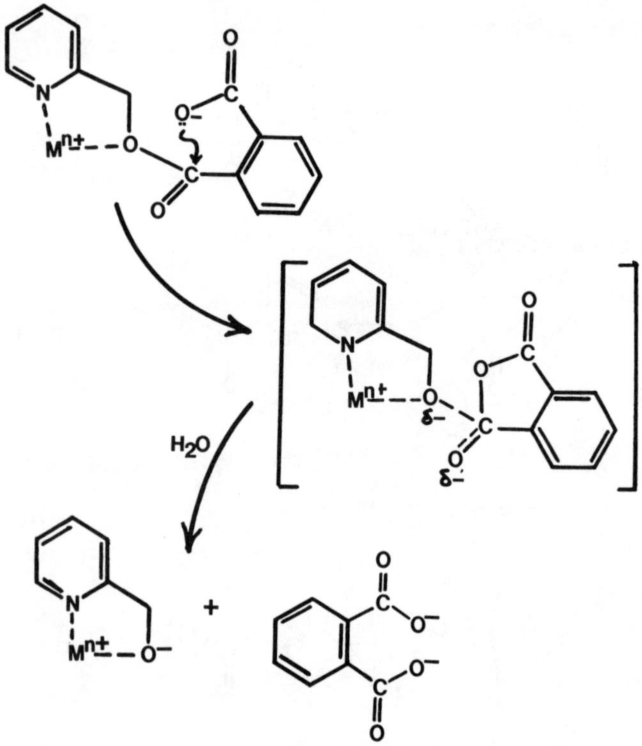

Figure 2 Ni(II), Cu(II), and Zn(II) catalysis of 2-pyridylmethyl hydrogen phthalate. A type 2 mechanism is operative: metal chelation generates a better leaving group.[13]

Picolinanilides can be used to illustrate how substituent effects determine which reaction step is rate limiting and affect the sensitivity to metal ion catalysis. Although compound I, shown in Figure 4, is capable of forming a five-membered metal chelate, addition of metals actually lowers overall hydrolysis rates. Formation of the metal chelate promotes N-H deprotonation, generating an amido complex which is more resistant to nucleophilic attack.[16] Compound II, in contrast, is quite sensitive to metal catalysis. N-Methyl substitution prevents formation of the amido complex, and the nitro substituents generate a more effective leaving group. As a consequence, nucleophilic attack becomes rate limiting. Polarization of the carbonyl bond (type 1 catalysis) and generation of the nucleophile Me-OH (type 4 catalysis) are both believed to contribute to the overall catalytic effect.[16]

For many amides and anilides, protonation of the leaving group (RNH^- or R_2N^-) facilitates breakdown of the tetragonal intermediate. Metal complexation may have an analogous effect. In order to test this hypothesis, anilides have been synthesized that bear auxiliary donor groups that facilitate metal coordination of the leaving group. Although metal catalysis has been demonstrated with these compounds,[17] alternative pathways of catalysis cannot be ruled out.

C. PHOSPHATE-CONTAINING ESTERS

Alkyl phosphate esters contain two distinct electrophilic sites: the phosphorus atom is a "hard" site (favoring reaction with OH^-), while the carbon in the alkyl linkage is a "soft" site (favoring reaction with H_2O). The relative importance of phosphorus- vs. carbon-

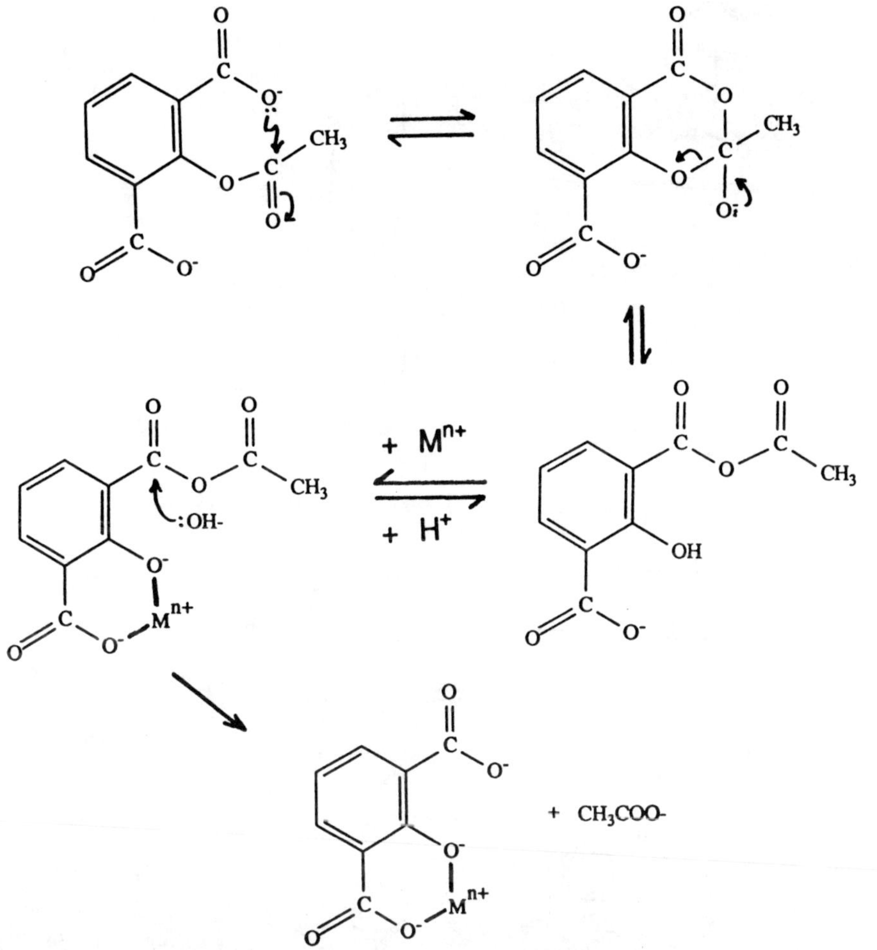

Figure 3 The ability of metal ions to catalyze the hydrolysis of 3-carboxyaspirin diminishes in the following order: Fe(III) > Al(III) >> Cu(II), Zn(II). A type 7 mechanism is operative: metal chelation blocks the reverse reaction.[11]

centered attack depends upon both pH and ester structure. If all three leaving groups are poor, for example, carbon-centered attack is the predominant mechanism.[18] Hydrolysis via phosphorus attack can proceed according to (1) a dissociative (S_N1) mechanism, involving a reactive three-coordinate metaphosphate intermediate, or (2) an associative mechanism, involving a five-coordinate intermediate or transition state.[19]

Metal-catalyzed hydrolysis of phosphate-containing esters is the subject of an excellent recent review by Hendry and Sargeson.[19] A number of different catalytic pathways have been documented. Figure 5, for example, presents a situation where an auxiliary donor group makes possible chelate formation between a metal ion and the leaving group, facilitating hydrolysis via a type 2 pathway.[20] An alternative "push-pull" mechanism is cited most frequently. It begins with attachment of a metal to the phosphoro-oxygen, which is a strong ligand donor group.[21] Increases in electrophilicity at the phosphorus atom accompanying metal coordination (type 1 pathway) are modest, however.[19] Most of the catalytic effect arises from the generation of a metal-bound hydroxo nucleophile,

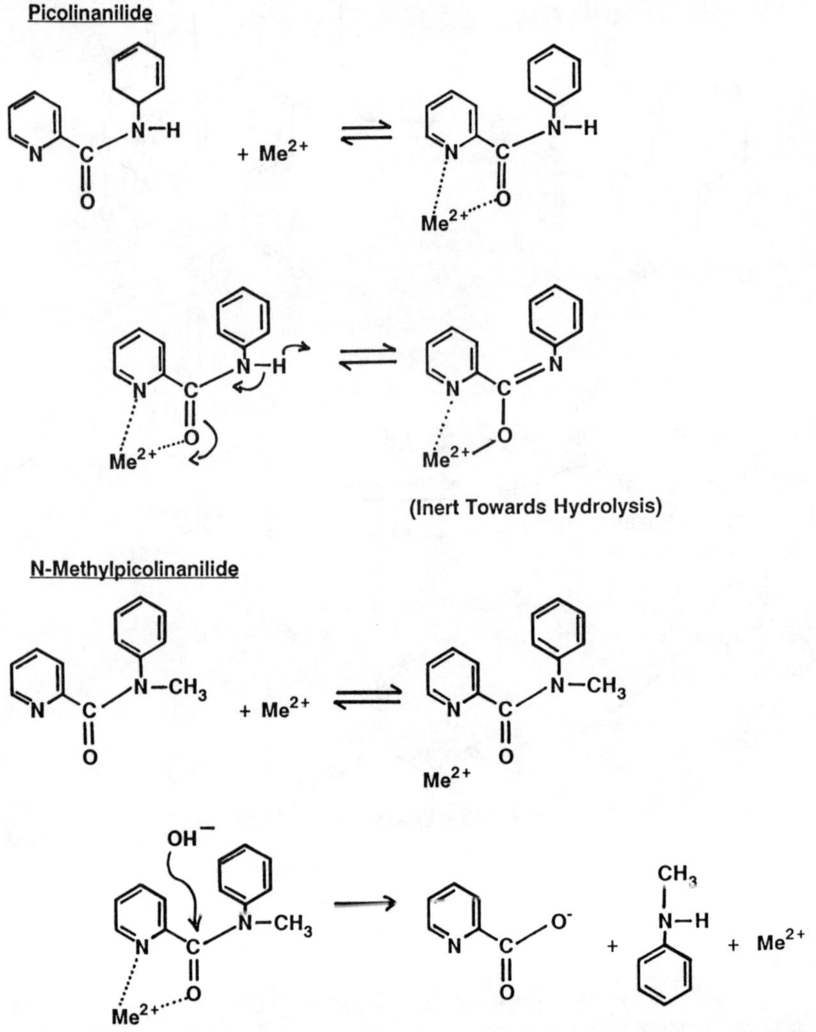

Figure 4 Chelation of metal ions by picolinanilide and by N-methylpicolinanilide, and subsequent nucleophilic attack by OH^-.[16]

which can attack the ester linkage via intramolecular reaction (type 3 and 4 pathways). Hoffmann[22] provides an excellent introduction to mechanisms of this kind.

D. OTHER HYDROLYZABLE COMPOUNDS

Our ability to discern the reaction mechanism and predict the magnitude of the catalytic effect depends upon the type of metal catalysis taking place. When the site of metal binding on an organic molecule is close to the site of nucleophilic attack, we can expect that some kind of catalysis or inhibition can take place. Activation of the nucleophile and other more complex reaction pathways are more difficult to evaluate.

Chemists have synthesized a series of compounds possessing various electrophilic substituents placed alongside auxiliary donor groups to facilitate metal binding. Hydration of 2-cyano-1,10-phenanthroline[23] and 2-cyanopyridine[24] to corresponding amides is

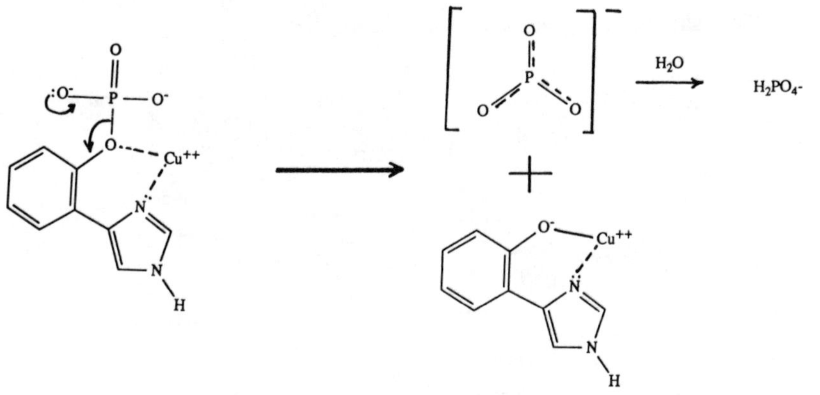

Figure 5 Cu^{II}-catalyzed hydrolysis of 2-(4(5)-imidazolyl)phenyl phosphate according to a type 2 mechanism (generation of a better leaving group.)[20]

susceptible to metal catalysis. Metal-catalyzed hydrolysis of anhydrides,[25] acetals,[26] and isothiocyanates[27] has also been observed.

E. PESTICIDE HYDROLYSIS INVOLVING DISSOLVED METAL IONS

Metal ion catalysis is most easily recognized when otherwise refractory compounds hydrolyze quickly upon addition of metal salts. By the mid 1950s, the sensitivity of thionophosphate pesticides such as Parathion and EPN (thionobenzene phosphoric acid *O*-ethyl-*O*-*p*-nitrophenyl ester) to metal ion catalysis was recognized.[28] Rate enhancements arising from the addition of 1.0×10^{-4} *M* Cu^{2+} were as high as 20-fold for Parathion (at pH 8.5) and 48-fold for EPN (at pH 8.2). Coordination of the sulfur donor group by Cu^{2+} activates the phosphorus center to nucleophilic attack (type 1 catalysis).

The phosphothionate pesticide Chlorpyrifos contains a pyridyl nitrogen capable of forming a six-membered chelate ring with appropriate metals, as shown in Figure 6.[29,30] Metal ion catalysis arises from the increased electrophilicity of the phosphorus atom in the chelate ring (type 1 catalysis[30]) and from possible leaving-group effects (type 2 catalysis). Although a number of metals have been observed to catalyze Chlorpyrifos hydrolysis (including Mg^{II} and Al^{III}), Cu^{II} is by far the most reactive. Recently, Mill and Mabey[31] calculated that Cu^{2+} concentrations would have to exceed 10^{-5} *M* in order for the metal-catalyzed pathway to surpass the rate of the uncatalyzed hydrolysis pathway, an occurrence which would seldom occur in natural waters.

Aldicarb, a systemic pesticide, is unusually sensitive to metal ion catalysis. It is important to understand the degradative pathways of Aldicarb, because of its appearance at excessive levels in groundwaters below agricultural lands.[32] As shown in Figure 7, Cu^{2+} forms a five-membered ring chelate with Aldicarb, which facilitates intramolecular

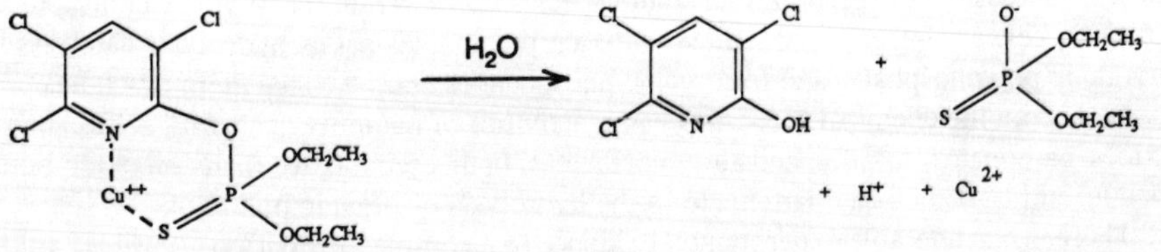

Figure 6 Cu(II)-catalyzed hydrolysis of chlorpyrifos. Type 1 and type 2 mechanisms are operative.[29,30]

Aldicarb decomposition involving Cu^{2+}:

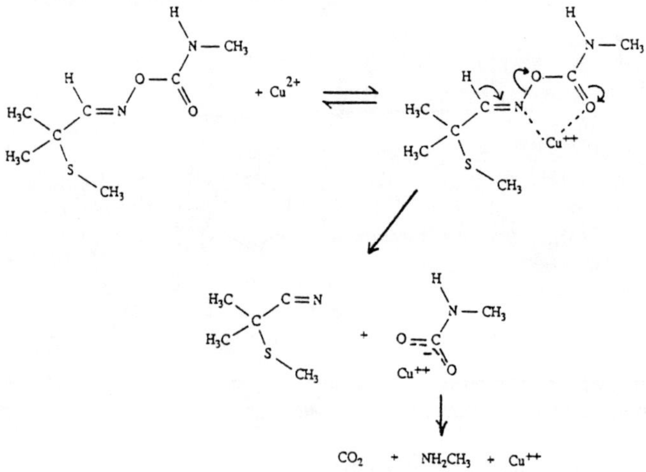

O-Acetyl-2-Pyridinecarboxaldoxime (also sensitive to catalysis):

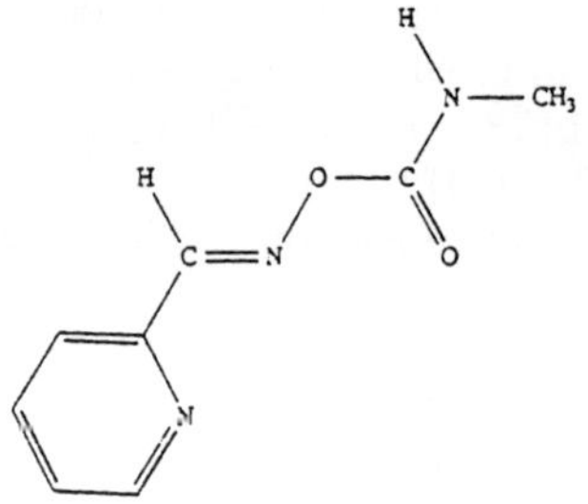

Figure 7 Carbamates sensitive toward metal catalysis.[32]

degradation. *O*-Acetyl-2-pyridinecarboxaldoxime, a structurally related compound susceptible to metal ion catalysis, is also shown in Figure 7.

F. CATALYSIS BY SURFACE-BOUND METAL IONS

Many potentially catalytic metals form low-solubility inorganic solids within the pH domain of natural waters. In order for a catalytic effect to be observed, surface-bound metals must be accessible to reaction with solute species. Research in the 1950s established that $La(OH)_3$ gels and other metal oxide/hydroxide precipitates act as hydrolysis catalysts for inorganic polyphosphates and for organic phosphate esters.[14,33] Since then, much has been learned about the chemical properties and reactivities of naturally occurring solids such as oxides, carbonates, sulfides, and aluminosilicates. In this section, we focus on oxide-bound metals, and how they can participate in the hydrolysis of organic pollutants.

There is considerable experimental evidence that complex formation equilibria among dissolved and surface-bound metal species are similar in many respects: surface complex formation constants for metal ions,[34] inorganic anions,[35,36] and organic ligands[37,38] increase in magnitude as the formation constants of analogous complexes in solution are increased. More recently, attempts have been made to compare reactivities of dissolved and surface-bound metal species. Notable achievements using this approach have been made in understanding the oxidation of surface-bound metals[39] and the dissolution of metal-containing minerals.[40]

In order to compare the catalytic reactivity of dissolved and surface-bound metals, we must investigate any unusual or unique characteristics of the oxide/water interface. A break or cleavage of an oxide under vacuum exposes lattice-bound metals that lack one or more ligands in their inner coordination sphere. The high Bronsted acidity of these unoccupied coordinative sites is responsible for the well-known hydrolytic reactivity of dehydrated and partially hydrated mineral surfaces.[41,42] The susceptibility of parathion to hydrolysis on clay surfaces has been well documented.[43-45] Much of this reactivity disappears upon hydration. In one study of the hydrolytic breakdown of 1-(4-methoxyphenyl)-2,3-epoxypropane on sodium smectite, for example, a maximum rate at 30% water content was decreased by more than a factor of two when the sodium smectite was placed in aqueous suspension.[46] The hydrolysis rate in suspension was only 2.7 times the rate in particle-free solution.

When abundant water is present, the coordinative requirements of surface-bound metals are met by adsorbed H_2O and OH^- molecules. Because of this "hydrous layer," inner-sphere coordination of solutes is a ligand exchange process, and solvent molecules are displaced in the process. Despite the diminished effect of Bronsted acidity upon addition of bulk water, dramatic effects of solids surfaces on hydrolysis rates in aqueous suspensions can still be observed. The hydrolysis of the insecticide phosmet, for example, was found to be 500 times faster in montmorillonite suspensions than in particle-free solution.[47] Zeolite suspensions have been observed to promote the hydrolysis of ethyl acetate.[48]

Surface catalysis is observed when all participating reactants adsorb to a significant extent, and when rate constants for reaction in the interfacial region are comparable to or exceed rate constants for reaction in homogeneous solution. Organic compounds possessing carboxylate groups adsorb relatively strongly at the oxide/water interface; surface complex formation models have been relatively successful at representing the effects of pH and medium composition on adsorption.[37,49,50] Recently, surface complex formation models have been used to represent the adsorption of organic compounds possessing phenolate, pyridyl, and amino substituents at the oxide/water interface.[51,52] At present, relatively little is known about the adsorption of functional groups directly pertinent to surface catalysis reactions (e.g., carbonyl groups within ester linkages).

Even when adsorption phenomena have been accounted for, relatively little is known about the nature (conformation and stoichiometry) of adsorbed species. This limits our ability to understand surface reactivity on a fundamental level. Based upon work that has already been completed, the following generalities can be made: (1) auxiliary donor groups that facilitate metal catalysis by metal ions in solution should also facilitate reactions involving surface-bound metals and (2) phenomena of only secondary importance in reactions of dissolved complexes (e.g., electrostatic and hydrophobic interactions) may play a much greater role in surface catalysis. Results from recent research, summarized below, help illustrate the nature and potential importance of surface catalysis.

We begin with an example of type 1 catalysis: the hydrolysis of phenyl picolinate (PHP), methyl picolinate (MEP), and related esters in TiO_2, FeOOH, and Al_2O_3 suspensions.[53,54] PHP and MEP are subject to metal ion catalysis in homogeneous solution. Appropriate metal ions chelate the heterocyclic nitrogen and carbonyl oxygen. The presence of the metal ion increases the partial positive charge at the carbonyl carbon, facilitating nucleophilic attack.[12] In an analogous manner, surface-bound metal ions should also be capable of chelating phenyl picolinate and methyl picolinate, thereby increasing their susceptibility to hydrolysis:

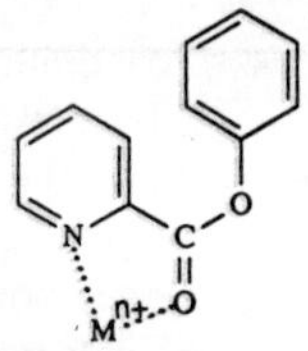

Chelation by
Dissolved Metal
Ions

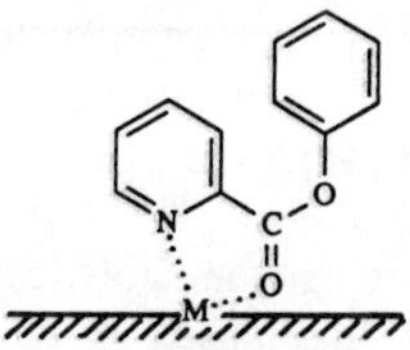

Chelation by
Surface-Bound
Metals

In the case of PHP, reaction progress can be monitored by measuring either loss of PHP or production of phenol. The sum ([PHP] + [Phenol]) remains constant in our experiments, indicating that competitive degradation processes are not important, and that the adsorption of either reactant or product is relatively low (below a few percent). As shown in Figure 8, PHP hydrolysis in particle-free blank solutions and in suspensions containing Al_2O_3 or SiO_2 is negligible during 100 h of reaction. Suspensions containing TiO_2 and FeOOH, however, cause a dramatic acceleration in PHP hydrolysis rate. Half-lives in the presence of TiO_2 are below 20 h. Similar results are obtained with MEP. It is important to establish that catalysis arises from reaction at the oxide surface, and not from release of soluble metals. As shown in Figure 9, removal of particles by filtration causes an immediate halt to any catalytic activity; the surface must be present for catalysis to occur.

The isomer phenyl isonicotinate was not subject to catalysis in our experiments, while PHP, MEP, and other esters with ligand donor groups in the *ortho* position were subject to catalysis. This is clear evidence that surface chelate formation is necessary for catalysis to occur.

The selectivity of hydrolyzable compounds to catalysis by different surfaces depends upon the nature of the auxiliary donor group. The pyridyl nitrogen of PHP and MEP encourages chelate formation with Ti^{IV} and Fe^{III} in preference to less polarizable Al^{III}. As a consequence, TiO_2 and FeOOH are effective hydrolysis catalysts, while Al_2O_3 is not. Oxygen donor groups, such as the phenolic group in phenyl salicylate and the oxygen heteroatom of methyl furanoate, are "harder" and more favorable to Al^{III}-chelate formation. As a consequence, all three surfaces catalyze the hydrolysis of these two "oxygen-donor" esters.

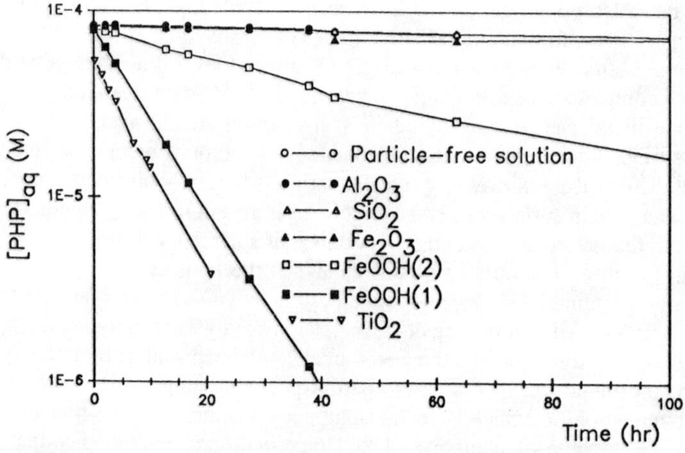

Figure 8 Effect of various metal oxides on the loss of phenyl picolinate (PHP) from solution via hydrolysis. All suspensions contained 10 g l^{-1} oxide, 1.0 m*M* acetate buffer (pH 5.0), and 50 m*M* NaCl.[53]

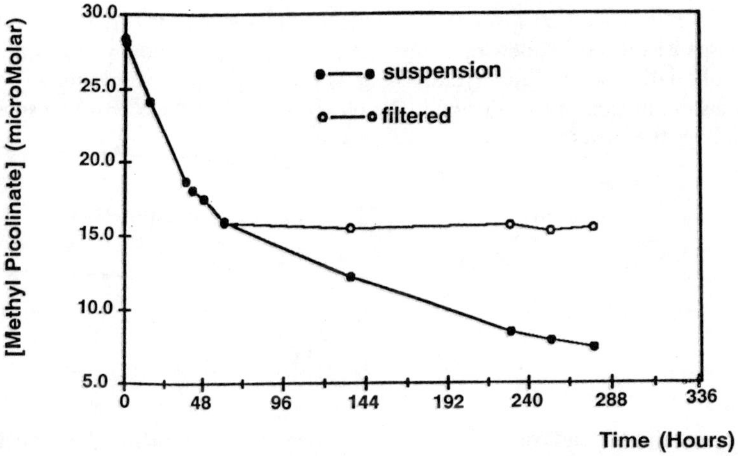

Figure 9 Proof that catalysis of methyl picolinate (MEP) hydrolysis occurs at the oxide surface. Removal of oxide particles by filtration halts the progress of the hydrolysis reaction (5 g l^{-1} TiO_2, pH 5.25) (unpublished data).

In particle-free solution, both PHP and MEP exhibit classic base-catalyzed hydrolysis behavior: reaction rate increases by an order of magnitude for every unit increase in pH, reflecting increases in the concentration of OH^-. Hydrolysis in the presence of TiO_2 is virtually independent of pH, and in the presence of FeOOH exhibits only a slight pH dependence (Figure 10). Activation of the ester linkage through surface chelate formation is apparently strong enough to facilitate reaction with the weak nucleophile H_2O, overshadowing the pH-dependent reaction with OH^-.

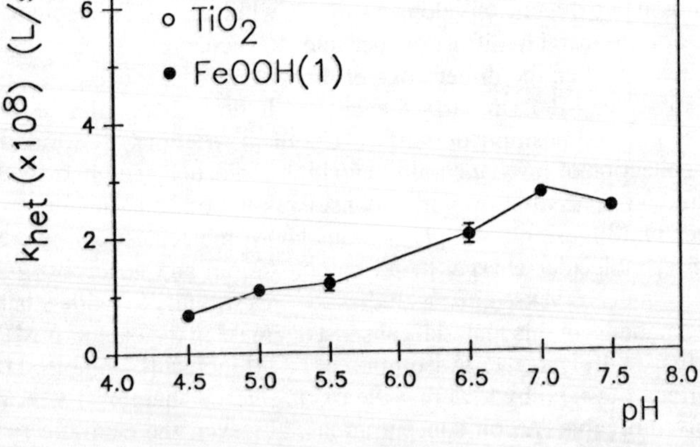

Figure 10 Effect of pH on k_{het}, the rate constant for surface-catalyzed hydrolysis of PHP normalized with respect to oxide loading.[53]

Next, we consider the hydrolysis of monophenyl terephthalate (MPT^-) in Al_2O_3 and TiO_2 suspensions. Positive-charged surfaces such as Al_2O_3 (below pH 8.6) and TiO_2 (below pH 6.4) accelerate the hydrolysis of MPT^- by an order of magnitude or more. Detailed examination of the reaction in the presence of Al_2O_3[55] has provided substantial evidence for the following catalytic mechanism:

$$>AlOH_2^+ + MPT^- \underset{k_{-1}}{\overset{k_1}{\rightleftharpoons}} >Al-MPT + H_2O \tag{3}$$

$$>Al\text{-}MPT + OH_d^- \xrightarrow{k_2} \text{products} \tag{4}$$

The subscript "d" indicates that the reaction rate is proportional to the concentration of hydroxide ion at the plane of closest approach to the surface (the innermost boundary of the diffuse layer).

In the case of MPT^-, the free, ionized carboxylate group provides the basis for substrate adsorption, through complex formation with surface-bound Al. The extent of MPT^- adsorption diminished as the pH was increased, as shown in Figure 11. This observation is consistent with the adsorption stoichiometry given in Reaction 3. The extent of MPT^- adsorption also decreased as the ionic strength was increased, indicating significant electrostatic interaction between the anionic ester and the positive-charged oxide surface (analogous to the adsorption of anions discussed by Hayes et al.[36]).

The anionic nucleophile (OH^-) is also accumulated at the positive-charged oxide surface. The pseudo first-order rate constant for MPT^- hydrolysis (k_h, shown in Figure 11) reflects the combined accumulation of both ester and nucleophile at the oxide surface; k_h reaches its highest value at the pH where the product $[>Al\text{-}MPT]\cdot[OH^-_d]$ is at a maximum.

Al_2O_3 and TiO_2-catalyzed hydrolysis of MPT^- exhibits elements of both type 3 and type 6 catalysis. Positive oxide surfaces accumulate both anionic reactants (OH^- and MPT^-), increasing their encounter frequency. Type 6 catalysis (electrostatic interactions favoring encounter between like-charged species) is typically more important for surface chemical reactions than for reactions in solution, because of the additive electrostatic effect arising from neighboring charged groups. Simultaneous attachment of substrate and nucleophile (type 3 catalysis) is also typical of surface chemical reactions. Adsorbed species diffuse in two dimensions along a surface, while dissolved species diffuse in three dimensions. For comparable diffusion coefficients, encounter frequencies between reactants are increased when the dimensions of the system are reduced.[56,57]

The sensitivity of MPT^- to surface-catalyzed hydrolysis enables us to examine the effect of medium composition on surface chemical reactions. Natural organic matter (NOM) is a collection of low-, medium-, and high-molecular-weight organic compounds whose nature and concentration varies considerably from one aquatic environment to another. Even within a single sample, various NOM molecules exhibit a wide range of molecular size, molecular charge, metal complexation, and adsorption characteristics. The Dismal Swamp NOM used in our studies adsorbs strongly to oxide surfaces. Addition of increasing amounts of this material caused a decrease in the extent of MPT^- adsorption and in the MPT^- hydrolysis rate. Adsorption of NOM molecules inhibited the reaction by blocking surface sites and by lowering the oxide surface charge.[58] Even at 30 mg C l^{-1}, NOM dissolved organic carbon concentrations, however, the catalytic reactivity of the surface was sufficient to yield rates of MPT^- hydrolysis that were substantially higher than rates measured in particle-free solutions; surface catalysis was still quite efficient.

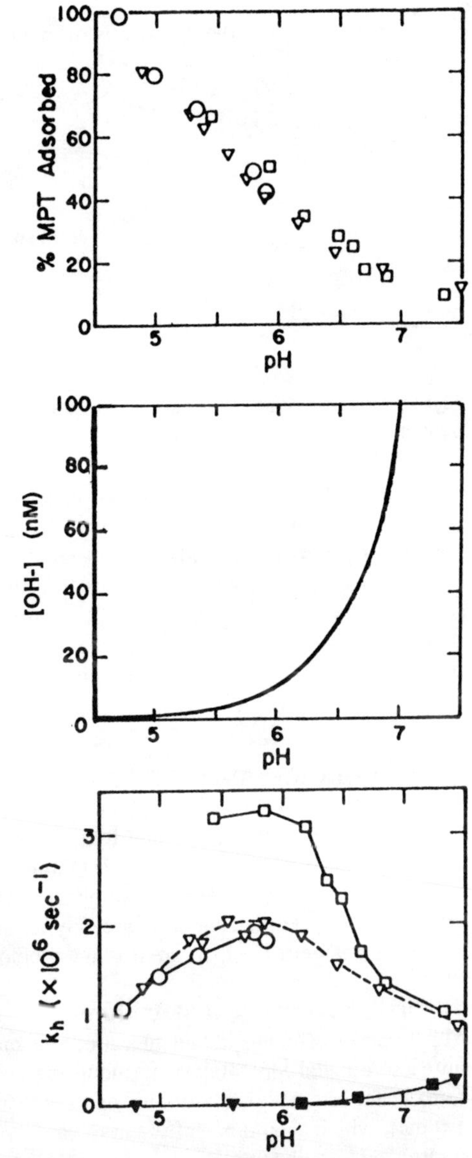

Figure 11 The hydrolysis of monophenyl terephthalate (MPT^-) is catalyzed by Al_2O_3 surfaces. The rate-limiting step is given by Reaction 3; the maximum in k_h at pH 5.5 reflects the maximum in the product [>Al-MPT^-][OH^-]. Buffers employed: 1.0 m*M* acetate (open circles); 1.0 m*M* 4-nitrophenol (open squares); a mixture of 1.0 m*M* acetate and 1.0 m*M* 4-nitrophenol (open triangles). The filled symbols represent k_h values measured in particle-free solutions.[55]

Susceptibility to surface-catalyzed hydrolysis is not restricted to simple laboratory chemicals; recent work in our laboratory has demonstrated that important pesticides are also susceptible to catalysis in oxide suspensions.[54] Catalysis of Chlorpyrifos hydrolysis by TiO_2, FeOOH, and Al_2O_3, for example, likely involves chelate formation between the surface-bound metal, the thionate sulfur, and the pyridinyl nitrogen.[54]

Recently, Stanton[59] reported that the fungicide oxycarboxin degrades when stored in borosilicate glass vessels. A surface-catalyzed reaction following a type 4 mechanism has been postulated, yielding the product 2 (vinylsulfonyl)-acetanilide (Figure 12). In alkaline solution, 2-(vinylsulfonyl)-acetanilide is formed by hydroxide ion attack at the oxathiin ring of the parent compound. Because of the low hydroxide ion concentration in neutral and acidic pH, hydrolytic ring cleavage is not observed in the absence of catalyst. The glass

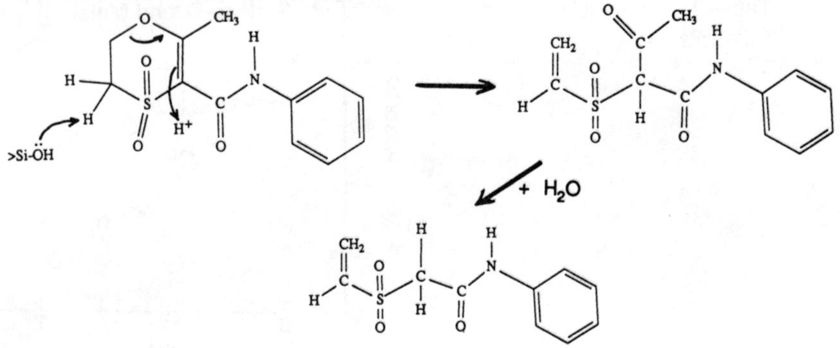

Figure 12 Hydrolytic decomposition of the fungicide oxycarboxin in the presence of glass surfaces.[59]

exerts its catalytic effect by providing sufficiently strong nucleophiles (surface-bound OH^- and O^{2-}) for attacking the oxathiin ring.[59] Quartz and aluminosilicate surfaces present in the aquatic environments may also be capable of catalyzing oxycarboxin hydrolysis.

In the case of surface-catalyzed hydrolysis reactions, the nature and relative significance of type 4 catalysis are poorly understood. New research is particularly needed in this area. Hoffmann[22] has recently reported that metal oxides catalyze the hydrolysis of *p*-nitrophenyl acetates via this mechanism.

IV. METAL CATALYSIS IN THE ENVIRONMENT

A. ABUNDANCE AND SPECIATION OF METALS

Table 1 lists the average crustal abundance of potential metal catalysts. Although Co, Ni, Cu, and Zn are well-documented hydrolysis catalysts, their natural abundance is low, suggesting that their environmental importance is limited. The light metals (Mg, Ca, and Al), and early members of the transition metals (Ti, Mn, and Fe) are much more abundant, but much less is known about their catalytic reactivity. Table 2 places metal species into reactivity categories, which are discussed below.

1. Dissolved Divalent Metal Ions

Mg(II) and Ca(II) are abundant cations in many aquatic environments, but relatively unreactive metal ion catalysts. Among the abundant elements, only Mn(II) and Fe(II) remain. Interstitial concentrations reported in anoxic sediments and aquifers reach 0.1 to 1.0 m*M*, which should be sufficient to cause an observed catalytic effect for susceptible hydrolyzable pollutants.

Mn(II) and Fe(II) are both classified as "borderline" metals by Pearson.[2] Fe(II) is more polarizable than Mn(II) and ranks higher in the Irving-Williams series, indicative of higher catalytic reactivity. In one study, ethyl glycinate-*N,N*-diacetic acid was found to be nine times more sensitive to catalysis by Fe(II) than by Mn(II).[60]

2. Al(III) and Fe(III) Inorganic Complexes

Relatively little is known about the catalytic reactivity of these two classic "hard" metal ions; the formation of hydroxo complexes, and solubility limitations under neutral and alkaline pH conditions make investigations of the reaction mechanism difficult. The high charge/radius ratio of these two metals makes them particularly effective catalysts for the hydrolysis of pyrophosphate, higher polyphosphates, and other "hard", anionic hydrolyzable species.[61] Many similarities are expected between the catalytic reactivity of Fe(III) and the more extensively studied but substitution-inert metal catalyst Co(III).

Table 2 Potentially Important Dissolved and Surface-Bound Metal Catalysts

	Set 1: Dissolved Divalent Metal Ions			
(Mg^{2+}, Ca^{2+})	Mn^{2+}	Fe^{2+}	(Co^{2+}, Ni^{2+})	(Cu^{2+}, Zn^{2+})
	Set 2: Al^{III}, Fe^{III} Hydroxo Complexes			
	Al^{III} Species		Fe^{III} Species	
Solubility limit $M(OH)_3$ (amorphous)	Al_T ≥50 μ*M* when the pH is ≤4.32		Fe^{III} ≥50 μ*M* when the pH is ≤2.64	
Hydroxo speciation at pH 4.0:	Al^{3+}	3.09×10^{-4} *M*	Fe^{3+}	1.01×10^{-9} *M*
	$AlOH^{2+}$	3.16×10^{-5}	$FeOH^{2+}$	6.50×10^{-8}
	$Al(OH)_2^+$	1.59×10^{-5}	$Fe(OH)_2^+$	2.06×10^{-7}
	$Al(OH)_3^o$	3.16×10^{-7}		
	$Al(OH)_4^-$	3.16×10^{-11}	$Fe(OH)_4^-$	2.59×10^{-15}
	Al_T	3.57×10^{-4}	Fe_T	2.72×10^{-7}
	Set 3: Al^{III}, Fe^{III} NOM Complexes			
1×10^{-3} *M* salicylates represents complexation by NOM (pH 4.00)	AlL^+	2.88×10^{-4} *M*	FeL^+	3.34×10^{-6} *M*
	AlL_2^-	6.71×10^{-7}	FeL_2^-	4.95×10^{-7}
	AlL_3^{3-}	3.12×10^{-13}	FeL_3^{3-}	9.43×10^{-13}
	Al_T	6.46×10^{-4}	Fe_T	4.11×10^{-6}
	Set 4: Hydrous Oxide Surfaces			
		$Al(OH)_3(s)$	$Fe(OH)_3(s)$	
$TiO_2(s)$		$AlOOH(s)$	$FeOOH(s)$	
		$Al_2O_3(s)$	$Fe_2O_3(s)$	
			$Fe_3O_4(s)$	
	Set 5: Alumino-Silicate Surfaces			
	Kaolinite		Mn+-Smectite	
			Fe-Rich Smectite	

Inorganic aluminum and iron speciation in the presence of excess $M(OH)_3(s)$ is presented in Table 2. $Al^{3+}(aq)$ and $Fe^{3+}(aq)$ dominate aqueous phase speciation at pH values below p^*K_1 (4.97 for Al(III), 2.19 for Fe(III)); dissolved hydroxo species dominate speciation at higher pH. Total dissolved metal concentrations (Al_T and Fe_T) drop precipitously as the pH is increased toward the neutral and alkaline range. The catalytic activity of surface-bound Al(III) and Fe(III) are dealt with separately.

3. Al(III) and Fe(III) Natural Organic Matter Complexes

In situations where organic chelating agents are present in a system, the catalytic effect of the metal is equal to or lower than the catalytic effect of the free metal ion.[62] Thus, although chelation by components of dissolved natural organic matter (NOM) may yield substantial increases in $Al^{III}{}_T$ and $Fe^{III}{}_T$, it is not known whether the catalytic rate will rise proportionately.

There has been much recent interest in the role of NOM in solubilizing aluminum, particularly in acidified aquatic environments.[63-65]. Although the exact structure of organic chelating groups cannot be known with certainty, hard oxygen-donor groups such as carboxylate ions and phenolate ions probably dominate Al(III) and Fe(III) binding. To serve as illustration, Table 2 presents results from a calculation involving the equilibration of 1.0 m*M* salicylate with excess $M(OH)_3(s)$. At pH 4.0, addition of salicylate yields a 1.8-fold increase in dissolved Al(III), and a 15-fold increase in dissolved Fe(III). Because of the tremendous potential of chelating compounds within NOM for mobilizing metals, the catalytic reactivity of the resulting complexes merits serious attention.

4. Hydrous Oxide Surfaces

Al^{III}-, Ti^{IV}-, and Fe^{III}-bearing mineral surfaces have already been observed to catalyze the hydrolysis of a small number of representative organic compounds. The identity of metal ions within an oxide surface is of primary importance; because of differences in complex formation capacity and polarizability among these three metals, their ability to catalyze the hydrolysis of representative organic compounds varies substantially.[53]

Even when the identity of the metal is kept constant, differences in the crystalline form of oxides undoubtedly influences reactivity. Fe^{III}, for example, occurs on such diverse natural surfaces as FeOOH (goethite, lepidocrocite), Fe_2O_3 (hematite, maghemite), $Fe(OH)_3$ (amorphous), and Fe_3O_4 (magnetite). Recent efforts to calculate surface complex formation constants based upon bond lengths and coordinative geometries[66,67] may provide insight into the reactivities of crystalline surfaces. Other phenomena contributing to catalysis by oxide surfaces are more difficult to assess, such as changes in the structure and ionic composition of water near a surface, and the effects of electrostatic and hydrophobic forces on reaction rates.

5. Aluminosilicate Surfaces

Mineral surfaces comprised of two or more metals are more complex than the simple hydrous oxides discussed above. Aluminosilicate surfaces can possess permanent charges arising from isomorphic substitution within either octahedral or tetrahedral layers. Edge sites (e.g., kaolinite) are often structurally and chemically different from planar sites. Smectites and other layered minerals contain interlamellar spaces; the hydration of these interlamellar spaces and their capacity to selectively accumulate or exclude solute ions are the key to understanding their unique or unusual reactivity toward organic pollutants (see References 41 and 42).

B. THE EFFECT OF SPECIATION ON CATALYSIS

Aquatic environments contain a vast array of constituent chemicals that affect the transformation and fate of organic pollutants. Speciation is the key to assessing these effects, and for predicting hydrolysis rates under realistic chemical conditions.

Johnson and Angelici[68] have provided a great deal of quantitative information about the metal-catalyzed hydrolysis of ethyl oxalate ($EtOx^-$). We use this information to explore the connection between speciation and catalysis. The overall reaction mechanism is given below:

$$\mathrm{EtOxH} = \mathrm{EtOx}^- + \mathrm{H}^+ \qquad \mathrm{pK_a} = 1.74 \tag{5}$$

$$\mathrm{EtOx}^- + \mathrm{M}^{2+} = \mathrm{EtOxM}^+ \qquad \mathrm{K_f} = [\mathrm{EtOxM}^+]/[\mathrm{EtOx}^-][\mathrm{M}^{2+}] \tag{6}$$

$$\mathrm{EtOx}^- + \mathrm{OH}^- \xrightarrow{k_{OH}} \mathrm{products} \qquad k_{OH} = 6.9 \times 10^{-5} M^{-1}\mathrm{s}^{-1} \tag{7}$$

$$\mathrm{EtOxM^+} + \mathrm{OH^-} \xrightarrow{k_M} \text{products} \tag{8}$$

$$\frac{d[\text{products}]}{dt} = \left(k_{OH} + k_M K_f\right)\left[OH^-\right]\left[EtOx^-\right] = k_h\left[EtOx^-\right] \tag{9}$$

Type 1 catalysis is operative here: the metal coordinates the carbonyl oxygen, activating the carbonyl carbon to nucleophilic attack. Johnson and Angelici[68] have stated that metal-ester complex formation is fast relative to subsequent attack by OH^-. We will assume that all metal complex formation reactions are in quasiequilibrium relative to OH^- nucleophilic attack.

Values of $\log K_f$ and k_M for three divalent metals are listed below (valid at 0.10 *M* ionic strength):[68]

	Cu^{2+}	**Zn^{2+}**	**Mg^{2+}**	
$\log K_f$	2.1	1.3	0.63	*(M^{-1})*
k_M	14.1	0.63	0.0030	*($M^{-1}\,s^{-1}$)*

We have calculated metal speciation in solutions containing a noncomplexing electrolyte (0.10 *M* $NaNO_3$) using the computer program MINEQL[69] as modified by Papelis et al.[70] Free metal concentrations ($[M^{2+}(aq)]$) are limited under neutral and alkaline pH by the formation of hydroxide solids and hydroxo species in solution:

$$M(OH)_2(s) + 2H^+ - M^{2+}(aq) + 2H_2O \qquad {}^{*c}K_{so} = \frac{\left[M^{2+}\right]}{a_{M(OH)_2}\left[H^+\right]^2} \tag{10}$$

$$M_T = \left[M^{2+}\right] + \left[MOH^+\right] + \left[M(OH)_2^o(aq)\right] + \left[M(OH)_3^-\right] \tag{11}$$

($^{*c}K_{so}$ is the concentration-scale solubility product for $M(OH)_2(s)$, written in terms of protons). Apparent first-order rate constants for ethyl oxalate hydrolysis (k_h) are shown in Figure 13 for 1.0×10^{-5} and 1.0×10^{-3} *M* total added Cu, Zn, and Mg. The smooth line with a slope of +1.0 represents OH^--catalyzed ethyl oxalate hydrolysis in the absence of added metals. The curve representing Mg lies alongside the metal-free line; catalysis by Mg is negligible in this case.

In acidic solutions, $\log k_h$ for Cu- and Zn-catalyzed hydrolysis increases with a slope of +1.0 as the pH is increased. Under these conditions, the amount of metal-ester complex formed is constant; the rate constant increase arises from increases in $[OH^-]$. Once a maximum is reached, however, $\log k_h$ actually decreases with a slope of −1.0 as the pH is increased. This curious behavior is caused by precipitation of the $M(OH)_2(s)$ solid phase at the pH corresponding to the maximum in $\log k_h$. Above this pH, $[M^{2+}(aq)]$ decreases by two orders of magnitude for every unit increase in pH, according to Equation 10. As a consequence, the product $[M^{2+}(aq)][OH^-]$ decreases by one order of magnitude for every unit increase in pH. Because $Cu(OH)_2(s)$ is less soluble than $Zn(OH)_2(s)$, the maximum in $\log k_h$ occurs at a lower pH. It is interesting to note that Zn is a better catalyst than Cu within a narrow pH range, because of its greater solubility.

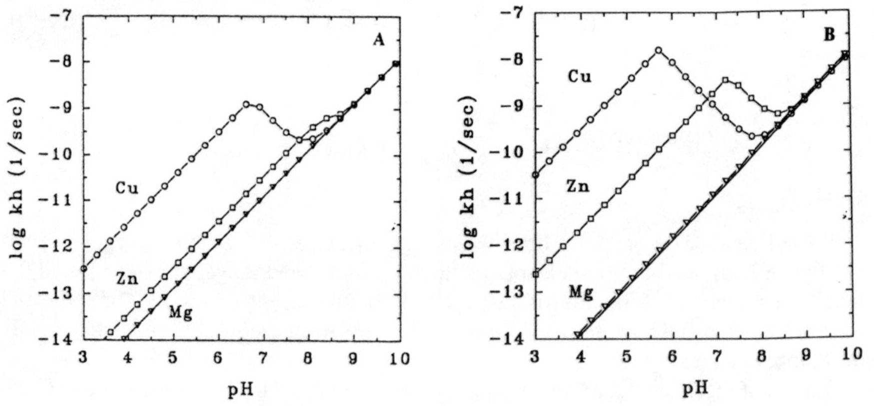

Figure 13 Apparent first-order rate constant k_h for the hydrolysis of 1.0×10^{-6} *M* ethyl oxalate in the presence of (A) 1×10^{-5} *M* and (B) 1×10^{-3} *M* Cu, Zn, and Mg. The smooth line with a 1:1 slope is the value of k_h in metal-free solution. (Based upon rate constants presented in Reference 68.)

We now examine the effect of adding naturally occurring ligands on the rates of metal-catalyzed hydrolysis. If solutions are undersaturated with respect to possible solid phases, adding ligand lowers free metal concentrations; speciation shifts from $[M^{2+}]$ toward metal-ligand complexes such as $[ML^{n+}]$. If a solid phase (e.g., $M(OH)_2(s)$) limits free metal solubility, adding ligand will have no effect on free metal concentrations; $[M^{2+}]$ is still governed by solubility equilibria such as Equation 10.

The inhibitory effect arising from the addition of oxalate and glycine is governed by the following equilibrium relationships:[7]

		Oxalate	Glycine	
$H_2L = HL + H^+$	pK_{a1}	1.25	2.35	(12)
$HL = L + H^+$	pK_{a2}	4.27	9.78	(13)
$Cu + L = CuL$	$\log K_1$	6.2	8.6	(14)
$Cu + 2L = CuL_2$	$\log \beta_2$	10.27	15.6	(15)

As shown in Figure 14, oxalate interferes with Cu-catalyzed hydrolysis of ethyl oxalate to a greater extent and over a wider pH than equimolar concentrations of glycine. Although glycine exhibits greater Cu-ligand complex formation constants than oxalate, it is substantially more basic. As a consequence, protons outcompete Cu for binding to glycine over a wider range in pH.

The example presented here was relatively simple. As new laboratory information becomes available, it would be useful to repeat calculations of this kind for pesticides and other more relevant hydrolyzable pollutants, and for Mn(II), Fe(II), and other more

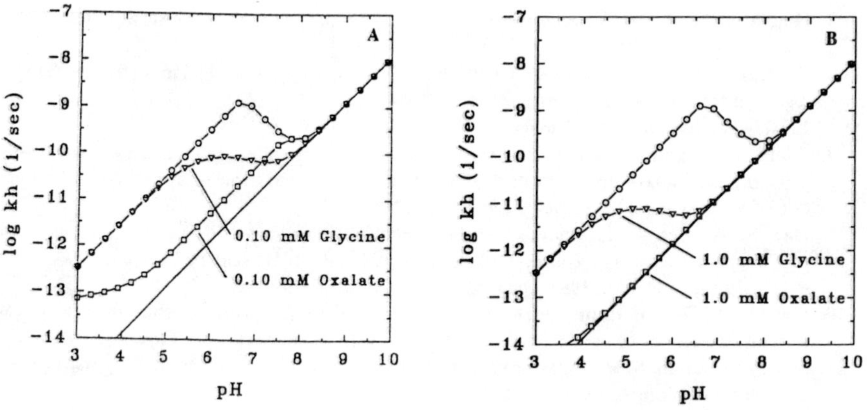

Figure 14 Cu complexation by oxalate or glycine lowers k_h for ethyl oxalate hydrolysis: (A) 0.10 m*M* added complexant; (B) 1.0 m*M* added complexant. The smooth line with a 1:1 slope is the value of k_h in metal-free solution. The open circles represent Cu-catalyzed reaction in the absence of complexing agent. Reaction Conditions: 1.0×10^{-5} *M* TOTCu, 1.0×10^{-6} *M* ethyl oxalate. (Based upon rate constants presented in Reference 68).

relevant metal catalysts. In addition, we should recognize that numerous mechanisms of metal catalysis exist which do not necessarily respond to changes in metal speciation in the same way.

V. SUMMARY AND CONCLUSIONS

Fundamental information regarding metal-ligand complex formation and mechanisms of metal-catalyzed hydrolysis provides insight into processes affecting the fate of organic pollutants in the environment. Although most of this information was derived from experiments in homogeneous solution, useful analogies can be made to reactions occurring at mineral/water interfaces. Basic understanding of reaction kinetics and mechanisms helps us to (1) identify organic pollutants most susceptible to metal catalysis, (2) identify naturally occurring metal species which should exhibit significant catalytic reactivity, and (3) evaluate how the speciation and complex composition of natural aquatic environments affect pathways and rates of pollutant hydrolysis.

ACKNOWLEDGMENTS

This work was supported by the Office of Exploratory Research of the U.S. Environmental Protection Agency (Grant R818894-01-1) The views and conclusions contained in this document are those of the authors and should not be interpreted as necessarily reflecting the official policies, either expressed or implied, of the U.S. Government.

REFERENCES

1. **Hoffmann, M. R.,** Trace metal catalysis in aquatic environments, *Environ. Sci. Technol.,* 14, 1061, 1980.
2. **Pearson, R. G.,** Acids and bases, *Science,* 151, 172, 1966.
3. **Evers, A., Hancock, R. D., Martell, A. E., and Motekaitis, R. J.,** Metal recognition in ligands with negatively charged oxygen donor groups. Complexation of Fe(III), Ga(III), In(III), Al(III), and other highly charged metal ions, *Inorg. Chem.,* 28, 2189, 1989.

4. **Houghton, R. P.,** *Metal Complexes in Organic Chemistry,* Cambridge University Press, Cambridge, England, 1979.
5. **Sigel, H. and McCormick, D. B.,** On the discriminating behavior of metal ions and ligands with regard to their biological significance, *Acc. Chem. Res.,* 3, 201, 1970.
6. **Emsley, J.,** *The Elements,* Clarendon Press, Oxford, 1989.
7. **Smith, R. M. and Martell, A. E.,** *Critical Stability Constants,* Plenum Press, New York, 1976.
8. **Barnum, D. W.,** Hydrolysis of cations. Formation constants and standard free energies of formation of hydroxy complexes, *Inorg. Chem.,* 22, 2297, 1983.
9. **Hay, R. W.,** Lewis acid catalysis and the reactions of coordinated ligands, in *Comprehensive Coordination Chemistry,* Vol. 6, Wilkinson, G., Gillard, R. D., and McCleverty, J. A., Eds., Pergamon Press, Oxford, 1987, chap. 61.4.
10. **Plastourgou, M. and Hoffmann, M. R.,** Transformation and fate of organic esters in layered-flow systems: the role of trace metal catalysis, *Environ. Sci. Technol.,* 18, 756, 1984.
11. **Suh, J. and Chun, K. H.,** Metal ion catalysis by blocking inhibitory reverse paths in the hydrolysis of 3-carboxyaspirin, *J. Am. Chem. Soc.,* 108, 3057, 1986.
12. **Fife, T. H. and Przystas, T. J.,** Divalent metal ion catalysis in the hydrolysis of esters of picolinic acid. Metal ion promoted hydroxide ion and water catalyzed reactions, *J. Am. Chem. Soc.,* 107, 1041, 1985.
13. **Fife, T. H. and Przystas, T. J.,** Transition-state effects in the divalent metal ion catalyzed hydrolysis of esters. Hydrolysis of 2-pyridylmethyl hydrogen phthalate, *J. Am. Chem. Soc.,* 102, 7297, 1980.
14. **Wilkins, R. G.,** *Kinetics and Mechanisms of Reactions of Transition Metal Complexes,* 2nd ed., VCH Publishers, Weinheim, Germany, 1991, chap. 6.
15. **Sayre, L. M.,** Catalysis of amide hydrolysis, *J. Am. Chem. Soc.,* 108, 1632, 1986.
16. **Reddy, K. V., Jacobson, A. R.. Kung, J. I., and Sayre, L. M.,** Divalent metal ion catalyzed hydrolysis of picolinanilides, *Inorg. Chem.,* 30, 3520, 1991.
17. **Przystas, T. J. and Fife, T. H.,** The metal-ion-promoted water- and hydroxide-ion-catalysed hydrolysis of amides, *J. Chem. Soc. Perkin Trans. 2,* 393, 1990.
18. **Gschwend, P., Schwarzenbach, R. P., and Imboden, D.,** *Environmental Organic Chemistry,* Wiley-Interscience, New York, 1993.
19. **Hendry, P. and Sargeson, A. M.,** Metal ion promoted reactions of phosphate derivatives, in *Progress in Inorganic Chemistry: Bioinorganic Chemistry,* Vol. 38, Lippard, S. J., Ed., Wiley-Interscience, New York, 1990, p. 201.
20. **Benkovic, S. J. and Dunikoski, L. K.,** An unusual rate enhancement in metal ion catalysis of phosphate transfer, *J. Am. Chem. Soc.,* 93, 1526, 1971.
21. **Gratzel, M.,** personal communication, 1992.
22. **Hoffmann, M. R.,** Catalysis in aquatic environments, in *Aquatic Chemical Kinetics,* Stumm, W., Ed., Wiley-Interscience, New York, 1990, chap. 3.
23. **Breslow, R., Fairweather, R., and Keana, J.,** Metal-catalyzed hydration of phenanthroline nitrile, *J. Am. Chem. Soc.,* 89, 2135, 1967.
24. **Belokon, Y. N., Tararov, V. I., Savel'eva, T. F., Vitt, S. V., Paskonova, E. A., Dotdayev, S. C., Borisov, Y. A., Struchkov, Y. T., Batasanov, A. S., and Belikov, V. M.,** Copper(II) ion promoted direct hydrolysis of 2-cyanopyridine to picolinic acid. Intramolecular catalysis by the coordinated N-β-hydroxyethyl group, *Inorg. Chem.,* 27, 4046, 1988.
25. **Breslow, R., McClure, D. E., Brown, R. S., and Eisenach, J.,** Very fast zinc-catalyzed hydrolysis of an anhydride. A model for the rate and mechanism of carboxypeptidase A catalysis, *J. Am. Chem. Soc.,* 97, 194, 1975.
26. **Fife, T. H. and Przystas, T. J.,** Divalent metal ion catalysis of acetal hydrolysis; effects of oxycarbocation stability and leaving group basicity, *J. Chem. Soc. Perkin Trans. 2,* 143, 1987.
27. **Satchell, D. P. N. and Satchell, R. S.,** The kinetics of soft metal ion promoted hydrolysis of phenyl isothiocyanates in aqueous solution, *J. Chem. Soc. Perkin Trans. 2,* 751, 1992.
28. **Ketelaar, J. A. A., Gersmann, H. R., and Beck, M. M.,** Metal-catalysed hydrolysis of thiophosphoric esters, *Nature (London),* 177, 392, 1956.
29. **Mortland, M. M. and Raman, K. V.,** Catalytic hydrolysis of some organic phosphate pesticides by copper(II), *J. Agric. Food Chem.,* 15, 163, 1967.
30. **Blanchet, P.-F. and St-George, A.,** Kinetics of chemical degradation of organophosphorus pesticides; hydrolysis of chlorpyrifos and chlorpyrifos-methyl in the presence of copper (II), *Pestic. Sci.,* 13, 85, 1982.

31. **Mill, T. and Mabey, W.,** Hydrolysis of organic chemicals, in *The Handbook of Environmental Chemistry, Vol. 2D: Reactions and Processes,* Hutzinger, O., Ed., Springer-Verlag, Berlin, 1988, 71.
32. **Bank, S. and Tyrrell, R. J.,** Copper(II)-promoted aqueous decomposition of aldicarb, *J. Org. Chem.,* 50, 4938, 1985.
33. **Butcher, W. W. and Westheimer, F. H.,** The lanthanum hydroxide gel promoted hydrolysis of phosphate esters, *J. Am. Chem. Soc.,* 77, 2420, 1955.
34. **Davis, J. A. and Hayes, K. F.,** *Geochemical Processes at Mineral Surfaces,* ACS Symp. Ser. 323, American Chemical Society, Washington, D.C., 1986.
35. **Sigg, L. and Stumm, W.,** The interaction of anions and weak acids with the hydrous goethite α-FeOOH) surface, *Colloids Surf.,* 2, 101, 1981.
36. **Hayes, K. F., Papelis, C., and Leckie, J. O.,** Modeling ionic strength effects on anion adsorption at hydrous oxide/solution interfaces, *J. Colloid Interface Sci.,* 125, 717, 1988.
37. **Kummert, R. and Stumm, W.,** The surface complexation of organic acids on hydrous γ-Al_2O_3, *J. Colloid Interface Sci.,* 75, 373, 1980.
38. **Stumm, W., Kummert, R., and Sigg, L.,** A ligand exchange model for the adsorption of inorganic and organic ligands at hydrous interfaces, *Croat. Chem. Acta,* 53, 291, 1980.
39. **Wehrli, B.,** Redox reactions of metal ions at mineral surfaces, in *Aquatic Chemical Kinetics,* Stumm, W., Ed., Wiley-Interscience, New York, 1990, chap. 11.
40. **Wehrli, B., Wieland, E., and Furrer, G.,** Chemical mechanisms in the dissolution kinetics of minerals; the aspect of active sites, *Aquat. Sci.,* 52, 3, 1990.
41. **Theng, B. K. G.,** Clay-activated organic reactions, in *International Clay Conference 1981,* van Olphen. H. and Veniale, F., Eds., Elsevier, Amsterdam, 1982.
42. **Voudrias, E. A. and Reinhard, M.,** Abiotic organic reactions at mineral surfaces, in *Geochemical Processes at Mineral Surfaces,* Davis, J. A. and Hayes, K. F., Eds., ACS Symp. Ser. 323, American Chemical Society, Washington, D.C., 1986, chap. 22.
43. **Yaron, B.,** Chemical conversion of parathion on soil surfaces, *Proc. Soil Sci. Soc. Am.,* 39, 639, 1975.
44. **Mingelgrin, U. and Saltzman, S.,** Surface reactions of parathion on clays, *Clays Clay Miner.,* 27, 72, 1979.
45. **Mingelgrin, U., Saltzman, S., and Yaron, B.,** A possible model for the surface-induced hydrolysis of organophosphorus pesticides on kaolinite clays, *Soil Sci.,* 41, 519, 1977.
46. **El-Amamy, M. M. and Mill, T.,** Hydrolysis kinetics of organic chemicals on montmorillonite and kaolinite surfaces as related to moisture content, *Clays Clay Miner.,* 32, 67, 1984.
47. **Sanchez-Camazano, M. and Sanchez-Martin, M. J.,** Montmorillonite-catalyzed hydrolysis of phosmet, *Soil Sci.,* 136, 89, 1983.
48. **Namba, S., Hosonuma, N., and Yashima, T.,** Catalytic application of hydrophobic properties of high-silica zeolites. I. Hydrolysis of ethyl acetate in aqueous solution, *J. Catal.,* 72, 16, 1981.
49. **Davis, J. A and Leckie, J. O.,** Effect of adsorbed complexing ligands on trace metal uptake by hydrous oxides, *Environ. Sci. Technol.,* 12, 1309, 1978.
50. **Balistrieri, L. S. and Murray, J. W.,** The influence of major ions of seawater on the adsorption of simple organic acids by goethite, *Geochim. Cosmochim. Acta,* 51, 1151, 1987.
51. **Zachara, J. M., Ainsworth, C. C., Cowan, C. E., and Schmidt, R. L.,** Sorption of aminonaphthalene and quinoline on amorphous silica, *Environ. Sci. Technol.,* 24, 118, 1990.
52. **Stone, A. T., Torrents, A., Smolen, J., Vasudevan, D., and Hadley, J.,** Adsorption of organic compounds possessing ligand donor groups at the oxide/water interface, *Environ. Sci. Technol.,* 27, 895, 1993.
53. **Torrents, A. and Stone, A. T.,** Hydrolysis of phenyl picolinate at the mineral/water interface, *Environ. Sci. Technol.,* 25, 143, 1991.
54. **Torrents, A.,** Hydrolysis of Organic Esters at the Mineral/Water Interface, Ph.D. thesis, Johns Hopkins University, Baltimore, 1992.
55. **Stone, A. T.,** Enhanced rates of monophenyl terephthalate hydrolysis in aluminum oxide suspensions, *J. Colloid Interface Sci.,* 127, 429, 1989.
56. **Hardt, S. L.,** Rates of diffusion controlled reactions in one, two, and three dimensions, *Biophys. Chem.,* 10, 239, 1979.
57. **Adam, G. and Delbruck, M.,** Reduction of dimensionality in biological diffusion processes, in *Structural Chemistry and Molecular Biology,* Rich, A. and Davison, N., Eds., W. H. Freeman, San Francisco, 1968, 199.

58. **Stone, A. T.,** The effect of Dismal Swamp dissolved organic matter on the adsorption and surface-enhanced hydrolysis of monophenyl terephthalate in aluminum oxide suspensions, *J. Colloid Interface Sci.*, 132, 81, 1989.
59. **Stanton, D. T.,** Glass-catalyzed decomposition of oxycarboxin in aqueous solution, *J. Agric. Food Chem.*, 35, 856, 1987.
60. **Leach, B. E. and Angelici, J.,** Metal-ion catalysis of the hydrolysis of some amino acid ester N,N-diacetic acids, *J. Am. Chem. Soc.*, 90, 2504, 1968.
61. **Watanabe, M., Matsuura, M., and Yamada, T.,** The mechanism of the hydrolysis of polyphosphates. VI. The effect of cations on the hydrolysis of pyro- and triphosphate, *Bull. Chem. Soc. Jpn.*, 54, 738, 1981.
62. **Hofstetter, R., Murakami, Y., Mont, G., and Martell, A. E.,** Influence of metal ions and of metal complexes on the hydrolysis of salicyl phosphate, *J. Am. Chem. Soc.*, 84, 3041, 1962.
63. **Campbell, P. G. C., Bisson, M., Bougie, R., Tessier, A., and Villeneuve, J.-P.,** Speciation of aluminum in acidic freshwaters, *Anal. Chem.*, 55, 2246, 1983.
64. **Pott, D. B., Alberts, J. J., and Elzerman, A. W.,** The influence of pH on the binding capacity and conditional stability constants of aluminum and naturally-occurring organic matter, *Chem. Geol.*, 48, 293, 1985.
65. **Backes, C. A. and Tipping, E.,** Aluminum complexation by an aquatic humic fraction under acidic conditions, *Water Res.*, 21, 211, 1987.
66. **Hiemstra, T., van Riemsdijk, W. H., and Bolt, G. H.,** Multisite proton adsorption modeling at the solid/solution interface of (hydr)oxides: a new approach. I. Model description and evaluation of intrinsic reaction constants, *J. Colloid Interface Sci.*, 133, 91, 1989.
67. **Hiemstra, T., de Wit, J. C. M., and van Riemsdijk, W. H.,** Multisite proton adsorption modeling at the solid/solution interface of (hydr)oxides: a new approach. II. Application to various important (hydr)oxides, *J. Colloid Interface Sci.*, 133, 105, 1989.
68. **Johnson, G. L. and Angelici, R. J.,** Metal-ion catalysis of ethyl oxalate hydrolysis, *J. Am. Chem. Soc.*, 93, 1106, 1971.
69. **Westall, J. C., Zachary, J., and Morel, F.,** *MINEQL: A Computer Program for the Calculation of Chemical Equilibrium Composition of Aqueous Systems,* Tech. Note No. 18, Ralph M. Parsons Laboratory, Massachusetts Institute of Technology, Cambridge, 1976.
70. **Papelis, C., Hayes, K. F., and Leckie, J. O.,** *HYDRAQL: A Program for the Computation of Chemical Equilibrium Composition of Aqueous Batch Systems Including Surface-Complexation Modeling of Ion Adsorption at the Oxide/Solution Interface,* Tech. Rep. No. 306, Department of Civil Engineering, Stanford University, Menlo Park, CA, 1988.

Chapter 23

The Role of Abiotic and Biotic Catalysts in the Transformation of Phenolic Compounds

J.-M. Bollag, C. Myers, S. Pal, and P. M. Huang

CONTENTS

I. INTRODUCTION

Xenobiotics bind to soil colloids through naturally occurring processes. These events may prove to be beneficial to the environment in that the toxic effect of hazardous xenobiotics can be neutralized through the incorporation of these compounds into soil humus. Studies have shown that the incorporation of xenobiotics into humus greatly reduces the movement of these chemicals caused by leaching.[1] Furthermore, the bound residues have been found to be less toxic and are less likely to interact with the biota.[2,3] Before the binding of xenobiotics to soil humus can be considered as a means of decontamination, it is necessary to obtain a thorough understanding of the binding mechanism and to evaluate the stability (persistence) and toxicity of the bound residues.

The following review focuses on the use of abiotic and biotic catalysts to mediate the cross-coupling of phenolic xenobiotics.

II. GENERAL PRINCIPLES OF OXIDATIVE COUPLING

The incorporation of phenolic compounds into humus is mediated by a variety of biological (i.e., microbial and plant enzymes) and abiotic catalysts (i.e., metal oxides, clay minerals, and organomineral complexes). Coupling reactions may also occur spontaneously in the presence of oxygen at neutral and alkaline pH. One of the most important reactions in the binding of phenolic derivatives to humus involves oxidative coupling.[4,5] In the first step of oxidative coupling reactions, susceptible aromatic compounds are oxidized to form unstable free radicals. The free radicals then go on to react with nearby molecules to form dimers and polymers. This reaction results in the formation of C-C and C-O bonds between phenolic species and C-C and C-N bonds between aromatic amines.[6]

With respect to phenolic compounds, the initial step involves the removal of an electron and hydrogen ion from the hydroxyl group, generating an alkoxy free radical. This intermediate then couples at positions *ortho* and *para* to the hydroxyl group (the *meta* position is nonreactive) to form a dimer. Figure 1, for example, shows this reaction with guaiacol as a reactant. Phenolic dimers are further oxidized to yield polymeric products, and if the potential of the oxidizing agent is high enough, C-C coupled dimers

0-87371-914-X/95/$0.00+$.50

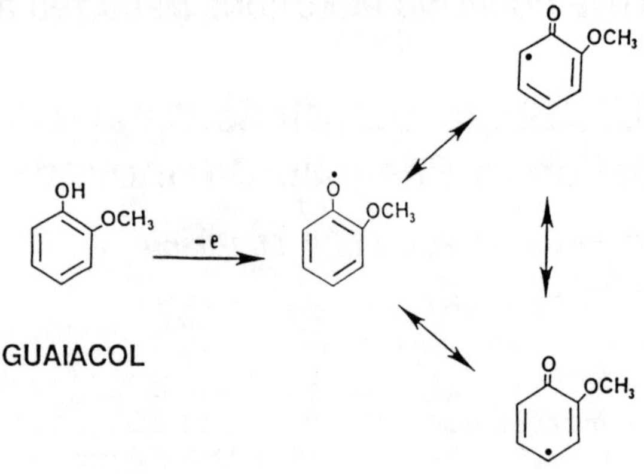

Figure 1 Formation of aryloxy radicals from guaiacol.

are oxidized to form extended quinones. Such a reaction pathway was demonstrated with guaiacol, whereby both enzymes and abiotic catalysts generated the same initial reaction products.[7]

III. OXIDATIVE COUPLING REACTIONS MEDIATED BY ABIOTIC AGENTS OR CATALYSTS

The oxidation and polymerization of phenolic compounds by various minerals has been reported by numerous investigators.[8-13] Shindo and Huang,[13] for instance, demonstrated that various Mn oxides and dioxides accelerated the conversion of diphenols, such as hydroquinone, resorcinol, and catechol, to humic acids. However, it is not clear whether all abiotic catalysts meet the definition of a catalyst. Catalysts are usually defined as materials which accelerate chemical reactions without themselves undergoing change. However, Twigg[14] pointed out that catalysts do change with use. In addition, Moore and Pearson[15] concluded that a catalyst is a substance that changes the rate of a desired reaction, regardless of the fate of the catalyst itself. The major difference between catalysts and oxidizing agents is whether the polyvalent atom which accepts the electron has the ability to mediate its transfer to a final electron acceptor (e.g., oxygen). For both abiotic and biotic agents, polyvalent atoms such as Cu, Mn, and Fe facilitate the transformation of phenolic compounds by acting as electron acceptors.[8-13] In particular, Cu behaves as a true catalyst when associated with a protein. An electron is transferred from the Cu-protein complex to an electron acceptor (oxygen) in a cyclic manner, irrespective of the oxidation state of the environment. However, when birnessite (MnO_2) serves as an electron acceptor, this compound is reduced to Mn^{2+} and may not be reconverted to its initial oxidized state. McBride[11] reported that in reactions in which birnessite catalyzed the oxidation of phenols, large quantities of Mn^{2+} accumulated in the solution. Even following long periods of incubation, large amounts of Mn^{2+} remained in the solution because Mn^{2+} oxidizes fairly slowly in the presence of dissolved O_2 except at very high pH.

Other abiotic factors have also been examined. For example, Wang and Huang[16] studied the catalytic activity of Ca-nontronite, Ca-bentonite, and Ca-kaolinite with respect to the capacity of these minerals to enhance the abiotic ring cleavage of pyrogallol. The amount of CO_2 released from the ring cleavage and the quantity of humic polymers formed were directly related and were dependent on the abiotic factor complex employed, such that catalytic activity was highest for Ca-nontronite > Ca-kaolinite > Ca-bentonite.

IV. OXIDATIVE COUPLING REACTIONS MEDIATED BY BIOTIC CATALYSTS

Degradation of xenobiotics in soil is largely due to the activity of microorganisms. Thus, it is not surprising that researchers have focused their efforts on using microorganisms and their enzymes to degrade xenobiotics that contaminate the environment. In 1949, Audus[17] first suggested the idea of using a microbial inoculant to enhance the degradation of 2,4-dichlorophenoxyacetic acid (2,4-D). Since that time, research has progressed to the level of isolating individual microbial enzymes and using these enzymes to catalyze the transformation of specific phenolic substrates.

Oxidoreductases are divided into two subclasses, the monophenol monooxygenases and the peroxidases. Included in the monophenol monooxygenase subclass are tyrosinases and laccases. The monophenol monooxygenases require bimolecular oxygen, but no coenzyme for activity, while the peroxidases require hydrogen peroxide for activity. Fungal laccases are used quite frequently because these enzymes appear to be similar in their activity regardless of their origin, and all are capable of catalyzing the transformation of phenolic substrates. However, quantitative differences in oxidation efficiencies have been observed among the various laccases. These variations are attributable to differences in pH optima and the nature (i.e., induced or noninduced) of the enzyme. The induced form of laccases are generally more active than the corresponding constitutive form.[18] This increase in activity may be necessary for the effective use of laccases for detoxification of an environment polluted by phenols.

Because of the complex and heterogeneous nature of soil, early experiments focused on studying the polymerization of a single phenolic compound in reactions catalyzed by isolated oxidoreductive enzymes. In several individual studies, the laccase from the soil fungus *Rhizoctonia praticola* was shown to polymerize different phenolic compounds, including naphtholic intermediates from carbamate insecticides,[19,20] methyl- or methoxy-substituted phenolic pesticides,[19-21] 1,5-dihydroxynaphthalene,[19] and 2,6-xylenol.[22] These experiments demonstrate the wide substrate specificity of laccase. Since a variety of phenolic compounds exist in the soil environment, polymerization is most likely to occur between different types of molecules. Accordingly, experiments have been performed to study the copolymerization of phenolic compounds. For instance, the formation of coupled hybrid phenolic compounds was demonstrated by incubating 2,4-dichlorophenol with vanillin in the presence of a laccase produced by *R. praticola.*[23] Three dimeric products were formed, as illustrated in Figure 2.

To study the incorporation of phenolic xenobiotics into humus, *in vitro* model systems were designed. In these investigations, phenolic substrates were incubated with isolated enzymes and individual components of soil humus, and the resulting products were separated using thin-layer chromatography (TLC) or high-performance liquid chromatography (HPLC). The structures of each of the various products were then characterized using mass spectrometry (MS). Bollag et al.[23] reported that 2,4-dichlorophenol could be cross-coupled with humic constituents such as orcinol, syringic acid, vanillic acid, and vanillin. In subsequent studies, various chlorinated phenols (4-chlorophenol, 2,4- and 2,6-dichlorophenol, 2,3,6- and 2,4,5-trichlorophenol, 2,3,5,6-tetrachlorophenol, and pentachlorophenol) were incubated with syringic acid and a laccase.[24] The resulting products were isolated and analyzed as described above, and two principal products were observed. One product consisted of quinonoid oligomers in which chlorophenols were bound by ether linkages to orthoquinoline products of syringic acid, and the other product was composed of phenolic oligomers in which chlorophenols were bound by ether linkages to decarboxylated products of syringic acid.

Sarkar et al.[25] conducted further studies to investigate the copolymerization of xenobiotics and humus under conditions which approximate the natural soil habitat. ^{14}C-Labeled 2,4-

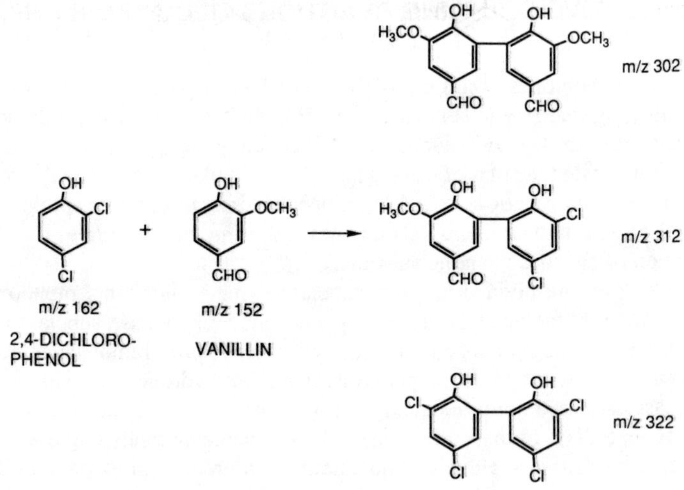

Figure 2 Oxidative coupling products resulting from the incubation of 2,4-dichlorophenol and vanillin with a laccase of the fungus *Rhizoctonia praticola.*[23]

dichlorophenol (^{14}C-2,4-DCP) was incubated with stream fulvic acid in the presence of phenoloxidases, and the binding of the chlorophenol was measured by determining the amount of radioactivity incorporated into fulvic acid. Chromatographic analysis indicated that although a large portion of the radioactivity remained in the solution, no unbound ^{14}C-2,4-DCP was present in the supernatant. Two types of polymers were formed during the coupling reaction: hybrid molecules containing fulvic acid and 2,4-DCP, and polymers of 2,4-DCP alone. Importantly, the presence of fulvic acid in the reaction mixture essentially doubled the removal of 2,4-DCP. In addition, the enzymatic coupling of this substrate was found to occur over a wide temperature and pH range. Such a lack of specificity in pH and temperature optima suggests that efficient enzymatic coupling of xenobiotics can occur in a variety of soil environments that undergo wide fluctuations in pH and temperature.

V. TRANSFORMATIONS OF PHENOLIC COMPOUNDS BY ENZYMES AND ABIOTIC AGENTS OR CATALYSTS

Various investigations have been performed to compare the catalytic activity of biotic and abiotic catalysts in the transformation of phenolic substances. The goal was to examine the structural similarities of the reaction products that resulted from biotic and abiotic transformation reactions. For instance, guaiacol was incubated with various oxidants in aqueous, buffered solutions. In each case, reaction conditions were controlled so that only the initial low-molecular-weight products were formed. With higher oxidant concentration, or longer reaction times, additional higher-molecular-weight compounds are formed. However, only the initially formed product structures were needed to draw conclusions about the mechanism of the formation of coupling products.

All reactions were conducted at 24°C in citrate/phosphate buffer at pH 7.3 with a guaiacol concentration of 1 m*M*. The catalyst was added, and at the specified time, the reaction was halted by bringing the solution to pH 2.5 with acetic acid. Boiled enzymes were used in control samples.

In separate reactions, manganese dioxide (0.025 g ml^{-1}) or bentonite-Wyoming clay (0.01 g ml^{-1}) were added to a reaction solution and the reaction was allowed to proceed

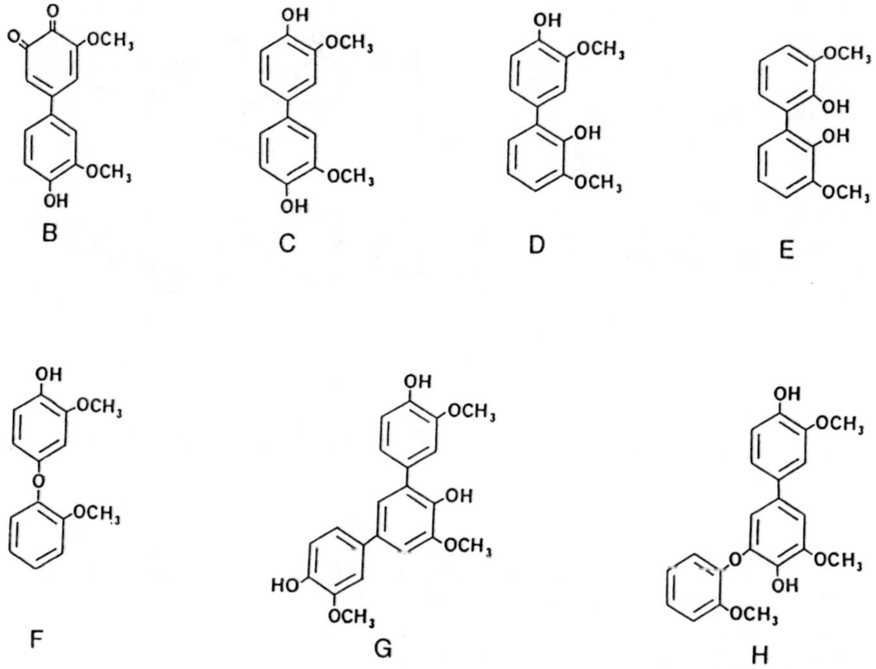

Figure 3 Structures of the oligomers formed following the one-electron oxidation of guaiacol.[7]

for 5 min. Tyrosinase was added to a reaction solution to give a final enzyme concentration of 0.41 tyrosinase units per milliliter (a tyrosinase unit is defined as the amount of enzyme that will cause an increase in absorption at 280 nm of 0.001 per minute at pH 6.5 at 25°C in a 3-ml reaction mixture containing L-tyrosine). Horseradish peroxidase-catalyzed reactions were performed with an enzyme concentration of 0.011 purpurogallin units per milliliter (a purpurogallin unit is defined as the amount of enzyme which forms 1.0 mg of purpurogallin, as determined from its absorption at 420 nm from pyrogallol in 20 s at pH 6.0 at 20°C). In the peroxidase reaction, hydrogen peroxide was added to the reaction mixture to a final concentration of 2.5 m*M*.

The activities of the laccases of *Rhizoctonia praticola* and *Trametes versicolor* were expressed as DMP (2,6-dimethoxyphenol) units, which is defined as the amount of enzyme that causes a change in absorbance of 1.0 unit per minute at 468 nm for a 3.5-ml aliquot containing 3.24 mol DMP at pH 4.2. The laccase of *R. praticola* was incubated with the guaiacol substrate at a concentration of 0.05 DMP units per milliliter at pH 7.3 for 45 min, and the laccase of *T. versicolor* at a concentration of 0.37 DMP units per milliliter at pH 5.5 for 2 min. To compare the activity of birnessite, an amount of birnessite was determined which corresponded to 1 unit of enzyme activity.

All reaction products were analyzed for the presence of guaiacol and guaiacol-derived oligomers using high-performance liquid chromatography (HPLC).[7] The reactions were limited by the concentration of the specific catalyst and the incubation time so that only the initially formed low-molecular-weight oligomers were produced.

The first seven products to appear following the peroxidase-catalyzed oxidation of guaiacol were structurally identified using mass spectrometry and proton nuclear magnetic resonance spectroscopy[7] (Figure 3). The guaiacol-derived product mixture was composed of five dimers (compounds B to F) and two trimers (compounds G and H) with

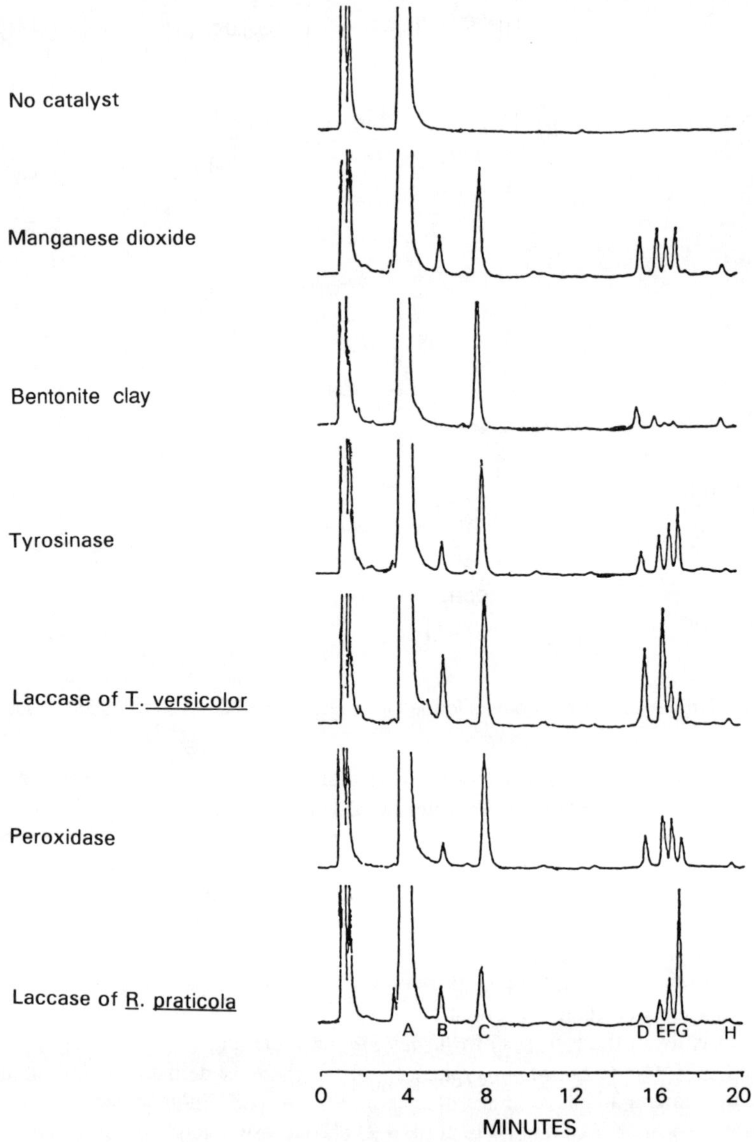

Figure 4 HPLC chromatograms of reaction solutions containing guaiacol and various oxidative agents. The product peak labels correspond to labels on the structures shown in Figure 3. Peak A represents guaiacol.[7]

quinoid and phenolic structural features. The same seven oligomers were produced using other biological and inorganic soil components as oxidizing agents. Manganese dioxide (birnessite), clay (bentonite), tyrosinase, and the laccases of *T. versicolor* and *R. praticola* each effected qualitatively equivalent transformations of guaiacol, as shown by the presence of the same seven product peaks in the respective HPLC chromatograms.[7] Figure 4 compares the chromatograms for the six oxidation reactions. All control samples of boiled enzymes gave chromatograms equivalent to that of unreacted guaiacol. The

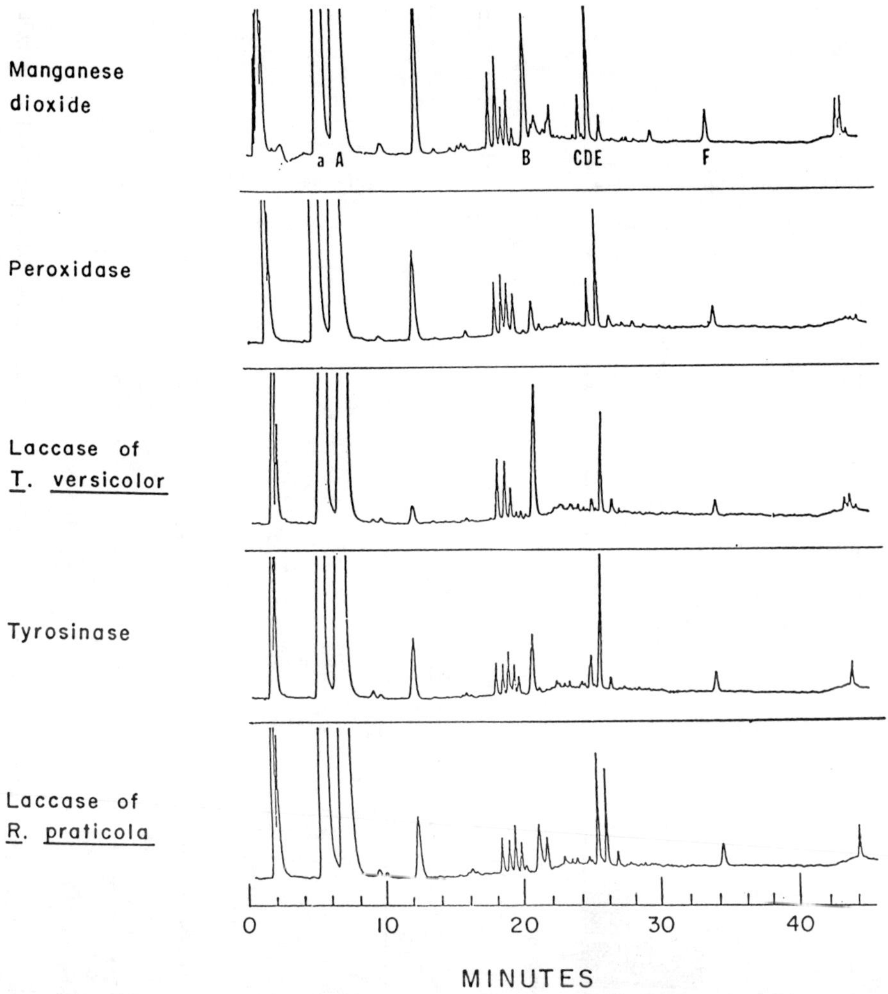

Figure 5 HPLC chromatograms of reaction solutions containing various oxidative agents as reactions initiators for the cooligomerization of guaiacol (peak a) and 4-chloroaniline (peak A). Compounds B through F were previously identified.[26]
B trimer [9-(4-chlorophenyl)-3,6-dimethoxy-1,2,7-trihydroxycarbazole]
C trimer [3-hydroxy-3′-methoxy-*N*-(4-chlorophenyl)-4-imino-4′-diphenoquinone]
D trimer [5-(4-chloroanilino)-3,3′-dimethoxy-4,4′-diphenoquinone]
E trimer [6-chloro-3-hydroxy-3′-methoxy-*N*-(4-chlorophenyl)-4-imino-4′-diphenoquinone]
F tetramer [3-(4-chloroanilino)-3′-hydroxy-5-methoxy-*N*-(4-chlorphenyl)-4′-imino-4-diphenoquinone]

qualitative similarity of the product mixtures for the six reaction solutions is evident. Note, however, that the products obtained in the clay-mediated oxidation of guaiacol do not include compound B, a dimer. It is likely that this dimer, which is an orange-colored quinone, was adsorbed to, or reacted with, the clay, as was suggested by the orange color of the clay residue. Nonetheless, the other oligomers (compounds C through H) were detected in the clay reaction solution.

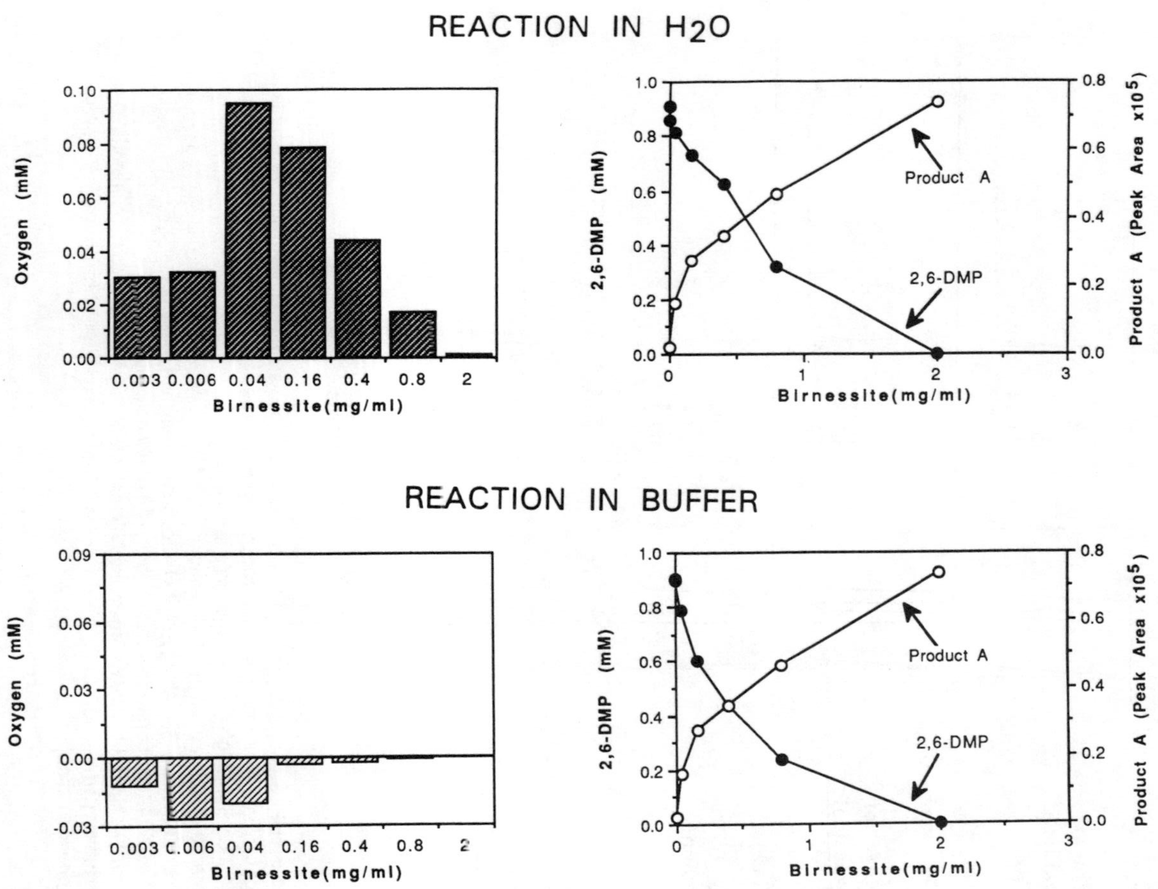

Figure 6 Transformation of 2,6-dimethoxyphenol in H_2O and phosphate buffer by various amounts of birnessite as determined by an oxygen monitor and HPLC analysis (incubation time: 30 min).

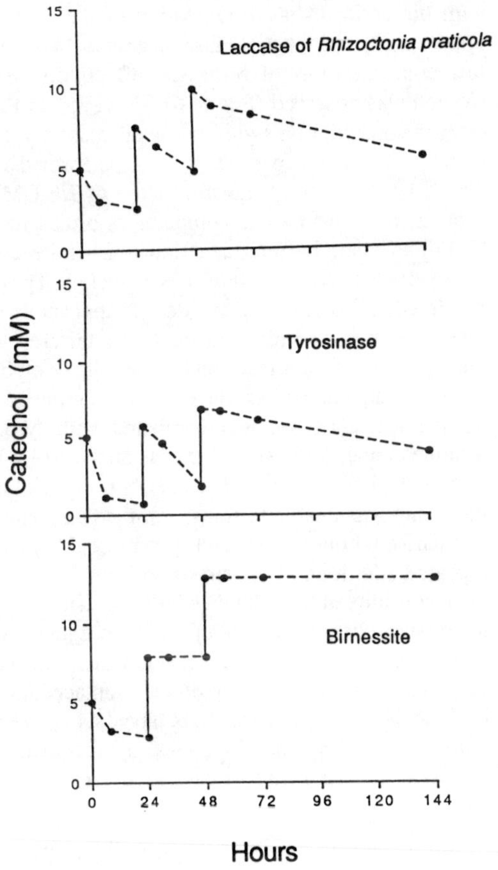

Figure 7 Transformation of catechol by laccase (0.4 units per ml), tyrosinase (0.1 units per ml), and birnessite (0.6 mg/ml) after repeated addition of substrate.

Although the same seven guaiacol-derived oligomers were formed in six different reaction solutions, the amounts of the respective oligomers varied (Figure 4). This phenomenon is most likely a result of different reaction conditions such as the pH of the reaction mixture or the rate of oxidation. Regardless of the amount of each oligomer that was formed, the fact remains that the same products were formed, and this is a direct result of guaiacol being in an environment where its oxidation potential was favorable for the loss of an electron from the phenolic oxygen.

In separate experiments, various inorganic and biological soil components were examined for their ability to initiate the cooligomerization of guaiacol and 4-chloroaniline.[26] The same five cooligomers produced in the *T. versicolor* laccase-mediated reaction were also found in the reaction solutions where manganese dioxide, peroxidase, laccase of *R. praticola,* and tyrosinase were used as oxidants. HPLC chromatograms of the five reaction product mixtures are shown in Figure 5.

VI. MECHANISM OF ABIOTICALLY AND BIOTICALLY GENERATED REACTIONS

Further research has been conducted to clarify and to compare the role of biotic and abiotic agents in the transformation of phenolic compounds. To this end, experiments were performed in which the substrate 2,6-dimethoxyphenol (2,6-DMP) was incubated with birnessite (MnO_2). Oxidation of the substrate was determined by measuring the uptake of O_2 and substrate disappearance. In reactions in which water served as a solvent, low concentrations of birnessite effectively oxidized 2,6-DMP, and consumption of oxygen was observed (Figure 6). However, as the concentration of birnessite increased, oxygen consumption significantly decreased, yet the oxidation of 2,6-DMP continued to increase. Similar experiments were performed using laccase and tyrosinase to oxidize 2,6-DMP. The rate of transformation of 2,6-DMP was considerably higher in reactions catalyzed by enzymes as compared to birnessite-mediated reactions.

In phosphate-buffered reactions, birnessite was shown to oxidize 2,6-DMP, but little or no oxygen consumption was detected. These results are in agreement with those previously reported by McBride.[12] It appears that anions, in particular oxyanions, in the phosphate-buffered reaction mixture interfere with the measurement of oxygen uptake during the oxidation reaction of phenolic compounds by birnessite.

In subsequent studies, the effect of continuous additions of the substrate catechol was determined. Catechol was incubated with tyrosinase, laccase, or birnessite, and the disappearance of the substrate was analyzed by HPLC. After 24 and 48 h, the catechol was replenished. During the first 8 h of incubation, birnessite oxidized catechol, yet birnessite was unable to transform the two subsequent catechol additions (Figure 7). In comparison, both laccase and tyrosinase retained their ability to transform catechol with repeated additions of the substrate.

The results of these studies indicate that, while both biotic and abiotic agents are able to oxidize phenolic compounds, these agents function differently in their capacity to mediate oxidative coupling. The enzymes behaved as catalysts and mediated the successful transfer of an electron to an electron acceptor in a cyclic manner, whereas it appears that the Mn ion of birnessite behaved as an electron acceptor and was reduced during oxidative coupling, the Mn ion was not reconverted to its initial state. Thus, it appears that birnessite functions in this case as an oxidizing agent.

VIII . CONCLUSIONS

The use of biotic and abiotic catalysts to degrade phenolic xenobiotics to less toxic metabolites and/or to incorporate phenolic derivatives into humus may represent an effective method to control soil and water pollution. In the soil, the polymerization/incorporation of hazardous xenobiotics not only reduces the amount of leachable compound, but also reduces the toxicity and bioavailability of these chemicals. In the detoxification of wastewaters, water-soluble pollutants can be converted to water-insoluble polymers which can easily be removed by filtration or sedimentation.[27,28]

Studies to examine the polymerization of phenolic pollutants and their derivatives have traditionally been carried out using laboratory models because the reaction conditions can be controlled and the results are more reproducible. However, results obtained with laboratory models cannot necessarily be extrapolated to predicting the fate of pollutants *in situ.* At the same time, since soil is a highly heterogeneous and physically, chemically, and biologically complex environment, it is unrealistic to conduct field studies without a means for standardizing reaction conditions. Nonetheless, future research must address the question of how to apply laboratory-based methodology to the natural habitat and must continue to verify that the bound pollutants are stable and nontoxic to plant, microbial, and animal life. In addition, further studies should be directed toward developing new methods for maximizing the binding process, e.g., through the use of immobilized enzymes, the addition of copolymerizing agents, or the use of abiotic catalysts.

REFERENCES

1. **Calderbank, A.,** The occurrence and significance of bound pesticide residues in soil, *Rev. Environ. Contam. Toxicol.*, 108, 71, 1989.
2. **McCarthy, J. F. and Jimenez, B. D.,** Reduction in bioavailability to bluegills of polycyclic aromatic hydrocarbons bound to dissolved humic matter, *Environ. Toxicol. Chem.*, 4, 511, 1985.
3. **Ogram, A. V., Jessup, R. E., Ou, L. T., and Rao, P. S. C.,** Effects of sorption on biological degradation rates of (2,4-dichlorophenoxy)acetic acid in soils, *Appl. Environ. Microbiol.*, 49, 582, 1985.
4. **Stevenson, F. J.,** *Humus Chemistry: Genesis, Composition, Reactions*, Wiley-Interscience, New York, 1982.
5. **Bollag, J.-M. and Loll, M. J.,** Incorporation of xenobiotics into soil humus, *Experientia*, 39, 1221, 1983.
6. **Sjoblad, R. D. and Bollag, J.-M.,** Oxidative coupling of aromatic compounds by enzymes from soil microorganisms, in *Soil Biochemistry*, Vol. 5, Paul, E. A. and Ladd, J. N., Eds., Marcel Dekker, New York, 1981, 113.
7. **Simmons, K. E., Minard, R. D., and Bollag, J.-M.,** Oxidative coupling and polymerization of guiaiacol, a lignin derivative, *Soil Sci. Soc. Am. J.*, 52, 1356, 1988.
8. **Shindo, H. and Huang, P. M.,** Role of Mn(IV) oxide in abiotic formation of humic substances in the environment, *Nature (London)*, 298, 363, 1982.
9. **Stone, A. and Morgan, J. J.,** Reduction and dissolution of manganese (III), and manganese (IV) oxides by organics. I. Reaction with hydroquinone, *Environ. Sci. Technol.*, 18, 450, 1984.
10. **Stone, A. and Morgan, J. J.,** Reductive dissolution of metal oxides, in *Aquatic Surface Chemistry*, Stumm, W., Ed., John Wiley & Sons, New York, 1987, 221
11. **McBride, M. B.,** Adsorption and oxidation of phenolic compounds by iron and manganese oxides, *Soil Sci. Soc. Am. J.*, 51, 1466, 1987.
12. **McBride, M. B.,** Oxidation of dihydroxybenzenes in aerated aqueous suspensions of birnessite, *Clays Clay Miner.*, 37, 341, 1989.
13. **Shindo, H. and Huang, P. M.,** Catalytic effects of manganese (IV), iron (III), aluminum, and silicon oxides on the formation of phenolic polymers, *Soil Sci. Soc. Am. J.*, 48, 927, 1984.
14. **Twigg, M. V.,** *Catalyst Handbook*, 2nd ed., Wolfe Publishing, London, 1989.
15. **Moore, J. W. and Pearson, R. G.,** *Kinetics and Mechanism*, 3rd ed., John Wiley & Sons, New York, 1988.
16. **Wang, M. and Huang, P. M.,** Pyrogallol transformations as catalyzed by nontronite, bentonite, and kaolinite, *Clays Clay Miner.*, 37, 525, 1989.
17. **Audus, L. J.,** The biological detoxication of 2,4-dichlorophenoxyacetic acid in soil, *Plant Soil*, 2, 31, 1949.
18. **Leonowicz, A., Trojanowski, J., and Orlicz, B.,** Induction of laccase in basidiomycetes: apparent activity of the inducible and constitutive forms of the enzyme with phenolic substrates, *Acta Biochim. Polon.*, 25, 369, 1978.
19. **Sjoblad, R., Minard, R., and Bollag, J.-M.,** Polymerization of 1-naphthol and related phenolic compounds by an extracellular fungal enzyme, *Pestic. Biochem. Physiol.*, 6, 457, 1976.
20. **Sjoblad, R. and Bollag, J.-M.,** Oxidative coupling of aromatic pesticide intermediates by a fungal phenol oxidase, *Appl. Environ. Microbiol.*, 33, 906, 1977.
21. **Bollag, J.-M., Shuttleworth, K. L., and Anderson, D. H.,** Laccase-mediated detoxification of phenolic compounds, *Appl. Environ. Microbiol.*, 54, 3086, 1988.
22. **Liu, S.-Y. and Bollag, J.-M.,** Enzymatic binding of the pollutant 2,6-xylenol to a humus constituent, *Water Air Soil Poll.*, 25, 97, 1985.
23. **Bollag, J.-M., Liu, S.-Y., and Minard, R. D.,** Cross-coupling of phenolic humus constituents and 2,4-dichlorophenol, *Soil Sci. Soc. Am. J.*, 44, 52, 1980.
24. **Bollag, J.-M. and Liu, S.-Y.,** Copolymerization of halogenated phenols with syringic acid, *Pestic. Biochem. Physiol.*, 23, 261, 1985.
25. **Sarkar, J. M., Malcolm, R. L., and Bollag, J.-M.,** Enzymatic coupling of 2,4-dichlorophenol to stream fulvic acid in the presence of oxidoreductases, *Soil Sci. Soc. Am. J.*, 52, 688, 1988.

26. **Simmons, K. E., Minard, R. D., and Bollag, J.-M.,** Oxidative co-oligomerization of guaiacol and 4-chloroaniline, *Environ. Sci. Technol.,* 23, 15, 1989.
27. **Klibanov, A. M., Alberti, B. N., Morris, E. D., and Felshin, L. M.,** Enzymatic removal of toxic phenols and anilines from waste waters, *J. Appl. Biochem.,* 2, 414, 1980.
28. **Maloney, S. W., Manem, J., Mallevialle, J., and Fiessinger, F.,** Transformation of trace organic compounds in drinking water by enzymatic oxidative coupling, *Environ. Sci. Technol.,* 20, 249, 1986.

Chapter 24

The Role of Abiotic Interactions with Humic Substances on the Environmental Impact of Organic Pollutants

N. Senesi and T. M. Miano

CONTENTS

I. INTRODUCTION

A pollutant or toxic chemical may be defined as any substance that is foreign to the natural ecosystem and, either directly or indirectly, adversely effects the natural physical, chemical, and biological equilibria and processes in the global environment or a portion of it.

Organic pollutants (OP) can reach the soil either by application on purpose, such as pesticides, or incidentally by wet or dry deposition from the vapor phase, or in forms of atmospheric particulates (i.e., fly ash) originating from waste chemicals and solid waste incineration. Pollutant chemicals can also reach the soil through municipal and industrial wastes, landfill effluents, and composts of various nature and source applied to the soil as fertilizers or amendments.

Once on the soil surface, the parent chemicals can either enter the soil or be transported to the aquatic environment by runoff into surface waters and by erosion of contaminated soils into streams. When in the soil, the behavior and fate of the pollutant is controlled by several factors, including the physical and chemical properties of the pollutant, and type and extent of interactions the pollutant experiences with various components of the soil system. These components are substantially inorganic, organic, and biological materials that interact differently with OP, depending on their intrinsic properties.

0-87371-914-X/95/$0.00+$.50

Several studies have suggested that OP show a greater affinity for organic than for mineral surfaces. The content and nature of soil organic matter (SOM) play, therefore, a major role in the performance of OP in soil. The most ubiquitous and widespread natural nonliving organic materials in soils are humic substances (HS). A number of physical, chemical, biochemical, and photochemical properties qualify these as privileged natural compounds in the interaction with organic chemicals. Humic substances may affect the fate of OP in soil in several ways, which include adsorption and partitioning, solubilization, hydrolytic degradation, and photodecomposition. These processes also have important implications in biodegradation and detoxication, bioavailability and phytotoxicity, bioaccumulation, and residue persistence and monitoring of OP in soil.

The main purpose of this review is to summarize and discuss molecular and mechanistic aspects of the most important modes of interaction between HS and some classes of OP, with emphasis on pesticides and their degradation products.

II. SOIL HUMIC SUBSTANCES

Humic substances consist of a physically and chemically heterogeneous mixture of relatively high-molecular-weight, yellow- to black-colored organic compounds of mixed aliphatic and aromatic nature, formed by secondary synthesis reactions (humification) during the decay process and transformation of biomolecules that originate from dead or living organisms and microbial activity.[1] These materials are exclusive of undecayed plant and animal tissues, their partial decomposition products, and the soil biomass. Approximately 60 to 70% of the total soil-organic carbon occurs in HS. Estimated levels of soil organic carbon on the earth surface occurring as HS are 30×10^{14} Kg.[1]

On the basis of their solubility in water at various pHs, HS are divided into two main fractions: (1) humic acids (HA), the portion that is soluble in dilute alkaline solution and is precipitated upon acidification to pH 2, and (2) fulvic acid (FA), the portion that is soluble at any pH value, even below 2.[1]

Humic and fulvic acids cannot be described by unique, chemically defined molecular structures. Both are operationally defined by a model structure constructed on the basis of available compositional, structural, functional, and behavioral data and containing the same basic structural units and the same types of reactive functional groups that are common to all the single, indefinitely variable and unknown molecules.[1] In Figure 1, "typical" model molecules proposed for soil HA and FA are presented.[1,2] The macromolecules bear chemically active functional groups such as carboxylic and phenolic hydroxyls, which render the humic polymer acidic. Hydrophilic as well as hydrophobic sites are present. The structure and composition of HA are more complex than those of FA, which generally feature more oxygen-containing functional groups (preferentially carboxylic), and lower molecular weight.

Physical properties of HA and FA, such as size and the conformation they assume in different conditions, are also very important factors in determining their reactivity and extent of adsorption or "trapping" of OP.

Humic materials are polydisperse and exhibit polyelectrolytic behavior in aqueous solution.[1,3,4] Surface activity is an important property of HS that promotes interactions especially with hydrophobic OP. The amphophilic character and surface activity increase with increasing pH as –COOH and –OH groups form more hydrophilic sites.[3] High pH values and a high concentration of HS depress the surface tension of water. This increases soil wettability and promotes interaction of HS with both hydrophobic and hydrophilic OP in solution. Humic and fulvic acids with their different size and conformation, more or less open, flexible, sponge-like structure may trap and even fix OP that can fit into the voids and holes.[3]

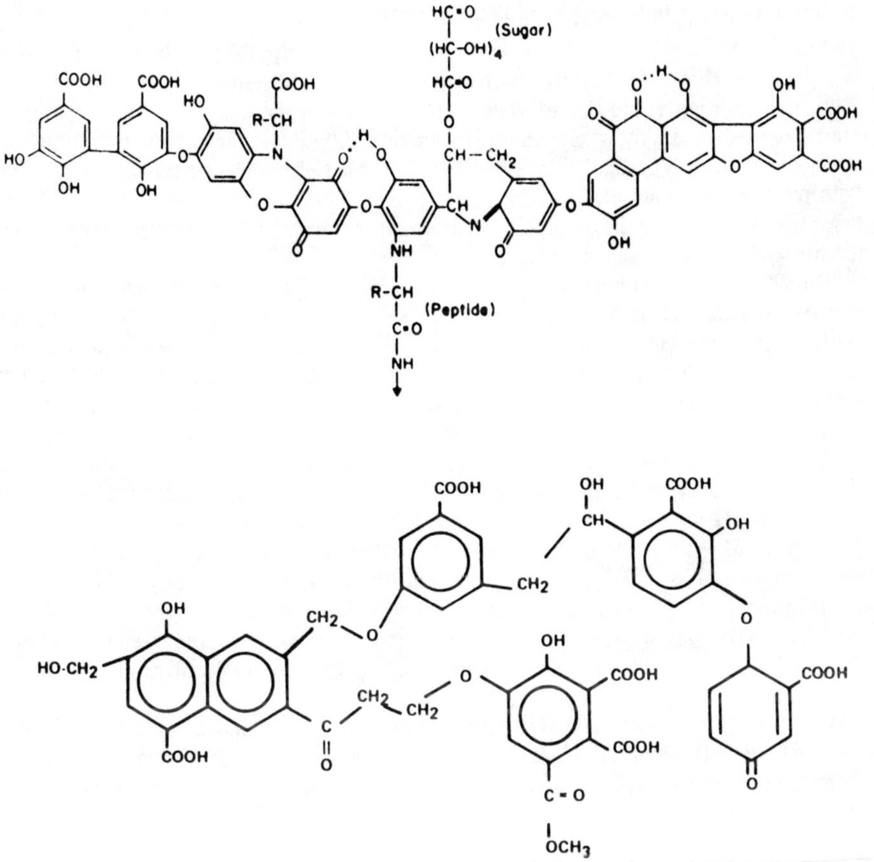

Figure 1 "Typical" model molecular structures proposed for soil HA[1] (top) and FA[2] (bottom).

Humic substances also contain relatively high amounts of stable free radicals, probably semiquinones, which can bind certain OP.[5] The increase in the free radical content of HA and FA in aqueous media with increasing pH or by visible-light irradiation, or in reducing conditions, enhances markedly the chemical and biochemical reactivity.[5]

III. GENERAL PROPERTIES OF ORGANIC POLLUTANTS

Organic pollutants in soil include a great variety of pesticides and their biological, chemical, and photochemical degradation products, aliphatic and aromatic derivatives of petrol and plastics, such as polynuclear aromatic hydrocarbons (PAH) and phthalic acid diesters (PAE), organic solvents, surfactants and detergents, etc.

These compounds feature a wide variety of physical, chemical, and biological properties and belong to widely differing chemical classes. Despite this, they can be classified into two groups, depending on whether their main interactions with soil components, especially with HS, may involve specific chemical or unspecific physical forces. Members of the former group are hydrophilic, ionic or ionizable, basic or acidic compounds; among those of the latter group are hydrophobic, nonionic, mainly nonpolar compounds.

A. CATIONIC AND BASIC COMPOUNDS

The most important and studied cationic pesticides are bipyridilium herbicides, diquat and paraquat, that are generally applied in the form of dibromide and dichloride salts. They exhibit high solubility in water and very low volatility. Diquat and paraquat photodecompose readily when exposed to sun or UV light, but are not photodecomposed when adsorbed onto particulate matter.[6] They are able to form charge-transfer complexes with phenols and other electron-donor molecules.[7] Other cationic pesticides of importance in agriculture are the fungicide phenacridane chloride, the germicide thyamine, and the plant-growth regulator phosphon.

Diquat and paraquat become partly inactivated in highly organic soils.[8,9] Sorption of diquat and paraquat on SOM is suggested to be the major factor responsible for the decrease in herbicide activity,[10] although these chemicals are still biologically active toward plants and microorganisms.[11-14] Phenacridane chloride and thyamine have the highest levels of adsorption on SOM, followed by phosphon, diquat, and paraquat.[11]

Amitrole and s-triazines are weakly basic compounds, easily ionizable by protonation, that are widely applied to soil as herbicides. Amitrole is soluble in water and behaves chemically as a typical aromatic amine. s-Triazines have a low water solubility that increases at pH values where strong protonation occurs, generally between pH 5.0 and 2.0. Structural modifications of the substituents significantly affect solubility at all pH values. Most s-triazines can be chemically hydrolyzed and partly decomposed by UV and IR radiation, including sunlight, in aqueous systems. They are, in general, relatively volatile. s-Triazines possess electron-rich nitrogen atoms that confer electron-donor ability, i.e., the capacity to interact with electron-acceptor molecules, giving rise to electron donor-acceptor (charge-transfer) complexes.

The biological activity and transport in soil of s-triazine and triazole herbicides is negatively correlated with the level of SOM due to adsorption phenomena.[11,12,15] Chemical hydrolysis and sunlight photodecomposition of s-triazines are also influenced by the presence of SOM.[16-18]

B. ACIDIC AND ANIONIC COMPOUNDS

Several classes of weakly acidic organic compounds are widely used in agriculture for their herbicidal or insecticidal action. They possess carboxyl or phenolic functional groups capable of ionizing in aqueous media to yield anionic species. Chlorinated aliphatic acids show the highest water solubility and strongest acidity. Substituted phenols such as dinitrophenols, dinoseb, dalapon, ioxynil, bromoxynil, and pentachlorophenol (PCP), and phenoxyalkanoic acids, such as 2,4-D and 2,4,5-T, have intermediate to low solubility in water. All these compounds are, however, highly water soluble when applied to soil in the common anionic, alkali salt formulations. With the exception of picloram and phenols, acidic pesticides are substantially nonvolatile. Other important acidic herbicides are the halogenated benzoic acids chloramben, dicamba, and TBA, and the pyridine derivative picloram.

The bioactivity and transport of acidic pesticides and their ester formulations correlate with the SOM content, even though the adsorption level of these herbicides is much lower than that of cationic and basic pesticides.[9,11] The persistence of acidic pesticides, measured in terms of biological activity as well as residual toxicity, is highest in soils rich in OM.[11,12,19] The leachability of picloram, amiben, and PCP is negatively correlated with OM,[19,20] whereas the efficacy of PCP decreases with increasing OM content due to increasing adsorption.

C. NONIONIC COMPOUNDS

Nonionic pesticides do not ionize significantly in aqueous systems and vary widely in their composition, physical properties, such as water solubility, polarity, and volatility, and chemical reactivity.

Chlorinated hydrocarbon insecticides are among the most widely known and studied nonionic pesticides. With the exception of lindane, all these compounds are insoluble in water. DDT is about ten times more insoluble than the other compounds of this family; thus, it is considered to be immobile in soil systems. Endrin, dieldrin, and aldrin show higher water solubility and are, therefore, slightly mobile in soils. The vapor pressure of chlorinated hydrocarbons varies widely from low (DDT, endrin, and dieldrin) to moderate (toxaphene and aldrin), high (chlordane and lindane), and very high (heptachlor). Volatilization of DDT from soils and other surfaces is, therefore, almost insignificant. Insecticidal activity, degradation, inactivation, leaching, and volatilization of several chlorinated hydrocarbons, including aldrin, dieldrin, endrin, endosulfan, lindane, heptachlor, DDT, toxaphene, and chlordane, decrease as the SOM content increases. The most prominent effect occurs in moist soils.[9,21,22] In particular, retention and inactivation of DDT is quite generally correlated with SOM content and especially with the humified fraction.[9] Higher levels of SOM also increase the persistence of DDT, lindane, and aldrin, which is higher in a muck soil than in a mineral soil.[23] Transport of DDT in forest soils has been associated with soil HA and FA fractions.[24] SOM is reported to be the principal means of deactivation of DDT as determined by a bioassay technique.[25]

Organophosphates, which include malathion and parathion insecticides and the herbicide glyphosate, are more toxic, but exhibit higher water solubility and vapor pressure and lower persistence in soil than chlorinated hydrocarbons. Malathion and parathion insecticides are easily chemically hydrolyzed and biodegraded by microorganisms in soil systems. The insecticidal activity of organophosphates and their adsorption in soils is correlated with SOM content.[12] The bioactivity of phorate decreases with increased SOM content,[26] while diazinon and parathion show the same relationships in moist soils, but not in dry soils.[27] Fonofos is persistent for more than 2 years in an organic soil, and its mobility and persistence in soils is suggested to be partly a function of adsorption on HS.[28,29]

Phenylcarbamates, or carbanilates, such as carbaryl, methiocarb, aldicarb, and carbofuran, generally exhibit low water solubilities; thus, they are almost immobile in soil systems. Chlorpropham and propham are readily volatilized from soil systems, but terbutol and carbaryl are not. Among the chemical reactions carbamates may undergo are ester and amide hydrolysis, N-dealkylation, and hydroxylation. Phenylcarbamate herbicides show a lower herbicidal activity in fine-textured soils than in coarse-textured ones. This appears to be related to the higher OM content of the former soils.[30] The phytotoxicity of chlorpropham is greatly reduced by OM added to the soil.[10] Vapor losses of propham and chlorpropham from moist soils decrease as the percentage of OM increases.[19]

Substituted urea herbicides, including fenuron, linuron, monuron, diuron, cycluron, and many others, generally exhibit low water solubility, with the exception of fenuron, and low volatility. These compounds can undergo hydrolysis, acylation, and alkylation reactions in soils and show electron donor properties, being able to form charge-transfer complexes with electron-acceptor molecules. The herbicidal activity, adsorption, and transport of several phenylureas in soils decrease as the SOM content increases.[12,31-34] Addition of OM to sandy soil significantly reduced the herbicidal activity of fluometuron and fenuron in a growth chamber study and in field experiments.[11]

Substituted dinitroanilines, including trifluralin, nitralin, benefin, butralin, and profluralin, show very low water solubilities and low volatility, with the exception of trifluralin. All these compounds are relatively immobile in soil systems. The herbicidal activity of substituted dinitroanilines, i.e., trifluralin and benefin, is reduced by SOM.[30,35-38] The mobility of these herbicides in muck soils is much less than in mineral soils, thus suggesting a partial adsorption of these compounds by organic soil colloids.[11] In greenhouse experiments, orizalyn was inactivated by SOM at a much lower level than trifluralin and nitralin due to much lower adsorption of orizalyn by SOM.[11]

The phenylamide herbicide diphenamid is moderately water soluble and not volatile. It probably behaves much like the acetanilides in aqueous and soil systems. The movement and bioactivity of diphenamid decreases as the SOM content increases.[31] Similarly, addition of SOM to model soil systems significantly reduces the herbicidal activity of diphenamid.[11]

Thiocarbamate and carbothioate herbicides generally exhibit low water solubility and high vapor pressures, thus being easily lost by volatilization from soil. The most important nonionic benzonitrile herbicide is dichlobenil, which has a low solubility in water and low vapor pressure; thus, it is relatively immobile in most soils. Among acetamide herbicides, CDAA has a relatively high water solubility and vapor pressure with respect to propachlor and alachlor. Increasing SOM content results in decreased bioactivity, movement, leaching, and volatilization of relatively volatile pesticides, which include thiocarbamates, acetamides, and benzonitriles such as EPTC, pebulate, cycloate, CDEC, CDAA, and dichobenil.[12,39-44] Higher levels of OM result in lower initial toxicity of many herbicides and lower losses through volatilization. Thus, these compounds are slightly less effective but persist longer in fine-textured soils.[11] Movement and adsorption of aldicarb, carbofuran, and pyrazone were shown to be positively and significantly correlated to SOM content.[45,46]

PAH represent a hazardous class of widespread contaminants produced in large quantities from antropogenic sources, including the combustion of fossil fuel, chemical manufacturing, petroleum refining, metallurgical processes, and some coal, oil shale, and tar sand conversion systems. PAH are neutral, nonpolar hydrophobic organic molecules, with the hydrophobicity increasing with molecular weight, i.e., with the number of benzene rings in the structure. As they have a hydrophobic nature, adsorption is very important in determining their fate in surface and subsurface water-soil/sediment systems. Aqueous concentrations, transport, and surface-associated chemical and biological degradation processes of PAH are highly dependent on adsorptive/desorptive equilibria with sorbents, e.g., HS, present in the system. In particular, liquid to solid phase partitioning can play a significant role in the adsorption phenomena on HS.

PAE are used mainly as plasticizers, and also as pesticide carriers and insect repellents, in dyes, cosmetics, and lubricants. PAE are lipophilic or lyophobic liquids of medium viscosity, practically immiscible with water, and have low vapor pressure. The isolation and identification of small amounts of PAE (0.03% of the dry weight) from soil HA and FA[47-49] indicates that the lyophylic HS can interact with lyophobic PAE to form water-stable products. Water-soluble FA may therefore mediate mobilization, transport, and immobilization of PAE in soil solutions and aquatic environments.

IV. ADSORPTION MECHANISMS

Adsorption represents probably the most important mode of interaction of OP with SOM. Adsorption can occur through physical-chemical binding by specific mechanisms and forces with varying degrees of strengths. These include ionic, hydrogen, and covalent bonding, charge-transfer or electron-donor acceptor mechanisms, dipole-dipole and Van der Waals forces, ligand exchange, and cation and water bridging. Adsorption of nonpolar (hydrophobic) organic compounds can better be described in terms of nonspecific, hydrophobic, or partitioning processes between soil water and the soil organic phase.

The type and extent of adsorption will depend on the amount and properties of both the chemical and the humic molecule besides other factors, including biological activities, that are beyond the scope of this review. The various chemical properties of the pollutant molecule, such as number and type of functional groups, acidic or basic character, polarity and polarizability, ionic nature and charge distribution, water solubility, and shape and configuration, will often result in several possible adsorption mechanisms that may operate in combination.

For any given chemical, a sequence of different mechanisms may be responsible for adsorption onto HS. The organic molecule may be sorbed initially by sites that provide the strongest binding, followed by progressively weaker sites as the stronger adsorption sites become filled. Once adsorbed, the chemical may be subject to other processes that can affect retention. For instance, some chemicals may further react to become covalently and irreversibly bound, while others may become only physically trapped in the humic matrix. Adsorption processes may thus vary from complete reversibility to total irreversibility, that is, the adsorbed chemical may be easily desorbed, with various degrees of difficulty, or not at all.

A. ADSORPTION ISOTHERMS

Construction and use of adsorption isotherms from equilibrium adsorption data has been employed by numerous researchers to describe the adsorption of OP on a solid matrix. In general, two equations are used to quantitatively describe the adsorption of the chemical onto SOM: the Freundlich equation (Equation 1) and the Langmuir equation (Equation 2).[50]

$$x/m = K\,C^{1/n} \tag{1}$$

where x/m is the ratio of adsorbate (OP) to SOM mass, C is the adsorbate concentration in solution at the equilibrium, and K and n are constants.

$$x/m = K_1\,K_2\,C/(1 + K_1\,C) \tag{2}$$

where K_1 and K_2 are constants and x/m and C have the same meaning as in Equation 1.

Adsorption of linuron, 2,4-D, picloram, and fonofos on HA is shown to follow a Freundlich-type isotherm.[29,51,52] Adsorption of cationic compounds, including paraquat and diquat, phosphon, hyamine, and chlordimeform, on SOM results in L-shaped or Langmuir-type isotherms characterized by a curvilinear response at all concentrations used, which seems to level off at a certain adsorption maximum.[11,53] This indicates specific adsorption on homogeneous sites. pH-dependent adsorption of s-triazines by organic soil colloids and HA also shows L-shaped isotherms, indicating that adsorbed species are in equilibrium with species in solution at each pH value.[54-56] Adsorption of chlordimeform and atrazine on HA also has the form of a Langmuir isotherm,[53,57] whereas the adsorption of terbutryne on HA fit both equations.[58] However, neither the Freundlich nor the Langmuir equation fit the data if the adsorption of an OP on SOM is predominantly due to an ion-exchange mechanisms, e.g., the adsorption of paraquat on HA.[59] In this case, only the Rothmund-Kornfeld equation fits the results satisfactorily.

In conclusion, carefully controlled adsorption measurements and the shapes of the related isotherms may provide information on the mechanism involved in the adsorption of OP to HS.

B. IONIC BONDING (CATION EXCHANGE)

Adsorption via ionic bonding, or cation exchange, applies only to those pesticides which are in the cationic form or can become cationic by protonation in solution. It involves ionized, or easily ionizable, carboxylic and phenolic hydroxyl groups of HS.

Infrared (IR)[53,59,60] and potentiometric-titration[61] data support ionic binding as the dominant mechanism for adsorption of diquat, paraquat, and chlordimeform by HS. Divalent cationic bipyridilium herbicides can react with two negatively charged sites on HS, e.g., two $-COO^-$ groups or a $-COO^-$ plus a phenolate ion (Figure 2a). Humic and fulvic acids appear to retain paraquat and diquat at levels that are considerably lower than the exchange capacity of the HS.[8] Thus, not all negative sites on HS seem to be positionally available to bind large

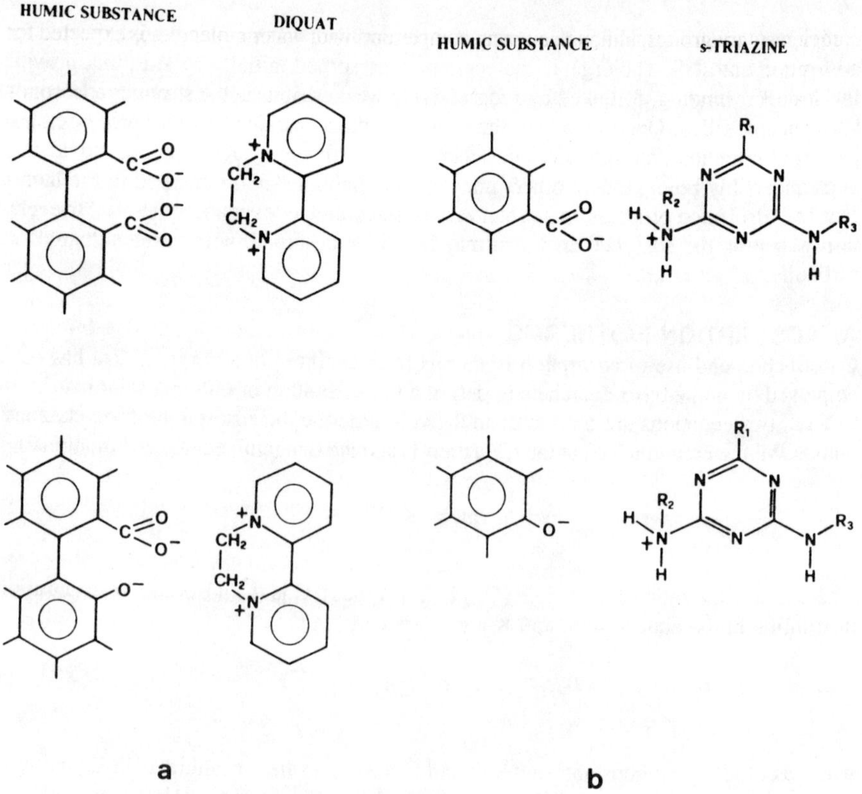

Figure 2 Model ionic binding between (a) HS and diquat and (b) HS and a protonated s-triazine.

organic cations, probably because of steric hindrance. Phosphon and phenacridane chloride are also reported to be adsorbed onto SOM through ionic bonding.[11]

Less basic pesticides such as s-triazines,[55] amitrole,[62] and dimefox[63] may become cationic through protonation, depending on their basicity and the pH of the system, which also governs the degree of ionization of acidic groups on the HS. Evidence that maximum adsorption of s-triazines on organic soils occurs at pH levels close to the pK_a of the herbicide is indicative of ion exchange.[55] However, the pH at the HS surface may be two orders of magnitude lower than that of the liquid phase[15]; thus, surface protonation of a basic molecule may occur even though the measured pH of the medium is greater than the pK_a of the adsorbate. IR studies of s-triazine-HA systems show that ionic bonding can occur between a protonated secondary amino group of the s-triazine and a carboxylate anion and, possibly, a phenolate group of the HA (Figure 2b).[64-71] Differential thermal analysis (DTA) data confirm IR results in showing an increased thermal stability of the HA upon its interaction with various s-triazines.[68,69] Structural features of s-triazines, such as the nature of the substituent in the 2-position and of the alkyl groups at the 4- and 6-amino groups, are shown to influence their basicity and, hence, reactivity with HS.[72,73] The higher reactivity of simazine with respect to atrazine and prometryne may be related to the smaller steric hindrance of the reactive N-H group of the former herbicide.[65]

C. HYDROGEN BONDING

The presence of numerous oxygen- and nitrogen-containing functional groups on HS renders highly probable the formation of H bonding for chemicals containing suitable complementary groups, although a strong competition with water molecules is expected for such sites.

Heat of formation, IR, and DTA data suggest the occurrence of one or more H bonds between HA and s-triazines, possibly involving C=O groups of HA and secondary amino groups of s-triazines (Figure 3a).[57,64,65,68]

Acidic pesticides such as chlorophenoxyalkanoic acids and esters, asulam, and dicamba can be adsorbed by H bonding onto humic substances at pH values below their pK_a through =COOH, =COOR, and similar groups (Figure 3b).[66,74,75]

H bonding plays an important role in the adsorption onto HS of several nonionic polar pesticides, including substituted ureas and phenylcarbamates, alachlor, metolachlor, cycloate, malathion, bromacil, and glyphosate,[11,62,69,74,76-79] and dialkylphtalates.[80]

D. CHARGE TRANSFER (ELECTRON DONOR-ACCEPTOR)

Humic macromolecules contain both electron-deficient structures, such as quinones, and electron-rich moieties, such as diphenols. This can account for their electron-donating and -accepting properties. Electron donor-acceptor mechanisms and formation of charge-transfer complexes will thus be possible with OP possessing electron donor or electron acceptor features.

Strong spectroscopic evidence exists for the formation of charge-transfer complexes between several HA and s-triazines (Equation 3), substituted ureas, and amitrole. The shift to the lower

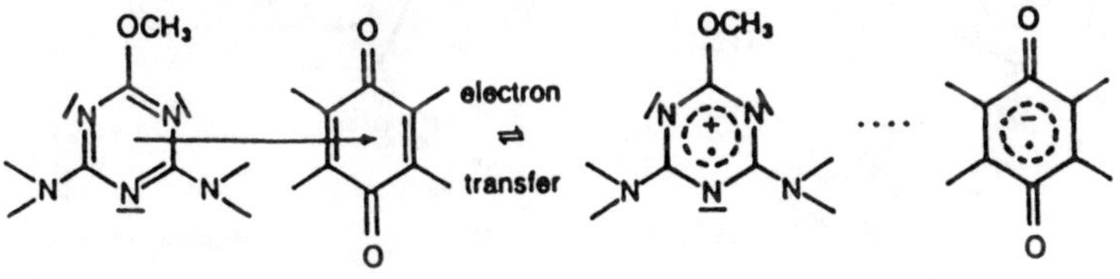

(3)

(s-triazine; electron–donor) (humic quinone; electron–acceptor) (radical cation and anion; charge-transfer complex)

wave number observed in the IR for the out-of-plane deformation vibration of the heterocyclic donor triazine is ascribed to the decreased electron density in the ring resulting from the formation of a complex with electron-deficient structures of HA.[68,71,81,82] The increase in free-radical concentration measured by electron spin resonance (ESR) in the interaction products of HA with s-triazine, substituted ureas, and amitrole is ascribed to the formation of semiquinone free radical intermediates in the single-electron donor-acceptor transfers occurring between the electron-rich amino or heterocyclic nitrogens of the herbicide and the electron-deficient, quinone-like structures in the HA.[62,68,69,71,76,82] Charge-transfer HA-s-triazine systems may also originate under the effect of light, which may induce an unpairing of electrons involved in the interaction and produce the observed increase in the ESR signal (photoinduced transfer).[50]

A number of structural and chemical properties of both the HA and the OP affect to some extent the efficiency of formation of the electron donor-acceptor system. The electron-acceptor tendency of the humic molecule seems to be related directly to the quinone content and inversely to the total acidity and the content of carboxyls and phenolic hydroxyls.[68,71] The presence of activating electron donors, such as a methoxyl group in the 2-position of the ring and isopropyl substituents on the amino groups, renders prometone the most efficient electron donor among s-triazines.[68,71] The absence of chlorine atoms on the phenyl ring of fenuron qualifies this urea as an electron donor stronger than ureas containing one or more deactivating electron-withdrawing chlorine atoms on the ring.[71]

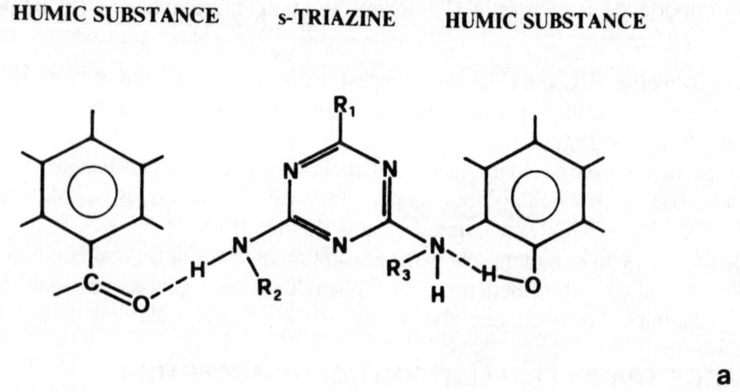

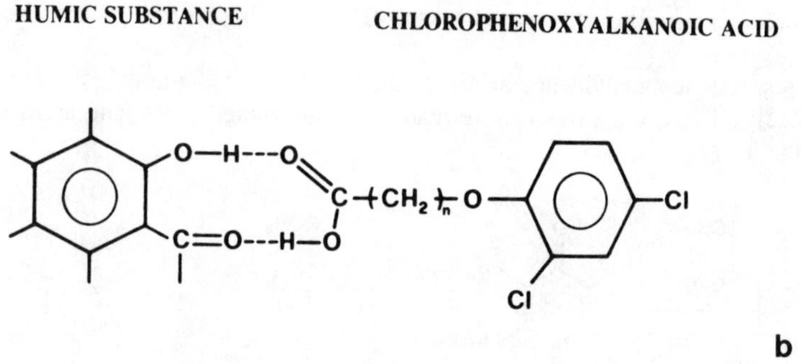

Figure 3 Model hydrogen bonding between (a) HS and a s-triazine and (b) HS and a chlorophenoxyalkanoic acid.

Infrared frequency shifts of the CH out-of-plane bending vibration of the electron acceptors paraquat, diquat, and chloridimeform, measured in their products of interaction with HS, are ascribed to the formation of charge-transfer complexes.[59,60] Ultraviolet-differential spectroscopy provides evidence of the formation of a charge-transfer complex between electron donor moieties of HS and the electron acceptor chloranyl (Equation 4).[83]

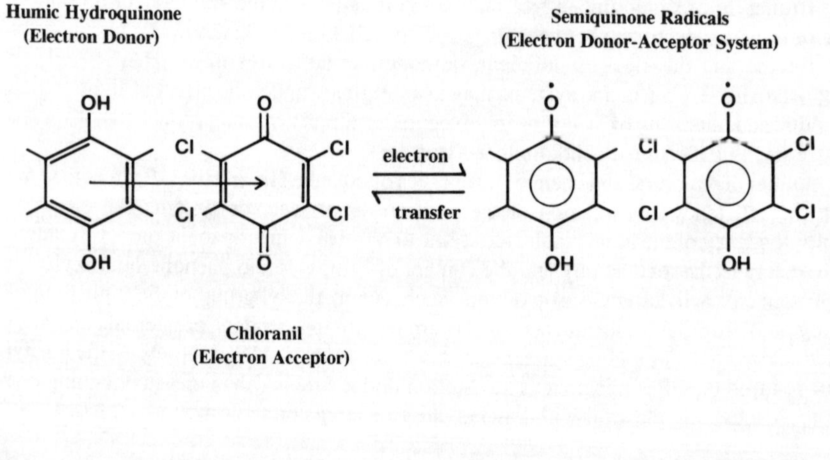

(4)

Other important OP such as DDT, dioxins, and polychlorobiphenyls (PCB) possess electron-accepting properties and may interact with HS by charge-transfer mechanisms similar to those previously described. Such a mechanism has been postulated to be responsible for the photolysis of DDT in the presence of HS.[84]

E. COVALENT BINDING

The presence of several types of reactive functional groups in HS offers multiple possibilities for their chemical reaction with suitable functional groups of the OP. These reactions are often mediated by chemical, photochemical, or enzymatic catalysts and result in the formation of covalent bonds, with incorporation of the pollutant or, more likely, its intermediates and products of degradation into the HS macromolecule.

Acylanilides, phenylcarbamates, phenylamide, phenylureas, dinitroaniline herbicides, nitroaniline fungicides, and organophosphate insecticides, such as parathion and methylparathion, are biodegraded in soil with the release of aromatic amines, such as chloroanilines. Similarly, catechols and phenols are released by the degradation of several classes of pesticides. These residues are subject to the formation of covalent bonds with SOM, with or without the mediation of microbial activity. Chloroanilines can be chemically bound to SOM by two possible mechanisms: nucleophilic addition to carbonyl and quinone groups and oxidative-coupling reactions to phenols (Equations 5 to 7).

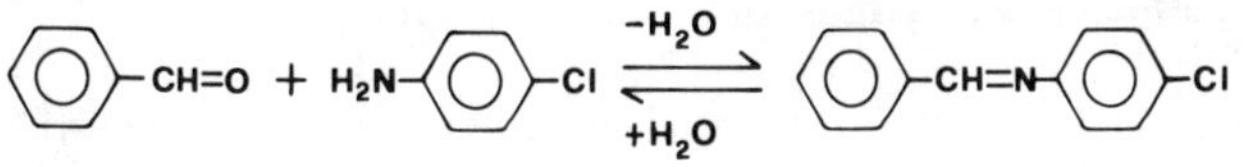

(5)

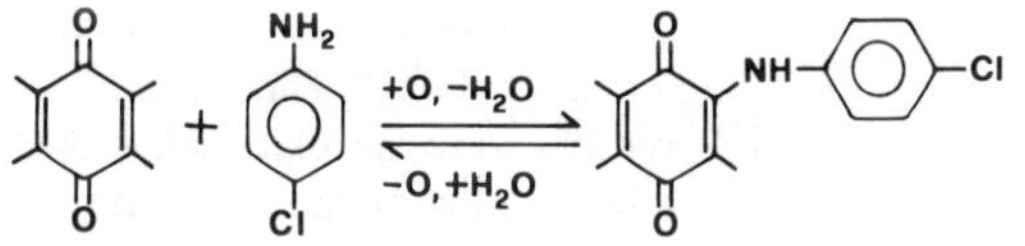

(6)

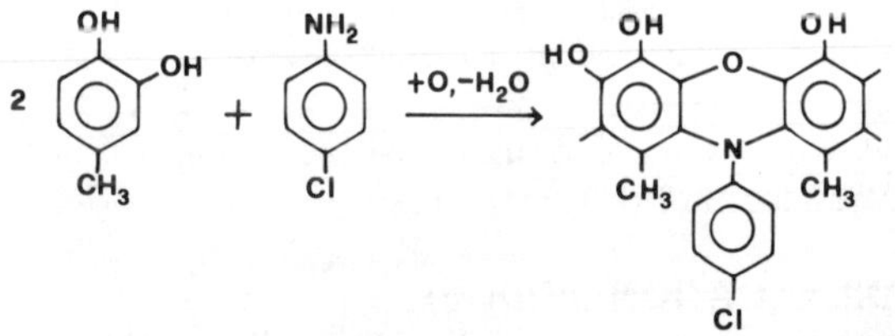

(7)

These reactions may lead to the formation of hydrolyzable (e.g., anil, a Schiff base, and anilinoquinone) or nonhydrolyzable (e.g., heterocyclic rings or ethers) bound forms.[85-87] The primary amines are suggested to interact with HS in a pathway consisting of an initial rapid, reversible formation of imine linkages with the HS carbonyl, followed by a slow, irreversible 1,4 nucleophilic addition to quinone-like rings, which is then followed by tautomerization and oxidation to yield an aminosubstituted quinone.[88] The slow reaction with HS is also given by secondary amines. Nucleophilic addition is suggested for the binding to HS of benzidine, α-naphthylamine, and *p*-toluidine.[89]

A marked quenching of free-radical concentration is measured by ESR spectroscopy in the interaction products of soil HA with water-dissolved chlorophenoxyalkanoic acids and esters. This suggests the occurrence of homolitic cross-coupling reactions between indigenous humic free radicals and phenoxy or aryloxy radical intermediates generated photochemically or by chemical or enzymatic catalysis in the initial, partial oxidative degradation of chlorophenoxyalkanoic compounds (Equation 8).[75,90] Indigenous inorganic catalysts (e.g. cupric and ferric ions)

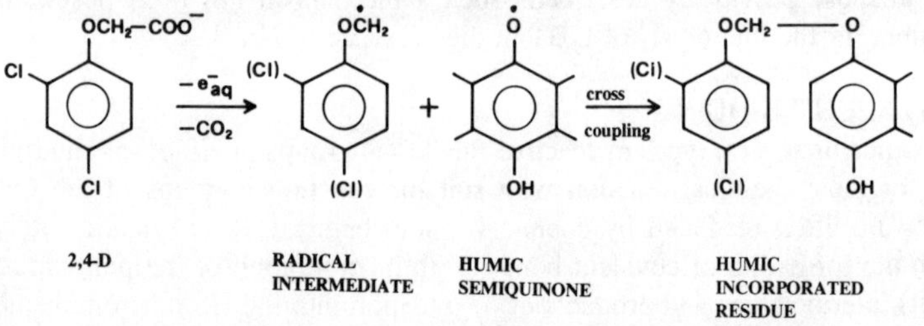

(8)

and residual enzymatic activity (e.g., due to phenoloxidases), both able to mediate chemically or biologically the oxidative degradation of these herbicides, are present in HS.[91] Phenoxy-type radicals may also be generated photochemically from phenoxyalkanoic compounds in the initial oxidation step of the nonbiological degradation process they undergo in aqueous solution and in the presence of light and air.[92,93] The coupling reactivity of HA free radicals with chlorophenoxy-derived radicals — assumed to be proportional to the extent of lowering of the residual free radical concentration in the interaction products — is negatively correlated with the carboxyl content and COOH-to-phenolic-OH ratio of the HA, and with the number of chlorine atoms on the phenoxy ring of the herbicide, which most likely interfere with cross-linking to the HA macromolecule.[75,94]

The significant increase of free radical concentrations and enlargement of ESR signal line widths measured in several HA-s-triazine interaction products[68,69,71] confirm the formation of covalent bonds, previously suggested on the basis of the increased adsorption of s-triazines on HA observed at high temperatures.[15] These results are attributed to the increased stabilization attained by free electrons on extended aromatic structures originated by the covalent binding of amino groups of the s-triazine to carbonyl and quinone groups of the HA.

Much evidence exists of the enzyme-catalyzed incorporation into HA polymers of chloro- and alkyl-substituted aniline residues in the presence of phenoloxidases,[95,96] and of chlorocatechol intermediates of the decomposition of 2,4-D and 2,4,5-T in the presence of peroxidase.[97] No evidence is available for the enzymatic binding and incorporation of triazine and phenylurea residues in HS.[98]

F. LIGAND EXCHANGE (CATION BRIDGE)

Adsorption by ligand exchange mechanisms involves the displacement of hydration water or other weak ligands partially holding a polyvalent metal ion associated to SOM by a suitable functional group of the pollutant. This group may act as ligand to the metal, with the formation of an inner-sphere complex. s-Triazines and anionic pesticides such as picloram are likely to bind HS by this mechanism (Equation 9).[99]

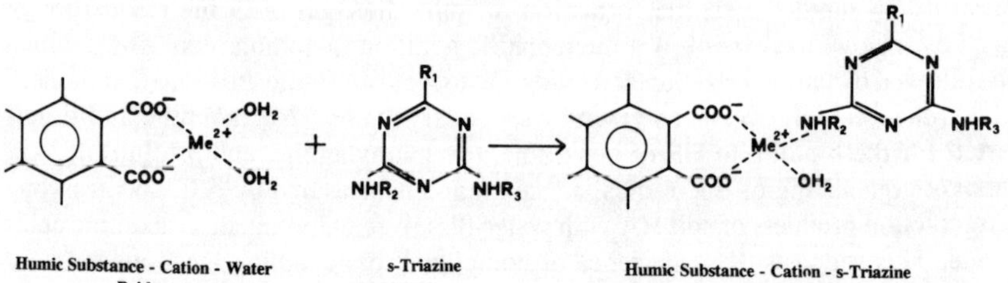

(9)

G. DIPOLE-DIPOLE AND VAN DER WAALS FORCES

Van der Waals forces consist of weak, short-range dipolar or induced dipolar attractions that operate, eventually in addition to stronger binding forces, in all adsorbent-adsorbate interactions. These forces assume, however, particular importance in the adsorption of nonionic and nonpolar pesticides. Since Van der Waals forces are additive, their contribution increases with the size of the pollutant molecule and with its capacity to adapt to the surface of the humic molecule.

Although scarce experimental evidence is available, Van der Waals forces are considered to be actively involved in the adsorption onto soil HS of large-size bipyridilium cations,[100] carbaryl and parathion,[101] alachlor and cycloate,[62] benzonitrile and DDT,[102] and several thiocarbamates, carbothioates, and acetanilides,[11] and to represent the principal adsorption mechanism for picloram and 2,4-D.[51,78]

H. HYDROPHOBIC ADSORPTION AND PARTITIONING

Hydrophobic adsorption on the surface or trapping within internal pores of the humic macromolecular sieve has been proposed as an important nonspecific mechanism for retention of nonionic, nonpolar OP that interact weakly with water. Hydrophobic active sites of HS include aliphatic side chains or lipid portions and aromatic lignin-derived moieties with high carbon content and bearing a small number of polar groups. Water molecules do not compete effectively with hydrophobic molecules for these sites.

Hydrophobic adsorption by SOM is suggested to be important for DDT and other organochlorine insecticides,[102] dialkylphthalates,[48,49] leptophos, methazole, norflurazon, oxadiazinon, butralin, and profluralin,[66] metolachlor,[78] chlorinated hydrocarbons,[102] picloram and dicamba,[51] 2,4-D,[74] parathion,[101] phenylcarbamates,[103] substituted anilines,[11] and PCB.[104] Hydrophobic association has also been proposed as a possible mechanism for adsorption of the s-triazines[105] and the phenylureas[52,106] by SOM.

Hydrophobic retention need not be regarded as an active adsorption mechanism and is often considered as a "partitioning" between water and a nonspecific organic phase. In contrast to adsorption, the term partitioning describes a process in which the adsorbate permeates, i.e., dissolves, into the network of the organic phase. Thus, partitioning is distinguished from adsorption by the homogeneous, aspecific distribution of the sorbed material throughout the entire volume of the organic phase.[107] Partitioning is modeled as an equilibrium process similar to that between two immiscible solvents, such as water and an organic solvent such as 1-octanol. Humic substances in both the solid and dissolved phase are thus considered as a nonaqueous solvent into which the hydrophobic pollutant can partition from water.

Thermodynamic arguments have been proposed, but may be too simplified to distinguish between partitioning and adsorption and may lead to erroneous conclusions.[108] The wide variation observed by several authors in the strength of interactions (expressed as partition coefficients) between hydrophobic OP and HS of different sources and nature is attributed to the variation in the chemical and structural composition of the latter.[109] This confirms that strong and "specific" chemical interactions are involved and should be accounted for in any interpretation of hydrophobic OP-HS association.[110]

V. SOLUBILIZATION EFFECTS

The actual water solubility of certain pollutants can be significantly modified by adsorption onto SOM, thus affecting the mobility and migration of pollutants in soil. The effect of adsorption on pollutant solubility in soil depends on whether the adsorption occurs on insoluble, immobile organic fractions such as HA, or on dissolved or suspended mobile fractions such as FA. Soil organic matter in the form of HA or FA can, therefore, either decrease or increase the actual solubility of pollutants in the soil liquid phase.

At a given concentration of dissolved HS, relatively water-insoluble, nonionic pollutants such as DDT, PCB, PAH, n-alkanes, and PAE, are most easily solubilized.[111-116] The increase in solubility may be the result of direct pollutant adsorption or partitioning, or of an overall increase in solvency.[117]

The effect of solubility enhancement decreases with the increasing intrinsic water solubility of the pollutant.[117] This trend has been observed for relatively water-insoluble higher alkanes in comparison with more water-soluble aromatic compounds[112] and for DDT vs. lindane.[118]

The observed pollutant-solubility enhancement has been described as possibly resulting from a partition-like interaction between solute and dissolved OM of high molecular weight. The latter is regarded as a "microscopic organic phase," similar to a micelle.[113] The effectiveness of this interaction appears to be largely controlled by a number of properties of the humic molecule such as molecular size, polarity, configuration, conformation, composition, total acidity, and surface activity.[113,114,116] Humic substances having high molecular weight, carbon content, and aromaticity, as well as large nonpolar volume and low oxygen and polar-group content and hydrophilicity, are more effective than low-molecular-weight OM in enhancing the water solubility of DDT and PCB.[117]

The magnitude of OP solubility enhancement is also dependent on the concentration of dissolved HS, solution pH, and temperature. The water solubilities of compounds such as DDT, some PCB, and higher alkanes are sensitive to low levels of dissolved HS; those of highly water-soluble OP are less so. The solubility enhancement is decreased if the pH is raised for DDT and PCB.[113,114] At alkaline pH, the hydration and concomitant repulsion of ionized acidic groups of the humic molecule would lead to more expanded conformations and thus decrease the extent of nonpolar regions available for interaction with the hydrophobic OP. A temperature decrease from 25 to 9°C increases twofold the binding constant of DDT to dissolved HS, thus increasing its apparent water solubility.[114]

All these results suggest that the solubility effect exerted by dissolved HS on hydrophobic OP cannot be completely and satisfactory explained on the basis of a simple partitioning model of interaction. Specific binding interactions should be duly accounted for in any general interpretation of such a complex process by considering suitable, additional factors and mechanisms that can better account for these effects.

VI. HYDROLYSIS CATALYSIS

Hydrolysis is an important degradative reaction that leads to products with solubility, volatility, reactivity, and other physical and chemical properties different from those of the parent pollutant molecule, thus affecting its behavior and fate in soil.

Humic substances, particularly in the dissolved phase, are able to exert a catalytic or inhibitory effect in the abiotic hydrolysis of a number of herbicides. In the presence of FA or HA, the acid hydrolysis rates of n-alkyl esters of 2,4-D and chloro-s-triazines are enhanced,[119,120] whereas that of alkaline hydrolysis of the n-octyl ester of 2,4-D (2,4-DOE) is lowered.[121] Sorption on HS has little or no effect on the rates of pH-independent (i.e., nonacid-base catalyzed) hydrolysis of some organophosphorothioate esters such as chlorpyrifos and diazinon, halogenated alkenes, aziridine derivatives, and others.[122]

The catalytic effects of HA and FA on the dechlorohydroxylation of the chloro-s-triazines simazine, atrazine, and propazine are attributed to specific interaction through

H bonding between the surface carboxylic groups of the humic molecule and the heterocyclic nitrogens of the triazine (Equation 10). This would

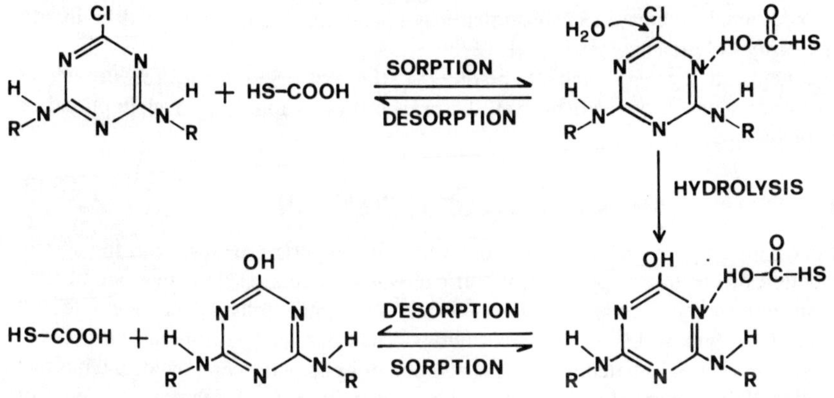

(10)

enhance the electron-withdrawing effect of the electron-deficient carbon bearing the chlorine atom, thus reducing the activation energy barrier for hydrolytic cleavage of the C-Cl bond and promoting its replacement by the weak nucleophile water (Equation 10).[57,120] This mechanism explains the correlation established between soil-catalyzed hydrolysis of chloro-s-triazines and the amount adsorbed by SOM.[18] Successively, a first-order kinetic reaction was measured with respect to the herbicide concentration for the hydrolysis of atrazine at pH 4 in an aqueous suspension of HA.[57] The catalytic effect of HA could depend not only on the number of effective acid groups, but also on the arrangement of these groups on the HA molecule.[57] Hydrolysis of atrazine in aqueous FA solution at pH ≤7 also follows first-order kinetics with respect to the herbicide concentration.[120] An increase in the FA concentration results in a higher hydrolysis rate constant and a shortened half-life, but has no effect on the activation energy, which, however, increases with an increase in the pH of the reaction mixture. A change in the pH of the FA solution would change the types and concentrations of acidic functional groups involved in the hydrolysis of atrazine, which in turn might affect the mechanism of hydrolysis as indicated by the change in the activation energy.[120] Undissociated carboxyl groups of SOM are suggested to be the only catalytically active sites, leading directly to catalyzed hydrolysis of the H-bonded atrazine, while phenolic groups have no catalytic effects.[55,123]

The retardation of base-catalyzed hydrolysis of 2,4-DOE by solution-phase HS is envisioned in terms of the changes in solution chemistry in the vicinity of the negatively charged (at neutral and alkaline pH values) surfaces of humic molecules.[121]

A tentative general mechanism has been proposed for the overall effects of HS on the hydrolysis kinetics of hydrophobic OP. The model is based on an analogy with micellar catalysis, with only a minor contribution from general acid-base catalysis.[124] The model has been successfully tested with experimental data obtained from the base-catalyzed hydrolysis of 2,4-DOE[121] and acid-catalyzed hydrolysis of atrazine.[57] Similar to anionic surfactants, the effects in the two cases are attributed to electrostatic stabilization of the transition state for the acid catalysis, in which the substrate becomes more positively charged, and to the destabilization of the transition state for base-catalyzed hydrolysis, in which the substrate becomes more negatively charged.[124] On the basis of this model, one would expect the effect on base-catalyzed hydrolysis to be small for parathion, which associates only weakly with HS, but large for compounds such as DDT, which associates strongly with them.[124] In conditions where much higher concentrations of HS are possible, i.e., in sewage sludge or in sediment and soil interstitial water, the impact of HS on

OP hydrolysis kinetics is predicted to be larger. Hydrolysis reactions of both parathion and DDT are expected, therefore, to be strongly retarded in these environments.

An additional, indirect mechanism that may operate in certain reactions is inhibition by HA of hydrolytic enzymes in soils.[125,126]

In conclusion, a general model to predict the effects of HS on the hydrolysis rates of OP does not seem to be available yet, the proposed ones appearing oversimplified or incomplete.

VII. PHOTOSENSITIZATION

Sunlight-induced photochemical transformations are important pathways in the abiotic degradation of OP in the top layer of soil and waters. Photoreactions may modify the physical and chemical properties of pollutants and significantly affect their fate and behavior in the bulk soil.[127] Light-induced transformations can be distinguished in direct photolysis processes, initiated by direct absorption of light by the compound, and indirect or sensitized photolysis, involving light absorption by natural "photosensitizers" or producers of photoreactants.

Humic substances can strongly absorb sunlight and may accelerate or induce by photosensitization a number of photoreactions that some OP undergo in the presence of artificial sensitizers such as riboflavin, methylene blue, rose bengal, acetone, benzophenone, etc., the occurrence of which is unlikely to occur under natural environmental conditions. Humic substances may thus behave as initiators of photoreactions, some of which involve reactive secondary products that have important implications in accelerating, increasing, or even determining light-induced transformation of nonabsorbing, photochemically stable OP.

Humic substances principally act as sensitizers or precursors for the production of highly reactive, short-lived species (so-called "photoreactants") such as the solvated electron, e^-_{aq}, singlet oxygen, $^1O_2$, superoxide anion, $O_2^{-\cdot}$, peroxy radicals, $RO_2\cdot$, hydrogen peroxide, H_2O_2, and redox-active species, including the excited state of the humic molecule and humic organic radicals. However, HS can also function as a scavenger of other phototransients such as hydroxy radicals, HO·.[128,129]

The formation of e^-_{aq} and humic-derived cation radicals in irradiated solutions of HS is suggested to result from photoejection of an electron from the excited states $^1HS^*$ or $^3HS^*$, according to:

$$^1HS \xrightarrow{\text{light}} {}^1HS^* \longrightarrow {}^3HS^* \tag{11}$$

$$^1HS^* \text{ or } {}^3HS^* \longleftrightarrow \left[HS^{+\cdot} + e^-\right] \underset{-H_2O}{\overset{+H_2O}{\rightleftharpoons}} HS^{+\cdot} + e^-_{aq} \tag{12}$$

The solvated electron is a powerful reductant that may react rapidly with electronegative xenobiotics such as chlorinated organics, e.g., dioxins, that are dehalogenated.[128,129]

The dominant reaction of solvated electrons is however, with O_2, to form $O_2^{-\cdot}$ (superoxide ion Equation 13) which, in turn, may disproportionate and react with water to form hydrogen peroxide (Equation 14). Excited (triplet-state) HS ($^3HS^*$ (Equation 10) can transfer energy directly to ground-state oxygen ($^3O_2$) to form singlet oxygen ($^1O_2$), which, in turn, may generate HA cation radicals and superoxide ion (Equation 15). The latter two species can also be formed directly by reaction of $^3HS^*$ with $^3O_2$. Photochemically derived HS cation radicals may generate organoperoxy cation radicals by reaction with $^3O_2$ (Equation 16).

$$e^-_{aq} + {}^3O_2 \longrightarrow O_2^- \cdot \tag{13}$$

$$2\,O_2^- \cdot \longrightarrow O_2 + O_2^{2-} \longrightarrow O_2 + H_2O_2 + 2OH \tag{14}$$

$$^3HS* + {}^3O_2 \longrightarrow {}^1HS* + {}^1O_2 \longrightarrow HS^+ \cdot + O_2^- \cdot \tag{15}$$

$$HSO_2^+ \cdot + {}^3O_2 \longrightarrow HSO_2^+ \cdot \tag{16}$$

Singlet oxygen and superoxide radicals are efficient, but selective photoreactants for the transformation of various OP. The occurrence of singlet oxygen is suggested to be of importance for the elimination of dissociated forms of phenolic compounds (e.g., chlorinated phenols), for cyclic dienes, and for sulfur compounds.[130] For example, HS are able to catalyze the photooxidation to the corresponding sulfoxide of the sulfide group of the thioether insecticides disulfoton, fenthion, methiocarb, and butocarboxim at the soil surface, but not of the sulfur group in methylthiotriazenes, thiocarbamates, or dithiolane insecticides.[127] Quantitative kinetic data show that photosensitized oxygenations of 2,5-dimethylfuran and the insecticide disulfoton in air-saturated natural water samples containing HS are at least one order of magnitude faster than those in distilled water.[93,131] The same photooxygenated products are obtained in both natural water samples and solutions of soil-derived HS. Furthermore, HS-derived peroxy radicals are important photooxidants for alkyphenols.[129] Up to half the triplet states of HS are estimated to have energies sufficiently high to photosensitize reactions of various OP, including polycyclic aromatic hydrocarbons, nitroaromatic compounds, polyenes, and diketones.[131]

The photoexcited humic parent macromolecule can also undergo direct photochemical reactions with some OP, thanks to the presence of conjugated structures such as keto and quinone groups. The three major possible pathways are energy transfer (or photosensitization), charge transfer, and photoincorporation.[128] The first pathway (see Equation 17 below) is a type of indirect photolysis, with HS in a photoexcited triplet state acting as sensitizer, that transfers energy to previously bound acceptor molecules and so catalyzes the degradation of OP that have low excited-state energies but cannot absorb sunlight themselves.[128] DDT is such a compound. The process is reversible, that is, HS can also act as an acceptor of energy from the excited OP, thus quenching the photodegradation of the latter.

$$\begin{array}{l} ^3\text{HS donor}^* + \text{OP acceptor} \rightleftharpoons (\text{HS}---\text{OP})^* \rightleftharpoons \text{HS} + \text{OP}^* \\ \qquad\qquad\qquad\qquad\qquad\qquad\qquad\qquad\qquad\quad \downarrow \\ \qquad\qquad\qquad\qquad\qquad\qquad\qquad\qquad\quad \text{OP photoproducts} \end{array} \tag{17}$$

Charge transfer, i.e., photoinduced electron transfer reactions, may occur from irradiated HS systems to polyaromatic electron acceptors such as PAH and paraquat.[50,128] HS can also photosensitize reactions involving hydrogen atom transfer, which are likely to involve triplet-state intermediates.[131] For example, HA are able to photosensitize reactions involving hydrogen transfer from the nitrogen of aniline to the sensitizer at much higher rates than those observed in the aniline photoreaction in distilled water.[93] In both photoreacting systems, azobenzene is a major product resulting from coupling of the anilino-free radical intermediates.[93]

ESR studies have suggested that visible and UV light irradiation of HS may enhance the indigenous free radical contents of HS,[132-134] which are highly susceptible to free radical-mediated interaction with OP. The ESR-monitored free radical increase observed in many donor-acceptor systems, such as HA-s-triazine and HA-urea herbicides, may be imputable to the unpairing of electrons originating from the formation of charge-transfer complexes under the effect of light.[68,69,76,135]

For example, photolysis of atrazine in water is more extensive, although initially retarded, in the presence of 0.01% dissolved FA (Equation 18).[136] Fulvic acid is able, in fact, to photosensitize the further degradation of the initially formed 2-hydroxy-atrazine to N-dealkylated products (Equation 18). UV-irradiation (254 nm) experiments conducted with prometryn show a first-order reaction in distilled water and in HA suspension at pH 3, while second-order reaction rate kinetics are observed in the presence of dissolved HA at pH 6 and 8 and FA at pH 3.6 and 8.[137] An additional, phytotoxic dealkylated product, 4-amino-6-(isopropylamino)-1-triazine, is detected when the photolysis of prometryn is performed in aqueous solutions of HA or FA.[137]

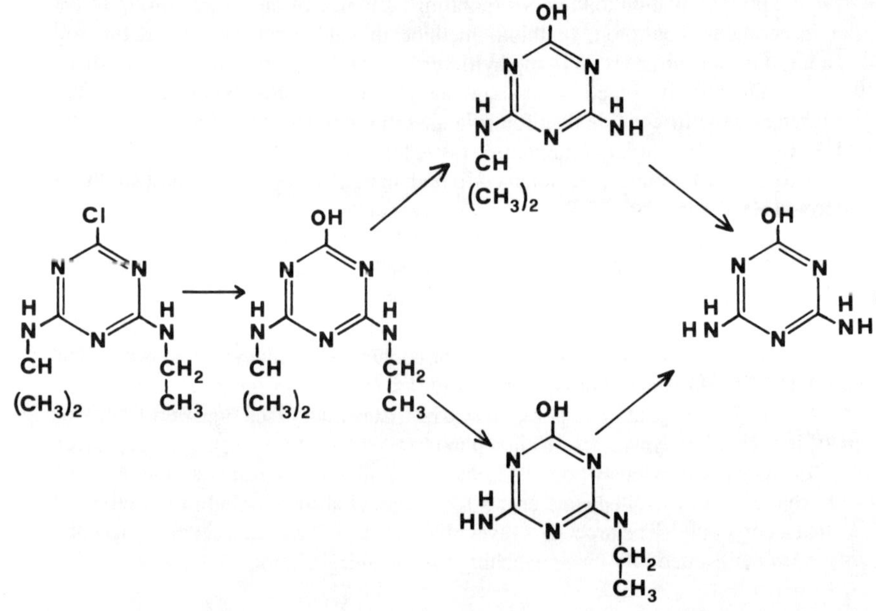

(18)

First-order kinetics are measured for the photolysis of TPB and its decomposition product diphenylborinic acid (DPBA) in the presence of HA and FA.[138] The sensitized-enhanced reaction rates for TPB are lower than for DPBA, with HA having a somewhat greater effect. Dissolved oxygen might compete with the organoborates in energy transfer reactions with the excited HS and thus act as a quenching agent.

Solutions of a commercial HA are significantly less active than natural HS in the photosensitization of 3,4-dichloroaniline, a metabolite of several important herbicides.[139] This is probably due to trapping of the reactive 3,4-DCA intermediates by the HA.

Direct photoincorporation of OP such as polychlorobenzenes into the humic macromolecule may also occur through radical combination or cycloaddition.[128]

VIII. SUMMARY AND CONCLUSIONS

Organic pollutants may interact with HS in the solid and dissolved phases in several ways; thus, interactions may influence to various extents the behavior and fate of the pollutants in soil. In particular, adsorption processes directly or indirectly control all the other processes in soil that affect OP by determining how much of the OP is solubilized and moves into the aqueous and gaseous phase, degrades, or is consumed by organisms. Humic substances can also catalyze some chemical degradation reactions such as hydrolysis, and influence photodegradation of OP. These processes may lead to the formation of reaction intermediates and products having physical and chemical properties different from the parent compound and behaving differently in soil.

Rarely, a unique process is involved in the interaction. More often, several processes with different mechanisms may occur, one or a few of which may dominate for a given pollutant under given conditions. The type and extent of interaction may change with time and may ultimately result in irreversible modification of the chemical and biological properties of the pollutant, and its speciation and partition in the soil-water-organism system. This may possibly lead to the formation of intermediates and products of degradation having a mobility, toxicity, and persistence which can be different from those of the parent molecule and those of products obtained in the absence of HS.

Complete immobilization of the pollutant may occur, with stable incorporation in the humic polymer or, alternatively, the pollutant can attach itself reversibly to organic soil fractions and eventually migrate with these, constituting a potential time-delayed source of contamination by a time-delayed release of toxic units. Experimental evidence of at least partial remobilization of pesticide residues has been obtained for dichloroaniline derivatives of propanil, intact methoxychlor, methylparathion, dinitroanilines, and methabenzthiazuron. However, the question remains unresolved and needs further research. Any simplistic representation of the interaction, such as depicting it as a partitioning, may therefore not be indicative of the phenomena actually involved in the interaction of most OP with SOM.

Further, adsorption of pesticides onto soil HS gives rise to a problem in the analytical (qualitative and quantitative) determination of pesticide residues in soil and water. Thus, it appears necessary to develop new procedures and methods which take into consideration these aspects and may bring about a solution.

New methods employing enzyme-catalyzed polymerization of OP into HS may be extremely useful for removing or minimizing pollutant concentrations in aquatic and terrestrial environments. As well, HS may feasibly act as naturally occurring photosensitizers to induce or accelerate a number of photodegradation processes which OP can undergo in the presence of added artificial sensitizers.

A better knowledge of the chemical nature and reactivity of HS, which are the major natural compounds interacting with OP in soil, and a better understanding of the mechanisms of their interactions with OP is expected to help in the development of a quantitative thermodynamic description and kinetic modeling of pollutant behavior and fate in soils. Advanced techniques that are available and show promise for the molecular and mechanistic investigation of these interactions include Fourier-transform IR, nuclear magnetic resonance, ESR and fluorescence spectroscopies, and differential scanning calorimetry.

Joint interdisciplinary research efforts thus appear necessary for the establishment of predictive measures to face and minimize pesticide pollution problems for soil, water, organisms, and the global environment.

REFERENCES

1. **Stevenson, F. J.,** *Humus Chemistry: Genesis, Composition, Reactions,* John Wiley & Sons, New York, 1982, 443.
2. **Langford, C. H., Gamble, D. S., Underdown, A. W., and Lee, S.,** Interaction of metal ions with a well characterized fulvic acid, in *Aquatic and Terrestrial Humic Materials,* Christman, R. F. and Gjessing, E. T., Eds., Ann Arbor Science, Ann Arbor, MI, 1983, 219.
3. **Schnitzer, M.,** Humic substances chemistry and reactions, in *Soil Organic Matter,* Schnitzer, M. and Khan, S. U., Eds., Elsevier, Amsterdam, 1978, 1.
4. **Hayes, M. H. B. and Swift, R. S.,** The chemistry of soil organic colloids, in *The Chemistry of Soil Constituents,* Greenland, D. J. and Hayes, M. H. B., Eds., John Wiley & Sons, New York, 1978, 179.
5. **Senesi, N.,** Application of electron spin resonance (ESR) spectroscopy in soil chemistry, *Adv. Soil Sci.,* 14, 77, 1990.
6. **Calderbank, A.,** The bipyridilium herbicides, *Adv. Pestic. Control Res.,* 8, 127, 1968.
7. **Haque, R. and Lilley, S.,** Infrared spectroscopic studies of charge-transfer complexes of diquat and paraquat, *J. Agric. Food Chem.,* 20, 57, 1972.
8. **Khan, S. U.,** Humic substances reactions involving bipyridilium herbicides in soil and aquatic environments, *Residue Rev.,* 52, 1, 1974.
9. **Khan, S. U.,** The interaction of organic matter with pesticides, in *Soil Organic Matter,* Schnitzer, M. and Khan, S. U., Eds., Elsevier, Amsterdam, 1978, 137.
10. **Scott, D. C. and Weber, J. B.,** Herbicide phytotoxicity as influenced by adsorption, *Soil Sci.,* 104, 151, 1967.
11. **Weber, J. B.,** Interaction of organic pesticides with particulate matter in aquatic and soil systems, *Adv. Chem. Ser.,* 111, 55, 1972.
12. **Weber, J. B. and Weed, S. B.,** Effect of soil on the biological activity of pesticides, *J. Ser. N. C. State Univ. Agric. Exp. Stn.,* 4087, 223, 1974.
13. **Tucker, B. V., Pack, D. E., and Ospenson, J. N.,** Adsorption of bipyridylium herbicides in soil, *J. Agric. Food Chem.,* 15, 1005, 1967.
14. **Tucker, B. V., Pack, D. E., Ospenson, J. N., Omid, A., and Thomas, W. D., Jr.,** Paraquat soil bonding and plant response, *Weed Sci.,* 17, 448, 1969.
15. **Hayes, M. H. B.,** Adsorption of triazine herbicides on soil organic matter, including a short review on soil organic matter chemistry, *Residue Rev.,* 32, 131, 1970.
16. **Armstrong, D. E., Chesters, G., and Harris, R. F.,** Atrazine hydrolysis in soil, *Soil Sci. Soc. Am. Proc.,* 31, 61, 1967.
17. **Jordan, L. S., Farmer, W. J., Goodin, J. R., and Day, B. E.,** Nonbiological detoxication of the s-triazine herbicides, *Residue Rev.,* 32, 267, 1970.
18. **Burkhard, N. and Guth, J. A.,** Chemical hydrolysis of 2-chloro-4, 6-bis(alkylamino)-1,3,5-triazine herbicides and their breakdown in soil under the influence of adsorption, *Pestic. Sci.,* 12, 45, 1981.
19. **Weed, S. B. and Weber, J. B.,** Pesticide-organic matter interactions, *J. Ser. N. C. State Univ. Agric. Exp. Stn.,* 3840, 39, 1974.
20. **Eliasson, L., Hallman, V., and Tolf, E.,** Leaching of picloram from different soils, *Sver. Skogsvardsforb Tidskr.,* 67, 491, 1969.
21. **Bowman, M. C., Schechter, M. S., and Carter, R. L.,** Behavior of chlorinated insecticides in a broad spectrum of soil types, *J. Agric. Food Chem.,* 13, 360, 1965.
22. **Farmer, W. J., Spencer, W. F., Shepherd, R. A., and Cliath, M. M.,** Effect of flooding and organic matter applications on DDT residues in soil, *J. Environ. Qual.,* 3, 343, 1974.
23. **Lichtenstein, E. P.,** Adsorption of some chlorinated hydrocarbon insecticides from soils into various crops, *J. Agric. Food Chem.,* 7, 430, 1959.
24. **Ballard, T. M.,** Role of humic carrier substances in DDT movement through forest soil, *Soil Sci. Soc. Am. Proc.,* 25, 145, 1971.
25. **Peterson, J. R., Adams, R. S., Jr., and Cutkamp, L. K.,** Soil properties influencing DDT bioactivity, *Soil Sci. Soc. Am. Proc.,* 53, 72, 1971.
26. **Kirk, R. E. and Wilson, M. C.,** The effect of soil type and moisture on germination and growth from wheat seed treated with phorate, *J. Econ. Entomol.,* 53, 771, 1960.
27. **Harris, C. R.,** Influence of soil type on the activity of insecticides in soil, *J. Econ. Entomol.,* 59, 1221, 1966.

28. **Khan, S. U., Hamilton, H. A., and Hague, E. C.,** Fonofos residues in an organic soil and vegetable crops following treatment of the soil with the insecticide, *Pestic. Sci.,* 7, 553, 1976.
29. **Khan, S. U.,** Adsorption of dyfonate (o-ethyl-s-phenyl-ethyl-phosphonodithioate) on humic acid, *Can. J. Soil Sci.,* 57, 9, 1977.
30. **Weber, J. B., Weed, S. B., and Waldrup, T. W.,** Effect of soil constituents on herbicide acidity in modified-soil field plots, *Weed Sci.,* 22, 454, 1974.
31. **Dubey, H. D., Sigafus, R. E., and Freeman, J. F.,** Effect of soil properties on the persistence of linuron and diphenamid in soils, *Agron. J.,* 58, 228, 1966.
32. **Obien, S. R., Suchisa, R. H., and Younge, Or.,** The effects of soil factors on the phytotoxicity of neburon to oats, *Weeds,* 14, 105, 1966.
33. **Douding, R. L. and Freeman, J. F.,** Residual phytotoxicity of fluomethuron in soils, *Weed Sci.,* 16, 226, 1968.
34. **Nash, R. G.,** Plant uptake of ^{14}C-diuron in modified soil, *Agron. J.,* 60, 177, 1968.
35. **Lambert, S. M., Porter, P. E., and Schieferstein, R. H.,** Movement and sorption of chemicals applied to the soil, *Weeds,* 13, 185, 1965.
36. **Hollist, R. L. and Foy, C. L.,** Trifluralin interaction with soil constituents, *Weed Sci.,* 19, 11, 1971.
37. **Menges, R. M. and Hubbard, J. L.,** Phytotoxicity of bensulide and trifluralin in several soils, *Weed Sci.,* 18, 244, 1970.
38. **Weber, J. B.,** Behavior of organic pesticides in soil, *Proc. Soil Sci. N. C.,* 14, 74, 1971.
39. **Ashton, F. M. and Sheets, T. J.,** The relationship of soil adsorption of EPTC to oats injury in various soil types, *Weeds,* 7, 88, 1959.
40. **Fang, S. G., Theisen, P., and Freed, V. H.,** Effect of water evaporation, temperature and rates of application on the retention of ethyl-N-N-di-n-propyl thiocarbamate in various soils, *Weeds,* 9, 569, 1961.
41. **Miller, C. W., Demoranville, I. E., and Charig, A. J.,** Persistence of dichlobenil in nonberry bogs, *Weeds,* 14, 296, 1966.
42. **Koren, E., Foy, C. L., and Ashton, F. M.,** Phytotoxicity and persistence of four thiocarbamates in five soil types, *Weed Sci.,* 16, 172, 1968.
43. **Koren, E., Foy, C. L., and Ashton, F. M.,** Adsorption, volatility and migration of the thiocarbamate herbicides in soil, *Weed Sci.,* 17, 148, 1969.
44. **Briggs, G. G. and Dawson, J. E.,** Hydrolysis of 2,6-dichlorobenzonitrile in soils, *J. Agric. Food Chem.,* 18, 97, 1970.
45. **Jamet, P. and Piedallu, M. A.,** Mouvement du carbofuran dans different types de sols, *Phytiatr.-Phytopharm.,* 24, 279, 1975.
46. **Jamet, P. and Piedallu, M. A.,** Etude de l'adsorption et de la desorption de la pyrazone (amino-5-chloro-4-phenyl-2,2 N pyridazinone-3) par different types de sols, *Weed Res.,* 15, 113, 1975.
47. **Ogner, G. and Schnitzer, M.,** Humic substances: fulvic acid-dialkylphthalate complexes and their role in pollution, *Science,* 170, 317, 1970.
48. **Matsuda, K. and Schnitzer, M.,** Reaction between fulvic acid, a soil humic material, and dialkylphthalates, *Bull. Environ. Contam. Toxicol.,* 6, 200, 1971.
49. **Khan, S. U. and Schnitzer, M.,** The retention of hydrophobic organic compounds by humic acid, *Geochim. Cosmochim. Acta,* 36, 745, 1972.
50. **Senesi, N. and Chen, Y.,** Interactions of toxic organic chemicals with humic substances, in *Toxic Organic Chemicals in Porous Media,* Gerstl, Z., Chen, Y., Mingelgrin, U., and Yaron, B., Eds., Springer-Verlag, Berlin, 1989, 37.
51. **Khan, S. U.,** Equilibrium and kinetic studies of the adsorption of 2,4-D and picloram on humic acid, *Can. J. Soil Sci.,* 53, 429, 1973.
52. **Khan, S. U. and Mazurkewitch, R.,** Adsorption of linuron on humic acid, *Soil Sci.,* 118, 339, 1974.
53. **Maqueda, C., Perez Rodriguez, J. L., Martin, F., and Hermosin, M. C.,** A study of the interaction between chlordimeform and humic acid from a typic chromoxevert soil, *Soil Sci.,* 136, 75, 1983.
54. **Weber, J. B., Weed, S. B., and Best, J. A.,** Displacement of diquat from clay and its phytotoxicity, *J. Agric. Food Chem.,* 17, 1075, 1969.
55. **Weber, J. B., Weed, S. B., and Ward, T. M.,** Adsorption of s-triazines by soil organic matter, *Weed Sci.,* 17, 417, 1969.
56. **Gaillardon, P., Calvet, R., and Tercé, M.,** Adsorption et desorption de la terbutryne par une montmorillonite-Ca et des acides humiques seul on en melanges, *Weed Res.,* 17, 41, 1977.

57. **Li, G. C. and Felbeck, G. T., Jr.,** A study of the mechanism of atrazine adsorption by humic acid from muck soil, *Soil Sci.,* 113, 430, 1972.
58. **Gupta, R. K., Raman, S., and Raman, K. V.,** Adsorption of terbutryne on humic acid, *J. Indian Soc. Soil Sci.,* 33, 255, 1985.
59. **Burns, I. G., Hayes, M. H. B., and Stacey, M.,** Some physico-chemical interactions of paraquat with soil organic materials and model compounds. II. Adsorption and desorption equilibria in aqueous suspension, *Weed Res.,* 13, 79, 1973.
60. **Khan, S. U.,** Adsorption of bipyridilium herbicides by humic acids, *J. Environ. Qual.,* 3, 202, 1974.
61. **Narine, D. R. and Guy, R. D.,** Binding of diquat and paraquat to humic acid in aquatic environments, *Soil Sci.,* 133, 356, 1982.
62. **Senesi, N., Padovano, G., Loffredo, E., and Testini, C.,** Interaction of amitrole, alachlor and cycloate with humic acids, in *Proc 2nd Int. Conf. Environmental Contamination,* CEP Cons., Edinburgh, 1986, 169.
63. **Grice, R. E., Hayes, M. H. B., and Lundie, P. R.,** Adsorption of organophosphorous compounds by soil constituents and by soils, in Proc 7th Br. Insecticide and Fungicide Conf., Vol. 11, 1973, 73.
64. **Sullivan, J. D. and Felbeck, G. T.,** A study of the interaction of s-triazine herbicides with humic acids from three different soils, *Soil Sci.,* 106, 42, 1968.
65. **Turski, R. and Steinbrich, A.,** Studies on the possibilities of binding herbicides of triazine derivatives by humic acids, *Pol. J. Soil Sci.,* 4, 120, 1971.
66. **Carringer, R. D., Weber, J. B., and Monaco, T. J.,** Adsorption-desorption of selected pesticides by organic matter and montmorillonite, *J. Agric. Food Chem.,* 23, 569, 1975.
67. **Senesi, N. and Testini, C.,** Adsorption of some nitrogenated herbicides by soil humic acids, *Soil Sci.,* 10, 314, 1980.
68. **Senesi, N. and Testini, C.,** Physico-chemical investigations of interaction mechanisms between s-triazine herbicides and soil humic acids, *Geoderma,* 28, 129, 1982.
69. **Senesi, N. and Testini, C.,** The environmental fate of herbicides: the role of humic substances, *Ecol. Bull. Stockholm,* 35, 477, 1983.
70. **Kalouskova, N.,** Kinetics and mechanisms of interaction of simazine with humic acids, *J. Environ. Sci. Health B,* 21, 251, 1986.
71. **Senesi, N., Testini, C., and Miano, T. M.,** Interaction mechanisms between humic acids of different origin and nature and electron donor herbicides: a comparative IR and ESR study, *Org. Geochem.,* 11, 25, 1987.
72. **Weber, J. B.,** Spectrophotometrically determined ionisation constants of 1,3-alkylamino-s-triazines and the relationships of molecular structure and basicity, *Spectrochim. Acta,* 23A, 458, 1967.
73. **Weber, J. B.,** Mechanism of adsorption of s-triazines by clay colloids and factors affecting plant availability, *Residue Rev.,* 32, 93, 1970.
74. **Khan, S. U.,** Interaction of humic acid with chlorinated phenoxyacetic and benzoic acids, *Environ. Lett.,* 4, 141, 1973.
75. **Senesi, N., Miano, T. M., and Testini, C.,** Role of humic substances in the environmental chemistry of chlorinated phenoxyalkanoic acids and esters, in *Chemistry for Protection of the Environment 1985,* Pawlowski, L., Alaerts, G., and Lacy, W. J., Eds., Stud. Environ. Sci. No. 29, Elsevier, Amsterdam, 1986, 183.
76. **Senesi, N. and Testini, C.,** Spectroscopic investigations of electron donor-acceptor processes involving organic free radicals in the adsorption of substituted urea herbicides by humic acids, *Pestic. Sci.,* 14, 79, 1983.
77. **Miano, T. M., Piccolo, A., Celano, G., and Senesi, N.,** Infrared and fluorescence spectroscopy of glyphosate-humic acid complexes, *Sci. Total Environ.,* 123, 83, 1992.
78. **Kozak, J.,** Adsorption of prometryn and metholachlor by selected soil organic matter fractions, *Soil Sci.,* 136, 94, 1983.
79. **Gaillardon, P., Calvet, R., and Gaudry, J. C.,** Adsorption de quelques phenylurées herbicides par des acides humiques, *Weed Res.,* 20, 201, 1980.
80. **Khan, S. U.,** Determining the role of humic substances in the fate of pesticides in the environment, *J. Environ. Sci. Health B,* 15, 1071, 1980.
81. **Müller-Wegener, U.,** Electron donor acceptor complexes between organic nitrogen heterocycles and humic acid, *Sci. Total Environ.,* 62, 297, 1987.

82. **Senesi, N.,** Free radicals in electron donor-acceptor reactions between a soil humic acid and photosynthesis inhibitor herbicides, *Z. Pflanzenernaehr. Bodenkd.*, 144, 580, 1981.
83. **Melcer, M. E., Zalewski, M. S., Hassett, J. P., and Brisk, M. A.,** Charge-transfer interaction between dissolved humic materials and chloranil, in *Aquatic Humic Substances. Influence on Fate and Treatment of Pollutants,* Suffet, I. H. and MacCarthy, P., Eds., Adv. Chem. Ser. No. 219, American Chemical Society, Washington, D.C., 1989, 173.
84. **Miller, L. L. and Narang, R. S.,** Induced photolysis of DDT, *Science,* 169, 368, 1970.
85. **Hsu, T. S. and Bartha, R.,** Hydrolysable and nonhydrolysable 2,4-dichloroaniline-humus complexes and their respective rates of biodegradation, *J. Agric. Food Chem.*, 24, 118, 1976.
86. **Adrian, P., Lahaniatis, E. S., Andreux, F., Mansour, M., Scheunert, I., and Korte, F.,** Reaction of the soil pollutant 4-chloroaniline with humic acid monomer catechol, *Chemosphere,* 18, 1599, 1989.
87. **Völkel, W., Choné, T., Mansour, M., and Andreux, F.,** Dégradation dans un sol brun de culture de 3,4-dichloroaniline libre ou liée à un monomère de type humique, in *C. R. XXIe Congrès Groupe Francais des Pesticides 1991,* Ecole Nationale Superieure d'Agronomie et des Industries Alimentaires–Institut Nationale Polytechnique de Lorraine, Nancy, 1991, 355.
88. **Parris, G. E.,** Covalent binding of aromatic amines to humates. I. Reactions with carbonyls and quinones, *Environ. Sci. Technol.*, 14, 1099, 1980.
89. **Graveel, J. G., Sommers, L. E., and Nelson, D. W.,** Sites of benzidine, α-naphthylamine and p-toluidine retention in soils, *Environ. Toxicol. Chem.*, 4, 607, 1985.
90. **Senesi, N., Testini, C., and Metta, D.,** Binding of chlorophenoxyalkanoic herbicides from aqueous solution by soil humic acids, *Proc. Int. Conf. "Environmental Contamination" 1984,* CEP Cons., Edinburgh, 1984, 96.
91. **Bollag, J. M., Liu, S. Y., and Minard, R. D.,** Cross-coupling of phenolic humus constituents and 2,4-dichlorophenol, *Soil Sci. Soc. Am. J.*, 44, 52, 1980.
92. **Crosby, D. G.,** Herbicide photodecomposition, in *Herbicides: Chemistry, Degradation and Mode of Action,* Vol. 2, Kearney, P. C. and Kaufmann, D. D., Eds., Marcel Dekker, New York, 1976, chap. 18.
93. **Zepp, R. G., Baughman, G. L., and Schlotzhauer, P. F.,** Comparison of photochemical behavior of various humic substances in water: sunlight-induced reactions of aquatic pollutants photosensitized by humic substances, *Chemosphere,* 10, 109, 1981.
94. **Senesi, N., Miano, T. M., and Testini, C.,** Incorporation of water dissolved chlorophenoxyalkanoic herbicides by humic acids of various origin and nature, in *Current Perspectives in Environmental Biogeochemistry,* Giovannozzi-Sermanni, G. and Nannipieri, P., Eds., Consiglio Nazionale delle Ricerche–Increments Produttirita Risorse Agricole, Rome, 1987, 295.
95. **Bollag, J. M., Minard, R. D., and Liu, S. Y.,** Cross-linkage between anilines and phenolic humus constituents, *Environ. Sci. Technol.*, 17, 72, 1983.
96. **Berry, D. F. and Boyd, S. A.,** Reaction rates between phenolic humus constituents and anilines during cross-coupling, *Soil Biol. Biochem.*, 17, 72, 1985.
97. **Stott, D. E., Martin, J. P., Focht, D. D., and Haider, K.,** Biodegradation, stabilization in humus and incorporation into soil biomass of 2,4-D and chlorocatechol carbons, *Soil Sci. Soc. Am. J.*, 47, 66, 1983.
98. **Bollag, J. M. and Loll, M. J.,** Incorporation of xenobiotics into soil humus, *Experientia,* 39, 1221, 1983.
99. **Nearpass, D. C.,** Adsorption of picloram by humic acids and humin, *Soil Sci.*, 121, 272, 1976.
100. **Burns, I. G., Hayes, M. H. B., and Stacey, M.,** Spectroscopic studies on the mechanisms of adsorption of paraquat by humic acid and model compounds, *Pestic. Sci.*, 4, 201, 1973.
101. **Leenheer, J. A. and Aldrichs, J. L.,** A kinetic and equilibrium study of the adsorption of carbaryl and parathion upon soil organic matter surfaces, *Soil Sci. Soc. Am. Proc.*, 35, 700, 1971.
102. **Pierce, R. H., Olney, C. E., and Felbeck, G. T.,** Pesticide adsorption in soils and sediments, *Environ. Lett.*, 1, 157, 1971.
103. **Briggs, G. G.,** Molecular structure of herbicides and their sorption by soils, *Nature (London),* 223, 1288, 1969.
104. **Strek, H. J. and Weber, J. B.,** Behavior of polychlorinated biphenyls (PCBs) in soils and plants, *Environ. Pollut. A*, 28, 291, 1982.

105. **Walker, A. and Crawford, D. V.,** The role of organic matter in adsorption of the triazine herbicides by soil, in *Isotopes and Radiation in Soil Organic Matter Studies,* Food and Agriculture Organization/International Atomic Energy Agency, Eds., International Atomic Energy Agency, Vienna, 1968, 91.
106. **Hance, R. J.,** Observations on the relationship between the adsorption of diuron and the nature of the adsorbent, *Weed Res.,* 5, 108, 1965.
107. **Chiou, C. T., Porter, P. E., and Schmedding, D. W.,** Partition equilibria of nonionic organic compounds between soil organic matter and water, *Environ. Sci. Technol.,* 17, 227, 1983.
108. **Mingelgrin, U. and Gerstl, Z.,** Reevaluation of partitioning as a mechanism of nonionic chemicals adsorption in soils, *J. Environ. Qual.,* 12, 1, 1983.
109. **Davis, W. M. and Delfino, J. J.,** Interaction among hydrophobic organic pollutants and humic substances, *Am. Chem. Soc. Div. Environ. Chem.,* 32, 7, 1992.
110. **Chin, Y.-P. and Weber, J. W., Jr.,** Estimating the effect of dispersed organic polymers on the sorption of contaminants by natural solids. I. A predictive thermodynamic humic substance-organic solute interaction model, *Environ. Sci. Technol.,* 23, 978, 1989.
111. **Wershaw, R. L., Burcar, P. J., and Goldberg, M. C.,** Interaction of pesticides with natural organic matter, *Environ. Sci. Technol.,* 3, 271, 1969.
112. **Boehm, P. D. and Quinn, J.,** Solubilization of hydrocarbons by the dissolved organic matter in seawater, *Geochim. Cosmochim. Acta,* 37, 2459, 1973.
113. **Chiou, C. T., Malcolm, R. L., Brinton, T. I., and Kile, D. E.,** Water solubility enhancement of some organic pollutants and pesticides by dissolved humic and fulvic acids, *Environ. Sci. Technol.,* 20, 502, 1986.
114. **Carter, C. W. and Suffet, I. H.,** Binding of DDT to dissolved humic materials, *Environ. Sci. Technol.,* 16, 735, 1982.
115. **Landrum, P. F., Nihart, S. R., Eadie, B. J., and Gardner, W. S.,** Reverse-phase separation method for determining pollutant binding to Aldrich humic acid and dissolved organic carbon of natural waters, *Environ. Sci. Technol.,* 18, 187, 1984.
116. **Shinozouka, N., Lee, C., and Hayano, S.,** Solubilizing action of humic acid from marine sediment, *Sci. Total Environ.,* 62, 311, 1987.
117. **Kile, D. E. and Chiou, C. T.,** Water-solubility enhancement of nonionic organic contaminants, in *Aquatic Humic Substances. Influence on Fate and Treatment of Pollutants,* Suffet, I. H. and MacCarthy, P., Eds., Adv. Chem. Ser. No. 219, American Chemical Society, Washington, D.C., 1989, 131.
118. **Caron, G., Suffet, I. H., and Belton, T.,** Effect of dissolved organic carbon on the environmental distribution of non polar organic compounds, *Chemosphere,* 14, 993, 1985.
119. **Struif, B., Weil, L., and Quentin, K. E.,** The behavior of herbicides, phenoxy acetic acids and their esters in waters, *Wom. Wasser,* 45, 53, 1975.
120. **Khan, S. U.,** Kinetics of hydrolysis of atrazine in aqueous fulvic acid solution, *Pestic. Sci.,* 9, 39, 1978.
121. **Perdue, E. M. and Wolfe, N. L.,** Modification of pollutant hydrolysis kinetics in the presence of humic substances, *Environ. Sci. Technol.,* 16, 847, 1982.
122. **Macalady, D. L., Tratnyek, P. G., and Wolfe, N. L.,** Influences of natural organic matter on the abiotic hydrolysis of organic contaminants in aqueous systems, in *Aquatic Humic Substances. Influence on Fate and Treatment of Pollutants,* Suffet, I. H. and MacCarthy, P., Eds., Adv. Chem. Ser. No. 219, American Chemical Society, Washington, D.C., 1989, 323.
123. **Gamble, D. S. and Khan, S. U.,** Atrazine hydrolysis in soil: catalysis by the acidic functional groups of fulvic acid, *Can. J. Soil Sci.,* 65, 435, 1985.
124. **Perdue, E. M.,** Association of organic pollutants with humic substances: partitioning equilibria and hydrolysis kinetics, in *Aquatic and Terrestrial Humic Materials,* Christman, R. F. and Gjessing, E. T., Eds., Ann Arbor Science, Ann Arbor, MI, 1983, 441.
125. **Malini de Almeida, R., Pospisil, F., Vockova, K., and Kutacek, M.,** Effect of humic acid on the inhibition of pea choline esterase and choline acyltransferase with malathion, *Biol. Plant,* 22, 167, 1980.
126. **Mulvaney, F. L. and Bremner, J. M.,** Use of p-benzoquinone for retardation of urea hydrolysis in soils, *Soil Biol. Biochem.,* 10, 297, 1978.
127. **Miller, G. C., Hebert, V. R., and Miller, W. W.,** Effect of sunlight on organic contaminants at the atmosphere-soil interface, in *Reactions and Movement of Organic Chemicals in Soils,* Sawhney, B. L. and Brown, K., Eds., SSSA Spec. Publ. No. 22, Soil Science Society of America, Madison, WI, 1989, 99.

128. **Cooper, W. J., Zika, R. G., Petasne, R. G., and Fischer, A. M.,** Sunlight-induced photochemistry of humic substances in natural waters: major reactive species, in *Aquatic Humic Substances. Influence on Fate and Treatment of Pollutants,* Suffet, I. H. and MacCarthy, P., Eds., Adv. Chem. Ser. No. 219, American Chemical Society, Washington, D.C., 1989, 333.
129. **Hoigné, J., Faust, B. C., Haag, W. R., Scully, F. E., Jr., and Zepp, R. G.,** Aquatic humic substances as sources and sinks of photochemically produced transient reactants, in *Aquatic Humic Substances. Influence on Fate and Treatment of Pollutants,* Suffet, I. H. and MacCarthy, P., Eds., Adv. Chem. Ser. No. 219, American Chemical Society, Washington, D.C., 1989, 363.
130. **Hoigné, J., Bader, H., and Nowell, L. H.,** Rate constants of OH radical scavenging by humic substances: role in ozonation and in a few photochemical processes for the elimination of micropollutants, *Am. Chem. Soc. Div. Environ. Chem.,* 27(1), 208, 1987.
131. **Zepp, R. G., Schlotzhauer, P. F., and Sink, R. M.,** Photosensitized transformations involving electronic energy transfer in natural waters: role of humic substances, *Environ. Sci. Technol.,* 19, 74, 1985.
132. **Senesi, N. and Schnitzer, M.,** Effect of pH, reaction time, chemical reduction and irradiation on ESR spectra of fulvic acid, *Soil Sci.,* 123, 224, 1977.
133. **Senesi, N. and Schnitzer, M.,** Free radicals in humic substances, in *Environmental Biogeochemistry and Geomicrobiology, Vol. II, The Terrestrial Environment,* Krumbein, W. E., Ed., Ann Arbor Science, Ann Arbor, MI, 1978, 467.
134. **Senesi, N., Chen, Y., and Schnitzer, M.,** The role of free radicals in the oxidation and reduction of fulvic acid, *Soil Biol. Biochem.,* 9, 397, 1977.
135. **Senesi, N. and Testini, C.,** Theoretical aspects and experimental evidence of the capacity of humic substances to bind herbicides by charge-transfer mechanisms (electron donor-acceptor processes), *Chemosphere,* 13, 461, 1984.
136. **Khan, S. U. and Schnitzer, M.,** UV irradiation of atrazine in aqueous fulvic acid solutions, *J. Environ. Sci. Health,* 3, 299, 1978.
137. **Khan, S. U. and Gamble, D. S.,** Ultraviolet irradiation of an aqueous solution of prometryn in the presence of humic materials, *J. Agric. Food Chem.,* 31, 1099, 1983.
138. **Mill, G. L. and Carter, D. J.,** Photochemical degradation of tetraphenylboron sensitized by dissolved organic matter in natural water, *Am. Chem. Soc. Div. Environ. Chem.,* 27(1), 146, 1987.
139. **Miller, G. C., Zisook, R., and Zepp, R.,** Photolysis of 3,4-dichloroaniline in natural waters, *J. Agric. Food Chem.,* 28, 1053, 1980.

Chapter 25

Adsorption Mechanisms and Abiotic Catalytic Transformations of Some Agrochemicals by Clay Minerals

G. G. Ristori and P. Fusi

CONTENTS

I. INTRODUCTION

The safe and efficient use of an agrochemical that can come into contact with soil, either following foliar treatments or by means of its direct application to soil, requires knowledge of its activity and of the processes that control its persistence in soil. The dynamics of agrochemicals in soil are related to various phenomena, such as immobilization, solubilization, translocation, and chemical and biological transformations of the molecules. Figure 1 shows a schematic representation of the fate of agrochemicals, in general, and of pesticides, in particular, when they come into contact with soil. The scheme clearly shows that the activity of these compounds and their fate are closely related to their concentration in the soil solution. Thus, factors that influence this concentration would also determine the dynamics of an agrochemical in soil. From this point of view, adsorption and desorption, as well as any transformations of the adsorbed molecules, are among the most important processes.

The general principles of the adsorption of agrochemicals by soil, as a whole, are useful in studying the relationships between the physicochemical characteristics of the molecules and their biological activity, persistence, and movement in soil. Data on adsorption by a single soil component (e.g., humic colloids, clay minerals, or oxides) are helpful in describing and explaining the mechanisms of adsorption.

This chapter describes only the importance of the adsorption of agrochemicals on the mineral fraction of soil in their abiotic transformation. It intentionally omits the role of organic matter, and the adsorption processes are not considered from thermodynamic and kinetic points of view.

II. ADSORPTION OF AGROCHEMICALS ON CLAY MINERALS

Clay minerals are capable of adsorbing organic molecules by different mechanisms, which have been described in several excellent articles.[1-6] The adsorbing capacity of clay minerals is the result of various factors, such as the type of exchangeable cations, the high specific surface area (e.g., 600 to 800 $m^2\ g^{-1}$ for montmorillonites), their structure, and their charge localization.

Among these factors, the nature of the exchangeable cations is of major importance, as these cations can not only compete for adsorption sites on clays with positively charged

0-87371-914-X/95/$0.00+$.50

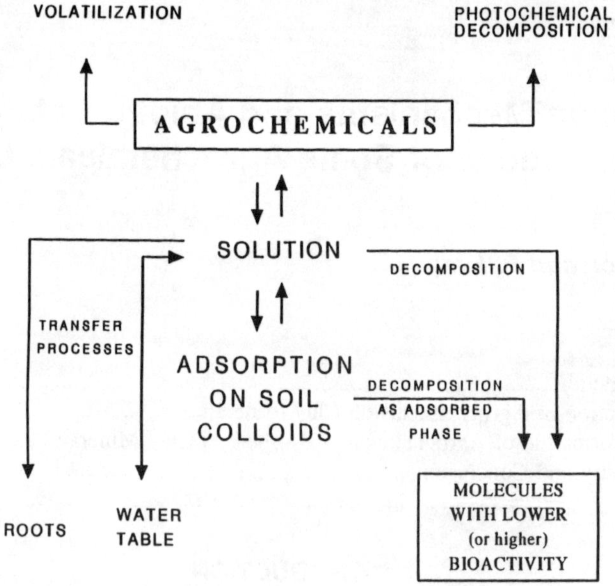

Figure 1 Dynamics of agrochemicals in soils.

organic molecules, but can also function directly as adsorption sites. The cations can also act indirectly by influencing water molecules in their immediate vicinity, as the polarizing power of the cation determines the degree of acidity of the solvation water. Based on spectroscopic techniques, the acidity of the clay mineral surface is greater than the bulk acidity of the same clay in water. Al and Fe cations can form hydroxides at the clay surface, which can result in an increase in the adsorbing capacity of the mineral. Amorphous oxides and hydroxides associated with clays are efficient absorbents, whereas crystallized oxides and hydroxides have a low adsorbing power.[7]

Adsorption is also affected by the chemical and physical properties of the molecules involved. The adsorption of agrochemicals is most often studied in water, and therefore interactions between an organic molecule and water are important in adsorption. Although adsorption from water simulates natural conditions, organic solvents are often employed. One of the advantages of using organic solvents is that adsorption is generally higher than in an aqueous medium, as organic molecules are adsorbed less from an aqueous environment because of the competition for adsorption sites with water molecules. In weakly polar organic solvents, which have little affinity for adsorption sites, there is little competition for these sites.

Nonaqueous solvents are also used when studying the nature of the bonds of agrochemical-clay complexes by IR spectroscopy, to minimize interferences by water. However, the chemistry of clay surfaces in water is different from that in a nonaqueous medium, and therefore the observations made in an organic medium cannot always be extrapolated to an aqueous system.[3]

The main types of binding mechanisms that occur at clay mineral surfaces can be summarized as follows.

1. Ionic bonds are established between a cationic organic molecule and the negative charge on a clay mineral. Examples of such molecules are the herbicides diquat, paraquat, protonated s-triazines, and amitrole.
2. Coordination bonds and ion-dipole bonds occur when exchangeable cations at the clay surface interact with nonionic polar molecules by coordination or ion-dipole bonding.

3. Hydrogen bonds are established between the water of hydration of exchangeable cations and polar molecules by means of a water bridge. Hydrogen bonds can also be established between the –NH and –OH groups or the N and O atoms of organic molecules and the -OH groups of clay minerals.
4. Charge transfer associations can result from electron transfer from the surface of the adsorbing clay mineral to the aromatic nucleus of the adsorbed organic molecule. This type of interaction has been demonstrated for the adsorption of dipyridilic cations, which, as indicated above, can also be adsorbed by ionic bonds.[8]
5. Protonation is the transfer of H^+ from water coordinated with exchangeable cations to H^+-acceptor groups or atoms on organic molecules.

III. TRANSFORMATION OF AGROCHEMICALS ADSORBED ON CLAY MINERALS

Organic molecules adsorbed on the surface of clay minerals undergo various types of reactions that can be subdivided as follows:

1. Reactions that lead to the formation of colored complexes by oxidation-reduction or acid-base interactions. An example is the behavior of benzidine, a colorless molecule that, when adsorbed on kaolinite or montmorillonite, is transformed into the corresponding yellow or blue cationic radical form.[9-11]
2. Polymerization reactions. Clay minerals catalyze the oxidative polymerization of phenolic compounds, such as pyrogallol, and can have an important role in the abiotic formation of humic substances.[12]
3. Degradation reactions. Clay minerals can activate organic reactions that lead to the transformation and/or degradation of adsorbed organic molecules.[13] A classic example of a degradation reaction catalyzed by clays is petroleum cracking. These reactions can be mediated by Lewis or Bronsted acids. Lewis acidic sites (electron acceptors) of clay minerals are Fe^{3+} and Al^{3+} ions located within the octahedral sheet and on the edges of clays, as well as exchangeable transition metal cations. The Bronsted acidity results from H^+ at exchange sites or from H^+ derived from the dissociation of water coordinated to exchangeable cations. The Bronsted type of acidity is responsible for most of the degradation reactions of agrochemicals adsorbed on clay minerals.

An interesting case of a hydrolytic degradation is shown by the herbicide asulam (4-aminobenzenesulfonylmethylcarbamate), which, when adsorbed on a homoionic Wyoming montmorillonite by cation exchange after protonation (regardless of the interlayer cation), is degraded under moderate heating (75 to 100°C) into various compounds. The nature of the compounds depends on the polarizing power of the cation.[14-16] With cations of high polarizing power, such as Al^{3+}, Cu^{2+}, and Mg^{2+}, asulam degrades into sulfanilic acid and carbamic acid. When the polarizing power of the cation is lower (e.g., Ba^{2+} and Ca^{2+}), the main degradation product is sulfanilamide, plus some sulfanilic acid and carbamic acid. With metals having a low polarizing power (e.g., K^+ and Na^+), sulfanilamide and the anion, 4-aminobenzenesulfonilcarbamate, are formed. With Li-montmorillonite, 4-aminobenzenesulfonilcarbamic acid and 4-aminobenzenesulfonilcarbamate, in addition to traces of sulfanilamide, were obtained, as the polarizing power of Li^+ is intermediate between that of other alkaline cations and alkaline earth elements.

The formation of sulfanilamide, instead of the sulfanilic and carbamic acids, was the result of the lack of a coordination between the C=O of asulam and the interlayer cation, especially with the alkaline metals. With Al^{3+} and Mg^{2+}, coordination was established on heating, with the consequent weakening and breakage of the S-N bond, which resulted in the formation of sulfanilic and carbamic acids. These results demonstrated that the increased "acidity" of the residual water, after heating, was responsible for the degradation of asulam. Although conditions of accentuated dehydration can occur in soil during periods of drought, the degradation of asulam can probably also occur under less drastic conditions, but over a longer period than the one employed in these experiments. Another consequence of heating asulam adsorbed on montmorillonite saturated with transition metals is the formation of two oxidation products, 4-nitrobenzenesulfonic and 4-nitrosobenzenesulfonic acids, in addition to sulfanilic and carbamic acids.[16]

The nature of the degradation products of asulam is also a function of the localization of isomorphous substitution in the clay relative to the amount of asulam degradation products formed after heating asulam-Mg-smectite complexes at 90°C (Table 1).[17] The data illustrate that the polarizing power of the Mg^{2+} ion (and, therefore, the acidity of the solvation water) depends on the localization of the layer charge derived from isomorphous substitution. In smectites with tetrahedral substitution, the negative charge is closer to the interlayer space, with the consequent reduction of the polarizing power of the cation. Therefore, asulam will be only weakly coordinated, and sulfanilamide is the principal degradation product. In contrast, in smectites with octahedral substitution, the influence of the layer charge on the cation is reduced, and the polarizing power of Mg^{2+} is essentially unaltered. Consequently, a coordination can be established with the C=O group of asulam, resulting in the subsequent degradation of the herbicide into sulfanilic and carbamic acids. Carbamic acid is unstable and is further degraded to CO_2 and NH_3.

The substitution of the amino group of asulam with a $-NO_2$ functional group results in a new molecule, nisulam, which retains herbicidal properties. nisulam is coordinated to the exchangeable cations of smectite by means of the $-NO_2$ group instead of the C=O group, because of the inductive and conjugative effects of the $-NO_2$ group. Therefore, nisulam, after heating, is degraded to 4-nitrobenzenesulfonamide, regardless of the nature of the interlayer cation and the characteristics of the clay mineral.[18]

Another example of hydrolytic degradation is shown by the fungicide fenarimol, which is adsorbed on montmorillonite by Van der Waals forces and hydrogen bonds involving the -OH groups and/or the nitrogens of the pyrimidine ring.[19] As a result of these weak interactions, the nature of the exchangeable cation does not appear to have a significant role in adsorption. When the complex is heated at 100°C for 8 h, the corresponding carbonium ion is formed as a result of a proton interaction between the solvation water of the exchangeable cation and the –OH groups of the molecule. If the complex is heated for 38 h, the molecule breaks at the level of the pyrimidine ring, with the consequent formation of a yellow dialdehyde. The weakness of the pyrimidine ring may be the result of the delocalization of the positive charge of the carbonium ion in positions 2, 4, and 6 of the pyrimidine ring, with a resulting weakening of the –C–N–C– bonds.

Another example of degradation is shown by the adsorption of the insecticide carbaryl on Al- and Cu-montmorillonite.[20] When the complexes are heated at 90°C for 30 h, the molecule is degraded into CO_2, *N*-methylammonium, 1-naphthol, and several polymerization products. However, when the molecule is adsorbed on Ca- or Na-montmorillonite, no degradation is observed, indicating that the greater polarizing power of Cu^{2+} and Al^{3+} than that of Ca^{2+} and Na^+ is involved in the degradation of carbaryl.

Nitrogen fertilizers also undergo catalytic degradation as a result of adsorption on smectite minerals. Mortland reported that the type of exchangeable cations on smectites influenced the degradation of urea.[21] Adsorbed oxamide ($NH_2COCONH_2$) was decomposed on all homoionic (Ca, Al, and Cu) smectite systems studied, although to different extents and through different mechanisms[22] (Table 2). Cu(II)-montmorillonite was very effective in degrading oxamide, mainly by the formation of coordination complexes between the N-H groups and Cu. These complexes are easily hydrolyzed by water. On Al-montmorillonite, the C=O group of adsorbed oxamide was protonated, and the molecule was degraded to carbamic acid, which is stable on a dry clay surface but

Table 1 Relative Amounts of Degradation Products of *Asulam* Formed After Heating Mg-Clay-*Asulam* Complexes at 90°C (Expressed as a Percent of Total Decomposition Products)[17]

Clay	Octahedral ions				Tetrahedral ions		Layer charge	Sulfanilic acid	Sulfanil-amide	4-Amino-Benzene sulfanilcarbamate
	Al	Fe	Mg	Li	Si	Al				
	(number of cations per half-unit cell)						(X^+)	(%)	(%)	(%)
Mont[a] Upton	1.52	0.16	0.33	—	3.91	0.09	0.39	77	15	8
Hectorite	0.02	—	2.70	0.24	3.98	0.02	0.32	71	21	8
Mont Chambers	1.42	0.17	0.40	—	3.90	0.10	0.53	70	20	10
Mont Clay Spur	1.52	0.16	0.34	—	3.88	0.12	0.40	69	21	10
Mont Otay	1.35	0.05	0.62	—	4.00	—	0.56	55	36	9
Mont Caffaro	1.38	0.55	0.13	—	3.55	0.45	0.40	5	65	30
Beidellite	1.12	0.42	0.71	—	3.48	0.52	0.48	25	60	15
Saponite	0.47	—	2.41	—	3.32	0.68	0.45	38	47	15
Nontronite	0.07	1.91	0.03	—	3.50	0.50	0.50	26	54	20

[a] Montmorillonite.

Table 2 Oxamide and Degradation Products Recovered After Wetting and Drying Treatments (% of Original Oxamide Content)[22]

Mont[a]-oxamide complexes and original oxamide content (μmol g^{-1})		Oxamide (%)	Oxamic acid (%)	Oxalic acid (%)	Decomposed fraction (%)
Cu-mont	NT[b]	25	0	1	74
(64.4)	2C[c]	7	2	3	88
	10C	2	5	3	90
Al-mont	NT	39	0	1	60
(59.0)	2C	34	1	3	62
	10C	30	1	2	67
Ca-mont	NT	71	11	4	14
(48.6)	2C	61	11	9	19
	10C	45	10	15	30

[a] Montmorillonite.
[b] No treatment.
[c] Wetting and drying cycle (100% RH, 80°C).

to CO_2 and NH_3 at high moisture levels. Ca-montmorillonite was less effective in degrading oxamide, and significant amounts of stable intermediate compounds, such as oxamic and oxalic acids, were progressively formed during wetting and drying cycles.

The foregoing discussion illustrates the importance of interactions between agrochemicals and mineral soil colloids in the transformation of the chemicals. Furthermore, the type of bond with which the organic molecules are adsorbed influences biodegradability, volatility, solubility, and biocide activity. Moreover, it must be emphasized that organic colloids are probably as important as clay minerals in the transformation of chemicals in soil. Clay-organo complexes, such as clay-humus complexes, can adsorb organic molecules that may not be adsorbed on pure clay minerals. For example, several chlorobenzene derivatives, which are adsorbed in only small amounts on clay minerals, are adsorbed in large quantities on montmorillonites on which quaternary alkylammonium cations, which give the clay surface hydrophobic rather than hydrophylic properties, have been adsorbed.[23] The greater the hydrophobic character of the cation saturating the clay, the greater the adsorption of pentachlorophenol from water.[24]

IV. SUMMARY AND CONCLUSIONS

The nature of the interactions between agrochemical and soil clay minerals depends on the chemical and physical characteristics of the molecules and the surface properties of clay minerals. The following binding mechanisms play an important role in agrochemical adsorption: (1) ionic bonds, (2) coordination bonds and ion-dipole bonds, (3) hydrogen bonds, (4) charge transfer associations, and (5) protonation. The water content of clay minerals is also very important for the adsorption of polar nonionic molecules because of a competition for adsorption sites between water and organic molecules. Another important factor to be considered is the polarizing power of the exchangeable cations of clay minerals, which determines in some cases the degree of acidity of solvation water. This allows the basic organic molecules to be protonated. The catalytic reactions occurring on clay surfaces substantially affect the persistence of agrochemicals in soils.

A better knowledge of the interactions between mineral and organic colloids of soil and agrochemicals or even of simpler organic molecules is necessary. Even if the latter molecules or their transformation products do not have biocidal activity, they can be useful as models for studying chemically similar agrochemicals, especially in relation to proper agronomic use and minimal environmental pollution by agrochemicals and their degradation products.

REFERENCES

1. **Greenland, D. J.,** Interaction between clays and organic compounds in soil. I. Mechanisms of interaction between clays and definite organic compounds, *Soil Fert.,* 28, 521, 1965.
2. **Bailey, G. W. and White, J. L.,** Adsorption of pesticides in soil, *Residue Rev.,* 32, 29, 1970.
3. **Mortland, M. M.,** Clay-organic complexes and interactions, *Adv. Agron.,* 22, 75, 1970.
4. **Cloos, P.,** Interaction entre les pesticides et la fraction minérale du sol, *Pedologie,* 22, 148, 1972.
5. **Theng, B. K. G.,** *The Chemistry of Clay-Organic Reactions,* Adam Hilgher, London, 1974.
6. **Cheng, H. H., Ed.,** *Pesticides in the Environment: Processes, Impacts, and Modeling,* Soil Science Society of America, Madison, WI, 1990.
7. **Terce, M. and Calvet, R.,** Role of aluminum in the adsorption of atrazine by clay minerals, in *Proc. Isr.-Fr. Symp. Behaviour of Pesticides in Soil,* Horowitz, M., Ed., Bet Dagan, Israel, 1975, 33.
8. **Haque, R., Lilley, S., and Coshow, W. R.,** Mechanism of adsorption of diquat and paraquat on montmorillonite surface, *J. Colloid Interface Sci.,* 33, 185, 1970.
9. **Weil-Malherbe, H. and Weiss, J.,** Color reactions and adsorption of some aluminosilicates, *J. Chem. Soc.,* 2164, 1948.
10. **Theng, B. K. G.,** Mechanism of formation of colored clay-organic complexes. A review, *Clays Clay Miner.,* 19, 383, 1971.
11. **Solomon, D. H., Loft, B. C., and Swift, J. D.,** Reaction catalyzed by minerals. IV. The mechanism of the benzidine blue reaction on silicate minerals, *Clay Miner.,* 7, 389, 1968.
12. **Wang, T. S. C., Huang, P. M., Chou, C.-H., and Chen, J.-H.,** The role of soil minerals in abiotic polymerization of phenolic compounds and formation of humic substances, in *Interactions of Soil Minerals with Natural Organics and Microbes,* Huang, P. M. and Schnitzer, M., Eds., Soil Science Society of America, Madison, WI, 1986.
13. **Grim, R. E.,** *Applied Clay Mineralogy,* McGraw-Hill, New York, 1962.
14. **Fusi, P., Ristori, G. G., and Malquori, A.,** Montmorillonite-asulam interactions. I. Catalytic decomposition of asulam adsorbed on H- and Al-clay. *Clay Miner.,* 15, 147, 1980.
15. **Ristori, G. G., Fusi, P., and Franci, M.,** Montmorillonite-asulam interactions. II. Catalytic decomposition of asulam adsorbed on Mg-, Ba-, Ca-, Li-, Na-, K- and Cs-clay. *Clay Miner.,* 16, 125, 1981.
16. **Fusi, P., Ristori, G. G., Franci, M.,** Oxidative processes related to degradation of asulam herbicide adsorbed on smectites, *Ann. Chim.,* 71, 443, 1981.
17. **Ristori, G. G., Fusi, P., and Cecconi, S.,** The influence of structural characteristics of smectites on catalytic decomposition of asulam, in *Proc. 6th Int. Clay Conf.,* Bologna-Pavia, 1981, 262.
18. **Fusi, P., Ristori, G. G., and Franci, M.,** Adsorption and catalytic decomposition of 4-nitrobenzenesulfonylmethylcarbamate by smectite, *Clays Clay Miner.,* 30, 306, 1982.
19. **Fusi, P., Ristori, G. G., Cecconi, S., and Franci, M.,** Adsorption and degradation of fenarimol on montmorillonite, *Clays Clay Miner.,* 31, 312, 1983.
20. **Fusi, P., Ristori, G. G., and Franci, M.,** Interactions of carbaryl with homoionic montmorillonite, *Appl. Clay Sci.,* 1, 375, 1986.
21. **Mortland, M. M.,** Urea complexes with montmorillonite: an infrared adsorption study, *Clays Clay Miner.,* 6, 143, 1966.
22. **Ristori, G. G., De Nobili, M., and Fusi, P.,** Non-biological degradation of oxamide adsorbed on homoionic montmorillonite, *Appl. Clay Sci.,* 6, 171, 1991.
23. **McBride, M., Mortland, M. M., and Pinnavaia, T. J.,** Adsorption of aromatic molecules by clay in aqueous suspension, *Adv. Environ. Sci. Technol.,* 3, 226, 1975.
24. **Boyd, S. A., Shaobai, S., Lee, J. F., and Mortland, M. M.,** Pentachlorophenol sorption by organo-clays, *Clays Clay Miner.,* 36, 125, 1988.

Chapter 26

Interactions Between Manganese Oxides and Multiple-Ringed Aromatic Compounds

G. Whelan, R. C. Sims, and I. P. Murarka

CONTENTS

I. INTRODUCTION

Biochemical and chemical biotic/abiotic oxidation, immobilization, and detoxification of recalcitrant organics, including polynuclear aromatic hydrocarbons (PAHs), may serve as a basis for *in situ* and at-grade treatment of certain organic wastes. Current research has focused on enzyme-catalyzed oxidation of pesticides and single-ringed aromatics.[1-4] For example, Bollag and Myers[5] concluded that biotic polymerization of xenobiotics in a humification process is possible because many of the degradation products of pesticides result in the formation of reactive intermediates with structures and/or functional groups similar to those found in natural humic material. Humic materials are natural organic substances that are common in the environment and are involved in the polymerization process with organic molecules.[6]

Recently, scientists have noted that abiotic-catalyzed polymerization may also represent an important aspect of humification.[7-11] Manganese-bearing silicates have demonstrated catalytic effects in enhancing the polymerization of polyphenols (e.g., hydroquinone).[11] Research is beginning to focus more on how abiotic catalysis affects the oxidation process (e.g., humification/polymerization) of organics and multiple-ringed aromatics.[12-16]

Research has demonstrated that aromatic compounds containing diol and dione functional groups tend to be susceptible to abiotic-catalyzed oxidation and also suggested that an increase in reactivity among diol compounds can be attributed to the ability of these compounds to form a corresponding quinone through a resonance-stabilized semiquinone radical.[17,18] Figure 1 presents a proposed relationship between naphthalenediol, semiquinone radical, naphthoquinone, semiquinone radical anion, and a semiquinone dianion.[19-22]

The objectives of this research were to investigate the nature of redox reactions in a simplified aqueous solution consisting of various multiple-ringed aromatic compounds and MnO_2 and to present a possible mathematical explanation of this abiotic process. Stoichiometric relationships between a dihydrodiol-modified PAH substrate (naphthalenediol) and Mn particles are reviewed. Manganese reductive dissolution and polynuclear aromatic oxidation were mathematically approached from experimental and theoretical perspectives. An initial set of mathematical expressions was developed based

0-87371-914-X/95/$0.00+$.50

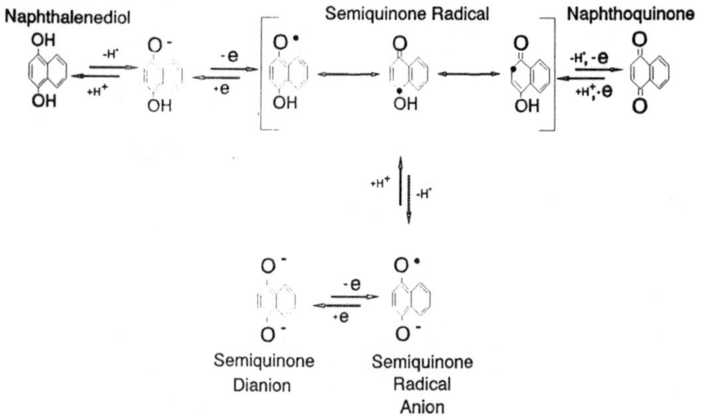

Figure 1 Proposed relationships between naphthalenediol, semiquinone radical, naphthoquinone, semiquinone radical anion, and semiquinone dianion.[19–22]

on experimental data; these expressions were then compared to theoretically derived equations based on stoichiometric and kinetic considerations.

Because of its oxidation-reduction (redox) potential, manganese oxides were used to oxidize the following five multiple-ringed aromatic compounds containing diol and dione functional groups under oxic conditions: 1,4-, 1,3-, and 2,3-naphthalenediol [$C_{10}H_6(OH)_2$]; 1,4-naphthoquinone ($C_{10}H_6O_2$); and 1,4-dihydroxy-9,10-anthracenedione [Quinizarin, $C_{14}H_6O_2(OH)_2$]. Figure 2 presents a schematic of each compound. In many experiments described herein, naphthalenediol has been chosen to illustrate relationships between the manganese oxides and organic substrates; any one of the other aromatic constituents could also have been chosen.

II. STOICHIOMETRY OF REDUCTIVE DISSOLUTION AND AUTOOXIDATION

Whelan and Sims[20,22] and Whelan[21] reviewed stoichiometric relationships between Mn(IV) and 1,4-naphthalenediol [i.e., 1,4-dihydroxynaphthalene, Nap-$(OH)_2$] in an oxic environment. If the reaction is assumed to proceed through an organic free radical, the following reaction describes the reductive dissolution of Mn(IV) and oxidation of naphthalenediol [Nap-$(OH)_2$] to form dissolved Mn(II) and a semiquinone radical (·O-Nap-OH) (see Figure 1).

$$\mathrm{Mn}^{\mathrm{IV}}_{\mathrm{O_2}}(\mathrm{s}) + 2\,\mathrm{Nap}-(\mathrm{OH})_2 + 2\,\mathrm{H}^+ = \mathrm{Mn}^{2+} + 2\left(\cdot\mathrm{O}-\mathrm{Nap}-\mathrm{OH}\right) + 2\,\mathrm{H_2O} \quad (1)$$

Stone and Morgan[19] have mechanistically described this reductive-dissolution/oxidation process in four steps, assuming inner-sphere complex formation between Nap-$(OH)_2$ and the oxide surface. Whelan and Sims,[20,22] Whelan,[21] and Figure 3 express these as follows:

1. Precursor-Complex Formation (Reductant Adsorption)

$$> \mathrm{Mn}^{\mathrm{IV}}(\mathrm{OH})_2 \;+\; 2\,\mathrm{Nap}-(\mathrm{OH})_2 \underset{k_{-1}}{\overset{k_1}{\longleftrightarrow}} > \mathrm{Mn}^{\mathrm{IV}}(\mathrm{O}-\mathrm{Nap}-\mathrm{OH})_2 \;+\; 2\,\mathrm{H_2O} \quad (2)$$

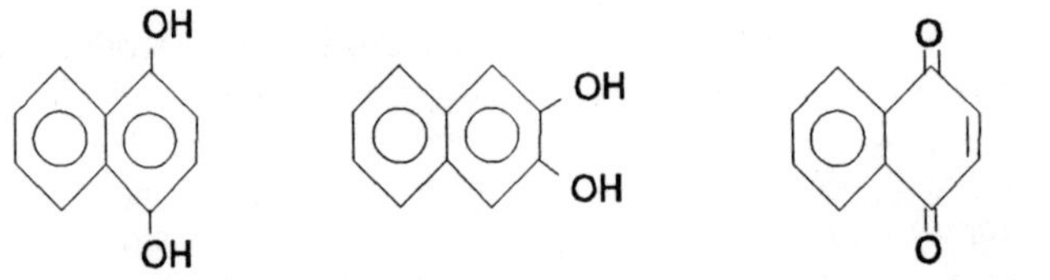

1,4-Naphthalenediol 2,3-Naphthalenediol 1,4-Naphthoquinone

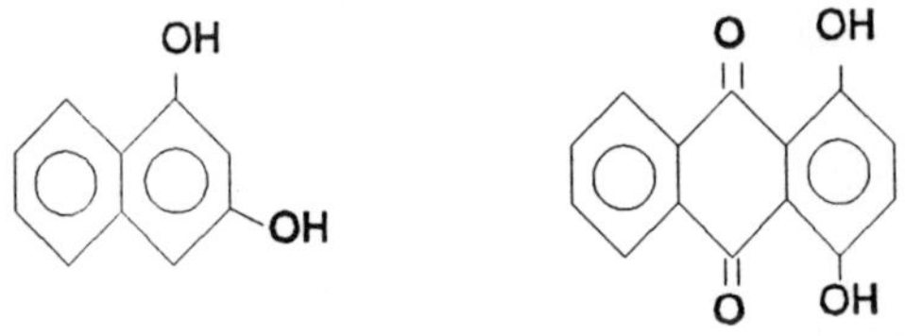

1,3-Naphthalenediol 1,4-Dihydroxy-9,10-anthracenedione (Quinizarin)

Figure 2 Schematic diagram of 1,4-, 2,3-, and 1,3-naphthalenediol, 1,4-naphthoquinone, and 1,4-dihydroxy-9,10-anthracenedione.[21,22]

where the symbol ">" refers to the oxide surface, and k_1 and k_{-1} are rate constants in the forward and reverse directions, respectively. Reductant adsorption, as presented in Figure 3, represents a schematic of the reductive-dissolution process, illustrates that naphthalenediol is a one-equivalent reductant, and should not be viewed as identifying the known ligand surface formation.

2. Electron Transfer

$$> Mn^{IV}(O - Nap - OH)_2 \underset{k_{-2}}{\overset{k_2}{\longleftrightarrow}} > Mn^{II}(\cdot O - Nap - OH)_2 \tag{3}$$

where k_2 and k_{-2} are rate constants in the forward and reverse directions, respectively.

3. Release of Oxidized Organic Product

$$> Mn^{II}(\cdot O - Nap - OH)_2 + H_2O \underset{k_{-3}}{\overset{k_3}{\longleftrightarrow}} > Mn^{II}OH_2 + 2(\cdot O - Nap - OH) \tag{4}$$

where k_3 and k_{-3} are rate constants in the forward and reverse directions, respectively, and ·O-Nap-OH is a radical. By noting that Mn(II) still resides on the oxide surface, the Mn(II) products of Equation 4 can also be written as follows, because the right- and left-hand sides of Equation 5 are equivalent:

$$> Mn^{II}OH_2 = > Mn^{IV}O_2 - (Mn^{II}OH_2) \tag{5}$$

where $>Mn^{IV}O_2$-($Mn^{II}OH_2$) represents the reduced metal complex on the Mn(IV) surface prior to Mn(II) release. Research has indicated that oxygen promotes oxidative coupling reactions of organics, creating dimers, trimers, and other less soluble, more surface-active oxidation products.[21-28] These oxidative-coupling reactions could account for the removal of radical products that are formed in the oxidation process.

4. Release of Reduced Metal Ion. Stone and Ulrich[29] noted that protons frequently assist in the metal-detachment step of dissolution reactions and that studies have generally found the number of protons involved to be equal to the valence of the detached metal (i.e., 2).[30] They further note that the actual number of protons involved in reductive dissolution is not known with certainty, because the presence of two or more oxidation states on the metal surface may alter the pH dependence of the metal-release step. The release of the reduced metal ion from the surface is expressed as follows:

$$> Mn^{IV}O_2 - \left(Mn^{II}OH_2\right) + 2\,H^+ \xrightarrow{k_4} > Mn^{IV}(OH)_2 + Mn^{2+} + H_2O \qquad (6)$$

where k_4 is a rate constant. In experiments to determine the effect of varying amounts of Mn^{2+} on MnO_2 dissolution rates, Stone and Morgan[15] found that the Mn^{2+} concentration in solution had no effect on initial rates of dissolution, although researchers have indicated that adsorption of Mn(II) may be an important process.[29] Therefore, one might conclude that Equation 6 is not rate limiting and can be considered to be unidirectional.

As shown schematically (Figure 3), the autooxidation of Mn^{2+} to Mn(IV) can be represented as[31]

$$Mn^{2+} + \tfrac{1}{2}O_2 + H_2O = Mn^{IV}O_2(s) + 2\,H^+ \qquad (7)$$

Two protons are produced in Equation 7, whereas Equation 1 indicates that two protons are consumed. Stumm and Morgan[31] note that the Mn^{2+} concentration decreases with time with an apparent autocatalytic effect. Benefield et al.[32] describe the autocatalytic oxidation of Mn(II) in the following manner:[31,33]

$$d\left[Mn^{2+}\right]/dt = -k_5\left[Mn_T\right]\left[P_{O_2}\right]\left[OH^-\right]^2\left[Mn^{2+}\right]\left[Mn_T - Mn^{2+}\right] \qquad (8)$$

where k_5 is a rate constant, Mn_T is the total manganese in the system, and $[P_{O_2}]$ is the partial pressure of oxygen. Wehrli[34] noted that thermodynamics indicate that manganese should be present in oxic waters as Mn(IV) oxides; however, experiments and analyses by Diem and Stumm[35] and Wilson[36] have indicated that Mn^{2+} is not oxidized by oxygen within several years. Wehrli[34] observed that recent experimental evidence had shown that mineral surfaces provide accelerated pathways for redox processes such as the oxygenation of metal ions and organic pollutants. He suggested that manganese oxygenation is accelerated by iron oxide particles by a factor of more than 10,000. Whelan and Sims[20,22] and Whelan[21] combined autooxidation with these four steps to schematically describe the reductive-dissolution/autooxidation process, as illustrated in Figure 3.

Whelan[21] also reviewed potential relationships between naphthalenediol and Mn(III) in an oxic environment, and Mn(IV) and Mn(III) in an anoxic environment. Although the anoxic environment is not discussed herein, the relationship between Mn(III) and naphthalenediol in an oxic environment, assuming free-radical complex formation, is as follows:

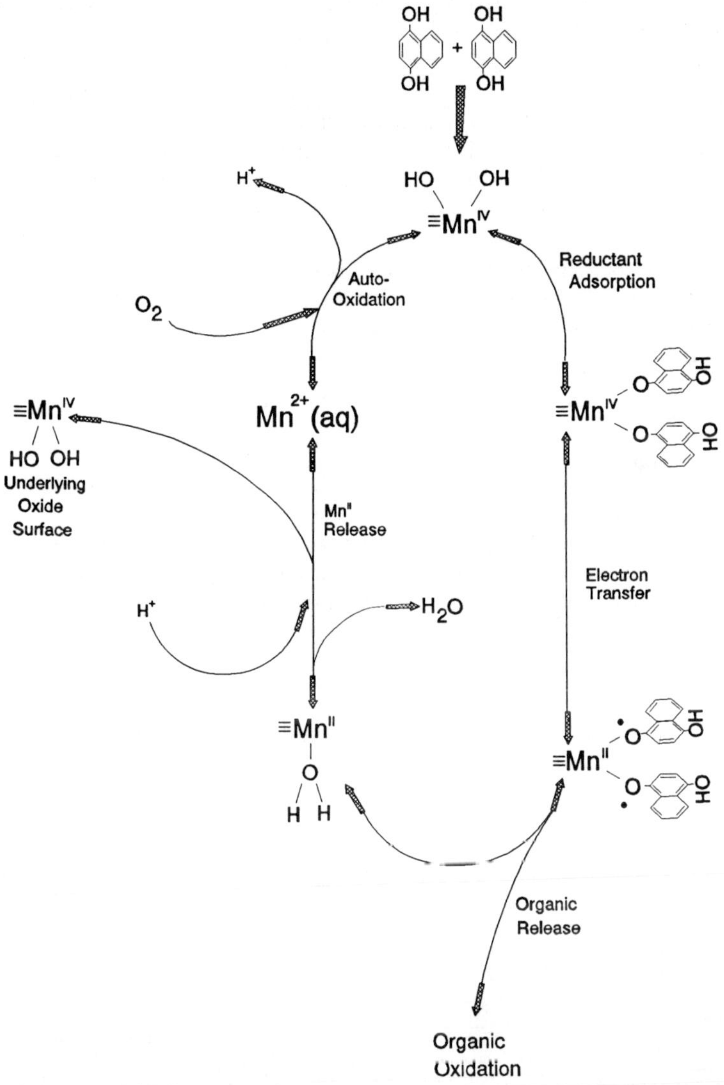

Figure 3 Schematic of the reductive-dissolution/autooxidation process.[19–22]

$$Mn^{III}OOH(s) + Nap-(OH)_2 + 2\,H^+ = Mn^{2+} + \cdot O-Nap-OH + 2\,H_2O \quad (9)$$

For both the oxic and anoxic cases using Mn(IV) and Mn(III), Whelan[21] demonstrated that a set of equations, similar to Equations 2 through 6, can be developed for inner-sphere complex formation.

III. KINETICS OF REDUCTIVE DISSOLUTION AND AUTOOXIDATION

The kinetics associated with reductive dissolution and autooxidation are reviewed for the surface interaction under oxic conditions of manganese(IV) dioxide and naphthalenediol, although the process is also applicable to the other constituents addressed herein. Assuming Equations 2 through 6 and Equation 8 are valid, the algorithms describing the kinetics of hypothetical reductive-dissolution and autooxidation reactions can be presented as illustrated by Figure 3. Assuming that the only species that contribute to the surface mass balance equation are $>Mn^{IV}(OH)_2$, $>Mn^{IV}(\cdot O\text{-}Nap\text{-}OH)_2$, $>Mn^{II}OH_2$, and $>Mn^{II}(\cdot O\text{-}Nap\text{-}OH)_2$, and that other competing anions and cations are not considered, the total moles of surface sites per liter of solution (S_T) can be written as follows:

$$S_T = \left[>Mn^{IV}(OH)_2\right] + \left[>Mn^{IV}(O-Nap-OH)_2\right] + \left[>Mn^{II}(\cdot O-Nap-OH)_2\right] + \left[>Mn^{II}OH_2\right] \tag{10}$$

Under the assumption that each reaction can be described as an elementary reaction, rate expressions can be proposed for $>Mn^{IV}(OH)_2$, $>Mn^{IV}(O\text{-}Nap\text{-}OH)_2$, $>Mn^{II}(\cdot O\text{-}Nap\text{-}OH)_2$, and $>Mn^{II}OH_2$, using Equations 2 through 6 and Equation 10:

$$d\left[>Mn^{IV}(OH)_2\right]/dt = -k_1\left[Nap(OH)_2\right]^2\left[>Mn^{IV}(OH)_2\right] + k_4\left[H^+\right]^2\left[>Mn^{II}OH_2\right] + k_{-1}\left[>Mn^{IV}(O-Nap-OH)_2\right] \tag{11}$$

$$d\left[>Mn^{IV}(O-Nap-OH)_2\right]/dt = k_1\left[Nap(OH)_2\right]^2\left[>Mn^{IV}(OH)_2\right] - \left(k_{-1}+k_2\right)\left[>Mn^{IV}(O-Nap-OH)_2\right] + k_{-2}\left[>Mn^{II}(\cdot O-Nap-OH)_2\right] \tag{12}$$

$$d\left[>Mn^{II}(\cdot O-Nap-OH)_2\right]/dt = k_2\left[<Mn^{IV}(O-Nap-OH)_2\right] - \left(k_{-2}+k_3\right)\left[>Mn^{II}(\cdot O-Nap-OH)_2\right] + k_{-3}\left[\cdot O-Nap-OH\right]^2\left[>Mn^{II}OH_2\right] \tag{13}$$

$$d\left[>Mn^{II}OH_2\right]/dt = k_3\left[>Mn^{II}(\cdot O-Nap-OH)_2\right] - \left\{k_{-3}\left[\cdot O-Nap-OH\right]^2 + k_4\left[H^+\right]^2\right\}\left[>Mn^{II}OH_2\right] \tag{14}$$

The rate expressions for the remaining nonsurface-constituent concentrations (i.e., $[Mn^{2+}]$, $[Nap(OH)_2]$, and $[\cdot O\text{-}Nap\text{-}OH]$) are as follows:

$$d\left[Mn^{2+}\right]/dt = k_4\left[H^+\right]^2\left[>Mn^{II}OH_2\right] - k_5\left[Mn_T\right]\left[P_{O_2}\right]\left[OH^-\right]^2 \left[Mn^{2+}\right]\left[Mn_T - Mn^{25}\right] \tag{15}$$

$$d\left[\text{Nap(OH)}_2\right]/dt = -2\,k_1\left[\text{Nap(OH)}_2\right]^2\left[>\text{Mn}^{\text{IV}}(\text{OH})_2\right] + 2\,k_{-1}\left[>\text{Mn}^{\text{IV}}(\text{O}-\text{Nap}-\text{OH})_2\right] \tag{16}$$

$$d\left[\cdot\,\text{O}-\text{Nap}-\text{OH}\right]/dt = 2\,k_3\left[>\text{Mn}^{\text{II}}(\cdot\,\text{O}-\text{Nap}-\text{OH})_2\right] - 2\,k_{-3}\left[\cdot\,\text{O}-\text{Nap}-\text{OH}\right]^2\left[>\text{Mn}^{\text{II}}\text{OH}_2\right] \tag{17}$$

IV. RATE-LIMITING MECHANISM OF REDUCTIVE DISSOLUTION

Stone and Morgan[19] note that "(m)ost reductive dissolution reactions of environmental and geochemical significance are ... controlled by surface chemical reactions" and that transport-controlled reductive dissolution reactions represent rare cases. If either electron transfer (Equation 3) or organic release (Equation 4) represents the rate-limiting step for Mn(IV) in an oxic environment, Equations 3 through 6 can be represented by an overall reaction:

$$>\text{Mn}^{\text{IV}}(\text{O}-\text{Nap}-\text{OH})_2 \xrightarrow{k_6} \text{Mn}^{2+} + \text{H}_2\text{O} + \text{oxidized products} \tag{18}$$

where k_6 is a rate constant in the forward direction. By performing a surface mass balance and assuming that the only species that contribute are $>\text{Mn}^{\text{IV}}(\text{OH})_2$ and $>\text{Mn}^{\text{IV}}(\text{O-Nap-OH})_2$ and other competing anions and cations are not considered, the surface mass balance equation reduces to:

$$S_T = \left[>\text{Mn}^{\text{IV}}(\text{OH})_2\right] + \left[>\text{Mn}^{\text{IV}}(\text{O}-\text{Nap}-\text{OH})_2\right] \tag{19}$$

The total manganese associated with the system (Mn_T) can be expressed as

$$\text{Mn}_T = \left[>\text{Mn}^{\text{IV}}(\text{OH})_2\right] + \left[>\text{Mn}^{\text{IV}}(\text{O}-\text{Nap}-\text{OH})_2\right] + \text{Mn}^{2+} \tag{20}$$

By combining Equations 2, 18, and 20, assuming electron transfer/organic release represents the rate-limiting step {i.e., $d[>\text{Mn}^{\text{IV}}(\text{O-Nap-OH})_2]/dt$ is at steady state}, and using an approach similar to that presented by Stone and Morgan,[19] the rate of change in Mn^{2+} concentration can be determined as follows:[21]

$$\frac{d\left[\text{Mn}^{2+}\right]}{dt} = \frac{\left\{k_6\left(\left[\text{Mn}_T\right]-\left[\text{Mn}^{2+}\right]\right)\right\}\left[\text{Nap}-(\text{OH})_2\right]^2}{\left\{\left(k_{-1}+k_6\right)/k_1\right\} + \left[\text{Nap}-(\text{OH})_2\right]^2} \tag{21}$$

with the total number of remaining sites at time t being $[\text{Mn}_T] - [\text{Mn}^{2+}]$. The exponent on the organic substrate [i.e., molar ratio (R)] is two, representing the number of moles of organic oxidized to moles of Mn(IV) reduced (refer to Equation 1). An expression similar to Equation 21, under oxic conditions, can also be developed for Mn(III) (e.g., MnOOH):[21,22]

$$\frac{d\left[Mn^{2+}\right]}{dt} = \frac{\left\{k_6\left(\left[Mn_T\right]-\left[Mn^{2+}\right]\right)\right\}\left[Nap-(OH)_2\right]}{\left\{\left(k_{-1}+k_6\right)/k_1\right\}+\left[Nap-(OH)_2\right]} \tag{22}$$

The exponent on the organic substrate is unity (refer to Equation 13). Under the conditions of a mixture of Mn(III/IV) particles under nonhomogeneous conditions, an expression for the rate of change of Mn^{2+} can be calculated as[21,22]

$$\frac{d\left[Mn^{2+}\right]}{dt} = \sum_{i=1}^{m} \frac{k_{6i}\left(\left[Mn_T\right]_i-\left[Mn^{2+}\right]_i\right)\left[Nap-(OH)_2\right]^{Ri}}{\left\{\left(k_{-1i}+k_{6i}\right)/k_{1i}\right\}+\left[Nap-(OH)_2\right]^{Ri}} \tag{23}$$

in which

$$\left[Mn_T\right] = \sum_{i=1}^{m}\left[Mn_T\right]_i \tag{24}$$

$$\left[Mn^{2+}\right] = \sum_{i=1}^{m}\left[Mn^{2+}\right]_i \tag{25}$$

where m is the number of different forms of manganese molecules (e.g., MnO_2, MnOOH, Mn_2O_3, δ-MnO_2, etc.) undergoing reductive dissolution, i is the index on the number of different forms of manganese molecules undergoing reductive dissolution, R is the ratio between the moles of organic oxidized to the moles of manganese reduced from Mn(IV/III) to Mn^{2+}, and all other terms retain their previous definition.

For Equation 21 to be qualitatively valid (and the more complex Equation 23), the temporal change of (1) the organic substrate would have to have a shape similar to that exhibited by a rectangular hyperbola but decreasing with time and (2) Mn^{2+} would have to have a shape similar to a rectangular hyperbola that increases with time (i.e., as the manganese is reduced to Mn^{2+}, the substrate is oxidized). Many mathematical expressions describe a shape similar to a rectangular hyperbola over at least a portion of the region, including Michaelis-Menton, C $(1 - e^{-kt})$, etc. A relationship for the substrate that describes a shape similar to a rectangular hyperbola but decreasing with time is:[21,22]

$$\left[Nap-(OH)_2\right] = S_m K_s/\left(K_s + t\right) \tag{26}$$

where S_m and K_s are constants. Stone and Morgan[15] noted that for manganese, the "initial dissolution rate is proportional to the amount of oxide loading, which suggests that the rate at any time during dissolution is proportional to the amount of remaining oxide ...." They described the initial dissolution rate with the following autocatalytic expression:

$$d\left[Mn^{2+}\right]/dt = k_{exp}\left(\left[Mn_T\right]-\left[Mn^{2+}\right]\right) \tag{27}$$

where k_{exp} is the experimentally determined rate constant. Because Equation 27 is only applicable during the initial phases of reductive dissolution when the rate of change of Mn^{2+} production is proportional to the oxide solids remaining, it does not necessarily adequately describe the rate of dissolution over longer time periods. To account for autocatalysis in the temporal rate of change of Mn^{2+} and the nonlinearity in this rate of change, Whelan[21] and Whelan and Sims[22] suggested a more appropriate expression for the time rate of change of Mn^{2+}:

$$d\left[Mn^{2+}\right]/dt = k_7\left(\left[Mn_T\right]-\left[Mn^{2+}\right]\right)/\left(K_m + t\right)^2 \tag{28}$$

where k_7 and K_m are constants. During initial times, the temporal rate of change of Mn^{2+}, as expressed by Equation 28, exhibits linearity with the remaining oxide. During other phases of the process, Equation 28 also exhibits the nonlinearity inherent in the data. By integrating Equation 28, when $[Mn^{2+}]$ equals zero at time zero for the initial condition, an expression for Mn^{2+} as a function of time can be developed:

$$\left[Mn^{2+}\right] = \left[Mn_T\right]\left\{1-\exp\left[-k_7\, t/K_m/\left(K_m + t\right)\right]\right\} \tag{29}$$

For Equation 29 to be valid, it should (1) have a shape similar to that exhibited by a rectangular hyperbola and (2) adequately describe the data. For example, (1) when time equals zero, $[Mn^{2+}]$ equals zero; (2) as time becomes large, the ratio of $[Mn^{2+}]/[Mn_T]$ approaches 1 - exp (k_7/K_m), not unity as a Michaelis-Menton-type rate expression would suggest. (3) Equation 29 has a shape similar to a rectangular hyperbola, and (4) the rate of change of Mn^{2+} is autocatalytic (see Equation 28). By combining Equation 29 with Equation 26, the following expression can be derived:

$$\frac{d\left[Mn^{2+}\right]}{dt} = \frac{\left\{k_7/\left(K_m - K_s\right)^2\right\}\left(\left[Mn_T\right]-\left[Mn^{2+}\right]\right)\left[Nap-(OH)_2\right]^2}{\left(K_s S_m\right)^2 + \left\{\left[Nap-(OH)_2\right]\left(K_m - K_s\right)\right\}^2 + 2\left[Nap-(OH)_2\right]K_s\, S_m\left(K_m - K_s\right)} \tag{30}$$

By noting that $(K_s\, S_m) >> \{2\, [Nap\text{-}(OH)_2]\, (K_m - K_s)\}$ should be valid over most of the range, Equation 30 reduces to a form that is the same as Equation 21:

$$\frac{d\left[Mn^{2+}\right]}{dt} = \frac{\left\{k_7/\left(K_m - K_s\right)^2\right\}\left(\left[Mn_T\right]-\left[Mn^{2+}\right]\right)\left[Nap-(OH)_2\right]^2}{\left\{K_s\, S_m/\left(K_m - K_s\right)\right\}^2 + \left[Nap-(OH)_2\right]^2} \tag{31}$$

Utilizing the theoretical development leading to the derivation of Equation 31, manganese oxides were used to oxidize five multiple-ringed aromatic compounds containing diol and dione functional groups under oxic conditions in an aqueous environment: 1,4-, 1,3-, and 2,3-naphthalenediol, 1,4-naphthoquinone, and Quinizarin.

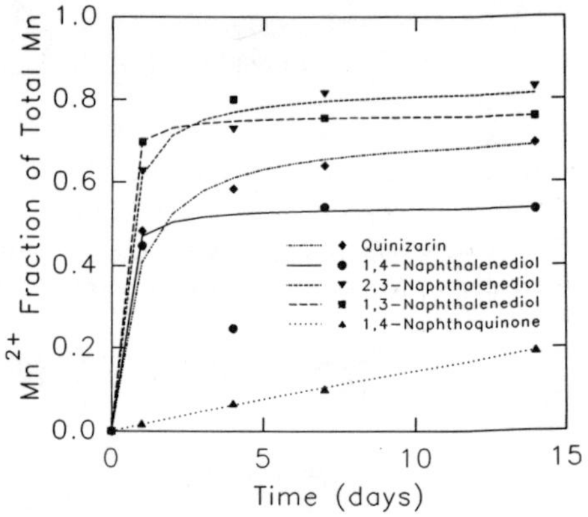

Figure 4 Nonlinear, least-squares regression plots associated with the fraction of total manganese as dissolved Mn^{2+} in a buffered system.[21,22,37]

The effectiveness of employing MnO_2 particles to oxidize these PAHs containing dihydrodiols and diones is best illustrated in the redox results presented in Figures 4 and 5. Figure 4 presents temporal variations in Mn^{2+} concentration by constituent, expressed as a fraction of the total manganese added to the reaction vessel. 1,4-Naphthalenediol on day 4 (Figure 4) was not employed in the analysis as it is a significant outlier. To indicate the degree of oxidation of the parent organics by the manganese particles, Figure 5 illustrates the fraction of the organic oxidized (i.e., molar ratio of organic oxidized to organic control) for 1,3- and 2,3-naphthalenediol and 1,4-naphthaquinone. For example, a 0.99 value in this figure indicates that 99% of the parent organic has been oxidized (either transformed or degraded); the possibility of sorption has been accounted for. Based on the data in Figures 4 and 5, Equations 28, 29, and 31 appear to describe the temporal variations in Mn^{2+} and the organic substrate as it relates to the reductive-dissolution process. These equations also satisfy the conditions for electron transfer/organic release as the rate-limiting step. Details of the experimental design and data are documented by Whelan.[21]

V. EFFECTS OF pH ON REDUCTIVE DISSOLUTION OF MANGANESE OXIDE PARTICLES

Equations 29 and 31 do not account for all relationships between reactants under varying environmental conditions (e.g., pH). Stone and Morgan,[19] Whelan,[21] and Whelan and Sims[22,28] have suggested that reductive dissolution of manganese oxides by organics is surface chemically controlled and not transport controlled. Therefore, assuming that electron transfer/release of aromatic organic is the rate-limiting step, Equations 3 through 6 can be condensed into the following expression:

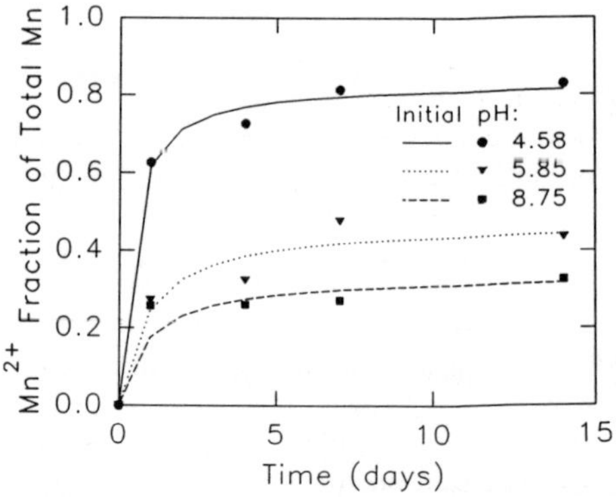

Figure 5 Nonlinear, least-squares regression plots associated with the fraction of total manganese as dissolved Mn^{2+} as a function of initial pH with 2,3-naphthalenediol.[21,22,37]

$$> Mn^{IV}(O - Nap - OH)_2 + n\,H^+ \xrightarrow{k_8} Mn^{2+} + H_2O + \text{oxidized products} \quad (32)$$

where k_8 is a rate constant in the forward direction and n represents the number of moles of protons consumed in Equation 32. By combining Equations 2, 20, and 32, and by using an approach similar to that presented by Stone and Morgan,[19] Whelan[21] derived the rate of change of Mn^{2+} as follows:

$$\frac{d[Mn^{2+}]}{dt} = \frac{k_8([Mn_T]-[Mn^{2+}])\{H^+\}^n[Nap-(OH)_2]^2}{\left(\left(k_{-1} + k_8\{H^+\}^n\right)/k_1\right) + [Nap-(OH)_2]^2} \quad (33)$$

where $\{H^+\}$ represents the proton activity with the total number of remaining sites at time t being $[Mn_T] - [Mn^{2+}]$. The organic term is squared because two moles of naphthalenediol are oxidized for each mole of manganese oxide reduced to Mn^{2+}.

Mathematically, Equation 33 describes the relationship between the rate of Mn^{2+} production, pH, and naphthalenediol concentration. If electron transfer/organic release is the rate-limiting step, then $k_1 >> k_8$, and k_{-1} is most likely much larger than $k_8\ \{H^+\}^n$, thereby reducing Equation 33 to

$$\frac{d[Mn^{2+}]}{dt} = \frac{k_8([Mn_T]-[Mn^{2+}])\{H^+\}^n[Nap-(OH)_2]^2}{(k_{-1}/k_1) + [Nap-(OH)_2]^2} \quad (34)$$

If $(k_{-1}/k_1) << [Nap\text{-}(OH)_2]^2$, which may be the case during the initial phases of the experiment when excess organic substrate is present, as illustrated by Stone and Morgan,[15] then Equation 34 reduces to

$$d[Mn^{2+}]/dt = k_8([Mn_T]-[Mn^{2+}])\{H^+\}^n \quad (35)$$

Stone and Morgan[15] proposed an empirical equation for hydroquinone:

$$d[Mn^{2+}]/dt = k_9([Mn_T]-[Mn^{2+}])\{H^+\}^m[QH_2] \quad (36)$$

where k_9 is a rate constant, m is a constant equaling 0.46, and QH_2 is hydroquinone. The power on QH_2 was determined to be unity; Stone and Morgan[15] mathematically demonstrated that the substrate hydroquinone is not squared because one mole of hydroquinone is oxidized for each mole of manganese dioxide reduced under anoxic conditions:

$$Mn^{IV}O_2(s) + QH_2 + 2\,H^+ = Mn^{2+} + Q + 2\,H_2O \quad (37)$$

where Q is benzoquinone. During the initial phases of the reductive dissolution process, Stone and Morgan[15] designed their experiments with hydroquinone in excess amounts; therefore, the hydroquinone concentration does not change significantly and can be considered in excess and a constant. As such, Equation 36 can be rewritten as follows:

$$d\left[Mn^{2+}\right]/dt = k_{10}\left(\left[Mn_T\right]-\left[Mn^{2+}\right]\right)\left\{H^+\right\}^m \tag{38}$$

in which

$$k_{10} = k_9\left[QH_2\right] \tag{39}$$

where k_{10} is a rate constant. When the organic substrate is in excess, Equations 35 and 38 have the same form. A similar type of analysis can be employed when the remaining organic substrate has been nearly completely oxidized {i.e., $(k_{-1}/k_1) >> [Nap\text{-}(OH)_2]^2$}; Equation 34 can be rewritten as follows:

$$d\left[Mn^{2+}\right]/dt = k_{11}\left(\left[Mn_T\right]-\left[Mn^{2+}\right]\right)\left\{H^+\right\}^n\left[Nap-(OH)_2\right]^2 \tag{40}$$

in which

$$k_{11} = \left(k_1/k_{-1}\right)k_2 \tag{41}$$

where k_{11} is a rate constant. Equation 40 is very similar to Equation 36.

Stone[26] performed a similar analysis using phenol when determining the exponent on the proton concentration by solving the following equation:

$$d\left[Mn^{2+}\right]/dt = k_{12}\left\{H^+\right\}^q\left[\text{phenol}\right]^p \tag{42}$$

where k_{12} is a rate constant and p and q are exponents. He solved for q and found that it ranged from <0.5 when a pH of 4 was approached to greater than unity for pHs greater than 6. Equation 42 is very similar in form to Equation 40.

Equation 32 describes the functional role that protons play in the reductive-dissolution process during the oxidation of aromatics. Protons are consumed in this process, and a decrease in the pH should increase the reductive-dissolution rate. Figures 6 and 7 present the effects that variations in pH have on the reductive dissolution of Mn(III/IV) to dissolved Mn(II) and oxidation of 2,3-naphthalenediol, respectively. These figures present the results of three different initial pH levels with all other conditions being equal: 4.58, 5.85, and 8.75.

Figure 6 presents the temporal variation in the dissolved manganese concentration by initial pH, expressed as a fraction of the total manganese added to the reaction vessel. Correspondingly, to indicate the degree of oxidation of the parent organics by the manganese particles, Figure 7 presents the fraction of the organic oxidized (i.e., molar ratio of organic oxidized to organic control). The results presented in Figures 6 and 7 complement each other and appear to follow the shapes suggested by Equation 33. Details of the experimental design and data are documented by Whelan.[21]

An analysis was performed to estimate the exponent n in the proton concentration in Equation 40. The condition associated with Equation 40 is chosen {i.e., $(k_{-1}/k_1) >> [Nap\text{-}(OH)_2]^2$}, as opposed to that associated with Equation 35 {i.e., $(k_{-1}/k_1) << [Nap\text{-}(OH)_2]^2$}, because most of the dissolved Mn(II) and naphthalenediol data collected represent long-term information (i.e., on the order of weeks) (see Figures 6 and 7). If Equation 35 were employed, short-term data (i.e., on the order of minutes) would have to have been collected. By combining Equations 28 and 40, the following expression can be derived

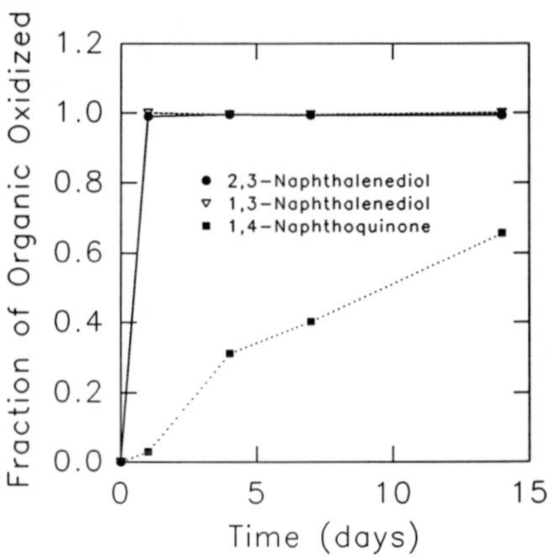

Figure 6 Fraction of the organic substrate oxidized during reductive dissolution.[21,22]

$$\frac{d\left[Mn^{2+}\right]}{dt} = \frac{k_7\left(\left[Mn_T\right]-\left[Mn^{2+}\right]\right)}{\left(K_m + t\right)^2} = k_5\left(\left[Mn_T\right]-\left[Mn^{2+}\right]\right)\left[Nap-(OH)_2\right]^2\left\{H^+\right\}^n \quad (43)$$

Linearizing Equation 43 gives

$$\text{Log}\left\{\text{RATIO}\right\} = \text{Log}\left\{k_5\left[Nap-(OH)_2\right]^2\right\} - n\,pH \quad (44)$$

in which

$$\text{RATIO} = k_7/\left(K_m + t\right)^2 \quad (45)$$

Day 14 results of Equation 44 are plotted in Figure 8 with a coefficient of determination (r^2) of 96%. The negative slope of the curve represents the exponent n and has a value of 0.4, which is similar to the value of 0.46 that Stone and Morgan[15] found for the oxidation of hydroquinone using reductive dissolution.

An analysis was also performed to determine the stoichiometric ratio between the moles of organic oxidized to the moles of Mn reduced from Mn(III/IV) to Mn^{2+}. Figure 9 presents the time-varying distribution of the molar ratio R with 95% confidence intervals for data associated with all analyses for each constituent presented herein (i.e.,

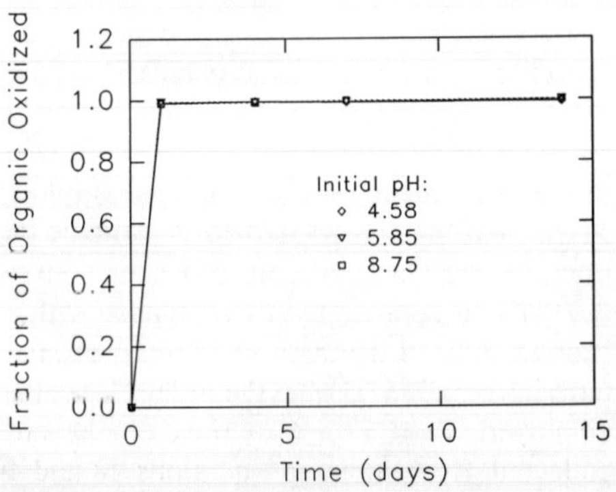

Figure 7 Temporally varying fraction of 2,3-naphthalenediol oxidized during reductive dissolution as a function of pH.[21,28]

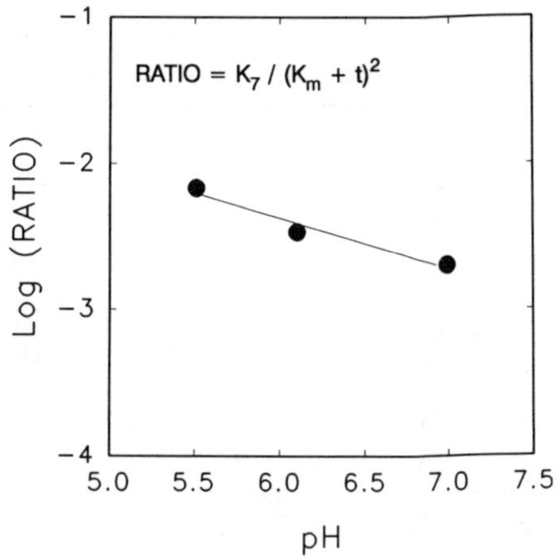

Figure 8 Log(RATIO) vs. pH.[21,28]

Sections IV and V), except Quinizarin, and for redox data not presented herein resulting from variations in concentration.[21,22,28] Substrate concentrations ranged from 4×10^{-5} to 4×10^{-3} *M* (8 to 800 mg l^{-1}) and produced results consistent with the theory presented herein.[21,28] Figure 9 illustrates that there is little variation in the mean molar ratios over time, ranging from 1.52 to 1.64, and there are no significant differences between the molar ratios at different time periods presented in Figure 4, based on confidence intervals.

VI. POLYMERIC PRODUCTS OF NAPHTHALENEDIOL FROM REDUCTIVE DISSOLUTION

Figure 3 and Equations 4, 18, and 32 suggest that if a free radical represented an intermediate step, one potential product of abiotic-catalyzed, reductive dissolution was an oxidized or "humified" polymer. The experimental precipitate observed in this study had a deep, dark-brown color similar to that exhibited by humic polymers. The redox reaction between MnO_2 particles and naphthalenediol resulted in the formation of such a precipitate. It was insoluble in dichloromethane, acetonitrile, and water at various pHs. Three analyses were performed to determine the origin of the material: infrared (IR) spectros-

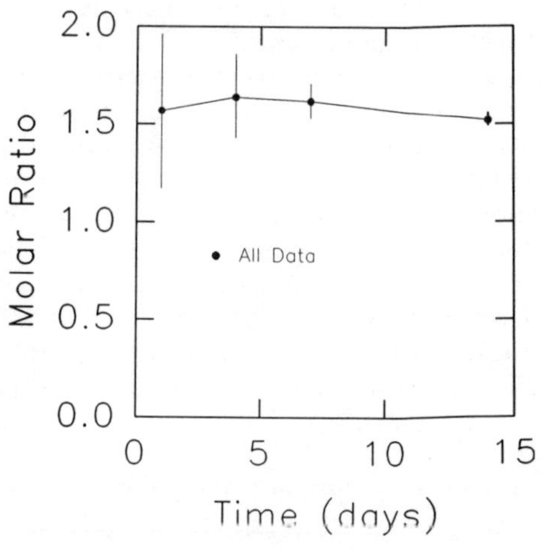

Figure 9 Temporal variation in the organic oxidized-to-manganese reduced molar ratio, with 95% confidence intervals.[21,22]

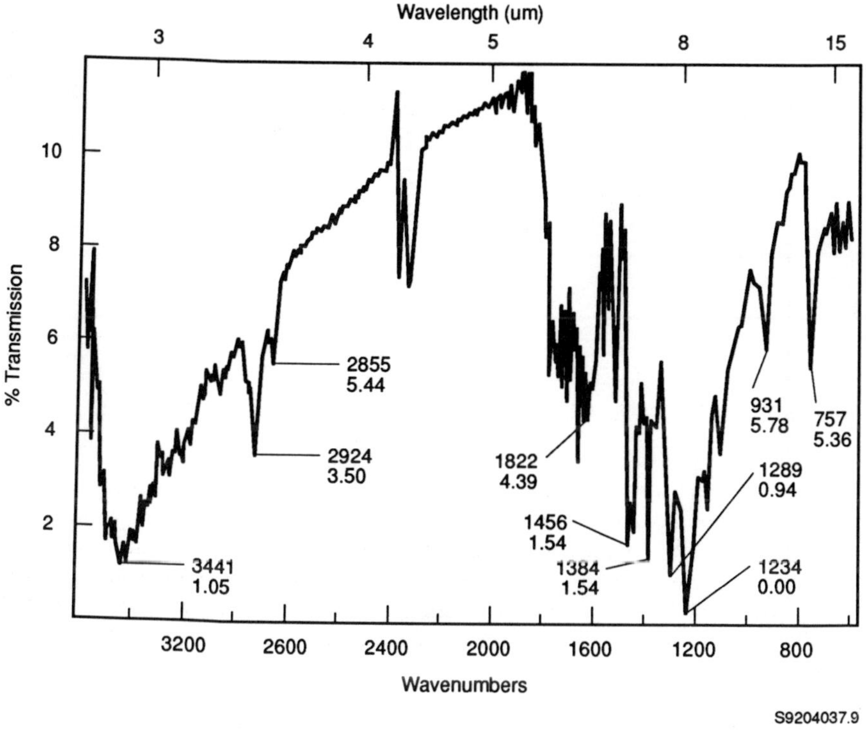

Figure 10 Infrared spectroscopy of naphthalenediol residue.[21,28]

copy; microelemental determination for carbon, oxygen, and hydrogen; and scanning electron microscopy using an energy-dispersive X-ray analysis (EDXA).

The IR results are presented in Figure 10. Even though the sample had been dried in a 103°C oven for 24 h, the material was so hydroscopic that water was still abundant in the spectral curve. Figure 10 presents the percent transmission vs. frequency in wave numbers over a range of 3600 to 600 cm^{-1}. Various bands occur near 3440, 2924, 2855, 1622, 1456, 1384, 1289, and 1234 cm^{-1}, and at other bands in the fingerprint region (i.e., <1400 cm^{-1}). Based on previous analyses by Stevenson,[6] this spectral curve appears to correspond to an organic compound similar in nature to that of a polymerized naphthalenediol product.[21,28]

The distribution of carbon, oxygen, and hydrogen by weight in naphthalenediol is 75.0, 20.0, and 5.0%, respectively. A microelemental analysis determined the composition based on the weight of the organic residue as 63.7, 25.2, and 4.2% of carbon, oxygen, and hydrogen, respectively, thereby accounting for 93.1% of the total mass; 6.9% is unaccounted for. Illustrative results of the EDXA are presented in Figure 11 and were used to identify the remaining 6.9% of the mass as that of chloride. Chloride appeared in residue samples but did not appear in background samples. The percentages by weight of the chloride in particle samples ranged from 2.5 to 8.0%. The chloride originates from the $NaClO_4$, which was used to fix the ionic strength of each solution. At a concentration of 10^{-3} *M,* $NaClO_4$ can supply plenty of available chloride. The 5.2% oxygen in excess of that present in the naphthalenediol (i.e., 20.0 vs. 25.2%) could be attributed to the perchlorate ion (ClO_4^-) or oxygen molecules that present in the oxic environment. Lee and Huang[38] note that Mn(III/IV) oxides are able to adsorb and polarize oxygen molecules, leading to the cleavage of the ring structure of polyphenols.

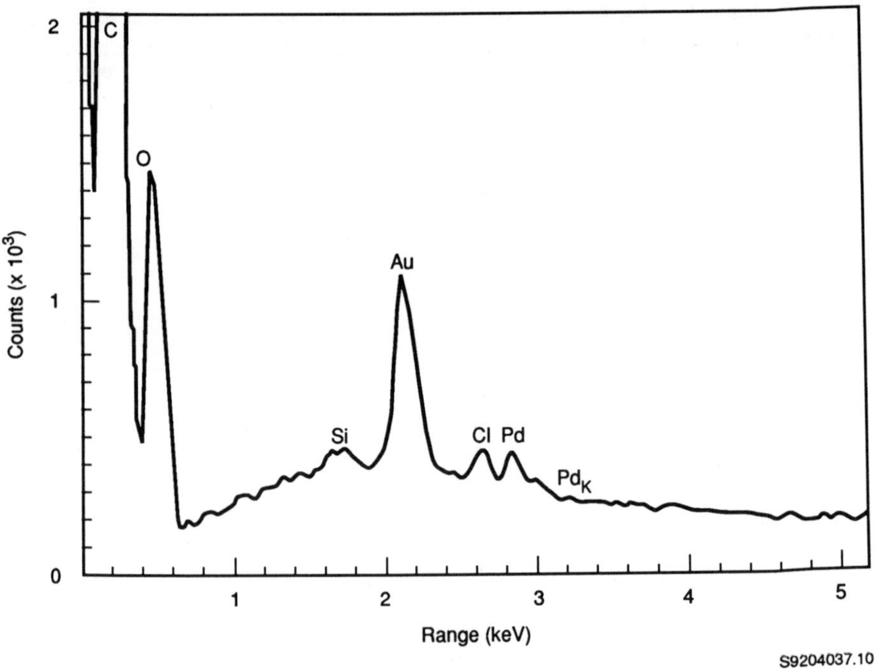

Figure 11 Energy-dispersive X-ray analysis of precipitate residue.[21,28]

VII. SUMMARY AND CONCLUSIONS

This research indicated that certain forms of PAH compounds (e.g., dihydrodiols and diones that form free-radical intermediates) are susceptible to oxidation and polymerization by manganese oxides. For example, all aromatic compounds exhibited varying degrees of oxidation over a 14-d period, resulting in 83, 76, 54, 70, and 20% of the manganese being reduced by 2,3-, 1,3-, and 1,4-naphthalenediol, Quinizarin, and 1,4-naphthoquinone, respectively. Correspondingly, after 14 d, 100% of the 2,3- and 1,3-naphthalenediol was oxidized, while approximately 65% of the 1,4-naphthalenediol was oxidized. Aromatics with diol functional groups were more successful at reducing manganese (and, hence, being oxidized) than those with only dione functional groups. The results also demonstrated that relatively insoluble aromatic compounds such as Quinizarin can be oxidized. Oxidation of Quinizarin resulted in a significant reduction of manganese particles from Mn(IV) to Mn(II) (i.e., 70%) over a 14-d period. Insoluble "humic-like" material precipitated during the reactions, indicating that a polymerization/humification process was occurring. Finally, the results suggested that electron transfer/organic release from the oxide surface is the rate-limiting step.

ACKNOWLEDGMENTS

Funding for this research was supplied by the U.S. Geological Survey (Award No. 14-08-0001-G1723) and the Electric Power Research Institute (EPRI) (Contract No. RP2879-09). Thanks are extended to Drs. J. J. Jurinak and D. K. Stevens of USU, J.-M. Bollag of Pennsylvania State University, A. T. Stone and K. L. Godfredtsen of Johns Hopkins University, and P. M. Huang of the University of Saskatchewan for providing technical information during the various phases of this work. Thanks are also extended to

Dr. S. V. Mattigod of Pacific Northwest Laboratory (PNL) for his technical review and comments. Appreciation is extended to Ms. J. Gephart of PNL for performing the editorial review of this manuscript. Funding for the editorial review was provided by PNL under contract DE-AC06–76RLO 1830.

REFERENCES

1. **Bollag, J.-M. and Liu, S.-Y.,** Biological transformation processes of pesticides, in *Pesticides in the Soil Environment,* Cheng, H. H., Ed., Soil Science Society of America, Madison, WI, 1990, chap. 6.
2. **Sarkar, J. M., Malcolm, R. L., and Bollag, J.-M.,** Enzymatic coupling of 2,4-dichlorophenol to stream fulvic acid in the presence of oxidoreductases, *Soil Sci. Soc. Am. J.,* 52, 688, 1988.
3. **Martin, J. P., Haider, K., and Linhares, L. F.,** Decomposition and stabilization of ring ^{14}C-labelled catechol in soil, *Soil Sci. Soc. Am. J.,* 43, 100, 1979.
4. **Liu, S.-Y., Freyer, A. J., Minard, R. D., and Bollag, J.-M.,** Enzyme-catalyzed complex formation of amino acid esters and phenolic humus constituents, *Soil Sci. Soc. Am. J.,* 49, 337, 1985.
5. **Bollag, J.-M. and Myers, C.,** Detoxification of aquatic and terrestrial sites through binding of pollutants to humic substances, *Sci. Total Environ.,* 117, 357, 1992.
6. **Stevenson, F. J.,** *Humus Chemistry: Genesis, Composition, Reactions,* John Wiley & Sons, New York, 1983.
7. **Wang, M. C. and Huang, P. M.,** Catalytic power of nontronite, kaolinite and quartz and their reaction sites in the formation of hydroquinone-derived polymers, *Appl. Clay Sci.,* 4, 43, 1989.
8. **Haider, K., Martin, J. P., and Filip, Z.,** Humus biochemistry, in *Soil Biochemistry,* Vol. 4, Paul, E. A. and McLanen, A. D., Eds., Marcel Dekker, New York, 1975, 195.
9. **Flaig, W., Beutelspacher, H., and Rietz, E.,** Chemical composition and physical properties of humic substances, in *Soil Components, Vol. 1, Organic Components,* Gieseking, J. E., Ed., Springer-Verlag, New York, 1975, 1.
10. **Shindo, H. and Huang, P. M.,** Catalytic effects of manganese(IV), iron(III), aluminum, and silicon oxides on the formation of phenolic polymers, *Soil Sci. Soc. Am. J.,* 48, 927, 1984.
11. **Whelan, G. and Sims, R. C.,** Abiotic immobilization/detoxification of recalcitrant organics, in *Proc. of SUPERFUND '90,* Hazardous Materials Control Research Institute, Silver Spring, MD, 1990, 820.
12. **Qiu, X. and McFarland, M. J.,** Bound residue formation in PAH contaminated soil composting using *Phanerochaete chrysosporium, Hazard. Waste Hazard. Mater.,* 8, 115, 1991.
13. **Abbott, C. A. and Sims, R. C.,** Use and efficiency of ethylene glycol monomethyl ether and monoethanolamine to trap volatilized [7-^{14}C] naphthalene and ^{14}C carbon dioxide, in *Proc. of SUPERFUND '89,* Hazardous Materials Control Research Institute, Silver Spring, MD, 1989, 23.
14. **Laha, S. and Luthy, R. G.,** Oxidation of aniline and other primary aromatic amines by manganese dioxide, *Environ. Sci. Technol.,* 24, 363, 1990.
15. **Stone, A. T. and Morgan, J. J.,** Reduction and dissolution of manganese(III) and manganese(IV) oxides by organics. I. Reaction with hydroquinone, *Environ. Sci. Technol.,* 18, 450, 1984.
16. **Stone, A. T. and Morgan, J. J.,** Reduction and dissolution of manganese(III) and manganese(IV) oxides by organics. II. Survey of the reactivity of organics, *Environ. Sci. Technol.,* 18, 617, 1984.
17. **Mihelcic, J. R. and Luthy, R. G.,** Degradation of polycyclic aromatic hydrocarbon compounds under various redox conditions in soil-water systems, *Appl. Environ. Microbiol.,* 54, 1182, 1988.
18. **Musso, H.,** Phenol coupling, in *Oxidative Coupling of Phenols,* Taylor, W. I. and Battersby, A. R., Eds., Marcel Dekker, New York, 1967, 1.
19. **Stone, A. T. and Morgan, J. J.,** Reductive dissolution of metal oxides, in *Aquatic Surface Chemistry,* Stumm, W., Ed., John Wiley & Sons, New York, 1987, 221.
20. **Whelan, G. and Sims, R. C.,** Oxidation of recalcitrant organics in subsurface systems, *Hazard. Waste Hazard. Mater.,* 9, 245, 1992.
21. **Whelan, G.,** Surface-Induced Oxidation of Multiple-Ringed Diol and Dione Aromatics by Manganese Dioxide, Ph.D. thesis, Utah State University, Logan, UT, 1992.
22. **Whelan, G. and Sims, R. C.,** Abiotic-catalyzed oxidation of naphthalenediol, naphthoquinone, and dihydroxyanthracenedione, *Water Res.,* submitted.

23. **Shindo, H. and Huang, P. M.,** Role of manganese(IV) oxide in abiotic formation of humic substances in the environment, *Nature (London),* 298, 363, 1982.
24. **LaKind, J. S. and Stone, A. T.,** Reductive dissolution of geothite by phenolic reductants, *Geochim. Cosmochim. Acta,* 53, 961, 1989.
25. **Shindo, H. and Huang, P. M.,** Catalytic polymerization of hydroquinone by primary minerals, *Soil Sci.,* 139, 505, 1985.
26. **Stone, A. T.,** Reductive dissolution of manganese(III/IV) oxides by substituted phenols, *Environ. Sci. Technol.,* 21, 979, 1987.
27. **Taylor, W. I. and Battersby, A. R.,** Eds., *Oxidative Coupling of Phenols,* Marcel Dekker, New York, 1967, vii.
28. **Whelan, G. and Sims, R. C.,** Abiotic-catalyzed oxidation of 2,3-naphthalenediol, *Water Res.,* submitted.
29. **Stone, A. T. and Ulrich, H.-J.,** Kinetics and reaction stoichiometry in the reductive dissolution of manganese(IV) dioxide and Co(III) oxide by hydroquinone, *J. Colloid. Interface Sci.,* 132, 509, 1989.
30. **Furrer, G. and Stumm, W.,** The coordination chemistry of weathering. I. Dissolution kinetics of δ-Al_2O_3 and BeO, *Geochim. Cosmichim. Acta,* 50, 1847, 1986.
31. **Stumm, W. and Morgan, J. J.,** *Aquatic Chemistry,* 2nd ed., John Wiley & Sons, New York, 1981.
32. **Benefield, L. D., Judkins, J. F., and Weand, B. L.,** *Process Chemistry for Water and Wastewater Treatment,* Prentice-Hall, New York, 1982.
33. **Morgan, J. J.,** Chemical equilibria and kinetic properties of manganese in natural waters, in *Principles and Applications of Water Chemistry,* Faust, S. D. and Hunter, J. V., Eds., John Wiley & Sons, New York, 1967, 561.
34. **Wehrli, B.,** Redox reactions of metal ions at mineral surfaces, in *Aquatic Chemical Kinetics: Reaction Rates of Processes in Natural Waters,* Stumm, W., Ed., John Wiley & Sons, New York, 1990, 311.
35. **Diem, D. and Stumm, W.,** Is dissolved Mn^{2+} being oxidized by O_2 in absence of Mn-bacteria or surface catalysts?, *Geochim. Cosmochim. Acta,* 48, 1571, 1984.
36. **Wilson, D. E.,** Surface and complexation effects on the rate of Mn(II) oxidation in natural waters, *Geochim. Cosmochim. Acta,* 44, 1311, 1980.
37. **Stevens, D. K.,** SANREG statistical computer program, unpublished, CEE Department, Utah State University, Logan, UT, 1992.
38. **Lee, J. S. K. and Huang, P. M.,** Photochemical effect on the abiotic transformations of polyphenolics as catalyzed by Mn(IV) oxide, in First Workshop of Working Group MO of the International Society of Soil Science on Impact of Interactions of Inorganic, Organic, and Microbiological Soil Components, Edmonton, Canada, 1992.

Chapter 27

Mobility and Persistence of Metolachlor and Terbuthylazine in Field Lysimeters

C. Marucchini, C. Vischetti, R. Giovanardi, and P. Ceccon

CONTENTS

I. INTRODUCTION

The degradation of groundwaters in rural areas is frequently attributed to agriculture. Experimental data useful for correlating agricultural practices with water quality are often inconsistent, owing to the time lag between the activity at the soil surface and the ultimate impact on the acquifer. Nevertheless, investigations on the transport of chemicals through the soil profile toward groundwater are needed to evaluate the risk of water pollution.

The mobility and persistence of agricultural chemicals, including pesticides, can be studied by using laboratory columns,[1] field lysimeters,[2,3] and field approaches.[4]

While laboratory column studies are usually far from real field conditions, field lysimeter studies can provide detailed data on pesticide mobility and persistence not normally obtained with field plot techniques, due to the impossibility of following the herbicide movement below the soil core sampling zone.

Water solubility and adsorption mainly affect the environmental fate of pesticides: in general, the higher the water solubility, the more weakly they are adsorbed, thus increasing the leacheability.[5] Metolachlor and terbuthylazine are two herbicides commonly used in Italy, particularly in corn crops. Terbuthylazine is a slightly water-soluble and strongly adsorbed compound, while metolachlor seem to be much more soluble and less adsorbed by the soil.[6]

This study compared the persistence and mobility of metolachlor and terbuthylazine where included in cultivated field lysimeters to evaluate the risk of polluting the water table. The data obtained from the experiment were compared with those obtained with the simulation model CALF[7,8] for predicting pesticide persistence and movement in soils.

II. MATERIALS AND METHODS

A. FIELD MANAGEMENT AND SAMPLING

The research was carried out in Udine (northeast Italy, 46° 03' N) in eight drainage lysimeters (1.6 m^2 in area, 1.5 m deep) buried at the center of 4×7 m plots and cultivated with a corn crop. Half of the lysimeters were filled with 0.5 m of 20-mm

0-87371-914-X/95/$0.00+$.50

Table 1 Main Properties of the Soil

Gravel (Ø >2 mm), %	12
Sand (0.05 < Ø <2 mm), %	49
Silt (0.002 < Ø <0.05 mm), %	23
Clay (Ø <0.002 mm), %	28
Field capacity, % v/v	24.5
Wilting point, % v/v	10.5
pH (H_2O)	8
Organic matter, g kg^{-1}	19
CEC, cmol kg^{-1}	16.5

sieved soil (Table 1) over a graveled bed, while the other half were filled with 1.0 m of the same soil. Preemergence herbicide treatment was applied on May 7 using a commercially formulated product (Primagran TZ, Ciba-Geigy) at the recommended rate of 6.6 kg ha^{-1} in 870 l ha^{-1} of water, thus distributing 1 kg ha^{-1} a.i. of terbuthylazine and 2 kg ha^{-1} a.i. of metolachlor. The collection of leachates at the bottom of the lysimeters was carried out once a day, whenever rainfall or irrigation induced drainage of water along the soil profile; after quantifying the drainage volume, a 200-ml sample of water was stored at –10°C before the analytical determinations.

Seven series of soil cores (200 g) were taken from the soil surface to a 0.5-m depth by 0.1-m steps, from May 6 to July 20, with the higher frequency close to the treatment.

B. ANALYTICAL PROCEDURE

1. Soil

Fifty grams of air-dried sieved soil sampled for each lysimeter at different depths (0 to 10, 10 to 20, 20 to 30, 30 to 40, and 40 to 50 cm) were mechanically shaken for 1 h in 100 ml of methanol. The suspension was centrifuged at 6000 g for 10 min and the supernatant filtered with extrarapid paper; this procedure was repeated twice.

The methanolic solution was diluted with 50 ml of H_20, extracted with chloroform (2 × 40 ml) in a separating funnel, evaporated at 40°C, and redissolved in 1 ml of methanol.

2. Water

Duplicate amounts of water samples (50 ml) were extracted twice with chloroform (50 + 50 ml), evaporated at 40°C, and redissolved in 1 ml of methanol.

The herbicides were determined with an HPLC apparatus (Perkin Elmer 410 Ser.) equipped with a C8 Supecoport column (4.6 mm i.d., 250 mm long) and with an UV detector (Perkin Elmer LC 95) at a wavelength of 254 nm. The mobile phase was 70% methanol and 30% water, and the flow rate 1 ml min^{-1}. Under these conditions, the retention times were 9.2 min for terbuthylazine and 11.2 min for metolachlor, while the sensitivity of the method was, respectively, 0.005 and 0.008 μg ml^{-1}.

III. RESULTS

Figure 1 reports the climatic data recorded during the trial; the mean air temperature was 18.3°C in May, 19.3°C in June, and 21.8°C in July. A 20-mm trickle irrigation was added soon after the herbicide application, inducing poor drainage on the following days.

Metolachlor and terbuthylazine showed different 50% disappearance times (DT_{50}, roughly 17 and 24 d, respectively), but each herbicide showed negligible differences between the two soil profiles (Figures 2 and 3). The same figures also show that the disappearance of the two herbicides followed a first-order kinetics. Laboratory trials

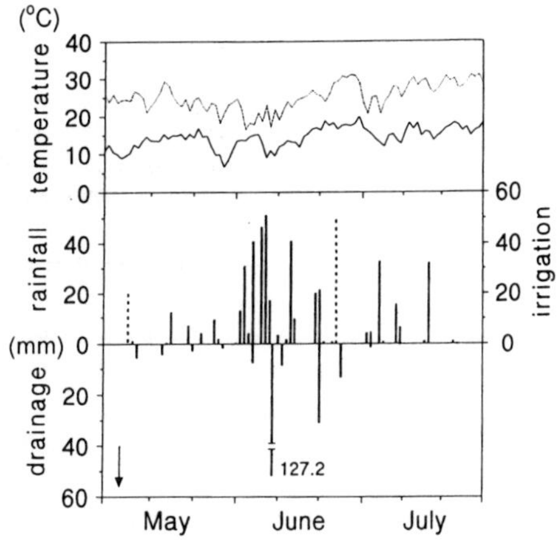

Figure 1 Weather data recorded during the experiment.

carried out in soils similar to that under experiment confirmed this pattern and the lower DT_{50} of metolachlor as compared to terbuthylazine.[9,10]

In spite of physicochemical differences between the two herbicides, no significant difference was noticed in their distribution in the soil profiles (Figures 4 and 5). Terbuthylazine was detected until July 20, while metolachlor disappeared within 58 d after the treatment (July 4). Neither herbicide was recovered at any depth below 40 cm. Within 58 d after the treatment, the herbicides had almost disappeared from the 0- to 10-cm soil layer.

The data obtained by water analysis showed significant concentrations of metolachlor, but any terbuthylazine was below the limits of sensitivity of the analytical procedure (Table 2). The recovery of metolachlor in drainage water started 6 d after the treatment and lasted roughly 1 month. At the end of this period (June 6), about 29 g ha^{-1} of metolachlor had leached through the 0.5-m lysimeters and 21 g ha^{-1} through the 1.0-m ones.

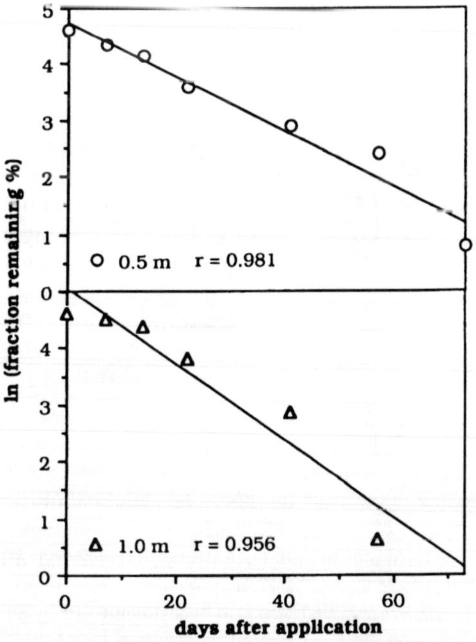

Figure 2 Disappearance of metolachlor in the two soil depths during the experiment.

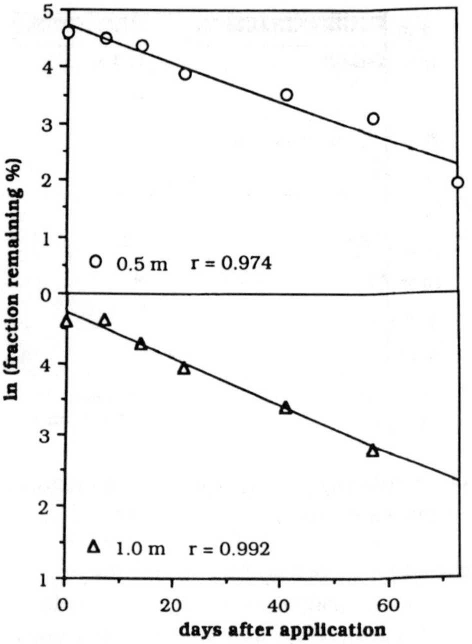

Figure 3 Disappearance of terbuthylazine in the two soil depths during the experiment.

IV. DISCUSSION

Under field lysimeter conditions, the two herbicides exhibited quite similar behavior.

The recovery of metolachlor and terbuthylazine from soil and water 30 d after application was about 25 and 40%, respectively. Provided that negligible amounts of both herbicides were taken up by the plants and lost by volatilization, it can be inferred that a large majority of the applied herbicide was degraded[11] and/or immobilized in soil as bound residue.[12] Additional studies on this topic are strongly needed to understand the environmental significance of herbicide disappearance.

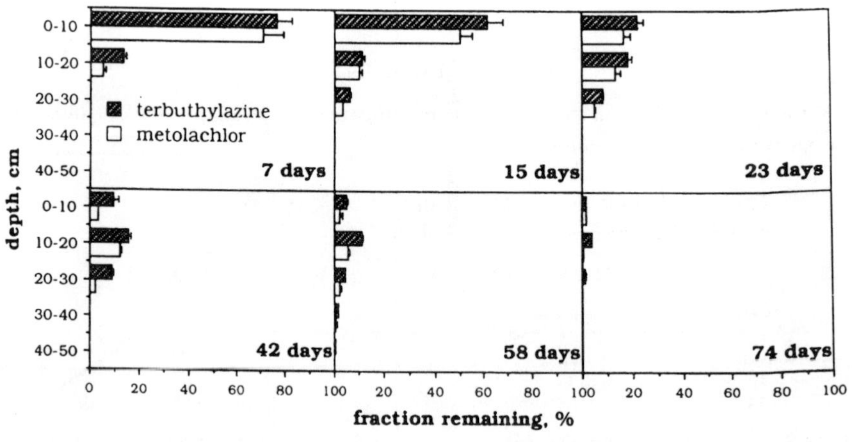

Figure 4 Distribution of terbuthylazine and metolachlor in the soil profile of 50-cm-depth lysimeters. Each data is the average of four replicates ± standard error.

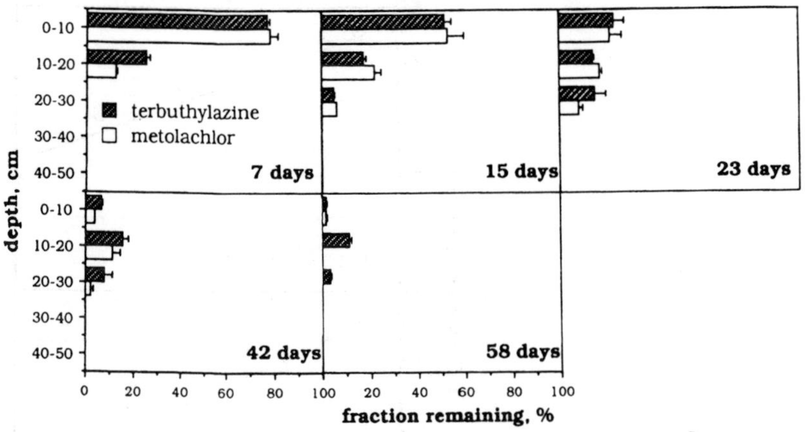

Figure 5 Distribution of terbuthylazine and metolachlor in the soil profile of 100-cm-depth lysimeters. Each data is the average of four replicates ± standard error.

According to Bowman,[4] terbuthylazine seemed to move deeper than metolachlor in the soil profile, but it is likely that it was due to the slower degradation of terbuthylazine.

The behavior of the herbicides differed mainly in the fraction recovered in water, even if less than 2% of the distributed metolachlor was leached. This is consistent with the water solubility and Freundlich constant, which suggested a higher affinity of terbuthylazine for the soil matrix than metolachlor.

Figure 6 contains the comparison between data simulated with the CALF model and those obtained in the experiment for the 50-cm-long lysimeters at different measuring dates. As shown, there is good agreement between the measured and simulated persistence; for metolachlor, the fit with first-order kinetics is weaker than for terbuthylazine (see Figures 2 and 3).

The data simulated with the CALF model for prediction of the fate of pesticides in the soil profile were in agreement with measured data, even if there was an underestimation of the herbicide movement, probably due to the presence of cracks through the soil profile.

V. CONCLUSIONS

Field studies were carried out to evaluate the mobility and persistence of metolachlor and terbuthylazine in drainage lysimeters.

Table 2 Herbicide Concentrations in Drainage Water as Affected by the Depth of the Soil Profile

	Metolachlor (μg l^{-1})		Terbuthylazine (μg l^{-1})	
DAA[a]	0.5 m	1.0 m	0.5 m	1.0 m
3	93	89	ND[b]	ND
16	178	168	8	ND
23	135	130	ND	ND
31	92	86	ND	ND

[a] Days after application.
[b] Below the detection limit.

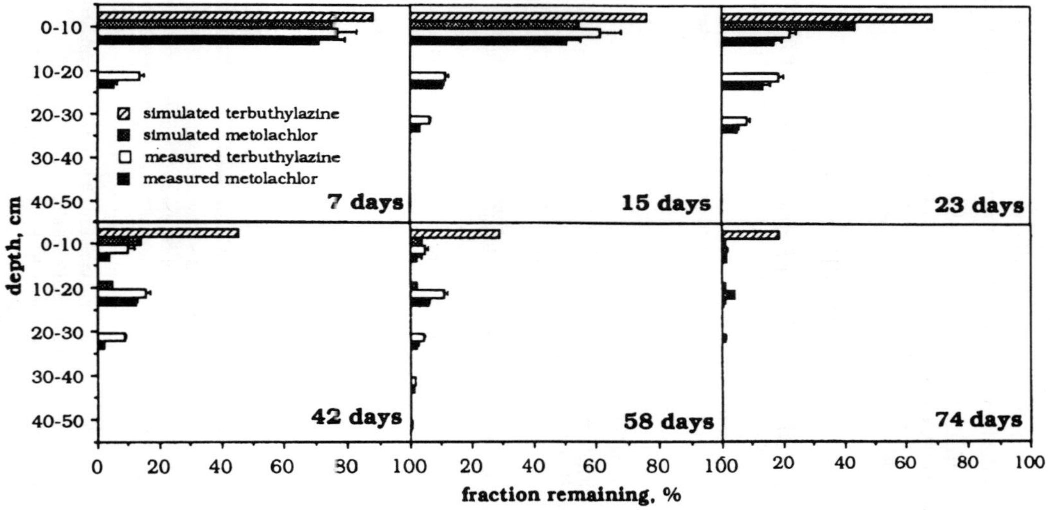

Figure 6 Comparison of the measured and simulated distribution of metolachlor and terbuthylazine in the soil profile of 50-cm-depth lysimeters.

The two herbicides showed different times to 50% loss: 17 d for metolachlor and 24 d for terbuthylazine. Both herbicides disappeared from soil following a first-order kinetics.

The data obtained by water analysis showed significant concentrations of metolachlor, but any terbuthylazine was below the limits of sensitivity of the analytical procedure.

In the performed experiment, metolachlor and terbuthylazine exhibited quite similar behavior, in spite of their different physicochemical properties.

The depth reached in the soil profile by the two herbicides was the same (30 cm); even in percolated water, only metolachlor was present.

Regarding the pesticide movement, generally, the experimental data showed a much deeper displacement for the two herbicides in comparison to the simulated data. However, simulation may be weak in this regard because it is based on average infiltration data only. It is fair to say, moreover, that at present it is virtually impossible to model "by-pass" flow.

ACKNOWLEDGMENTS

This research was supported by the National Research Council of Italy, Special Project RAISA, Subproject no. 2, Paper no. 1103. The authors gratefully acknowledge Mr. Alberto Coletti for his help in analyses.

REFERENCES

1. **Vischetti, C. and Businelli, M.,** Evaluation of a simulation model for prediction of chlorsulfuron fate in an Umbrian soil, *Sci. Total Environ.,* 123/124, 581, 1992.
2. **Bergstrom, L.,** Use of lysimeters to estimate leaching of pesticides in agricultural soils, *Environ. Pollut.,* 67, 325, 1990.
3. **Bowman, B. T.,** Mobility and persistence of metolachlor and aldicarb in field lysimeters, *J. Environ. Qual.,* 17, 689, 1988.
4. **Carsel, R. F.,** Comparison of pesticide root zone model predictions with observed concentrations for the tobacco pesticide metalaxyl in unsaturated zone soils, *Environ. Toxicol. Chem.,* 5, 345, 1986.
5. **Weber, J. B. and Miller, C. T.,** Organic chemical movement over and through soil, in *Reaction and Movement of Organic Chemicals in Soils,* Sawhney, B. L. and Brown, K., Eds., Soil Science Society of America, American Society of Agronomy, Madison, WI, 1989, 305.
6. **Chesters, G., Simsiman, G. V., Levy, J., Alhajjar, B. J., Fathulla, R. N., and Harkin, J. M.,** Environmental fate of alachlor and metolachlor, *Rev. Environ. Contam. Toxicol.,* 110, 1, 1989.
7. **Nicholls, P. H., Walker, A., and Baker, R. J.,** Measurement and simulation of the movement and degradation of atrazine and metribuzin in a fallow soil, *Pestic. Sci.,* 12, 484, 1982.
8. **Walker, A.,** Evaluation of a simulation model for prediction of herbicide movement and persistence in soil, *Weed Res.,* 27, 143, 1987.
9. **Zimdahl, R. L.,** Degradation of three acetanilide herbicides in soil, *Weed Sci.,* 30, 545, 1982.
10. **Vischetti, C. and Businelli, M.,** Simulazione del movimento e della persistenza di 4 erbicidi s-triazinici nei suoli umbri, *Ann. Fac. Agrar. Perugia,* 54, 217, 1990.
11. **Bouchhard, D. C., Lavy, T. L., and Marx, D. B.,** Fate of metribuzin, metolachlor, and fluometuron in soil, *Weed Sci.,* 30, 629, 1982.
12. **Calderbank, A.,** The occurrence and significance of bound pesticide residues in soil, *Rev. Environ. Contam. Toxicol.,* 108, 71, 1988.

Chapter 28

Soil-Pesticide Interactions and Their Impact on the Volatilization Process

W. F. Spencer, M. M. Cliath, and S. R. Yates

CONTENTS

I. INTRODUCTION

Volatilization is one of the dominant processes that controls the movement of applied pesticides into the general environment and their lifetime and effectiveness within the target area. Volatilization is defined as the loss of chemicals from surfaces in the vapor phase, e.g., vaporization followed by movement into the atmosphere. The potential volatility of a chemical is related to its inherent vapor pressure, but actual vaporization rates depend on the environmental conditions and all other factors that control the behavior of the chemical at the solid-air-water interface.[1] Volatilization of soil-applied pesticides is strongly dependent on the nature and rate of the interactions between the applied pesticide and the soil constituents. Vapor pressures of pesticides are greatly decreased by their interactions with soil, mainly resulting from adsorption. This chapter enumerates and discusses the factors affecting volatilization from soil, with particular emphasis on soil pesticide interactions and their impact on the volatilization process. Three comprehensive reviews of pesticide volatilization were recently published.[2-4]

A. FIELD-MEASURED VOLATILIZATION LOSSES

Field measurements using microclimate techniques have shown that volatilization rates from moist soil surfaces can be very large, with losses approaching 90% within 3 d, even for chemicals with vapor pressure less than 0.133 Pa (10^{-3} mmHg).[5] Table 1, taken from Glotfelty et al.,[6] illustrates this along with some factors affecting vapor losses. Volatilization rates were measured by the aerodynamic method, in which pesticide volatilization losses are calculated from pesticide concentration, wind speed, and temperature gradients obtained simultaneously in the atmosphere over the treated field.[10] Measured volatilization rates were dependent upon the nature of the pesticides and how they were used. Shallow soil incorporation and dry soil surfaces greatly restricted volatilization losses.

Similar results have been reported by others. For example, Cliath and co-workers[11] applied EPTC, a water-soluble herbicide, directly in surface irrigation water, where the

0-87371-914-X/95/$0.00+$.50

Table 1 Summary of Pesticide Volatility Studies[6]

Mode of application	Percentage lost	Time elapsed	Data source
Trifluralin (vapor pressure = 1.1 ×10^{-4} mmHg at 25°C)			
Soil incorporation to			
2.5 cm	22	120 d	White et al.[7]
7.5 cm	3.4	90 d	Taylor[5]
Surface, dry soil	2–25[a]	50 h	Experiment 3[b]
Surface, moist soil	50	3–7.5 h	Experiments 2 and 1
	90	2.5–7 d	Experiments 2 and 1
Heptachlor (vapor pressure = 3 ×10^{-4} mmHg at 25°C)			
Soil incorporation to			
7.5 cm	7	167 d	Taylor et al.[8]
Orchard grass	90	7 d	Taylor et al.[9]
Soil surface, dry	14–40[a]	50 h	Experiment 3
Soil surface, moist	50	6 h	Experiment 1
	90	6 d	Experiment 1
Lindane (vapor pressure = 6.3 ×10^{-5} mmHg at 25°C)			
Soil surface, dry	12	50 h	Experiment 3
Soil surface, moist	50	6 h	Experiment 2
	90	6 d	Experiment 2
Chlordane (vapor pressure = 1 ×10^{-5} mmHg at 25°C)			
Soil surface, dry	2	50 h	Experiment 3
Soil surface, moist	50	2.5 d	Experiment 1
DCPA (vapor pressure = 2.5 ×10^{-5} mmHg at 25°C)			
Soil surface, moist	2	34 h	Experiment 1

[a] Range between integrated flux and soil loss.
[b] Experiment numbers refer to data obtained in (1) 1985 or (2) 1977 Beltsville site), or (3) 1978 (Salisbury site).

chemical would be expected to move rapidly into the soil. However, the amount of EPTC volatilized from water and wet soil was equal to 73.6% of the amount applied — 28.4% volatilized in 6 h from the irrigation water and 45.2% from wet soil within 46 h after the water infiltrated (Table 2).

B. THE VOLATILIZATION PROCESS

Volatilization rates of pesticides from nonadsorbing surfaces are dependent only upon vapor pressure and the rate of movement of the chemical away from the evaporating surface. The actual rates of loss, or the proportionality constant relating vapor pressure to the volatilization rate, are almost entirely dependent upon external conditions that affect movement away from the evaporating surface, such as surface roughness, wind speed, air turbulence, etc.[1,2,11] The rate that a pesticide moves away from the surface is diffusion controlled. Close to a solid surface, there is relatively no vertical movement of air, and vaporized substances are transported from the surface through the stagnant air boundary layer (or laminar flow layer) only by diffusion. The actual rate of mass transfer away from the surface by diffusion will be proportional to the diffusion coefficient and to the vapor density at the evaporating surface.

When a pesticide is mixed into the soil, it vaporizes at a rate which depends mainly upon three processes: (1) its rate of movement away from the surface, as above, (2) its

Table 2 Amounts of EPTC in Runoff, and Volatilized from Water and Wet Soil, During and After a Flood Irrigation Application to Alfalfa[11]

EPTC	kg ha^{-1}	% of total applied
Applied in irrigation water (ave 2.17 ppm)	3.04	100.0
Runoff in tailwater (ave 1.70 ppm)[a]	0.21	7.0
Volatilized from water[b]	0.86	28.4
Volatilized from wet soil[c]	1.38	45.2
Total volatilized	2.24	73.6
Total lost	2.45	80.6

[a] EPTC concentration in runoff varied from 1.97 to 1.44 ppm during volatilization measurements.
[b] Volatilized between 1445 and 2045 h on 25 May.
[c] Volatilized between 2045 h on 25 May and 1845 h on 27 May.

effective vapor pressure at the surface or within the soil, and (3) its rate of movement to the vaporizing surface. Soil-incorporated pesticides volatilize at a greatly reduced rate dependent not only on the equilibrium distribution between the air, water, and soil matrix as related to vapor pressure, solubility, and adsorption, but also on their rate of movement to the soil surface. The initial volatilization rate of soil-incorporated pesticides will be a function of the vapor pressure of the chemical at the soil surface as modified by adsorption. The small fraction of the pesticide that remains on the soil surface after mixing is readily lost, and volatilization then becomes dependent on the rate of movement of the chemical to the soil surface by diffusion and convection in evaporating water compared with the rate of movement through the stagnant air boundary layer.

When water evaporates from the soil surface, the suction gradient produced results in appreciable upward movement of water to replace that which evaporated, and any pesticides in the soil solution will move toward the surface by mass flow (convection) with the evaporating water. The phenomenon called the "wick effect" was experimentally demonstrated by Spencer and Cliath[12] to accelerate the volatilization of organic chemicals. Volatilization rates due to mass flow or convection can be estimated from the water flux and the concentration of the chemical in the soil water. The dissolved concentration of chemical in the soil water can be calculated from an adsorption coefficient for the specific soil or estimated from K_{oc} values. When water is not evaporating, the volatilization rate depends on the rate of movement of the chemical to the soil surface by diffusion, and diffusion equations can be used to predict changes in chemical concentration within the soil and the rate of volatilization at the soil surface.[13] Most of the factors affecting volatilization rates will control one or more of the three processes outlined above.

II. FACTORS CONTROLLING PESTICIDE VOLATILIZATION

The factors controlling pesticide volatilization are grouped under four headings: (1) properties of the chemical, (2) soil environment, (3) weather or climatic factors, and (4) management and usage factors under a farmer's control.

A. PROPERTIES OF THE CHEMICAL

Properties include (1) vapor pressure, (2) solubility, (3) air water-partition coefficient or Henry's law coefficient, K_H, (4) soil water-partition coefficient or K_{oc}, (5) diffusion coefficient in air, (6) diffusion coefficient in water, and (7) degradation rate or $t_{1/2}$.

In a series of papers, Jury et al.[14-17] described and applied a screening model for assessing the relative volatility and persistence of pesticides and other trace organics in soil based on their physicochemical properties. In lieu of discussing the effect of individual chemical properties, some of their interacting effects on volatilization are shown as output from the above screening model. Figure 1 shows the volatilization output from the screening model for 20 pesticides using input data for Henry's law constant, adsorption coefficient, and degradation rate.[18] The relative magnitude of the calculated volatilization rates and their change with time depended upon the physicochemical properties of the pesticides and water evaporation rates. Water evaporation had a considerable effect on the volatilization of some chemicals and not on others. Also, volatilization of some pesticides decreased with time and volatilization of others increased with time when water was evaporating. The model output indicated that the volatilization behavior of a pesticide is controlled mainly by the ratio of its solution to the vapor concentration or Henry's law constant (K_h), which determines the extent to which the air boundary layer restricts volatilization from soil. The predictions indicate that chemicals with a sufficiently low K_H ($<<2.65 \times 10^{-5}$) will move to the surface in evaporating water faster than they can volatilize through the air boundary layer. Consequently, under evaporative conditions, their concentration increases at the soil surface and volatilization rate increases.

Laboratory and field experiments have confirmed these predictions with respect to the importance of Henry's law constant in controlling relative volatilization and surface accumulation. In the laboratory, this was verified by measuring volatilization from soil columns of two chemicals with widely different K_H values: prometon, with $K_H = 1.0 \times 10^{-7}$, and lindane, with $K_H = 1.33 \times 10^{-4}$. Figure 2 shows measured and predicted prometon vapor flux vs. time with and without water evaporating.[19] Extremely good agreement was observed between measured and predicted vapor flux values both with and without water evaporating. With water evaporating, the volatilization of prometon increased from 0.03 $\mu g/cm^{-2}$ d during the first 5 h to 1.24 $\mu g\ cm^{-2}$ d after 14 d, with water evaporating at 0.55 cm^{-1} d. This is a 41-fold increase in volatilization rate over the 14-d period due to accumulation of prometon at the soil surface as water evaporated. Measured distribution in the soil column after 14 d showed that prometon accumulated at the soil surface with water evaporating. Concentrations in the surface 0.3 cm increased from 10 to 81 $\mu g\ g^{-1}$. In the case of lindane volatilization under the same conditions, good agreement was observed between measured and predicted values, but volatilization decreased with time, with or without water evaporating.[19] After 14 d with water evaporating, lindane was distributed throughout the column with only a slight decrease in concentration at the surface instead of an accumulation as was observed with prometon.

Volatilization and movement of four herbicides as related to their physical and chemical properties was measured under field conditions.[20] Soil samples taken from a depth of 0 to 1 cm with a 3-cm-diameter soil plug sampler at various times after application were analyzed to follow herbicide concentration changes as related to water application or drying periods. The field was irrigated twice after the application irrigation, with soil drying occurring between the irrigations. With triallate ($K_H = 7.90 \times 10^{-4}$), concentrations continually decreased following application, whether the soil was drying or being wetted with irrigation water. Prometon ($K_H = 1.0 \times 10^{-7}$) reacted as expected: concentrations in the surface 1 cm markedly increased while the soil was drying. Following the first irrigation, prometon concentrations increased from 15.5 to 61.5 $\mu g\ g^{-1}$; following the second irrigation at day 10, concentrations increased from 5.0 to 27.9 $\mu g\ g^{-1}$; and after the third irrigation, the concentrations increased from 1.3 to 16.0 $\mu g\ g^{-1}$ over the final 13-d drying period. The other two herbicides with intermediate K_H values reacted in an intermediate fashion. Their concentrations increased only slightly during the drying periods.

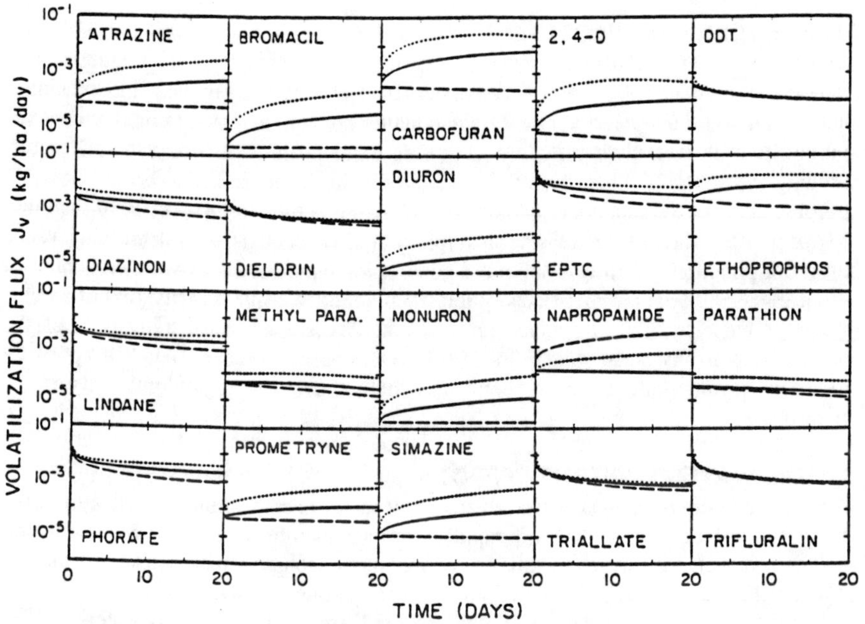

Figure 1 Volatilization flux vs. time for water evaporation rate.[18] E = 0 (dashed line), 2.5 mm d^{-1} (solid line), and 5.0 mm d^{-1} (dotted line), with L = 10 cm, $\theta = 0.3$, $f_{oc} = 0.0125$, and d = 4.75 and 2.375 mm, respectively, where L is depth of incorporation, θ is water content, f_{oc} is organic carbon fraction, and d is boundary layer thickness.

The phenomenon of pesticides accumulating at the soil surface following convective movement in evaporating water has several interesting ramifications. First, the mechanism could greatly enhance the volatilization of chemicals with low K_H, formerly considered to be essentially nonvolatile from wet soils. Second, the mechanism could significantly increase the amounts of some chemicals available for photolysis or for runoff into surface water bodies with rainfall or irrigation water.

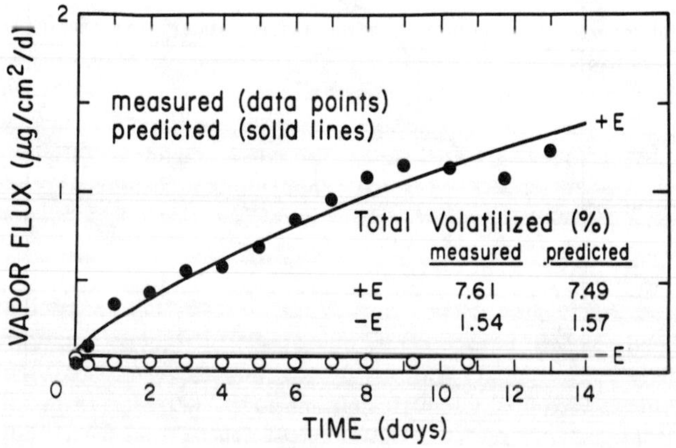

Figure 2 Volatilization of prometon from San Joaquin sandy loam with (+E) and without (–E) water evaporating.[19]

B. SOIL PHYSICAL FACTORS

Soil environmental factors include (1) soil organic matter content, (2) soil texture or clay content, (3) soil pH, (4) bulk density or porosity, (5) soil water content and its distribution, and (6) soil water evaporation rate. Organic matter content, soil texture or clay content, and soil pH all exert their influence over volatilization through adsorption effects on reducing the solution concentration and vapor density in the soil. For weakly polar or nonionic chemicals, the amount of soil organic matter is the most important of the three soil properties related to increasing adsorption and, consequently, to decreasing vapor density or potential volatility. With more polar or ionic molecules, clay minerals play an increasingly important role in adsorption and volatility. Bulk density or porosity is important because of its influence on vapor and nonvapor phase movement of the chemical and also because of its influence on water evaporation rates. Air-filled porosity is especially important for those chemicals whose movement is predominantly in the vapor phase.

C. SOIL WATER CONTENT EFFECTS

Soil water content effects far outweigh any other factor in controlling volatilization from soil. Pesticides vaporize much more rapidly from wet than from dry soils. Measurements of vapor pressures in soil at various water contents conclusively demonstrated that the greater vaporization from wet than from dry soils is due mainly to an increased vapor pressure resulting from displacement of the pesticide from soil surfaces by water.[21-24] This is illustrated in Figure 3, where the relative vapor density of dieldrin is plotted against its concentration in soil at various water contents.[21] In moist soil, the vapor density increased rapidly with the soil concentration of dieldrin, but in very dry soil, the vapor concentration was almost nil even at 100 ppm dieldrin. To put this in perspective, 3.9% water where vapor density is still high is equal to 90 atm of suction, which is much dryer than the wilting point, and 2.1% water in this soil is equivalent to less than one molecular layer of water. When water content is reduced below this level, adsorption sites on the soil mineral phase, on which the pesticide molecules are more strongly bound, become active so that fugacity, and hence vapor density of the pesticide, is greatly reduced. Similar

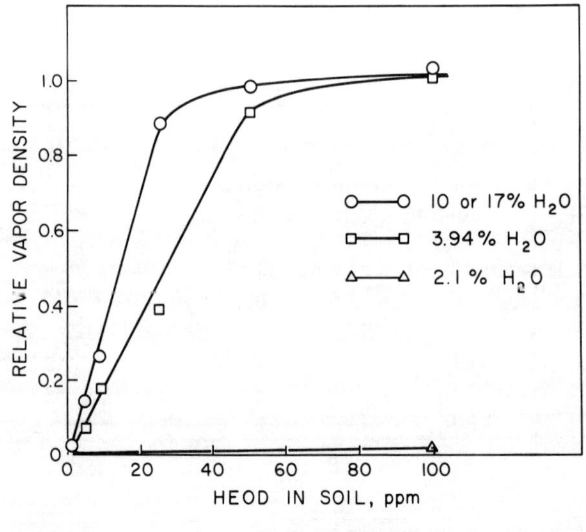

Figure 3 Relative vapor density of dieldrin (HEOD) vs. concentration in Gila silt loam as affected by soil water content.[21]

relationships were found to hold for other pesticides in other soils. The drying effect was largely reversible in response to changes in water content, in that vapor pressures increased to near their original values upon rewetting the air-dry soil.

The magnitude of the so-called "drying effect" as expressed by the ratio of vapor density in wet to vapor density in dry soil depends on the pesticide and the soil. The ratio of wet to dry vapor density for dieldrin in five soils was approximately 100.[21] The ratio of wet to dry vapor density in Gila silt loam for lindane[22] was 10, for DDT[23] 100, and for trifluralin[24] approximately 3000.

The mechanistic roles of soil humus and soil minerals and their contribution to soil sorption of nonionic pesticides in wet and dry soils are the same as reported by Chiou and co-workers[25] when they measured adsorption of the insecticides parathion and lindane from hexane onto dry and partially hydrated soil. They observed much greater adsorption of parathion and lindane from hexane on dry than on moist soil. Chiou and Shoup[26] measured vapor adsorption of organic compounds onto soil at various relative humidities. Again, increasing soil-water content greatly decreased the vapor adsorption of compounds such as benzene and various chlorobenzenes. They ascribed both the lesser adsorption of organic compounds from hexane and lesser vapor adsorption onto moist soil to the adsorptive displacement by water of organics adsorbed on the mineral matter.[26] Our observations[21-24] and theirs are consistent with Chiou's[27] concept of the soil as a dual sorbent in which the mineral fraction of the soil functions as a conventional solid adsorbent and the organic matter functions as a partition medium. In aqueous systems, adsorption on mineral matter is suppressed by water and the uptake by soil consists primarily of solute partitioning into the organic matter. In dry soils, it is the mineral fraction that strongly adsorbs the nonionic organic compounds.

In the field, soils can rapidly form a dry surface layer that greatly suppresses pesticide volatilization. Confirmation of the soil water content effects on volatility in the field was reported by Harper et al.[28] and by Glotfelty et al.[6] Harper et al.[28] measured trifluralin volatilization losses from a treated soybean field along with soil moisture and temperature changes throughout the growing season. Figure 4 shows diurnal trifluralin volatilization rates as affected by soil water content, water flux, and microclimate effects during one 24-h period.[28] This was a warm, sunny day with high soil water loss rates resulting in the water content in the surface 0.5 cm decreasing during midday to about one molecular layer of water. The pesticide flux began to increase in the morning but fell very rapidly when the soil water content in the surface 0.5 cm dropped to low levels. The water flux from the soil continued to increase, indicating that temperature and other conditions were conducive to high evaporative losses, but trifluralin flux remained low. Volatilization flux exhibited a much stronger dependence on soil water content than on temperature, water flux, or wind speed. On other days when the water content of the soil did not decrease because of poor evaporative conditions, the trifluralin flux showed a marked temperature dependence and increased during the day as temperature increased. These changes in volatilization flux are consistent with laboratory-measured factors affecting the vapor density of trifluralin.[24]

Glotfelty et al.[6] reported the expected diurnal changes in the volatilization flux of four pesticides from a moist silt-loam soil with maximum flux intensities reached near noon when insolation was highest. This pattern is usually observed in soils that contain sufficient water to maintain a moist soil surface throughout the day when net radiation is highest. Conversely, Glotfelty et al.[6] reported that volatilization was greatly reduced near noon from a sandy soil where sufficient drying occurred at the surface to restrict vapor density. During both evenings, before and after midday, the sandy soil surface was remoistened from dew and the upward movement of water, which reactivated adsorbed pesticides and resulted in an increased volatilization in the morning and the evenings.

These field observations confirm the primary importance of the moisture content of the surface soil layer in controlling the volatility of soil residues. The effect is primarily

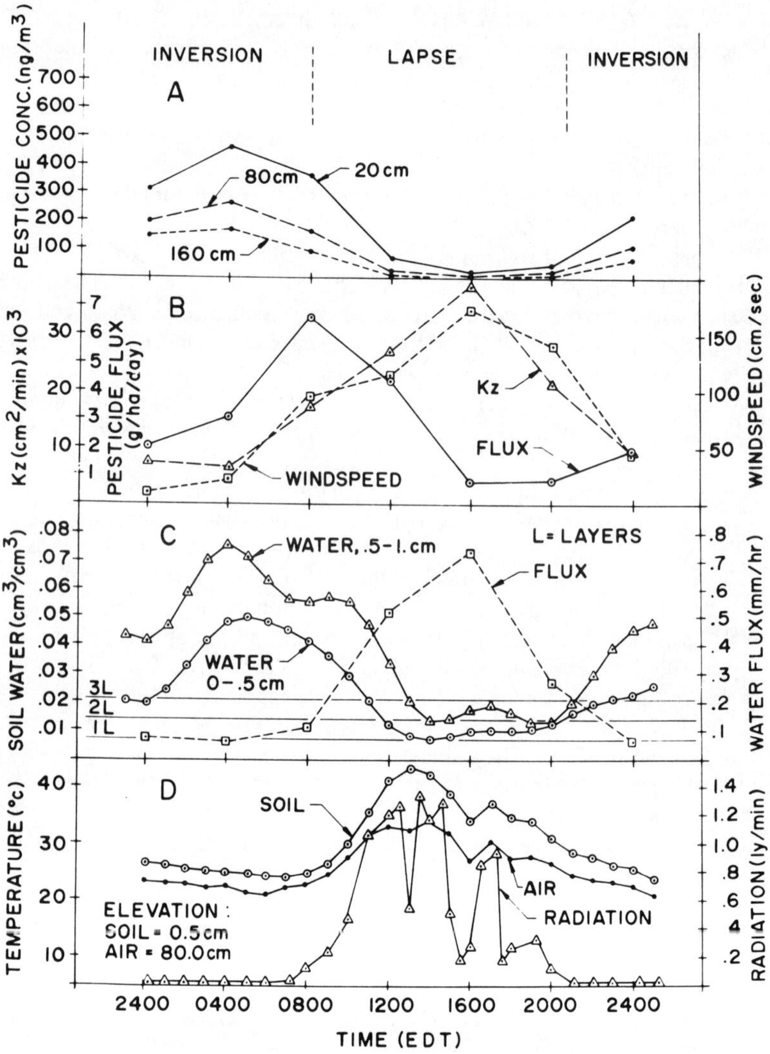

Figure 4 Diurnal trifluralin volatilization flux as affected by soil water contents, water flux, temperature, and wind speed.[28] K_z is a transfer coefficient.

because of the reduction in the fugacity of the residues by the increased adsorption in dry soil. A lesser contributing factor is the fact that enhancement of volatilization by the "wick effect" is also greater in moist than in dry soils.

D. WEATHER OR CLIMATIC FACTORS

Weather or climatic factors include (1) temperature, (2) relative humidity, (3) wind speed, (4) rainfall, and (5) insolation. Temperature influences volatilization rates mainly through its effect on vapor pressure. Vapor pressures of most pesticides increase three to four times for each 10°C increase in temperature. Temperature may also influence volatility through its effect on the movement of the chemical to the surface by diffusion or mass flow in evaporating water, or through its effect on the soil water adsorption/desorption equilibrium. For these effects, increases in temperature are usually associated with increases in volatilization rate. However, this may not always be the case, since an increase in temperature is also associated with an increase in the drying rate of the soil surface, thereby possibly decreasing vapor density and resulting in less volatilization than at the lower temperature, as explained in the previous section.

Humidity is important because of its effect on the water evaporation rate and the water content of the surface layer of soil. Wind speed and insolation also affect the water evaporation rate, as well as the pesticide movement from the soil surface. All other things being equal, volatilization rate will increase with increasing wind speed or increasing insolation. Rainfall, of course, greatly affects the water status of the soil and usually moves pesticide downward away from the soil surface.

E. MANAGEMENT AND USAGE FACTORS

Management and usage factors under control of the farmer include (1) rate of application, (2) method of application, (3) formulation, and (4) tillage practice. These factors can greatly affect volatilization rate. The rate and method of application, for example, soil surface vs. incorporation, in large part control the initial concentration of the pesticide at the soil surface. With a 1 kg ha^{-1} surface application of a pesticide to bare soil, the surface layer of about 1 mm penetrated by the spray may contain as much as 100 μg g^{-1}. With many pesticides, such a high soil concentration in moist soil will produce a vapor pressure or fugacity equal to that of the pure pesticide without soil. For example, saturated vapor densities in moist Gila silt loam were reached at soil concentrations of 25, 55, 15, and 73 μg g^{-1} for dieldrin,[21] lindane,[22] DDT,[23] and trifluralin,[24] respectively. In contrast to surface applications, incorporating the pesticide into the soil to a depth of 7.5 cm reduces the concentration of a 1 kg ha^{-1} application in the soil to about 1 μg g^{-1}. This means an approximate 100-fold reduction in fugacity and a 100-fold increase in the average path length to the soil surface.

The major effects of formulation are probably most evident during application due to drift and volatilization from droplets, but the formulation can also influence the penetration of the applied pesticide into the soil. For example, one difference worth noting is the difference between liquid and granules. Liquid sprays alight on all exposed areas, most of which are crests of soil clods. Granules will, in general, bounce off high places and settle in crevices. If so, granules, on average, will be several millimeters lower than spray residues, be better sheltered, and have a larger solid angle available for diffusion into the soil. Even if the soil is mechanically disturbed to allow incorporation, smooth granules have a greater tendency to settle downward than the disturbed soil clods.

The effect of microencapsulation on the volatilization of chlorpropham was documented in a field experiment by Turner et al.[29] The specific volatilization rate of the encapsulated formulation was between 12 and 26% of the conventional EC formulation during the first 8 d. Their results suggest that encapsulation techniques offer real possibilities for reducing the environmental impact of some pesticides.

Tillage practices can greatly affect pesticide volatilization rates. Glotfelty[30] recently summarized the effects of tillage practices on pesticide volatilization and degradation, with particular emphasis on the effects of conservation tillage practices. Conservation tillage is a soil management technique that emphasizes reduced frequency and intensity of tilling the soil and results in a preservation of some type of cover on the soil surface as a mulch. Compared to conventional tillage practices, conservation tillage may change greatly both the manner in which pesticides are used and the properties of the soil in ways that significantly affect pesticide volatilization. For example, the upper soil layers in a no-till field are generally wetter, cooler, and higher in organic matter than comparable fields conventionally tilled. The plant debris or mulch layer in no-till surface soil plays a key role in creating these conditions which all greatly affect volatilization. The mulch layer also can intercept a portion of applied pesticide and prevent it from reaching the soil surface. It will be more rapidly lost from the mulch layer than from bare soil.[31] The potential impact of tillage practices upon volatilization needs to be more clearly established.

III. SUMMARY AND CONCLUSIONS

We have gained considerable insight into the processes involved in volatilization and factors affecting volatilization rates.[2] Of course, the intrinsic properties of individual pesticides are important, but data from laboratory and field experiments indicate that the volatilization rate of a specific pesticide in soil is controlled mainly by soil conditions. Meteorological variables, such as temperature, radiation, and wind speed, are important insofar as they control the soil conditions, but soil conditions, and particularly soil moisture levels, are critical in controlling the supply of available pesticide at the surface, which in turn controls the volatilization rate.

There is much work yet to be done. Even though we conclude that soil moisture content effectively controls the adsorption and fugacity of pesticide residues, the factors controlling changes in moisture content in surface layers of soil are not well understood. The ability to understand and predict drying and rewetting in surface soils of various textures under a range of management practices is an essential step in improving the prediction of pesticide volatilization rates. Models need to be developed that will predict the rates of volatilization of pesticides from target areas over extended periods. Such models must include the effects of soil drying on the adsorption and reduction in fugacity of residues and take into account all the other factors that are important in volatilization, including the balance between evaporation and upward movement of soil water and the effects of the range of soil management practices, including reduced tillage. Such models will require climatic or weather data, and the precision of the output will depend on the availability of this information over short time periods.

REFERENCES

1. **Spencer, W. F., Farmer, W. J., and Cliath, M. M.,** Pesticide volatilization, *Residue Rev.,* 49, 1, 1973.
2. **Taylor, A. W. and Spencer, W. F.,** Volatilization and vapor transport processes, in *Pesticides in the Soil Environment: Processes, Impacts, and Modeling,* Cheng, H. H., Ed., Soil Science Society of America, Madison, WI, 1990, 213.
3. **Taylor, A. W. and Glotfelty, D. E.,** Evaporation from soils and crops, in *Environmental Chemistry of Herbicides,* Grover, R., Ed., CRC Press, Boca Raton, FL, 1989, 89.
4. **Glotfelty, D. E. and Schomburg, C. J.,** Volatilization of pesticides from soil, in *Reactions and Movement of Organic Chemicals in Soils,* Sawhney, B. L. and Brown, K., Eds., Soil Science Society of America, Madison, WI, 1989, 181.
5. **Taylor, A. W.,** Post-application volatilization of pesticides under field conditions, *J. Air Pollut. Control Assoc.,* 28, 922, 1978.
6. **Glotfelty, D. E., Taylor, A. W., Turner, B. C., and Zoller, W. H.,** Volatilization of surface-applied pesticides from fallow soil, *J. Agric. Food Chem.,* 32, 638, 1984.
7. **White, A. W., Jr., Harper, L. A., Leonard, R. A., and Turnbull, J. W.,** Trifluralin volatilization from a soybean field, *J. Environ. Qual.,* 6, 105, 1977.
8. **Taylor, A. W., Glotfelty, D. E., Glass, B. L., Freeman, H. P., and Edwards, W. M.,** The volatilization of dieldrin and heptachlor from a maize field, *J. Agric. Food Chem.,* 24, 625, 1976.
9. **Taylor, A. W., Glotfelty, D. E., Turner, B. C., Silver, R. E., Freeman, H. P., and Weiss, A.,** Volatilization of dieldrin and heptachlor residues from field vegetation, *J. Agric. Food Chem.,* 25, 542, 1977.
10. **Parmele, L. H., Lemon, E. R., and Taylor, A. W.,** Micrometeorological measurement of pesticide vapor flux from bare soil and corn under field conditions, *Water Air Soil Pollut.* 1, 433, 1972.

11. **Cliath, M. M., Spencer, W. F., Farmer, W. J., Shoup, T. D., and Grover, R.,** Volatilization of S-ethyl N, N-dipropylthiocarbamate from water and wet soil during and after flood irrigation of an alfalfa field, *J. Agric. Food Chem.,* 28, 610, 1980.
12. **Spencer, W. F. and Cliath, M. M.,** Pesticide volatilization as related to water loss from soil, *J. Environ. Qual.,* 2, 284, 1973.
13. **Mayer, R., Letey, J., and Farmer, W. J.,** Models for predicting volatilization of soil-incorporated pesticides, *Soil Sci. Soc. Am. Proc.,* 38, 563, 1974.
14. **Jury, W. A., Spencer, W. F., and Farmer, W. J.,** Behavior assessment model for trace organics in soil. I. Model description, *J. Environ. Qual.,* 12, 558, 1983.
15. **Jury, W. A., Farmer, W. J., and Spencer, W. F.,** Behavior assessment model for trace organics in soil. II. Chemical classification and parameter sensitivity, *J. Environ. Qual.,* 13, 567, 1984.
16. **Jury, W. A., Spencer, W. F., and Farmer, W. J.,** Behavior assessment model for trace organics in soil. III. Application of screening model, *J. Environ. Qual.,* 13, 573, 1984.
17. **Jury, W. A., Spencer, W. F., and Farmer, W. J.,** Behavior assessment model for trace organics in soil. IV. Review of experimental evidence, *J. Environ. Qual.,* 13, 580, 1984.
18. **Jury, W. A., Spencer, W. F., and Farmer, W. J.,** Use of models for assessing relative volatility, mobility, and persistence of pesticides and other trace organics in soil systems, in *Hazard Assessment of Chemicals, Current Developments,* Vol. 2, Saxena, J., Ed., Academic Press, New York, 1983, 1.
19. **Spencer, W. F., Cliath, M. M., Jury, W. A., and Zhang, L.-Z.,** Volatilization of organic chemicals from soil as related to their Henry's law constants, *J. Environ. Qual.,* 17, 504, 1988.
20. **Spencer, W. F. and Cliath, M. M.,** Movement of pesticides from soil to the atmosphere, in *Long Range Transport of Pesticides,* Kurtz, D. A., Ed., Lewis Publishers, Chelsea, MI, 1990, 1.
21. **Spencer, W. F., Cliath, M. M., and Farmer, W. J.,** Vapor density of soil-applied dieldrin as related to soil-water content and dieldrin concentration, *Soil Sci. Soc. Am. Proc.,* 33, 509, 1969.
22. **Spencer, W. F. and Cliath, M. M.,** Desorption of lindane from soil as related to vapor density, *Soil Sci. Soc. Am. Proc.,* 34, 574, 1970.
23. **Spencer, W. F. and Cliath, M. M.,** Volatility of DDT and related compounds, *J. Agric. Food Chem.,* 20, 645, 1972.
24. **Spencer, W. F. and Cliath, M. M.,** Factors affecting vapor loss of trifluralin from soil, *J. Agric. Food Chem.,* 22, 987, 1974.
25. **Chiou, C. T., Shoup, T. D., and Porter, P. E.,** Mechanistic roles of soil humus and minerals and adsorption on nonionic organic compounds from aqueous and organic solutions, *Org. Geochem.,* 8, 9, 1985.
26. **Chiou, C. T. and Shoup, T. D.,** Soil sorption of organic vapors and effects of humidity on sorptive mechanism and capacity, *Environ. Sci. Technol.,* 19, 1196, 1985.
27. **Chiou, C. T.,** Roles of organic matter, minerals, and moisture adsorption on nonionic compounds and pesticides by soil, in *Humic Substances in Soil and Crop Sciences: Selected Readings,* MacCarthy, P., Clapp, C. E., Malcolm, R. L., and Bloom, P. R., Eds. American Society of Agronomy, Madison, WI, 1990, 111.
28. **Harper, L. A., White, A. W., Jr., Bruce, R. R., Thomas, A. W., and Leonard, R. A.,** Soil and microclimate effects on trifluralin volatilization, *J. Environ. Qual.,* 5, 236, 1976.
29. **Turner, B. C., Glotfelty, D. E., Taylor, A. W., and Watson, D. R.,** Volatilization of microencapsulated and conventionally applied chlorpropham in the field, *Agron. J.,* 70, 933, 1978.
30. **Glotfelty, D. E.,** The effects of conservation tillage practices on pesticide volatilization and degradation, in *Effects of Conservation Tillage on Groundwater Quality: Nitrates and Pesticides,* Logan, T. J., Ed., Lewis Publishers, Chelsea, MI, 1987, 169.
31. **Whang, J. M., Schomburg, C. J., Glotfelty, D. E., and Taylor, A. W.,** Volatilization of fonofos, chlorpyrifos, and atrazine from conventional and no-till surface soils in the field, *J. Environ. Qual.,* 22, 173, 1993.

Chapter 29

Factors Affecting the Movements, Reactions, and Biotransformations of Xenobiotics

F. Andreux, M. Schiavon, C. Munier-Lamy, M. Mansour, and I. Scheunert

CONTENTS

I. INTRODUCTION: HOW TO CHARACTERIZE "SOIL POLLUTION" DUE TO PESTICIDES?

Contamination of waters has become one of the main concerns of modern societies during the last 2 decades. The presence in water catchments of organic xenobiotics, especially plant protection substances and their metabolites, has been reported in numerous instances. Once pesticide substances are applied to the soil-plant system, they can be casually, sometimes chronically, transferred to groundwaters, through the soil water phase. Whether these substances are used for agricultural or other purposes, they induce a special case of pollution that could be called *intentional* or *inevitable* pollution. Their presence in surface and groundwaters therefore indicates that the capacity of the soil to retain or degrade them is not unlimited, especially in the case of intensive and repeated treatments. In other words, *soil pollution precedes water pollution.*

It seems important to reassess what soil pollution is, especially in the case of pesticides. In Europe, a *legal definition* of drinking water pollution has been provided by the EEC, which would limit the maximal acceptable concentration of pesticide to 0.1 μg l^{1-} for each individual pesticide, and 0.5 μg l^{1-} for the sum of all pesticides present.[1] In soils, such a quantitative definition is possible only if the total of pesticides present in a given soil layer can be determined, and if the main exchange reaction parameters

0-87371-914-X/95/$0.00+$.50

between soil solid phase and soil solution are known. Thus, the pesticide concentration in the soil solution, extracted with standard methods, can be taken into account. A combination of qualitative and quantitative definitions can be proposed:[2]

1. Biocidal pollution, which corresponds to the so-called accidental pollution, is defined as the concentration above which a toxic action on living organisms is detected. To be detected, it requires biological assays.[3]
2. Chemical pollution is based on methodological considerations, especially the extractability of the molecule with organic solvents and the accuracy of detection methods.
3. Ecological pollution is referred to when the presence of micropollutants is known or suspected, but the molecules are not detectable by the above methods. However, these micropollutants can be occasionally released and detected, and even exhibit toxicity. In the case of soils, this definition is tightly related to the possibility of formation of "unextractable" or "bound" residues of pesticides in soils of different types and genesis.[4,5]

The behavior of pesticides and their derivatives in soils can be described as a function of (1) their chemical nature, (2) their interactions with soil components, and (3) the main soil characteristics, which are either internal (e.g., texture, structure, and exchange capacity) or external (e.g., climatic, anthropogenic actions, and management). Most studies have dealt with the first two aspects, whereas relationships to soil types are commonly restricted to pH, texture, and clay and soil organic matter (SOM) contents. It seemed, therefore, reasonable to look for a more comprehensive scheme of pesticide-soil interactions that would take into account soil factors as a whole, rather than successive single analytical characteristics.

The main purpose of this chapter is to document these interactions with pertinent examples taken from both field and laboratory experiments. Emphasis is placed on those chemicals which have the highest tendency to bind to humic substances (HS), especially phenols, anilines, and triazines. Mainly classical pesticides are considered, and new molecules used at doses of a few kilograms per square kilometer, such as sulfonylurea, are mentioned only casually.

II. FACTORS INFLUENCING THE MOVEMENT OF PESTICIDES IN SOIL WATER

A. SOIL DEPTH AND HYDRODYNAMICS

Less than 20 years ago, plant protection substances were considered to interact mainly with surface soil components, and their behavior in deeper soil layers was considered of secondary importance. Our current knowledge about pesticide transport in soils is clearly summarized by Wierenga,[6] who stated that "...although soil scientists have studied pesticides in soil for at least two decades, this has not prevented movement of some pesticides to the groundwater." One reason why the presence of pesticides has not been frequently established in subsoils is probably the lack of adequate sampling, extraction, and detection methods. Recent results clearly indicate that it is no longer feasible to restrict the study of pesticide behavior to the upper soil layer, but to consider the whole soil profile, or better, the whole vadose zone.

The transport of pesticides by soil water can occur at different soil depths, either superficially through erosion and runoff, or laterally and vertically through leaching. In general, the transported pesticide is mainly in the soluble form, and that in the suspended sediment is present at a higher concentration than in the whole soil.[7-9] The proportions of pesticides recovered in soil water depend on factors other than the characteristics of the molecules and the soils, especially the applied doses, and the time elapsed since their application. In the case of atrazine, numerous data are available for both runoff and vertical leaching. Table 1 indicates that losses by runoff are generally higher than those

Table 1 Evaluation of Transport of Atrazine by Runoff and Leaching

Applied doses (kg ha^{-1})	Runoff (%)	Duration (years)	Leaching (%)	Ref.
0.6 – 9.1	1.73 – 2.98	1		Hall et al.[7]
3.4	2.5 – 5.4	2		Ritter et al.[8]
2.2; 4.5	4.9 – 5.3	1		Hall[9]
Variable	0.2 – 5.7	3		Tripplett et al.[10]
1.45 – 4.03	0.2 – 1.9	3.5		Léonard et al.[11]
2.3	1	1		Wu[12]
1.25		1	0.85 – 1.94 (clayey soil)	Schiavon & Jacquin[13]
1.25		1	0.005 – 0.54 (silty soil)	Schiavon & Jacquin[13]
1.4		1	0.15	Muir & Baker[14]
1.2		1	0.25 – 0.80 (clayey soil)	Von Stryk & Bolton[15]
1.2 – 2.5		1	0.004 – 0.01 (silty soil)	Chevreuil & Chesterikoff[16]
1.4		1	2.0 (clayey soil)	Snégaroff[17]
1.0		2	2.0 (sandy soil)	Scheunert & Korte[18]
?		1	0.038 (silty soil)	Southwick et al.[19]

by leaching, with maximal values of 5.3 and 2% per year removed from the soil by the respective processes. In agricultural soils treated with standard doses of different pesticides, the concentration of dieldrin was 2.0 μg kg^{-1} at 1.35 m,[20] and that of atrazine 0.26 μg kg^{-1} at 1.80 m,[21] whereas atrazine, alachlor, metolachlor, and simazine reached concentrations of 0.80, 1.60, 1.20, and 10.20 μg kg^{-1}, respectively, at 2.30 m.[22] Because the SOM content and biodegradability generally decrease from the surface to the underlaying horizons, pesticide residues that can reach deep levels represent a potential of several kilograms per hectare, and can be transferred to the groundwater, regardless of the molecular type and soil porosity.

Lateral movements of pesticides can be determined, using tubing or tile drains, under natural conditions or in combination with rain simulation.[10,14,15] Various approaches are used to study pesticide leaching at different soil depths, and under field conditions. One of the most common methods for nutrient studies, using suction cup samplers, has seldom been applied to pesticides, probably because most of these molecules can be retained on the cup wall surface.[23] Assays using stainless and aluminum oxide devices were reported by Baumann et al.,[24] using various pesticide molecules. In other instances, authors dug sampling wells,[21] isolated undisturbed soil cores, or installed *in situ* soil-packed columns.

Numerous outdoor experiments with soil columns of different sizes have been carried out. The problem of whether lysimeter columns were representative of actual field conditions frequently arose, and was extensively discussed by Führ and his group, especially in the case of metamitron and metabenzthiazuron.[25] Using ^{14}C-labeled molecules established that not only the parent pesticide molecules, but also their transformation products (i.e., "metabolites") can migrate toward and throughout the subsoil layers. Atrazine and its main metabolites (deethylatrazine, deisopropylatrazine, diaminoatrazine, and hydroxyatrazine) were added by Schiavon,[26] to a "Sol brun lessivé" in 0.60-m-high lysimeter columns. After 1 year, he observed that about 50% of the ^{14}C derived from atrazine was retained in the top 0.12 m of the columns, following which the percentage decreased exponentially with depth. The three dealkylated molecules were more mobile, and the dechlorinated atrazine was less mobile, than the active molecule: percentages of ^{14}C recovered in the 0- to 0.24-m plough layer were 87% for atrazine, and between 83.4% (deethylatrazine) and 99.6% (hydroxyatrazine) for the metabolites. However, these molecules did not migrate freely throughout soil pores, as indicated by the low proportions of ^{14}C from the three dealkylated molecules which were extractable in methanol-water (80:20 v/v) solution. Contrarily, hydroxylation of atrazine did not hinder extraction with this solvent, although hydroxyatrazine gave rise to stronger adsorption in soil.[27]

Column devices have also been applied to study the effect of varying tillage systems.[10] Fermanich and Daniel[28] recently used 1-m-high lysimeters to compare the mobility and persistence of the insecticides ^{14}C-carbofuran and chlorpyrifos in a loamy sand soil, and under maize plants. These authors carefully concluded that no-till, when compared with conventional ploughing, induced chemical, biological, and/or physical changes in soils, and that these changes affected the adsorption and dissipation processes of the insecticides to a greater extent than the water movement itself.

B. RAINFALL AND SEASONAL CHANGES

Seasonal follow-up of pesticide residues dissolved in natural waters can give an idea of their average concentration, and of the risks of contamination of a given area. High contents of triazines were reported in rivers of Central Europe,[29] in which 708 analyses showed that atrazine, simazine, terbumeton, and terbuthylazine were present at concentrations higher than 0.1 $\mu g\ l^{-1}$, and frequently 0.4 $\mu g\ l^{-1}$. A recent survey of 38 molecules carried out in France[30] showed that 99.73% among about 2000 measurements from 11 groundwater sites and two surface water sites yielded concentrations lower than the limit of 0.1 $\mu g\ l^{-1}$, and that these molecules were not detectable in 99.06% of the cases.

In groundwaters, the fluctuations in the concentrations of pesticides cannot always be related to external pulses, owing to the imperfect knowledge about their transfer functions through the phreatic network. Thus, no balance sheet can be established for a given molecule. A weekly survey carried out in 1992 by Portal et al.[31] on a banned water catchment of eastern France showed that the concentrations of atrazine remained almost constant and close to the limit value of 0.1 $\mu g\ l^{-1}$, whereas the concentrations of deethylatrazine varied from five to ten times those of atrazine. These authors considered that the concentration of triazinic residues in catchments located in maize crop areas had been increasing during the last 20 to 25 years, and that the present concentrations corresponded to a kind of "equilibrium" value. This was not the case for more recently applied and more degradable substances, such as isoproturon, which presented fluctuating values, without a clear link to the local rainfall.

Lysimeter field devices used over long time periods can supplement catchment measurements, and allow one to (1) determine the amounts of pesticides released at any soil depth during and beyond the first year of their application; (2) establish the relationship between rainfall and leaching within the same periods, and (3) compare different soils submitted to rigorously similar conditions.

The kinetics of release of solutes can be evaluated in terms of either concentrations or percentages of added active substance. Figure 1 shows the kinetics of leaching of conversion products of ^{14}C-pendimethalin from a Podzol during the first year following application.[32] As previously established with phenylurea during 12 years,[33] the most common feature for leaching curves was an increasing concentration of the effluents recovered during the first year, toward a maximum after 1 or 2 years, then a decrease that followed approximately an exponential law of time.

In this kind of experiment, the interactions of the pesticide with the successive soil layers are unknown, and analysis of the effluents only gives an idea of the resulting processes. To study the impact of rainfall events on the leaching of pesticides from the

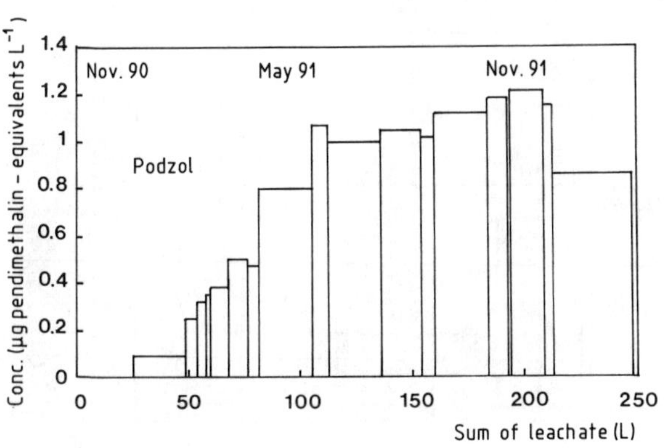

Figure 1 ^{14}C residues in water effluents from a Podzol following application of ^{14}C-pendimethalin (mean values from three lysimeters).[32]

topsoil layer, an experiment was carried out by Schiavon et al.[34] The behavior of ^{14}C ring-labeled atrazine in 0.06-m-high microcolumns packed with surface soil from three adjacent agricultural soils — a rendzina, a brown soil, and a pelosol — was compared during 1 year, starting in April. The annual rainfall was about 0.45 m, with maximum amounts in August, December, and February. The cumulated proportions of ^{14}C recovered after 1.5, 3, 6, and 12 months increased with time, to maximal values ranging between 35 and 63%, in the order: rendzina < brown soil ≤ pelosol. Figure 2[35] shows that in the brown soil, the proportions of leached ^{14}C residues were much higher after 45 d than in the other soils, whereas the increase of leaching with time was much lower. This suggested that the brown soil had a much higher initial porosity, and that compaction increased during the experiment to a larger extent in this soil than in the two others. This was clearly evidenced by comparing soil bulk densities between the start of the experiment and after 1 year (Table 2). A similar effect was proposed by Achik and Schiavon[36] to explain the faster dissipation of carbofuran in a clayey soil than in a loamy-clay soil, in spite of the lower adsorption capacity of the latter.

The proportions of leached ^{14}C-equivalent atrazine were used by Demon et al.[37] to construct model curves that took into account the total amount of ^{14}C *(tec)* remaining extractable in water and organic solvents after leaching (Figure 3). The *tec* values decreased exponentially as a function of the number of rainfall events, which allowed one to obtain a mathematical relationship between the proportion of leached ^{14}C and the distribution of rainfall events independently from the volume of rain. Based on the fact that the proportion of atrazine leached out in 1 year from the upper soil layer (0.06 m) of the brown soil did not exceed 55%, and assuming that all transfer parameters were constant in the whole plough layer, the annual leaching that can be extrapolated by reiteration would attain 6.0% out of the 0- to 0.30-m layer, and 0.36% out of the 0- to 0.60-m layer.

The advantage of this kind of model is to allow an interseasonal follow-up of pesticide leaching since the application time, by taking into account the extractability, that is to say the nature of the binding of the pesticide remaining in the column, in relation to the rhythms of the rainfalls.

C. WATER-SOLUBLE SOIL ORGANIC MATTER

The water solubility of xenobiotics is often considered as the major factor in their transport through the soil water phase. The strength of the interactions between a given molecule and the soil solid phase largely depends on the chemical nature of this molecule,

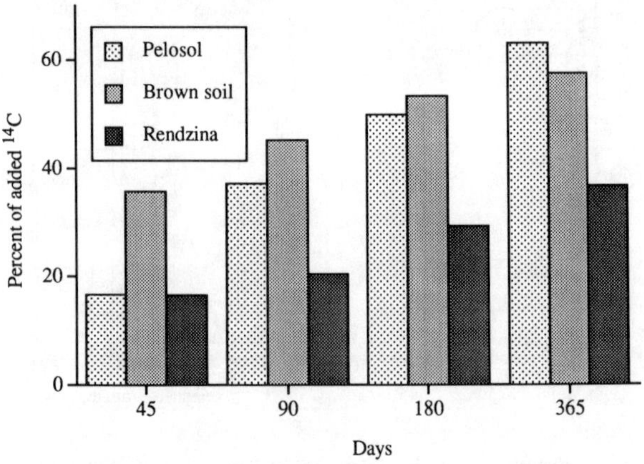

Figure 2 Changes with time of cumulated ^{14}C residues in water leachates from three soils exposed to outdoor conditions after application of ^{14}C ring-labeled atrazine.[35]

especially its polarity and water solubility.[38] Conversely, the lower the solubility in water, the higher is the solubility in hydrophobic organic solvents. The most classical example of this behavior is the empirical relationship between the partition coefficient in octanol-water (K_{ow}) and the solubility in water. A linear relationship between the neperian logarithms of these values should fit an ideal straight line, as indicated by the equations shown in Table 3. Chiou et al.[39] have confirmed such a relationship for a wide range of organics. The slight difference between the empirical and theoretical curves gives a measure of Log γ_0*, which is the solute activity coefficient in the octanol phase saturated with water.

How can water-soluble SOM enhance the solubility of pesticides in water? The answer to this question is probably of chief importance to determine the impact on pesticide movements of humic substance (HS) molecules present in natural water. Due to the presence of these molecules, the concentration of pesticides in the water phase could be higher than their theoretical solubility.[39] Recently, Enfield and Bengtsson[42] have developed a three-phase model to represent the effect of a supposed organic macromolecule on the relative mobility of a supposed hydrophobic compound, as a function of K_{ow}. According to this model, which makes several simplifying assumptions, the relative mobility of hydrophobic molecules with log $K_{ow} \geq 6$ (e.g., DDT) could be doubled by a concentration of HS macromolecule of 100 mg l^{-1}, whereas a concentration of 10,000 mg l^{-1} would be necessary to double the mobility of molecules with $K_{ow} \approx 3$ (e.g., atrazine, toluene).

In fact, the extent and stability of the reaction products between pesticide residues and water-soluble SOM largely vary from one pesticide to another. In their lysimeter study with ^{14}C-atrazine, Schiavon[26] and Schiavon et al.[43] observed that the proportion of unidentified residues quickly decreased during a 1-year experiment, and they finally identified seven

Table 2 Main Analytical Characteristics of the Three Soils Used in the Comparison of Atrazine Leaching Under Outdoor Conditions[34]

Soil	pH	Clay (g kg^{-1})	Org. C (g kg^{-1})	$CaCO_3$ (g kg^{-1})	CEC (cmol kg^{-1})	Bulk density (g cm^{-3}) 0 d	Bulk density (g cm^{-3}) 365 d
Brown soil	6.2	290	14	0	14.2	1.30	1.56
Pelosol	7.3	500	13	≤10	22.1	1.09	1.56
Rendzina	7.6	300	66	225	33.0	1.02	1.02

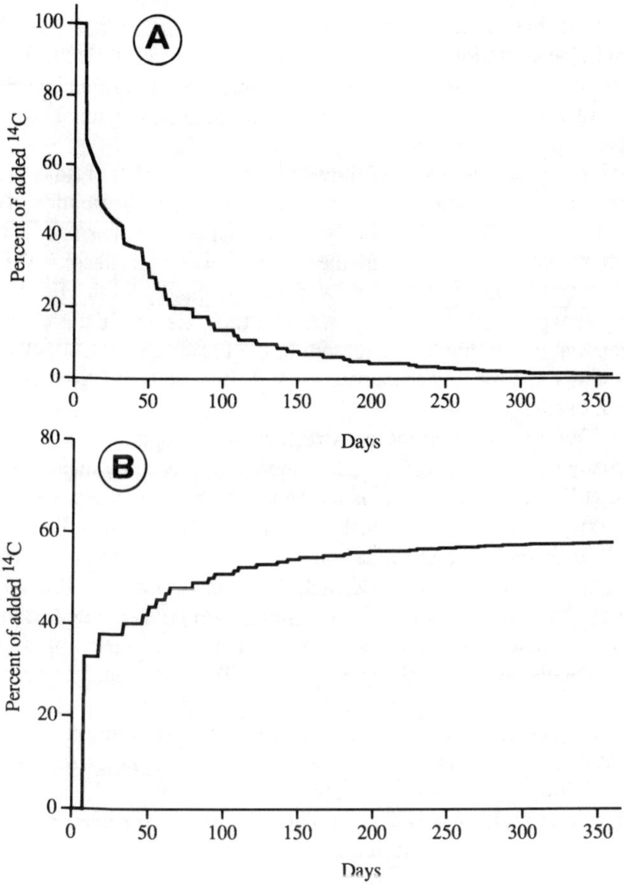

Figure 3 Model curves describing the changes with time of percentages of (A) total extractable ^{14}C residues *(tec)* and (B) cumulated ^{14}C residues leached from a brown soil exposed to outdoor conditions in 0.06-m-high columns, after application of ^{14}C ring-labeled atrazine.[37]

leacheable conversion products. In the case of buturon, Reiml et al.[33] recovered less than 30% of identified conversion products in the leachates 12 years after application to the soil. The effect of HS can be appreciated when not only unbound pesticide residues, but also the fraction extractable together with soil HS is considered. In the same way as the distribution of HS fractions varies with soil types, the distribution of pesticide residues among HS fractions would also be expected to differ from one soil to another. For example, Kloskowski and Führ[44] followed the distribution of ^{14}C-hydroxysimazine in chemical fractions of a

Table 3 Correlation Equations Between Octanol-Water Partition Coefficient (K_{ow}) and Solubility in Water (S_w) of a Variety of Organic Solutes

Equation	Ref.
$\text{Log } K_{ow} = -0.86 \cdot \text{Log } S_w + 0.71$	Chiou et al.[39]
$\text{Log } K_{ow} = -0.54 \cdot \text{Log } S_w + 0.65$	Karickhoff et al.[40]
$\text{Log } K_{ow} = -0.68 \cdot \text{Log } S_w + 4.25$	Hasset et al.[41]

"Parabraun" soil from Germany during a 102-d incubation in laboratory conditions. They observed a rapid incorporation of ^{14}C into SOM fractions, especially in the fulvic acid (FA) and humic acid (HA) fractions. The incorporation into the FA fraction tended to decrease with time, contrary to incorporation in the nonextractable humin fraction, which increased faster than that in the HA fraction.

The behavior of pesticides can be followed for longer periods of time, and preferably under outdoor conditions. Figure 4 shows the variations in the distribution of ^{14}C-atrazine among humic fractions of the brown soil, rendzina, and pelosol after 1.5, 3, and 6 months. These variations were in agreement with the above mentioned decrease with time of water leaching.[35] The extractable ^{14}C-atrazine was located mainly in the FA fraction at the beginning of the experiment, then decreased with time, whereas the proportions present in the HA fraction, and in humin, increased. In the brown soil, the slightly more acidic pH and coarser texture probably accounted for higher amounts of FA-bound atrazine than in the two other soils.

The role of FA as carrier in the coextraction of hydrophobic residues needs to be clarified, as many modern functional studies about HS, especially those using ^{13}C-NMR, have shown that FA have a lower aromaticity and a higher amount of hydrophilic groups than other fractions.[46] It is possible that transient complexes, including weak physical bonds as well as hydrogen and ionic bonds, are involved.[47] In the case of atrazine, solubilization by FA could be related to hydrolytic dechlorination, which was shown to be promoted by the interaction through H bonding with undissociated external –COOH groups of the FA molecule.[48] In a further step, charge-transfer complexes with more aromatic components, as described by Senesi and Testini,[49] could be formed. In other cases, especially with $-NH_2$-bearing residues, the incorporation of pesticides into HS polymers through stable covalent binding with phenolic or quinoid rings was demonstrated experimentally.[50-52] Based on the study of polymers synthesized from catechol in the presence of low concentrations of atrazine and its metabolites, Andreux et al.[53] concluded that these molecules were predominantly incorporated in FA-type fractions, except diaminoatrazine, which was bound to HA-type fractions.

III. FACTORS INFLUENCING THE RETENTION OF PESTICIDES IN THE SOIL SOLID PHASE

A. SOIL SORPTION CAPACITY

The proportions of organic xenobiotics dissolved in water are generally very low with respect to the total amounts applied and retained by the soil solid phase. Sorption processes were extensively reviewed by Weber and Miller.[54] One of the most classical parameters used to characterize the partition of pollutants between the soil solid surface and the interstitial water phase is the *equilibrium sorption capacity index* (K_f, K_{om}, or K_{oc}), as given by the Freundlich fitting of the adsorption isotherm equation, as follows:

$$x/m = K_f \cdot C_e^{1/n} \qquad (1)$$

where x/m is the amount of solute adsorbed per unit weight of adsorbent, C_e the equilibrium concentration, and 1/n the sorption constant, which characterizes the shape of the isotherm curve.[55]

In the case of equipartitioning between the liquid and solid phases, $1/n \approx 1$, and K_f becomes K_d, the *distribution sorption constant,* which allows one to compare the adsorption of different chemicals on a given soil. The K_{om} and K_{oc} value can be obtained by dividing K_f or K_d by the SOM content, or C content, of the adsorbent, respectively. They can be considered a measurement of the contribution of hydrophobic forces to sorption,

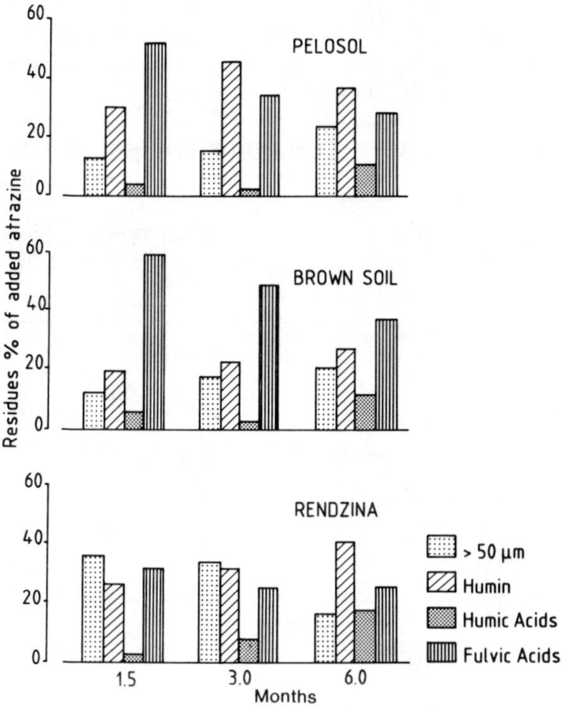

Figure 4 Distribution of ^{14}C-atrazine residues in humus fractions of three soils incubated under outdoor conditions.[34,45]

at least in the case of nonpolar or slightly polar molecules. They generally increase with decreasing water solubility (S_w). There are numerous examples of such variations. For instance, Hasset et al.[41] and Hasset and Banwart,[56] on a range of more than 100 molecules, obtained a linear regression curve as follows:

$$\text{Log } K_{OC} = -0.62(\pm 0.024) \cdot \text{Log } S_w + 3.95 \tag{2}$$

In a similar way, an empirical correlation between the sorption coefficient (K_{oc}) and octanol-water partition coefficient (K_{ow}) has been frequently reported.[40,57,58] The regression equations correspond to parallel straight lines for the different categories of molecules. The intercepts for log K_{ow} = 0 give K_{oc} values of 0.61 for hydrophobic molecules, 1.0 for s-triazines, and 11.6 for phenylurea and carbamates.

B. SOIL pH AND EXCHANGEABLE CATIONS

Apart from the octanol-water partition coefficient, several soil characteristics can affect sorption processes. It is often difficult to isolate one soil property in order to determine its impact on the sorption capacity of organic molecules. One possible method is to recalculate adsorption isotherms from that obtained at a given pH, by using the pH-dependent variation of the adsorption of the same amount of molecule added. In order to eliminate as much as possible other factors, such as "clay mineral content" and "SOM content," these determinations were carried out by Barriuso (in Reference 34) in the case of atrazine adsorbed on the humin fraction of the three already mentioned agricultural

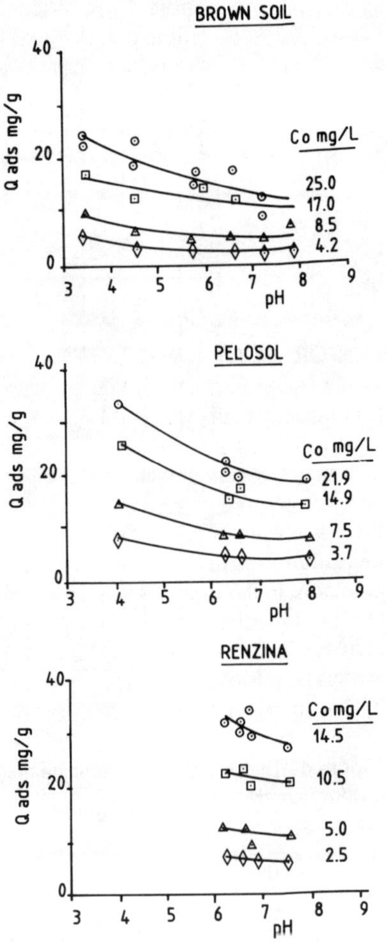

Figure 5 Effect of pH values on the adsorption *(Qads)* of atrazine on the humin fraction from three different soils.[34] Co is the initial concentration of the atrazine solution.

soils from eastern France. Figure 5 shows that the adsorption of atrazine decreased with increasing pH. In the rendzina soil, the low amounts adsorbed were due to the presence of calcium carbonate, and adsorption was enhanced after decarbonatation of the humin material. By constructing adsorption isotherms of ^{14}C-atrazine, it was clearly confirmed that K_{oc} values decreased with increasing pH (Table 4).

The effect of exchangeable cations on the adsorption of atrazine was studied by Bertin et al.[59] in the case of purified HAs from a rendzina soil and a synthetic polymer from catechol, each of them being saturated with different cations. As shown in Figure 6, the isotherm curves did not reach a plateau, due to the limiting solubility of the adsorbate. However, in both cases, the slopes of the respective curves indicated that the influence of the saturation cations on K_{oc} values followed the order:

$$Ca^{2+} < H^{+} < Al^{3+} < Fe^{3+}$$

These results were probably indicative of an increasing hydrophobicity of the HS molecules, as a consequence of the increasing stability of the complexes formed with these cations.

Table 4 K_{oc} Values Obtained From Adsorption Isotherms of Atrazine at Different pH Values on the Humin From a Pelosol[34]

pH	K_{oc}	n	r^2
4	4.54	0.747 ± 0.090	0.98
5	2.44	0.805 ± 0.047	0.98
6	1.79	0.835 ± 0.035	0.99
7	1.36	0.871 ± 0.024	1.00
8	1.29	0.846 ± 0.038	0.99

C. SOIL CLAY CONTENT

Under experimental conditions, clayey soils are frequently shown to have higher adsorption capacity than other soils.[60] In the case of carbofuran adsorbed on two cultivated soils which differed in their clay content but had the same low SOM content, Achik and Schiavon[36] observed that the clayey soil (52% clay) yielded a higher slope and a higher partition coefficient than the loamy clayey soil (24.2% clay). Similar differences between these two soils were observed in the presence of other molecules, especially atrazine and metsulfuron-methyl.[34] However, the influence of clay minerals on adsorption is detectable only in the case of low SOM content. When this amount is too low with respect to that of clay minerals, the latter can develop a specific adsorption capacity, with a large increase in K_{oc}, as shown by Hasset et al.[41] for α-naphthol. Such an influence may not always be positive, especially in the case of negatively charged molecules, as pointed out by Barriuso and Calvet,[61] who suggested the possibility of a repulsive effect against 2,4-D added to a Vertisol.

Apart from clay minerals, the clay-size fractions of soils generally contain variable proportions of noncrystalline to poorly crystalline inorganic components. Such components, especially Al and Fe oxides, were shown to provide sites that govern the degree of adsorbtivity of atrazine by soils.[62]

D. SOIL ORGANIC MATTER CONTENT

The prominent role of SOM in the adsorption of hydrophobic molecules has been known for at least 2 decades. In the case of triazines, such a role has been demonstrated since 1968 by Walker and Crawford.[63] In general, linear correlations were found between K_d and SOM (or C) contents,[41,55] with slopes that were characteristic of the pesticide. In the above mentioned adsorption experiments using ^{14}C ring-labeled atrazine and three different soils, Barriuso (in Reference 34) established that the rendzina soil, which contained about six times more SOM than the two other soils, had K_d values that were about six times higher. Adsorption isotherms (Figure 7) illustrated the differences between the three soils in the case of atrazine, as well as of a sulfonylurea (metsulfuron-methyl), which showed a similar discrimination, although with smaller amounts adsorbed. In the latter case, the specific effect of the clay fraction was more marked than with atrazine, resulting in relatively higher K_{oc} values for the brown soil and the pelosol than for the rendzina (Table 5).

E. CHEMICAL NATURE OF SOIL ORGANIC MATTER

In studies about soil sorption capacity, most authors consider SOM as a whole, regardless of the differences in its chemical nature. The importance of water-soluble SOM in the transport of pesticides through the soil solution has already been mentioned (see Section II.C). The mechanisms of abiotic interactions between xenobiotics and SOM

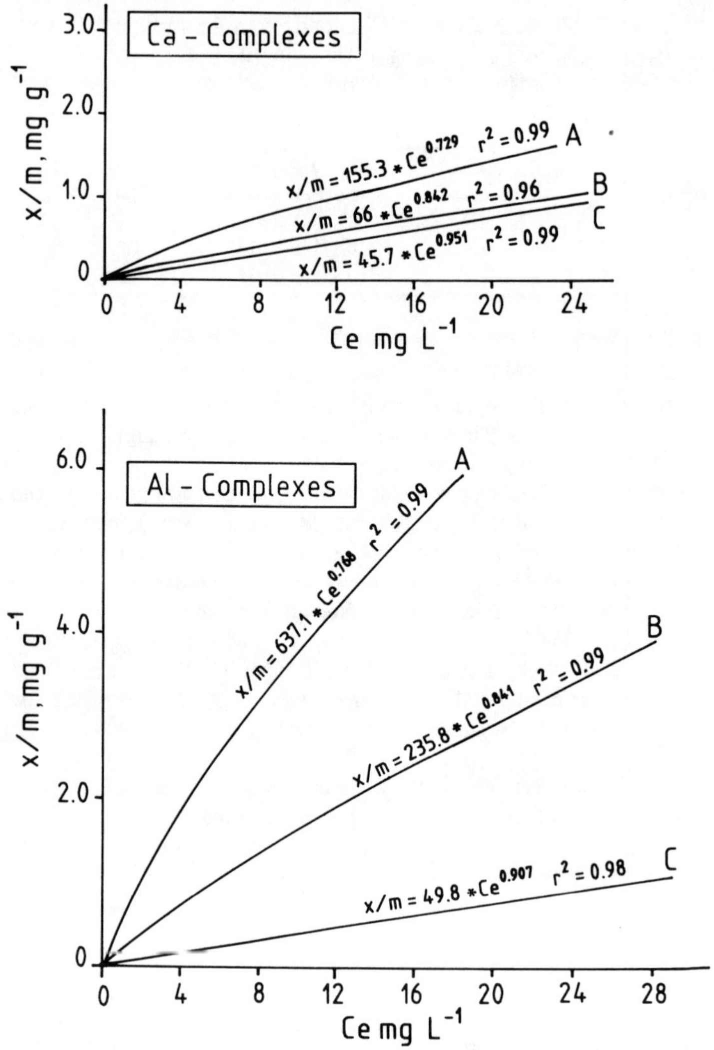

Figure 6 Adsorption isotherms of atrazine on Ca^{2+} and Al^{3+} complexes of two humic-like polymers and one humic acid.[59] (A) catechol polymer; (B) catechol + glycine polymer; (C) rendzina humic acid.

have been reviewed by Senesi and Miano.[47] When persistent pesticides are involved, increasing incorporation of the residues into more polymeric and more aromatic materials can occur with elapsing time. One of the most active mechanisms of this incorporation is oxidative copolymerization with polyphenolic materials, as pointed out by numerous authors in the case of substituted phenols[50,64] and anilines.[51,65-67] Such reactions can be enhanced by either mineral catalysts (e.g., clays, metal oxides) or enzymes.[52] Even in the case of substituted secondary amines, especially triazines, the possible formation of covalent bonds through nucleophilic addition to quinonic structures of HS in water medium (20 mg: 20 mg: 100 ml water) was discussed by Senesi and Testini.[49] By preparing synthetic polymers from catechol in the presence of low concentrations of atrazine and its metabolites (1 mg: 20 mg to 1 mg: 160 mg: 100 ml

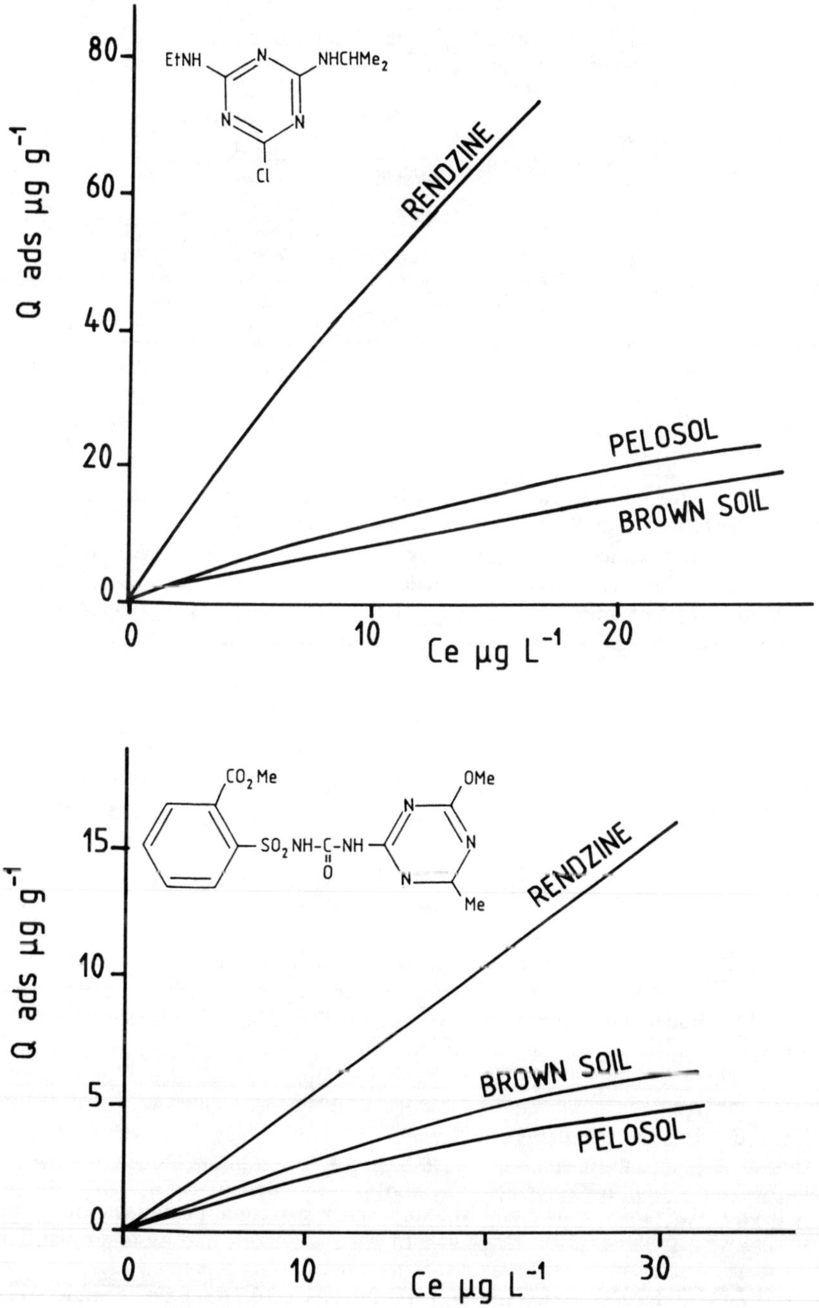

Figure 7 Adsorption isotherms of atrazine (above) and metsulfuron-methyl (below) on three soils.[34]

water), Andreux et al.[53] showed evidence of covalent bonds only in the case of diaminoatrazine, which was found in HA-type fractions. This suggests that in natural conditions, structural modifications of the pesticide contribute to enhanced incorporation of pesticide derivatives into soil components.

Table 5 K_{oc} Values Obtained From Adsorption Isotherms of Atrazine and Metsulfuronmethyl on Three Different Soils[34]

Soil type	K_{oc}	n	r^2
		Atrazine	
Rendzina	1.06	0.846 ± 0.009	0.99
Brown soil	0.85	0.878 ± 0.008	0.99
Pelosol	1.35	0.816 ± 0.007	0.99
		Metsulfuron methyl	
Rendzina	0.11	0.870 ± 0.105	0.99
Brown soil	0.15	1.017 ± 0.077	0.98
Pelosol	0.18	0.870 ± 0.105	0.99

In the formation of such irreversible bonds, the *time factor* is of prevalent importance. Desorption studies have shown that the bonds formed during adsorption assays were partly irreversible, as indicated by the classical phenomenon of hysteresis. Desorption kinetics of pesticides can be modeled with different mathematical representations. One of them assumes that adsorption sites belong to two compartments which follow a linear and an exponential distribution, respectively, according to Equation 3:

$$Q_{ads} = K_1 \cdot C_e + Q_{adsn} - Q_{adsn} \cdot \exp(-K_2 \cdot C_e) \qquad (3)$$

where Q_{ads} is the total amount adsorbed, K_1 and K_2 are the respective adsorption constants of the first and second compartments, and Q_{adsn} is the amount adsorbed in the second compartment. A third compartment is the water solution, in which C_e is the equilibrium concentration of the pollutant. It was established that the proportion of strongly bound atrazine (i.e., that defined by the exponential distribution) was a function of SOM content.[2]

When the soil was incubated for several weeks or months, irreversibly adsorbed residues predominated. In order to localize these bound residues, combined grain-size and chemical fractionation of the soil were carried out. There are only few works that have applied physical fractionation techniques to soils enriched with labeled pesticides. Clay-size (i.e., 2μm) fractions are generally the most active in binding pesticide residues.[62] Again with atrazine applied to the same three soils, Barriuso et al.[68] showed drastically different distributions (Figure 8). A tendency toward a bimodal distribution with soil particle-size, especially in the rendzina soil, means that imperfectly decomposed plant residues associated to the sand-size (i.e., 200- to 2000–μm) fractions were also able to bind pesticide residues.

IV. FACTORS INFLUENCING THE DEGRADATION OF FREE AND SOIL-BOUND PESTICIDES

The degradation of pesticides in soils is a complex phenomenon, for three main reasons: (1) pesticides are present at low concentration, (2) they are distributed very quickly among the various soil fractions, which differ from each other in terms of microbial habitats and interactions with the substrates, and, (3) they are diversely resistant to chemical or biological degradation and frequently require specific soil conditions or adaptation of soil organisms to these conditions.

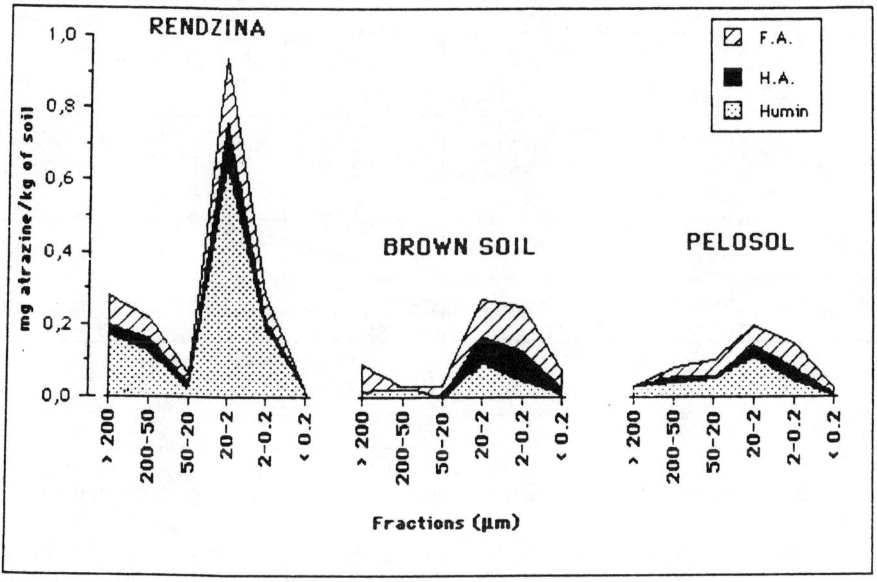

Figure 8 Grain-size fractionation of a rendzina and a brown soil incubated during 6 months with ^{14}C ring-labeled atrazine.[68]

A. SOIL BIOLOGICAL ACTIVITY

In soils, the major part of degradation reactions is performed biotically, either by microorganisms or by extracellular enzymes present in soil. This applies to transformation reactions as well as to mineralization and the formation of bound residues. Plants and soil fauna may also contribute to the transformation of xenobiotics in soil; the conversion products formed within their bodies reach the soil either by excretion or by decay of the organisms after their death. The bulk of degradation reactions is due to a small fraction within the microflora only ("competent microflora"). This is reflected by the observation that total biomass in soil is not always positively correlated with the degradation rates of pesticides.[69] Thus, the influence of soil factors on the degradation of organic xenobiotics is exerted, for the main part, not directly, but through those microorganisms that perform the degradation of the respective chemical. Recent literature surveys have been published by Stotzky,[70] Alexander and Scow,[71] and Scheunert.[72] Although zero-order degradation kinetics of organic substrates were reported in several instances, most processes can be described by nonlinear curves, especially first- or higher-order reactions, as:

$$-d_S/d_t = k \cdot S^n \tag{4}$$

where S is the substrate concentration, k the rate constant, and n the order constant. In a complex environment like soil, the degradation kinetics can be affected by numerous factors, such as climate, physicochemical effects, and soil management.

B. TEMPERATURE

Increasing temperature generally increases the degradation rates of xenobiotics.[69,72] In abiotic degradation, the rates of chemical reactions increase with increasing temperature. The influence of temperature on biological degradation is more complex than on abiotic degradation, and includes that on biomass level, microbial composition, and microbial

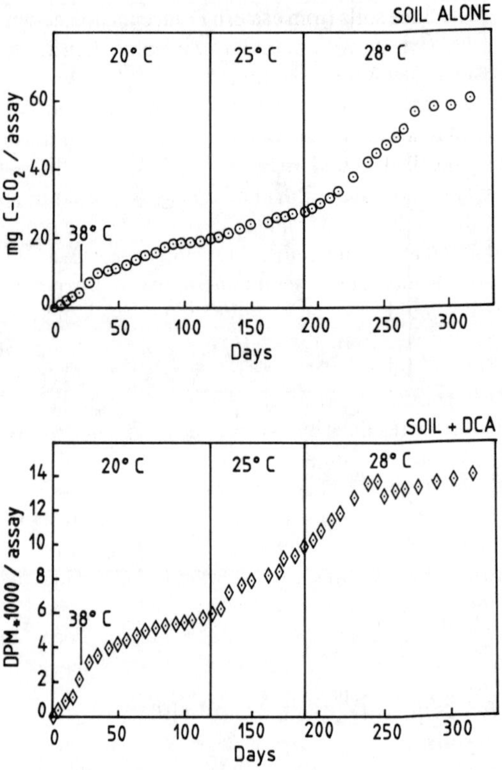

Figure 9 Consequences of stepwise increases of incubation temperature on the total soil respiration (above) and biodegradation of 3,4-dichloroaniline, DCA, (below).[75]

activity. Influence on microbial composition implies effects in favor or to the disadvantage of microbial strains capable of degrading the xenobiotics in question. Microbial activity has an optimum temperature. Thus, mineralization rates of the thiocarbamate herbicides diallate and triallate increased between 4 and 22°C[73] but were higher at 20°C than at 33 or 44.5°C.[74]

Temperature fluctuations can deeply affect biodegradation kinetics of highly resistant substrates. Such an effect was studied in the case of the degradation of ^{14}C-labeled 3,4-dichloroaniline (DCA) in a brown soil by Völkel et al.,[75] during about 1 year. Figure 9 shows that stepwise increases of incubation temperature resulted in stronger increases in the degradation of DCA than of the native SOM. When a 38°C temperature was applied at the beginning of the experiment, thermophilic organisms were selected, which seemed to degrade preferentially the labeled DCA.

C. MOISTURE CONTENT

Under field conditions, the positive effect of higher temperature on biomineralization may be superimposed by the negative effect of lower humidity. Some organic compounds can persist in soils for years, in spite of being degradable by soil microorganisms. This means that they are unavailable, and that desorption into the aqueous phase by rewetting is a prerequisite for their biodegradation. This is known as the process of "aging" of pesticide residues. Aged residues are generally less desorbable from soil than freshly added pesticides, as shown for simazine by Scribner et al.[76] The generally positive influence of soil moisture on biomineralization rates[69] can probably be explained by two factors: (1) the solubilization of pesticides, and thus their bioavailability to microorganisms, increases with increasing water content, and (2) the microbial biomass passes through an optimal value with increasing water content.[77]

The three soils from eastern France incubated with ^{14}C-labeled atrazine[34] were exhaustively extracted with water and methanol, then air dried and stored. After 12- to 22-month periods, they were submitted to a second treatment with the same solvents.[78] The patterns of extraction curves differed according to aging time and soil type (Figure 10). At least five to nine extractions were necessary, which means that in spite of a 1-year incubation, the so-called bound residues were not totally stable, and that drying and rewetting probably contributed to their reextraction. As for adsorption kinetics, the amounts of reextractable bound residues were lower in the rendzina soil, higher in the brown soil, and intermediate in the pelosol. Organic matter was the predominant factor in the rendzina soil, whereas clay mineral content was in the pelosol. In the brown soil, neither of these factors could account for protection of the bound residues. It was found that more $^{14}CO_2$ was evolved from the soil in the presence than in the absence of synthetic nutrient medium, indicating that the residues could be partly degraded by cometabolism reactions. Such a conclusion was also reached by Khan and Behki[79] in a study of the degradation of atrazine residues by two species of *Pseudomonas*. In the Bernoux study,[78] unidentified *Bacillus* and actinomycetes were involved, and evidence of simultaneous enhancement of iron-reducing activity was shown using α–α-dipyridyl staining in liquid medium.

Outdoor experiments with ^{14}C-atrazine in the same three soils showed that after 6 months, the fraction of bound residues was the most modified, when compared with extractable fractions. In the brown soil, using the rainfall-based model designed by Demon et al.,[37] it was confirmed that a large proportion of bound residues disappeared, probably as a consequence of degradation processes (Figure 11). The extent of degradation was much lower in other soils, probably because larger amounts of more stable native SOM were present.

D. SUNLIGHT RADIATION

Photochemical degradation pathways can occur in all phases and compartments of the environment, such as the atmosphere, aquatic systems, and soils. It has been shown that light can penetrate from 0.001 to 0.01 m in soils, depending on soil texture, compaction, etc. Thus, solar radiations can make an appreciable contribution to the transformation and degradation of organics which may be present on soil surfaces.[80] Abiotic transformations of organic chemicals in soil are to be investigated with particular consideration to the dynamic and catalytic effects, in relation primarily to the interactions of these chemicals with the environment. When they are in contact with soil, one must distinguish photodegradation in the adsorbed, solid, or aqueous phase of the soil. In the case of agrochemicals, the nature of formulation components must also be taken into consideration.

The complexity of ecosystems substantially impedes the study of such reactions. Single reactions, therefore, require conditions which are simplified, but as close as possible to those of various ecosystems, to make possible a rational interpretation of experimental results. In photochemical reactions, two different effects must be considered: *direct photolysis* of a foreign substance which absorbs UV light in the spectral region of sunlight (i.e., λ >290 nm) and *indirect photolysis,* in which the light energy is absorbed by other solutes and is either transmitted to the foreign chemical under consideration (sensitization by HS) or leads to the formation of species that enter into a chemical reaction with the foreign chemical, as shown in Table 6.

Both clay mineral and HS are known to produce oxygen-reactive species (singlet and atomic oxygen, superoxide anion radicals, and hydroxyl radicals), and to contribute in that way to the enhancement of photodegradation of chemicals which either do not absorb appreciably or do not absorb all UV light of λ >290 nm.[81] Nevertheless, in some cases, quenching effects can be observed, while photodegradation of chemicals can take place in the presence of HS showing the lower energy level of the exited state of the HS species

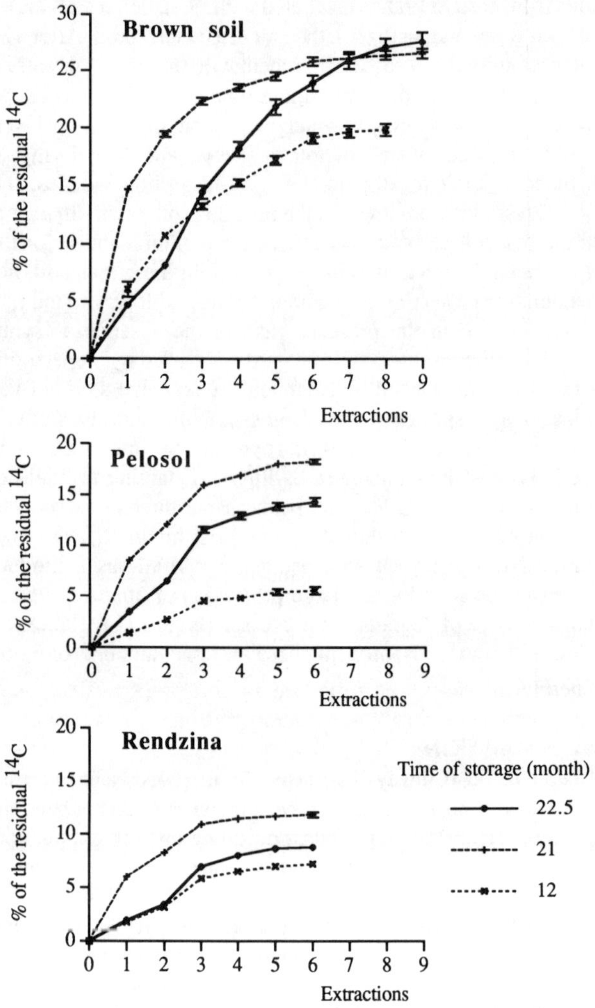

Figure 10 Availability of ^{14}C-labeled bound residues of atrazine, as affected by time of storage.[78]

compared to that of the environmental chemical.[82] Photochemical processes on soil surfaces are complex transformations which have to be taken into consideration, as are all the above mentioned reactions occurring in the soluble phase. Furthermore, combined photochemical and thermal effects were shown to be of capital importance in degradation processes.[83]

E. SOIL SOLID PHASE

If one attempts to correlate the degradation rates of xenobiotics with the composition of the soil solid phase, it turns out to be very difficult to consider one soil property in isolation from others. As soil is a complex medium, pesticides may interact with soil components and compete for adsorption or (bio)degradation. Abiotic degradation reactions that are catalyzed by clay minerals could be dependent on both the level and

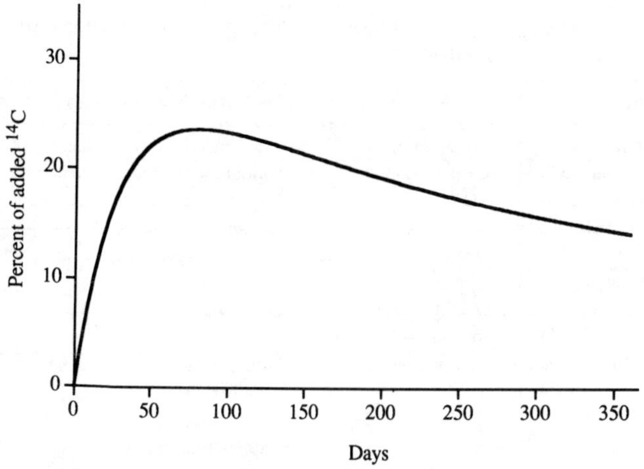

Figure 11 Disappearance with time of ^{14}C-atrazine bound-residues, as calculated from an outdoor experiment in a brown soil.[37]

composition of the clay fraction.[84,85] The same applies to abiotic oxidations promoted by metal oxides in soil.[84] In some cases, direct influences of inorganic soil components on biotic degradation rates were observed, e.g., orientation of the microbial synthesis of oxidative enzymes by sands or clay minerals,[86] or the activation of insolubilized enzymes. However, in nearly all cases, the influence of soil properties on biotic degradation reactions is their combined action on soil microorganisms in general and especially on the "competent microflora." For example, the pH value as a result of carbonate content and the acidic functions of HS is generally positively correlated with soil microbial activity. Organic C content could positively influence both the biomass and activity of microorganisms. However, since it enhances the adsorption capacity of soils, it thus decreases the availability of xenobiotics to the soil microorganisms. Detailed discussion about the physicochemical factors in the influence of soil clay minerals on both SOM quality and bacterial survival can be found in the recent work by Theng and Orchard.[87]

When both aged and freshly added pesticides were present, the latter were generally degraded preferentially. A comparison can be made with the behavior of biological substrates such as amino acids, proteins, or sugars. For these molecules, adsorption and binding to clay minerals, the most commonly proposed mechanisms to explain their decreased or prevented degradation by soil microorganisms, have been extensively reviewed by Stotzky.[70] In the presence of reactive SOM, especially phenolics, incorporation into stable humic-like polymers is the prevailing factor, as mentioned above (see Section III.E). In the case of pesticides, interaction products with phenolics and quinones during the oxidative polymerization represent an adequate chemical model for the formation of unextractable (bound) residues.[50,51,64-67]

When exposed to soil microorganisms, these polymeric bound residues exhibited higher stability than the free added pesticide. In the case of polymer-bound chlorophenol, Dec and Bollag[88] showed that the release of ring C in the form of CO_2 was very slow, and did not exceed 1% of the total bound material after more than 10 weeks. Similarly, by reacting 3,4-dichloroaniline with catechols, Völkel et al.[67] obtained quinonic reaction products which had the structures shown in Figure 12. When such an addition compound was incubated for more than 1 year in an agricultural soil, it was converted into CO_2 with a tenfold slower rate than the corresponding 3,4-dichloroaniline. From these results, the authors inferred that this kind of reaction product could occur under natural conditions, once added to the soil.

Table 6 Photoprocesses Affecting Natural Organic Matter and Pesticides in Water Phase

Substrates	Products	Probable mechanisms and effects
Natural organic chromophores	$\cdot C + \cdot HO_2$ $C+ + \cdot O_2-$	Transfer (H, electron, energy) to O_2; initiation of numerous reactions
Petroleum hydrocarbon (RH, ArH)	ROO ·	Oxygen addition; oxidized organic radicals
Pesticides (disulfoton)	Disulfoton sulfoxide	Singlet oxygen; changes in properties and toxicity
Herbicides (2,4-D; 2,4,5-T)	Oxidation and reduction products	Direct photolysis; complex
Pentachlorophenol	Phenols, quinones, acids	Direct photolysis; changes in properties and toxicity
PCBs, DDT	Complex	Direct or sensitized photolysis; dehydrochlorination

Although there is still a lack of comparison between various kinds of soils, it is quite probable that the rates and reactions of incorporation of a given xenobiotic compound among HS will follow a similar pattern in all soil types. This was observed by Scheunert et al.,[89] who established that ^{14}C ring-labeled 4-chloroaniline bound to different soil fractions was degraded into $^{14}CO_2$ to different extents when reincubated in a clay loam soil from Germany (Table 7). Again in this case, it was noticed that the FA fractions were less stable, and the HA fractions more stable, against both bio- and photodegradation.

F. SOIL MANAGEMENT AND PLANT COVER

Soil management and plant cover affect degrading soil microflora via effects on organic C, pH, humidity, temperature, etc. Thus, forest soils having high organic C content and low pH values exhibit a lower degradation capacity toward some xenobiotics than a meadow soil with lower organic C content and a higher pH value.[90] Even when the persistence of the chemical in soil is low, changes in the natural soil biological equilibria can be noticed once the chemical is added. This was the case with the fungicide flutriafol, which was applied at different doses to a brown soil by Bordes.[91] Figure 13 shows that this molecule interfered differently with soil cellulolytic populations, according to the amount added. At low doses, only small and temporary effects were noticed, whereas at higher doses, an inhibition phase was observed, followed by an enhancement phase, which corresponded to the selective stimulation of cellulolytic bacteria.

Secondary effects on crops can be feared, especially in the absence of detoxification processes. However, aged pesticide residues can be bioavailable, without noticeable phytotoxicity. In the above mentioned work, Scheunert et al.[89] observed that bound residues of chloroaniline present in soil HS could be taken up by a barley plant. The amounts were not high (>0.3% of ^{14}C-labeled molecule), but were the same percent order

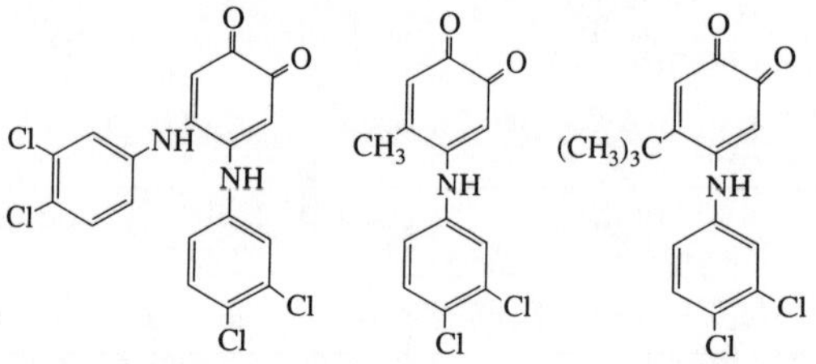

Figure 12 Products of oxidative addition of 3,4-chloroaniline to (1) catechol, (2) 5-methylcatechol, and (3) 5-ter.butylcatechol.[67]

Table 7 Concentration of Soil-Bound Residues of 4-Chloroaniline ^{14}C in Soil Fractions, and Their Mineralization and Photodegradation[89]

Soil fraction	Conc. of bound ^{14}C (μg g^{-1})[a]	Aerobic mineralization (28 d, % $^{14}CO_2$)	Photodegradation (λ>290 nm)
Inorganic fractions	0.53	1.73	0.59
Humic acids	30.70	1.21	0.08
Humin	6.97	4.13	1.18
Fulvic acids	5.95	4.98	1.43
Total soil	1.50	2.18	1.39
Free 4-chloroaniline	None	9.04	26.06

[a] Microgram equivalent to the parent compound per gram of dry soil fraction.

as for degradability. These amounts were maximal for FA and minimal for HA. Similar results were obtained with bound residues of ^{14}C ring-labeled atrazine added to Italian ray grass plants in pot experiments.[92] In this case, plant absorption occurred with residues obtained from natural HA, as well as with those prepared from synthetic humic polymers.

V. SUMMARY AND CONCLUSIONS

On the basis of selected examples, the behavior of xenobiotics, mostly pesticides, was related to the main soil characteristics that can reflect the effects of different soil formation conditions.

Soil physical parameters, especially density and porosity, together with parent rock and topography, strongly control the movements of pesticides in the soil water phase, through both leaching and runoff. One important aspect to be considered is soil depth, since interactions between the pesticides and the soil matrix are different in surface and subsurface soil. Emphasis should be placed on the latter aspect, as organic-bound xenobiotic solutes can reach the water table, and little information is available about their dynamics through deeper soil layers, as well as in the vadose zone.

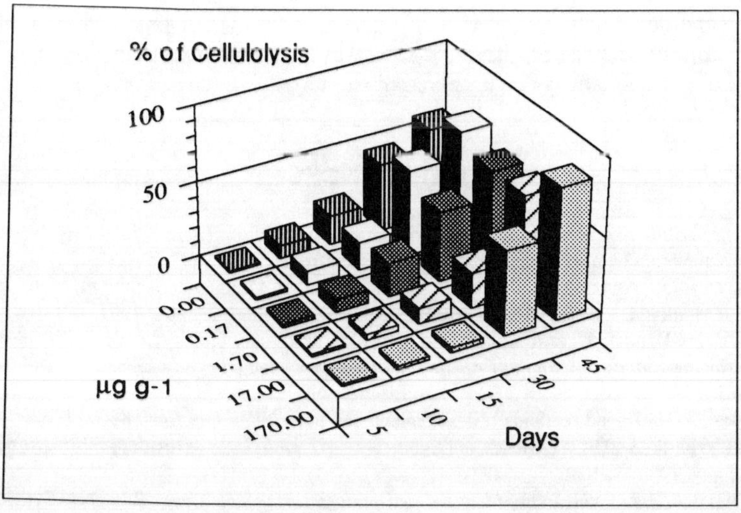

Figure 13 Impacts of increasing doses and time of application of the fungicide flutriafol on cellulolytic activity in a maize culture soil.[91]

A large array of *chemical parameters* are responsible for the retention of xenobiotics in the soil solid phase. Soil sorption capacity can be quantified and described by different mathematical equations which show the existence of different energy levels or "compartments" in the retention of these molecules. Soil pH and exchangeable cations affect mainly the size of these compartments, whereas the soil matrix, especially the nature and content of clay minerals and SOM, seem to be more decisive in the retention energy. Covalent bonds between xenobiotics and SOM are the most stable ones, especially those involving phenolic moities. However, apart from some synthetic models, little is known about the chemical reactions involved in the retention of organic xenobiotics in soils.

Regarding degradation processes, discussion focused mainly on *the factors that determine biological degradation, with respect to abiotic degradation.* Temperature and moisture conditions, as the main soil climatic factors, also affect the kinetics of the formation and degradation of pesticide-bound residues. Since they determine the rates of accumulation and decomposition of humus, they have a prominent impact on the binding of pesticides to humus. Although limited to the very upper soil layer, photolytic reactions can assist other degradation processes. Adsorption and the binding of xenobiotics to the soil solid phase frequently limit or delay biological degradation. However, the mechanisms of such protective interactions still require study. Soil management and plant cover were not specifically examined, although most studies dealt with agricultural soils. One has to keep in mind that soil management influences biological activity, especially microbial and genetic diversity. As a monospecific cover results in a selected and adapted rhizospheric microflora, emphasis should be placed on the search for new species growing in natural environments, and that would exhibit specific properties in the degradation of or the resistance to certain pesticide types.

Finally, the *time factor* was shown to play a major role in most processes. Many authors have placed emphasis on adsorption reactions, which describe interactions occurring at the very beginning of the addition of the pesticide to soil. In the case of persistent pesticides, *aging processes* are also important, and are responsible for the formation of nonextractable (bound) residues. Their formation is generally intense and rapid in soils with high SOM content, at least in the case of hydrophobic molecules. When experimental studies are run, it is important to consider long-term incubations, preferentially under outdoor conditions, and during at least one annual cycle. In this way, the seasonal behavior of pesticides can be observed, as well as the possibility of late release of bound residues, that cannot always be observed in shorter experiments.

REFERENCES

1. **Commission des Communautés Européennes.** Directive "Qualité de l'eau destinée à la consommation humaine," no. 80/778/CEE, Paramètre 55, 1988.
2. **Barriuso, E., Andreux, F., Schiavon, M., and Portal, J. M.,** Intérêts et limitations des méthodes de séparation des micropolluants organiques des sols, *Sci. Sol,* 29, 301, 1991.
3. **Lefèbvre-Drouet, E. and Calvet, R.,** Méthodes de dosage biologique des herbicides, in *Les Herbicides: Mode d'Action et Principes d'Utilisation,* Scalla, R., Ed., INRA, Versailles, France, 1981, 337.
4. **Khan, S. U.,** Bound pesticide residues in soil and plants, *Residues Rev.,* 84, 1, 1982.
5. **Calderbank, A.,** The occurrence and significance of bound pesticide residues in soil, *Rev. Environ. Contam. Toxicol.,* 108, 71, 1989.
6. **Wierenga, P. J.,** Solute transport, in *Future Developments in Soil Science Research,* Boersma, L. L., Elrick, D. E., Corey, R. B., Cheng, H. H., Tucker, T. C., Rutledge, E. M., Unger, P. W., Foster, N. W., Kincheloe, S., and Hsu, P. H., Eds., Soil Science Society of America, Madison, WI, 1987, 23.
7. **Hall, J. K., Pawlus, M., and Higgins, E. R.,** Losses of atrazine in runoff water and soil sediment, *J. Environ. Qual.,* 1, 172, 1972.

8. **Ritter, W. F., Johnson, H. P., Lovely, W. G., and Molnau, M.,** Atrazine, propachlor, and diazinon residues on small agricultural watersheds, *Environ. Sci. Technol.,* 8, 38, 1974.
9. **Hall, J. K.,** Erosional losses of s-triazine herbicides, *J. Environ. Qual.,* 3, 174, 1974.
10. **Triplett, G. B., Conner, B. J., and Edwards, W. M.,** Transport of atrazine and simazine in runoff from conventional and no-tillage corn, *J. Environ. Qual.,* 7, 77, 1978.
11. **Leonard, R. A., Langdale G. W., and Fleming, W. G.,** Herbicide runoff from upland piedmont watersheds. Data and implications for modelling pesticide transport, *J. Environ. Qual.,* 8, 223, 1979.
12. **Wu, T. L.,** Dissipation of the herbicides atrazine and alachlor in a Maryland corn field, *J. Environ. Qual.,* 9, 459, 1980.
13. **Schiavon, M. and Jacquin, F.,** Etude de la présence d'atrazine dans les eaux de drainage, in *C. R. Journ. d'Etudes Herbicides,* Columa, Versailles, France, December 13 to 14, 1973, 35.
14. **Muir, D. C. and Baker, B. E.,** Detection of triazine herbicides and their degradation products in tile-drain water from fields under intensive corn (maize) production, *J. Agric. Food Chem.,* 24, 122, 1976.
15. **Von Stryk, F. G. and Bolton, E. F.,** Atrazine residues in tile-drain water from corn plots as affected by cropping practices and fertility levels, *Can. J. Soil Sci.,* 57, 249, 1977.
16. **Chevreuil, M. and Chesterikoff, A.,** Etude du transfert des pesticides sur un bassin versant de la Brie. Modalités et essai de bilan, *C. R. Séances Acad. Agric. Fr.,* 65, 835, 1979.
17. **Snégaroff, J.,** La pollution des eaux par les triazines herbicides dans le secteur du marais charentais, *Phytiatr. Rev. Fr. Méd. Pharm. Vég. Phytophar.,* 28, 249, 1979.
18. **Scheunert, I. and Korte, F.,** Interactions in the fate of chemicals in terrestrial systems, *Ecotoxicol. Environ. Saf.,* 9, 385, 1985.
19. **Southwick, L. M., Willis, G. H., Bengston, R. L., and Lormand, T. J.,** Atrazine and metolachlor in subsurface drain water in Louisiana, *J. Irrig. Drain Eng.,* 116, 16, 1990.
20. **Halleberg, G. R.,** Pesticide pollution of groundwater in the humid United States, *Agric., Ecosyst. Environ.,* 26, 299, 1989.
21. **Wehtje, G. R., Spalding, R. F., Burnside, O. C., Lowry, S. R., and Leavitt, J. R.,** Biological significance and fate of atrazine under aquifer conditions, *Weed Sci.,* 31, 610, 1983.
22. **Huang, L. Q. and Frink, C. R.,** Distribution of atrazine, simazine, alachlor, and metholachlor in soil profiles in Connecticut, *Bull. Environ. Contam. Toxicol.,* 43, 159, 1989.
23. **Smith, C. N. and Carsel, R. F.,** A stainless-steel soil solution sampler for monitoring pesticides in the vadose zone, *Soil Sci. Soc. Am. J.,* 50, 263, 1986.
24. **Baumann, T., Merkel, B., Ruppert, T., and Niessner, R.,** Problems related to pore water sampling for pesticide determination, Lehrstuhl f. Hydrogeologie, Hydrochemie und Umwelt analyt., T.U., München, Germany, unpublished paper.
25. **Kubiak, R., Führ, F., Mittelstaedt, W., Hansper, M., and Steffens, W.,** Transferability of lysimeter results to actual field situations, *Weed Sci.,* 36, 514, 1988.
26. **Schiavon, M.,** Studies of the leaching of atrazine, of its dechlorinated derivatives, and of hydroxyatrazine from soil, using ^{14}C ring-labelled compounds, under outdoor conditions, *Ecotoxicol. Environ. Saf.,* 15, 46, 1988.
27. **Goswamy, K. P. and Green, R.,** Simultaneous extraction of hydroxyatrazine, atrazine and ametryne from some Hawaiian soils, *Soil Sci. Am. Proc.,* 3, 702, 1973.
28. **Fermanich, K. J. and Daniel, T. C.,** Pesticide mobility and persistence in microlysimeter soil columns from a tilled and no-tilled plot, *J. Environ. Qual.,* 20, 195, 1991.
29. **Hörmann, W. D., Tournayre, J. C., and Egli, H.,** Triazine herbicide residues in Central European streams, *Pestic. Monit. J.,* 13, 128, 1979.
30. **My, J.,** Qualité des eaux et pesticides, *Phytoma,* 422, 20, 1990.
31. **Portal, J.-M., Babut, M., and Schiavon, M.,** Etat de contamination des eaux d'un captage de Lorraine, in Proc. *Colloque Phyt'eau,* Versailles, October 22 to 23, 1992, in press.
32. **Dörfler, U., Scheunert, I., and Schroll, R.,** Lysimeterversuche zum Transportverhalten von ^{14}C-markierten Pflanzenschutzmitteln im Boden, *DVLW Schriftenreihe Wasser,* in press.
33. **Reiml, D., Scheunert, I., and Korte, F.,** Leaching of conversion products of (^{14}C) buturon from soil during 12 years after application, *J. Agric. Food Chem.,* 37, 244, 1989.
34. **Schiavon, M., Barriuso, E., Portal, J.-M., Andreux, F., Bastide, J., Coste, C. M., and Millet, A.,** Étude du Devenir de Deux Substances Organiques Utilisées dans les Sols, l'une Massivement (l'Atrazine) et l'Autre à l'Etat de Trace (le Metsulfuron-Méthyl), à l'Aide de Molécules Marquées au ^{14}C. Rep. Contract No. 7219, SRETIE, Ministère de l'Environnement, France, 1990.

35. **Demon, M., Schiavon, M., Portal, J.-M., and Munier-Lamy, C.,** Seasonal dynamic of atrazine in three soils under outdoor conditions, *Chemosphere,* submitted.
36. **Achik, J. and Schiavon, M.,** Carbofuran transfer and persistence in drained agricultural soils related to their structure and adsorption properties, *Ecotoxicol. Environ. Saf.,* 18, 83, 1989.
37. **Demon, M.,** unpublished data, 1991.
38. **Scheunert, I. and Klein, W.,** Predicting the movement of chemicals between environmental compartments, in *Appraisal of Tests to Predict the Environmental Behaviour of Chemicals,* Shehan, P., Korte, F., Klein, W., and Bourdeau, P., Eds., SCOPE 25, John Wiley & Sons, New York, 1985, 285.
39. **Chiou, C. T., Malcolm, R. L., Brinton, T. L., and Kile, D. E.,** Water solubility enhancement of some organic pollutants and pesticides by dissolved humic and fulvic acids, *Environ. Sci. Technol.,* 20, 502, 1986.
40. **Karickhoff, S. W., Brown, D. S., and Scott, T. A.,** Sorption of hydrophobic pollutants on natural sediments, *Water Res.,* 13, 241, 1979.
41. **Hasset, J. J., Banwart, W. L., and Griffin, R. A.,** Correlation of compound properties with sorption characteristics of nonpolar compounds by soils and sediments: concepts and limitations, in *Environmental and Solid Wastes,* Francis, C. W. and Auerbach, S. I., Eds., Butterworths, Boston, 1983, 161.
42. **Enfield, C. G. and Bengtsson, G.,** Macromolecular transport of hydrophobic contaminants in aqueous environments, *Ground Water,* 26, 64, 1988.
43. **Schiavon, M., Portal, J.-M., and Andreux, F.,** Données actuelles sur les transferts d'atrazine dans l'environnement, *Agronomie,* 12, 129, 1992.
44. **Kloskowski, R. and Führ, F.,** Behaviour of {ring-^{14}C} hydroxy-simazine in a Parabraun soil, *Chemosphere,* 14(11/12), 1913, 1985.
45. **Andreux, F., Schiavon, M., Bertin, G., Portal, J.-M., and Barriuso, E.,** The usefulness of soil fractionation methods in studies about the behaviour of pollutants in soils, *Toxicol. Environ. Chem.,* 31-32, 29, 1991.
46. **Hatcher, P. G., Breger, I. A., Dennis, L. W., and Maciel, G. E.,** Solid-state ^{13}C-NMR of sedimentary humic substances: new revelations on their chemical composition, in *Aquatic and Terrestrial Humic Materials,* Christman, R. F. and Gjessing, E. T., Eds., Ann Arbor Science, Ann Arbor, MI, 1983, 37.
47. **Senesi, N. and Miano, T.,** The role of abiotic interactions with humic substances on the environmental impact of organic pollutants, in *Environmental Impact of Soil Component Interactions,* Vol. 1, *Natural and Anthropogenic Organics,* Huang, P. M., Berthelin, J., Bollag, J.-M., McGill, W. B., and Page, A. L., Eds., Lewis Publishers, Chelsea, MI, 1995.
48. **Khan, S. U.,** Kinetics of hydrolysis of atrazine in aqueous fulvic acid solution, *Pestic. Sci.,* 9, 39, 1978.
49. **Senesi, N. and Testini, C.,** Theoretical aspects and experimental evidence of the capacity of humic substances to bind herbicides by charge-transfer mechanisms (electron-donor-acceptor processes), *Chemosphere,* 13, 461, 1984.
50. **Bollag, J.-M., Liu, S.-Y., and Minard, R. D.,** Cross-coupling of phenolic humus constituents and 2,4-dichlorophenol, *Soil Sci. Soc. Am. J.,* 44, 52, 1982.
51. **You, I. S., Jones, R. A., and Bartha, R.,** Evaluation of a chemically defined model for the attachment of 3,4-dichloroaniline to humus, *Bull. Environ. Contam. Toxicol.,* 29, 476, 1982.
52. **Bollag, J. M., Pal, S., and Huang, P. M.,** Role of abiotic and biotic catalysts in the transformation of phenolic compounds, in *Environmental Impact of Soil Component Interactions,* Vol. 1, *Natural and Anthropogenic Organics,* Huang, P. M., Berthelin, J., Bollag, J.-M., McGill, W. B., and Page, A. L., Eds., Lewis Publishers, Chelsea, MI, 1995.
53. **Andreux, F., Portal, J.-M., Schiavon, M., and Bertin, G.,** The binding of atrazine and its dealkylated derivatives to humic-like polymers derived from catechol, *Sci. Total Environ.,* 117/118, 207, 1992.
54. **Weber, J. B. and Miller, C. T.,** Organic chemical movement over and through soil, in *Reactions and Movement of Organic Chemicals in Soils,* Sawhney, B. L. and Brown, K., Eds., SSSA Spec. Publ. No. 22, Soil Science Society of America, Madison, WI, 1989, 305.
55. **Hamaker, J. W. and Thompson, J. M.,** Adsorption, in *Organic Chemicals in the Soil Environment,* Vol. 1, Goring, C. and Hamaker, J., Eds., Marcel Dekker, New York, 1972, 51.

56. **Hasset, J. J. and Banwart, W. L.,** The sorption of nonpolar organics by soils and sediments, in *Reactions and Movement of Organic Chemicals in Soils,* Sawhney, B. L. and Brown, K., Eds., SSSA Spec. Publ. No. 22, Soil Science Society of America, Madison, WI, 1989, 31.
57. **Briggs, G. G.,** Theoretical and experimental relationships between soil adsorption, octanol-water partition coefficients, water solubilities, bioconcentration factors, and parachor, *J. Agric. Food Chem.,* 29, 1050, 1981.
58. **Brown, A. W. A. and Flagg, E. W.,** Empirical prediction of organic pollutant sorption in natural sediments, *J. Environ. Qual.,* 10, 382, 1981.
59. **Bertin, G., Schiavon, M., Portal, J.-M., and Andreux, F.,** Contribution to the study of nonextractable pesticide residues in soils: incorporation of atrazine in model humic acids prepared from catechol, in *Diversity of Environmental Biogeochemistry,* Berthelin, J., Ed., Elsevier, Amsterdam, 1991, 105.
60. **Stevenson, F. J.,** *Humus Chemistry, Genesis, Composition, Reactions,* John Wiley & Sons, New York, 1982.
61. **Barriuso, E. and Calvet, R.,** Soil type and herbicides adsorption, *Int. J. Environ. Anal. Chem.,* 46, 117, 1992.
62. **Huang, P. M., Grover, R., and McKercher, R. B.,** Components and particle size fractions involved in atrazine adsorption by soils, *Soil Sci.,* 138, 20, 1984.
63. **Walker, A. and Crawford, D. V.,** The role of organic matter in adsorption of the triazine herbicides by soil, in *Isotopes and Radiation in Soil Organic Matter Studies,* International Atomic Energy Agency, Vienna, 1969, 91.
64. **Sarkar, J. M., Malcolm, R. L., and Bollag, J.-M.,** Enzymatic coupling of 2,4-dichlorophenol to stream fulvic acid in the presence of oxidoreductases, *Soil Sci. Soc. Am. J.,* 52, 688, 1988.
65. **Bollag, J.-M., Minard, R. D., and Liu, S.-Y.,** Cross-linkage between anilines and phenolic humus constituents, *Environ. Sci. Technol.,* 17, 72, 1983.
66. **Adrian, P., Andreux, F., Viswanathan, R., Freitag, D., and Scheunert, I.,** Fate of anilines and related compounds in the environment. A review, *Toxicol. Environ. Chem.,* 29-30, 109, 1989.
67. **Völkel, W., Choné, T., Mansour, M., Scheunert, I., and Andreux, F.,** MnO_2-induced addition of 3,4-dichloroaniline to catechol rings: a model for the formation of pesticide bound residues in humus, in *Proc. 6th Intern. I.H.S.S. Meeting,* Monopoli, Italy, September 21 to 25, 1992, in press.
68. **Barriuso, E., Schiavon, M., Andreux, F., and Portal, J. M.,** Localization of atrazine non-extractable (bound) residues in soil size fractions, *Chemosphere,* 22(12), 1131, 1991.
69. **Scheunert, I., Dörfler, U., and Adrian, P.,** Biomineralisierung von Umweltchemikalien in verschiedenen Böden in Abhängigkeit von jahreszeitlichen Bedingungen, in *Landwirtschaft im Spannungsfeld von Belastungfaktoren und gesellschaftlichen Ansprüchen,* Kongreßband 1990 Berlin, VDLUFA Verlag, Darmstadt, 1990, 633.
70. **Stotzky, G.,** Influence of soil mineral colloids on metabolism processes, growth, adhesion, and ecology of microbes and viruses, in *Interactions of Soil Minerals with Natural Organics and Microbes,* Huang, P. M. and Schnitzer, M., Eds., Soil Science Society of America, Madison, WI, 1986, 310.
71. **Alexander, M. and Scow, K. M.,** Kinetics of biodegradation in soil, in *Reactions and Movement of Organic Chemicals in Soils,* Sawhney, B. L. and Brown, K., Eds., SSSA Spec. Publ. No. 22, Soil Science Society of America, Madison, WI, 1989, 243.
72. **Scheunert, I.,** Transformation and degradation of Pesticides in soil, in *Chemistry of Plant Protection,* Vol. 8, Ebing, W., Ed., Springer Verlag, Berlin, 1992, 23.
73. **Anderson, J. P. E.,** Einfluss von Temperatur und Feuchte auf Verdamfung, Abbau und Festlegung von Diallat im Boden, *Z. Pflanzenkr., Pflanzenschutz,* 7, 141, 1975.
74. **Anderson, J. P. E.,** Herbicide degradation in soil: influence of microbial biomass, *Soil Biol. Biochem.,* 16, 483, 1984.
75. **Völkel, W., Choné, T., Mansour, M., and Andreux, F.,** Dégradation dans un sol brun de culture, de la 3,4-dichloroaniline libre ou liée à un monomère de type humique, in *Réactions et Biotransformations des Xénobiotiques,* 21st Congr. Groupe Français des Pesticides, May 22 to 23, 1991, Nancy-Brabois, 1992, 355.
76. **Scribner, S. L., Benzing, T. R., Sun, S., and Boyd, S. A.,** Desorption and bioavailability of aged simazine residues in soil from a continuous corn field, *J. Environ. Qual.,* 21, 115, 1992.

77. **Hartmann, A. and Simmeth, I.,** Einfluß des Wasserpotentiale auf mikrobielle Aktivitäten von Böden, in *Landwirtschaft in Spannungsfeld von Belastungfaktoren und gesellschaftlichen Ansprüchen,* Kongreßband 1990 Berlin, VDLUFA Verlag, Darmstadt, 1990, 671.
78. **Bernoux, M.,** Contribution expérimentale à l'étude de la stabilité des résidus liés d'atrazine, DEA de Sciences Agronomiques, ENSAIA-INPL, Nancy, 1991, 38.
79. **Khan, S. U. and Behki, R. M.,** Effects of *Pseudomonas* species on the release of bound ^{14}C residues from soil treated with (^{14}C) atrazine, *J. Agric. Food Chem.,* 38(11), 2090, 1990.
80. **Mansour, M. and Korte, F.,** Abiotic degradation pathways of selected xenobiotic compounds in the environment, in *Studies on Environmental Science, 29, Chemistry for Protection of the Environment,* Pawlowski, L. and Alaerts, G., Eds., Elsevier, Amsterdam, 1986, 257.
81. **Mansour, M., Feicht, E., and Méallier, P.,** Improvement of the photostability of selected substances in aqueous medium, *Toxicol. Environ. Chem.,* 20-21, 139, 1989.
82. **Choudry, G. G.,** Humic substances: structural aspects and photophysical, photochemical and free radical characteristics, in *The Handbook of Environmental Chemistry,* Vol. 1, Hützinger, O., Ed., Springer-Verlag, Berlin, 1984, 1.
83. **Lee, J. S. K. and Huang, P. M.,** Photochemical effect on the abiotic transformations of polyphenolics as catalyzed by Mn(IV) oxides, in *Environmental Impact of Soil Component Interactions,* Vol. 1, *Natural and Anthropogenic Organics,* Huang, P. M., Berthelin, J., Bollag, J.-M., McGill, W. B., and Page, A. L., Eds., Lewis Publishers, Chelsea, MI, 1995.
84. **Wolfe, N. L., Mingelgrin, U., and Miller, G. C.,** Abiotic transformations in water, sediments and soil, in *Pesticides in the Soil Environment: Processes, Impacts, and Modelling,* Cheng, H. H., Ed., Soil Science Society of America, Madison, WI, 1990, 103.
85. **Ristori, G. G. and Fusi, P.,** Adsorption mechanisms and nonbiological catalytic transformations of agrochemicals by clay minerals, in *Environmental Impact of Soil Component Interactions,* Vol. 1, *Natural and Anthropogenic Organics,* Huang, P. M., Berthelin, J., Bollag, J.-M., McGill, W. B., and Page, A. L., Eds., Lewis Publishers, Chelsea, MI, 1995.
86. **Filip, Z. K. and Claus, H.,** Effects of soil minerals on the microbial formation of enzymes and their possible use in remediation of chemically polluted sites, in *Environmental Impact of Soil Component Interactions,* Vol. 1, *Natural and Anthropogenic Organics,* Huang, P. M., Berthelin, J., Bollag, J.-M., McGill, W. B., and Page, A. L., Eds., Lewis Publishers, Chelsea, MI, 1995.
87. **Theng, B. K. G. and Orchard, V. A.,** Interactions of clays with microorganisms and bacterial survival in soil: a physicochemical perspective, in *Environmental Impact of Soil Component Interactions,* Vol. 1, *Natural and Anthropogenic Organics,* Huang, P. M., Berthelin, J., Bollag, J.-M., McGill, W. B., and Page, A. L., Eds., Lewis Publishers, Chelsea, MI, 1995.
88. **Dec, J. and Bollag, J.-M.,** Microbial release and degradation of catechol and chlorophenols bound to synthetic humic acids, *Soil Sci. Soc. Am. J.,* 53, 1366, 1988.
89. **Scheunert, I., Ter Meer-Bekk, C., and Korte, F.,** Distribution and biodegradability of ^{14}C-residues bound in various soil fractions after treatment of the soil with model ^{14}C-chemicals, in *Quantification, Nature and Bioavailability of Bound ^{14}C-Pesticide Residues in Soil, Plants and Food,* International Atomic Energy Agency, Vienna, 1986, 31.
90. **Dörfler, U., Adrian, P., and Scheunert, I.,** Prüfung der biologischen Abbaukapazität von Böden in Abhängigkeit von jahreszeitlichen Bedingungen. Umweltforschungsplan des Bundesministers für Umwelt, Naturschutz und Reaktorsicherheit, Forschungsbericht 10602052, im Auftrag des Umweltbundesamtes, Berlin, 1990.
91. **Bordes, O.,** Influence du flutriafol sur l'activité cellulolytique de la microflore d'un sol brun, DAA de Sciences Agronomiques, ENSAIA-INPL, Nancy, 42, 1991.
92. **Bertin, G., Schiavon, M., and Pottier, C.,** Plant bioavailability of "natural" and "model" humic acid-bound (^{14}C) atrazine residues, *Toxicol. Environ. Chem.,* 26, 203, 1989.

Chapter 30

Effects of Soil Minerals on the Microbial Formation of Enzymes and Their Possible Use in Remediation of Chemically Polluted Sites

Z. K. Filip and H. Claus

CONTENTS

I. INTRODUCTION

Extensive agricultural and industrial activities may result in a flow of different xenobiotics into soil and underground environments. If composed of recalcitrant molecules, xenobiotics can accumulate and become harmful to these environments and their inhabitants. Some of them are toxic. Numerous studies show that different microorganisms or some enzymes are capable of *in vitro* transformation or degradation of several xenobiotic compounds, such as pesticides, detergents, and other organic pollutants. Phenoloxidases are active in transforming different phenolic substances of natural and anthropogenic origin *in vitro*[1-4] and in soils.[2,5-9] Phenols also have been recognized as main contaminants to groundwater by the Environmental Protection Agency (EPA). Questions arise in this respect, whether soil components such as clays and other solids may affect the microbial formation and activity of phenoloxidases. Answers to these questions can be of importance when considering a possible use of phenoloxidases in the remediation of chemically polluted sites.

II. MICROBIAL FORMATION OF PHENOLOXIDASES AS AFFECTED BY SOIL ADSORBENTS

Many white-rot fungi and actinomycetes can produce phenoloxidases such as laccase (EC 1.10.3.2), tyrosinase (EC 1.14.18.1), and peroxidase (EC 1.11.1.7). In our experiments,[10] we used *Polyporus versicolor* and *Pleurotus ostreatus* strains producing laccases, and *Streptomyces*

0-87371-914-X/95/$0.00+$.50

Table 1 Characteristics of Some Phenoloxidases Used in the Experiments[41]

Phenoloxidase	Molecular weight (kDa)	Isoelectric point	Isoenzymes	Other features	Stability range pH
Tyrosinase from *Streptomyces eurythermus*	16	6.4	1–2	Contains Cu	5.0–10 max = 8.0
Tyrosinase from *Agaricus bisporus*	110	4.8	Tetramer	Contains Cu	4.0–10 max = 7.0
Peroxidase from horseradish	40	5.3–7.3	7	Glycoprotein, contains Fe	3.0–10 max = 6.0
Laccase from *Pleurotus ostreatus*	47	3.5	3–6	Inducible glycoprotein, contains Cu	2.5–10 max = 6.0
Laccase from *Polyporus versicolor*	61	3.0	2	Inducible glycoprotein, contains Cu	2.0–10 max = 6.0

eurythermus and *Streptomyces michiganensis* strains producing tyrosinases. The characteristics of the specific enzymes and a peroxidase from horseradish are summarized in Table 1.

Bentonites (2:1 clays) and kaolinites (1:1 clays) of different cation-exchange capacities and homoionic to H^+, Na^+, Ca^{2+}, Cu^{2+}, or Al^{3+} served as typical soil mineral adsorbents in the experiments. Other solids used were bentonite complexed with a commercial humic acid, quartz sand, and porous glass (Table 2).

In our previous studies, clay additions to the cultures of some yeasts, soil fungi, and actinomycetes markedly accelerated and increased growth, especially in well-aerated cultures.[11-13] The influence of solids (1% w/v) in batch cultures of *S. michiganensis, P. ostreatus* and *P. versicolor* is demonstrated in Table 3. The addition of kaolinite resulted in a slight decrease of the total protein production (15 to 20%). The enzyme activities, however, were sometimes enhanced, particularly the laccase from *P. versicolor,* which increased up to 50%. The presence of different bentonites and of a bentonite-humus complex in the culture media resulted in a strong reduction of phenoloxidase activity. In the *S. michiganensis* cultures, for instance, protein contents were reduced by 20 to 34%, but tyrosinase activity ceased almost completely. A significant decrease in enzyme activities was obtained also for the laccases from *P. ostreatus* and *P. versicolor.* Again, the bentonite-humus complex inhibited growth and enzyme activities in these cultures. Quartz sand had no significant effect on the behavior of *S. michiganensis* and *P. ostreatus,* but it enhanced the production of laccase in *P. versicolor* up to 1.3-fold. The addition of porous glass had little effect on the microbial growth and the activity of laccases, but the tyrosinase activity of *S. michiganensis* was considerably enhanced in the presence of this solid. In cultures of *S. eurythermus* with increasing concentrations of bentonite from 0.01 to 0.5% (w/v), the microbial growth (protein content) remained almost unaffected, while there was a continuous decrease in both the bulk and specific tyrosinase activity. This indicates that the bentonite inhibited not only the activity, but also the production of tyrosinase by *S. eurythermus.*

The addition of homoionic kaolinites (1% w/v) exerted different effects on the growth of microorganisms and the activity of enzymes (Table 4).[10] The growth of *S. michiganensis*

Table 2 Clays and Other Solids Used in the Experiments on the Formation and Adsorption of Phenoloxidases[10-13]

Solids	Characteristics and origin	Cation-exchange capacity (MEQ 100^{-1}g)
	Bentonites (2:1 clays)	
B1	Native aluminum silicate (Sigma, FRG)	81
B2	Natural Ca-bentonite from Hallertau (Bayern), min. contents of montmorillonite 90% (Südchemie, FRG)	96
B3	Similar to B2 but min. contents of montmorillonite 60%	61
	Kaolinites (1:1 clays)	
K1	Distribution of particle size: >40 µm ca. 15%, >20 µm ca. 40%, <20 µm ca. 60% (Merck, FRG)	6.4
K2	Well crystallized clay (KG a1, Ga) (Clays Minerals Society, U.S.; specific surface ca. 10 m^2 g^{-1}	4.7
Quartz sand (QS)	Material from a groundwater aquifer of the Senckenberger Forest (near Hamburg, FRG); particle size: < 2 mm	—
Porous glass (PG)	Porosity 55%; pore size 125–315 µm (Schott, FRG)	—
Humic acid (HA)	Commercial product, (Roth, FRG)	—

was inhibited by the H^+-, Na^+-, Ca^{2+}-, and Cu^{2+}-saturated clays by about 20%, and the inhibition amounted to 30% for Al^{3+}-kaolinite. Similarly, the activity of tyrosinase was diminished. On the other hand, the growth of *P. versicolor* was less affected, and the production of laccase was enhanced. With Cu^{2+}-kaolinite, the increase amounted to 155%. Increasing the concentration of the Cu^{2+}-kaolinite (1 to 4% w/v) also increased the activity of tyrosinase as well as the yield of protein in the *S. eurythermus* cultures. The behavior of phenoloxidase producers in sand cultures may have some resemblance to that expected in natural environments (Figure 1). There was a significantly higher activity of phenoloxidases in sand autoclaved before inoculation than in nonsterile sand. This phenomenon may be due to competitive effects of microorganisms indigenous to the groundwater sand. Their presence was clearly indicated by the increase of fluorescein diacetate hydrolysis in noninoculated sand with nutrients added. In sand cultures of *P. ostreatus,* the growth of the indigenous microorganisms appeared to completely inhibit the production of laccase. Similar observations were made by Leonowicz and Bollag[14] for laccase producers in soil cultures.

III. TRANSFORMATION OF DIFFERENT XENOBIOTICS BY PHENOLOXIDASES

A. PEROXIDASE

Bordeleau and Bartha[6] studied the oxidative coupling of 43 substituted anilines using a peroxidase from the soil fungus *Geotrichum candidum.* All anilines monosubstituted by F^-, Cl^-, Br^-, or I^- and also methylated anilines were oxidized and transformed to colored

Table 3 Production of Protein and Activity of Phenoloxidases in the Presence of Solids (as % of Controls)[10]

	Solids[a]							
Subject	**K1**	**K2**	**B1**	**B2**	**B3**	**B1+ HA**	**QS**	**PG**
	Streptomyces michiganensis							
Protein content	85	88	66	69	80	76	92	73
Tyrosinase activity	71	102	0.3	0.6	2.2	0.6	94	149
	Pleurotus ostreatus							
Protein content	87	79	75	82	ND	62	116	93
Laccase activity	102	78	34	42	ND	56	81	91
	Polyporus versicolor							
Protein content	96	96	78	77	97	60	119	100
Laccase activity	147	109	6.0	10	54	6.0	131	83

[a] For abbreviations, see Table 2.

polyaromatic compounds. Nitroanilines were not transformed, however. Dichlorinated isomers, but not 2,6-dichloroaniline, were also oxidized and coupled into azobenzenes. Similar products were formed from chloromethylanilines, but not from 2-chloro-6-methylaniline. From trichloroanilines involved in the tests, only a 2,3,4-isomer was transformed by peroxidase; 2,3,5-trimethylaniline was oxidized to an azo compound. No transformation was observed with tetra- and pentasubstituted fluoroanilines.

Berry and Boyd[15] studied the catalytic oxidation of several monosubstituted anilines and phenols brought about by a horseradish peroxidase. They found electron-donating substituents to increase the oxidation, and nitrosubstituents to cause an inhibition of an enzymatic oxidation.

Table 4 Production of Protein and Activity of Phenoloxidases in the Presence of Homoionic Kaolinite (as % of Controls)[10]

	Solids				
Subject	**H^+**	**Na^+**	**Ca^{2+}**	**Cu^{2+}**	**Al^{3+}**
	Streptomyces michiganensis				
Protein content	79	78	80	82	68
Tyrosinase activity	32	71	76	78	60
	Streptomyces eurythermus				
Protein content	86	92	93	91	70
Tyrosinase activity	69	109	68	161	72
	Pleurotus ostreatus				
Protein content	74	88	82	158	82
Laccase activity	112	119	80	93	91
	Polyporus versicolor				
Protein content	71	86	97	100	81
Laccase activity	130	128	122	155	133

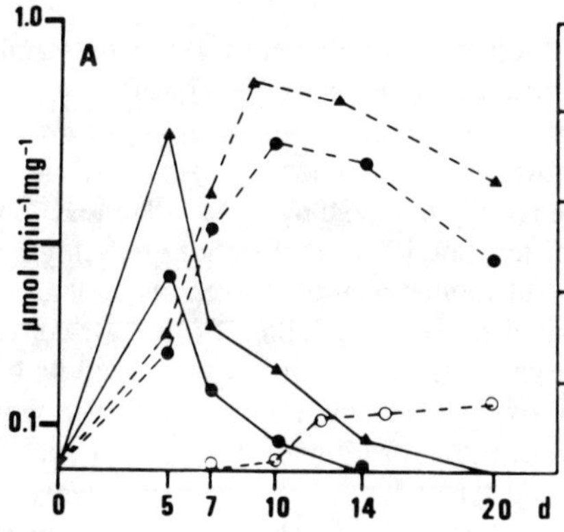

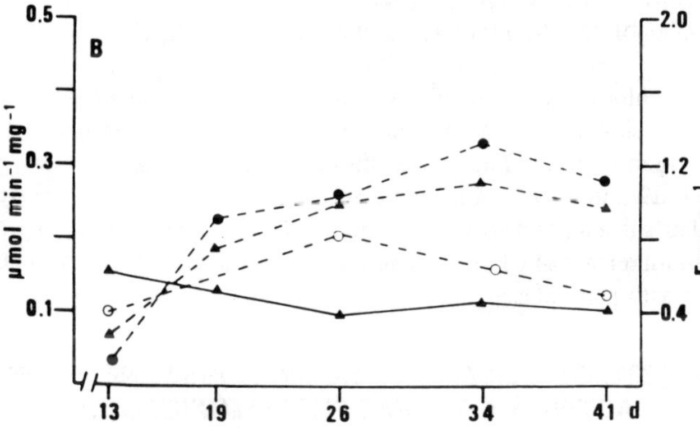

Figure 1 Activity of tyrosinase from *S. eurythermus* (A) and laccase from *P. ostreatus* (B) in sand cultures.[10] ○ , nonsterile, supplemented with nutrients; ● , nonsterile, inoculated and supplemented with nutrients; ▲ , sterile, inoculated and supplemented with nutrients; ______, specific enzyme activity; - - - - - fluorescein diacetate hydrolysis.

The fungus *Phanerochaete chrysosporium* exhibits versatile peroxidase activities. These include depolymerization and degradation of lignins and also oxidation of several anthropogenic xenobiotics, such as polycyclic aromatic hydrocarbons,[16-18] dibenzo(p)dioxines,[17] and chlorophenols.[19,20] It was also demonstrated that in cultures of *P. chrysosporium,* several organic pollutants can be fully mineralized. This includes chlorinated phenols from wastewaters of the cellulose processing industry,[21] PCBs, dioxins, DDT, Lindan and benzo(a)pyrene,[22-24] chlorinated anilines,[25] pentachlorophenol,[26] and phenanthren.[27] The involvement of a peroxidase complex from *P. chrysosporium* in the individual oxidation reactions or processes has to be elucidated.

B. LACCASE AND TYROSINASE

A laccase from *Rhizoctonia praticola* and *Trametes versicolor,* and also a soil extract containing a laccase were capable of coupling diverse phenols and lignin degradation products (e.g., vanillin, vanillic acid, and ferulic acid) with xenobiotics such as chlorophenols, naphthalene, and halogenated anilines under laboratory experiments.[1-4,7,28-31,34] Several authors also observed cross-coupling reactions between individual xenobiotics using mass spectroscopy.[1-4,28,32,33] A laccase from *T. versicolor* was capable of oxidizing not only phenols, but also halogenated and alkyl- or alkoxyl-substituted anilines.[35]

In our laboratory, we tested the oxidation of 49 phenols and aromatic amines by phenoloxidases.[36-38] Thirty-eight compounds (78%) were oxidized by the laccase from *P. versicolor* and 7 (14%) by tyrosinase from *Agaricus bisporus* (Table 5). Anisol was the only substrate which was weakly oxidized by tyrosinase and not by laccase. Alkyl-substituted phenols were rapidly oxidized by laccase. No reaction occurred with bulky molecules such as 2,6-di-tert-butyl-4-methylphenol, probably because of some steric hindrance. Alkyl-chloro-substituted phenols (seven compounds) were oxidized by the laccase. Of 16 halogenated phenols, including F- and Br-containing substitutes, 11 were oxidized by the laccase. Reaction velocity decreased with increasing degree of halogen substitution. No oxygen consumption was observed with pentachlorophenol. No reaction occurred with nitro-substituted phenols. No synergistic reactions were observed when laccase and tyrosinase were added simultaneously to substrate solutions. With *para*-substituted phenols, reaction velocity decreased in the sequence: amino > isopropyl > fluoro > chloro > nitro residues. Substituents in *meta* positions, e.g., 3-chlorophenol and 3-aminophenol, mostly exerted negative effects on reaction velocity.

Oxidation of chlorinated compounds was often accompanied by a different degree of dehalogenation. Using tyrosinase, chloride was released from some substrates even in the absence of oxygen consumption. For example, from 4-chlorophenol and 4-chloro-3-methyl-phenol, 29 and 47% of Cl^- bound in the molecule was released, respectively. These and other substances also were dechlorinated effectively by laccase and peroxidase. A maximum release of Cl^-, i.e., 64%, was measured when 4-bromo-2-chlorophenol was incubated with peroxidase.

IV. EFFECTS OF CLAYS AND HUMIC SUBSTANCES ON THE ACTIVITY OF SOME PHENOLOXIDASES

To evaluate a possible use of phenoloxidases as a tool in the remediation of chemically polluted soil and underground sites, it is necessary to estimate their behavior toward the most common and important soil constituents such as clays and humic substances. Humic substances appear to exert different effects on different kinds of peroxidases. Serban and Nissenbaum[39] reported a higher resistance against proteolysis and heat by a peroxidase bound to a humic acid. The enzyme was probably only incorporated into the humic matrix, and it remained active. However, horseradish peroxidase also was found to form ionic, hydrogen, and covalent bonds with soil humic acids.[40] The enzyme immobilization resulted in a decrease in its activity and an increase in a resistance to adverse physicochemical conditions (pH). It has been demonstrated that the adsorption of phenoloxidases was strongly enhanced by decreasing pH.[36-38,41] In the presence of bentonites, laccase activity was rapidly reduced at pH <5.0 and disappeared completely at pH 3.0, indicating that all the enzyme was adsorbed. In the presence of kaolinites, the decrease of free laccase activity was slower, and some activity still remained at pH 2.0. Similar results were obtained for the other enzymes. Maximum adsorption occurred at pH values near the isoelectric points (IEPs) of the phenoloxidases.

Enzyme adsorption significantly correlated with the cation exchange capacity (CEC) of clays: at the adequate pH, phenoloxidases were strongly adsorbed on bentonites and a bentonite-humus complex, often resulting in total inhibition of enzyme activity (Table 6). The kaolinites were less effective. Quartz sand exerted similar effects as kaolinites, whereas porous glass had no effect.[36-38,41] The binding capacities of the homoionic clays were estimated from the adsorption isotherms. About 0.5 mg of peroxidase and 1 to 2 mg of tyrosinase (from *A. bisporus)* were adsorbed by 1.0 mg of bentonite. Only 10 μg of tyrosinase (from *A. bisporus)* were adsorbed by 1.0 mg of kaolinite. All phenoloxidases

Table 5 Enzymatic Oxidation of Some Substituted Phenols and Anilines[38]

Compound	Oxygen consumption ($nmol\ O_2\ min^{-1}\ ml^{-1}$) Laccase from *Polyporus versicolor*	Oxygen consumption ($nmol\ O_2\ min^{-1}\ ml^{-1}$) Tyrosinase from *Agaricus bisporus*
Alkyl-Substituted Phenols		
2,6-Dimethylphenol	9,600	1,800
2,4,6-Trimethylphenol	5,200	0
2,3,5-Trimethylphenol	3,000	0
2,3,6-Trimethylphenol	3,000	0
2-Isopropylphenol	3,700	1,400
3-Isopropylphenol	4,100	0
4-Isopropylphenol	4,400	0
2,6-Diisopropylphenol	5,200	0
2,4-Di-tert.butylphenol	700	0
2,5-Di-tert.butylphenol	0	0
2,6-Di-tert.butyl-4-ethylphenol	0	0
Substituted Chlorophenols		
2-Chloro-6-methylphenol	6,600	0
2-Chloro-5-methylphenol	2,200	0
4-Chloro-2-methylphenol	3,500	0
4-Chloro-3-methylphenol	1,500	0
2-Chloro-4,5-dimethylphenol	6,600	0
2,4-Dichloro-6-methylphenol	5,900	0
2,2-Methylen-bis-3,4,6-trichlorophenol	400	0
Halogen-Substituted Phenols		
2-Chlorophenol	3,800	0
3-Chlorophenol	400	0
4-Chlorophenol	900	0
2,4-Dichlorophenol	1,600	0
2,6-Dichlorophenol	2,500	0
2,3-Dichlorophenol	200	0
3,4-Dichlorophenol	0	0
2,5-Dichlorophenol	0	0
2,4,6-Trichlorophenol	2,300	0
2,4,5-Trichlorophenol	300	0
2,3,4-Trichlorophenol	0	0
2,3,5-Trichlorophenol	0	0
2,3,5,6-Tetrachlorophenol	700	0
Pentachlorophenol	0	0
4-Fluorophenol	1,600	0
4-Bromo-2-chlorophenol	2,200	0
Arylamines, Nitrophenols		
2-Aminophenol	41,000	15,000
3-Aminophenol	3,100	0
4-Aminophenol	8,500	900
2,5-Dimethylaniline	2,600	0
N-Methylaniline	4,400	0
N,N-Dimethylaniline	5,900	0
3-Aminobenzo-trifluoride	400	0
2-Nitrophenol	0	0
3-Nitrophenol	0	0

Table 5 Enzymatic Oxidation of Some Substituted Phenols and Anilines[38] (Continued)

Compound	Oxygen consumption ($nmol\ O_2\ min^{-1}\ ml^{-1}$) Laccase from *Polyporus versicolor*	Tyrosinase from *Agaricus bisporus*
Others Phenolics		
Phenol	1,000	8,800
2-Phenylphenol	600	0
Anisol	0	400

except tyrosinase from *A. bisporus* lost their activity when adsorbed on bentonite. The bound tyrosinase retained 2 to 11% of the free enzyme activity. With kaolinite, the retained activity was 20 to 50% for tyrosinase from *A. bisporus* and about 15% for laccase from *P. versicolor.*

Attempts were made to desorb the complexed phenoloxidases by increasing pH and ionic strength.[36-38,41] The enzymes were found to be bound more strongly by bentonite than by kaolinite: above the IEP of the individual enzyme, about 45% of the tyrosinase activity became desorbed from the kaolinites but only 15% from the bentonites. Specific activities of desorbed phenoloxidases were distinctly reduced as compared with the controls. By increasing the ionic strength, we would have expected an enzyme desorption due to an ion exchange process. However, a change of the salt concentration from 0.2 to 1.0 *M* NaCl did not influence desorption of the enzymes.

Table 6 Maximum Adsorption of Phenoloxidases on Solids at Adequate pH Values (as % of Controls)[41]

Phenoloxidase	pH	Solids[a] B1	B2	B3	K1	K2	B1+HA	QS
Tyrosinase from *Streptomyces eurythermus*	7.0	97	93	91	0.0	3.0	97	0.0
Tyrosinase from *Agaricus bisporus*	5.0	99	100	100	63	75	99	41
Peroxidase from horseradish	4.0	100	97	87	50	36	100	0.0
Laccase from *Pleurotus ostreatus*	3.5	100	11	81	36	50	100	57
Laccase from *Polyporus versicolor*	3.0	100	79	42	39	30	100	35

[a] For abbreviations, see Table 2.

V. POSSIBLE USE OF PHENOLOXIDASES IN REMEDIATION OF CHEMICALLY POLLUTED SITES

There are three ways to use enzymes to eliminate organic pollutants: (1) application of enzyme preparations, (2) stimulation of *in situ* enzyme production by autochthonous microorganisms, and (3) inoculation of a polluted site by specific enzyme-producing microbial strains.

Some examples of these procedures are briefly discussed here.

A. APPLICATION OF ENZYME PREPARATION

A possible technical use of peroxidases has been recognized especially for wastewater treatment. When incubating wastewater from the coal-processing industry with horseradish peroxidase and hydrogen peroxide, up to 99% of phenols could be floculated in a wide concentration and pH range. Other contaminants, e.g., PCBs, were coprecipitated.[42,43] However, some harmful by-products can be formed in this way. Duguet et al.[44] and Maloney et al.[45] identified dioxins and chlorinated dibenzofurans in the supernatant after 2,4-dichlorophenol was precipitated by a peroxidase. In some cases, instead of peroxidase, tyrosinase could be used as a cheaper means, for it does not require the presence of hydrogen peroxide.[46] The substrate spectrum for this enzyme is narrower, however, and its irreversible inactivation during the reaction was also observed. The latter disadvantage could be overcome by the immobilization of tyrosinase on a solid carrier. Letts and Chase[47] and Vilanova et al.[48] observed an increase in the stability of tyrosinase immobilized on cellulose, collagen, or glass particles. Laccase from *Coriolus versicolor* and a horseradish peroxidase were quickly inactivated when applied to phenolic wastewaters in experiments by Davis and Burns.[49] The stability of both enzymes could be significantly raised, however, through their copolymerization with L-tyrosin and immobilization on alginate particles. An interesting use of a phenoloxidase was demonstrated by May and Li.[50] The enzyme was immobilized in a semipermeable liquid membrane, which allowed phenols to react with the enzyme but prevented the transformed products from returning in the liquid phase. Such a technique could be of interest when speculating on the remediation of polluted groundwater aquifers using phenoloxidases. Sarkar et al.[51] were able to stabilize some oxidoreductases and hydrolases on soil and clay particles (bentonite, kaolinite). The laccase bound to bentonite was fully active, while tyrosinase lost 25% of its activity. The soil-binding capacity correlated positively with the clay content. The immobilized enzymes were more resistant to temperature and pH changes.

The differences in the substrate specificity and the behavior of the individual phenoloxidases in the presence of organic and inorganic soil constituents indicate that specific investigations probably are necessary in order to determine if enzyme preparations can be used in the bioremediation of polluted sites.

B. STIMULATION OF *IN SITU* ENZYME PRODUCTION BY AUTOCHTHONOUS MICROORGANISMS

Bordeleau and Bartha[5] reported a positive correlation between a phenoloxidase activity originated in soil bacteria, actinomycetes, and fungi and the transformation of 3,4-dichloroaniline to tetrachloroazobenzene. The addition of nutrients to the soil increased both the peroxidase activity and the transformation rate of the chemical compound. A specific stimulation of the phenoloxidase activities could perhaps be expected by a soil amendment with some lignin derivates and kaolinite. Berry and Boyd[52] enriched a soil sample contaminated with 3,3-dichlorobenzidine (DCB), a carcinogenic compound, in peroxidase, hydrogen peroxide, glucose, vanillic acid, and ferulic acid. A strong decrease in the DCB concentration occurred in the ferulic acid- and hydrogen peroxide-enriched soil. Ferulic acid is known to induce the synthesis of laccases very effectively.[53]

C. INOCULATION OF A POLLUTED SITE BY SPECIFIC ENZYME-PRODUCING MICROBIAL STRAINS

It is possible to spread spores or mycelia of phenoloxidase-producing fungi at a polluted site. Such an experiment was done by Lamar and Dietrich[54] with *P. chrysosporium* and *Phanerochaele sordida.* When sterile soil contaminated with pentachlorophenol (PCP, 250 to 400 $\mu g\ g^{-1}$) was inoculated with fungal conidia, between 88 and 91% of PCP disappeared within about 6 weeks, probably as a result of its oxidative coupling with some components of the soil organic matter. Only peat was added to the soil as a carbon source (no inorganic nutrient), and the temperature was lower than optimum for fungal growth. The inoculum was effective because competition with autochthonous soil microorganisms did not occur, which can inhibit phenoloxidase production, as shown in some of our experiments.[10]

VI. SUMMARY AND CONCLUSIONS

The physiological importance of phenoloxidases for their microbial producers is apparently based on the transformation of toxic components of the substrate (e.g., plant-related phenolic compounds) into some nontoxic polymers. Lyr[55] and Dupius et al.[56] showed that simple as well as chlorinated phenols can be detoxified by laccase, tyrosinase, and peroxidase, due to a dechlorination and inactivation of hydroxyl groups through coupling mechanisms. Similarly, Bollag et al.[3] demonstrated a detoxification of 2,6-xylene, 4-chloro-2-methylphenol, and *p*-cresol in cultures of *R. praticola* by a laccase. Besides chemical transformation, a precipitation of polymers might contribute to the final effect. Also, in soil and groundwater environments, modification and detoxification rather than mineralization of pollutants by enzymatic activities should be expected. In soils, detoxification may often include the incorporation of pollutants in or coupling with humic substances. The long-term behavior of those complexes, and especially their stability, are almost unknown, although some promising observations have already been made. Dec and Bollag[57] have shown that humic-xenobiotic complexes obtained after oxidative coupling seem relatively stable. Only small amounts of the xenobiotics were released with time, which furthermore became completely mineralized by soil microflora and abiotic factors. In conclusion, enzymatic activities may offer an alternative to a microbiological type of remediation, especially when pollutants consist mainly of phenols, halogenated phenols, or anilines and in environments which contain kaolinite rather than bentonite as a prevalent clay fraction. No heavy amendments of nutrients are necessary and no undesirable accumulation of biomass *in situ* occurs. Phenoloxidases may be of special interest in this respect, since they offer several advantages: (1) their yields are relatively high and their isolation from bulk cultures is not complicated, (2) the screening for specific producers is easy, (3) they have a broad substrate spectrum, and (4) they show a high degree of stability and activity, especially after immobilization. However, no complete mineralization of organic pollutants by phenoloxidases can be expected, for it normally requires a gradual sequence of different specific enzymes. A peroxidase complex formed by *P. chrysosporium* may offer some hope in this respect; however, further extensive investigations are necessary, which should include the ecotoxicological characteristics of enzyme transformation products and their behavior in soil and underground environments.[58-60] In agrosystems, the possible use of enzymes, especially immobilized ones, to detoxify pesticide-contaminated sites deserves special consideration.[61]

REFERENCES

1. **Bollag, J. M., Liu, S. Y., and Minard, R. D.**, Cross-coupling of phenolic humus constituents and 2,4-dichlorophenol, *Soil Sci. Soc. Am. J.*, 44, 52, 1980.
2. **Bollag, J. M., Minard, R. D., and Liu, S. Y.**, Cross-linkage between anilines and phenolic humus constituents, *Environ. Sci. Technol.*, 17, 72, 1983.
3. **Bollag, J. M., Shuttleworth, K. L., and Anderson, D. H.**, Laccase-mediated detoxification of phenolic compounds, *Appl. Environ. Microbiol.*, 54, 3066, 1988.
4. **Bollag, J. M., Sjoblad, R. D., and Minard, R. D.**, Polymerization of phenolic intermediates of pestides by a fungal enzyme, *Experientia,* 33, 1564, 1977.
5. **Bordeleau, L. M. and Bartha, R.**, Biochemical transformations of herbicide-derived anilines in culture medium and in soil, *Can. J. Microbiol.,* 18, 1857, 1972.
6. **Bordeleau, L. M. and Bartha, R.**, Biochemical transformations of herbicide-derived anilines: requirements of molecular configuration, *Can. J. Microbiol.,* 18, 1873, 1972.
7. **Sjoblad, R. D. and Bollag, J. M.**, Oxidative coupling of aromatic compounds by enzymes from soil microorganisms, in *Soil Biochemistry,* Vol. 5, Paul, E. A. and Ladd, J. N., Eds., Marcel Dekker, New York, 1981, chap. 3.
8. **Filip, Z. and Preusse, T.**, Phenoloxidierende Enzyme — ihre Eigenchaften und Wirkungen im Boden, *Pedobiologia,* 28, 133, 19885
9. **Claus, H. und Filip, Z.**, Phenoloxidierende und andere Enzyme als Mittel zur Umwandlung organischer Schadstoffe im Boden- und Grundwasserbereich, *Forum Städte-Hyg.,* 42, 214, 1991.
10. **Claus, H. and Filip, Z.**, Effects of clays and other solids on the activity of phenoloxidases produced by some fungi and actinomycetes, *Soil Biol. Biochem.*, 22, 483, 1990.
11. **Haider, K., Filip, Z., and Martin, J. P.**, Einfluß von Montmorillonit auf die Bildung von Biomasse und Stoffwechselprodukten durch einige Mikroorganismen, *Arch. Microbiol.,* 73, 201, 1970.
12. **Filip, Z., Haider, K., and Martin, J. P.**, Influence of clay minerals on growth and metabolic activity of Epicoccum nigrum and Stachybotrys chartarum, *Soil Biol. Biochem.,* 4, 135, 1972.
13. **Martin, J. P., Filip, Z., and Haider, K.**, Effect of montmorillonite and humate on growth and metabolic activity of some actinomycetes, *Soil Biol. Biochem.,* 8, 409, 1976.
14. **Leonowicz, A. and Bollag, J. M.**, Laccases in soil and the feasibility of their extraction, *Soil Biol. Biochem.,* 19, 237, 1987.
15. **Berry, D. F. and Boyd, S. A.**, Oxidative coupling of phenols and anilines by peroxidases: structure-activity relationships, *Soil Sci. Soc. Am. J.,* 48, 565, 1984.
16. **Haemerli, S. D., Leisola, M. S. A., Sanglard, D., and Feichter, A.**, Oxidation of benzo(a)pyrene by extracellular ligninases from Phanerochaete chrysosporium, J. *Biol. Chem.,* 261, 6900, 1986.
17. **Hammel, K., Kalyanaraman, B., and Kirk, T. K.**, Oxidation of polycyclic aromatic hydrocarbons and dibenzo (p)-dioxins by Phanerochaete chrysosporium ligninase, *J. Biol. Chem.,* 261, 16948, 1986.
18. **Sanglard, D., Leisola, M. S. A., and Fiechter, A.**, Role of extracellular ligninases in biodegradation of benzo(a)pyrene by Phanerochaete chrysosporium, *Enzyme Microbiol. Technol.,* 8, 209, 1986.
19. **Hammel, K. E. and Tardone, P. J.**, The oxidative 4-dechlorination of polychlorinated phenols is catalyzed by extracellular fungal lignin peroxidase, *Biochemistry,* 27, 6563, 1988.
20. **Aitken, M. D., Venkatadri, R., and Irvine, R. L.**, Oxidation of phenolic pollutants by a lignin degrading enzyme from the white-rot fungus Phanerochaete chrysosporium, *Water Res.,* 23, 443, 1989.
21. **Huynh, V. B., Chang, H. M., Joyce, T. W., and Kirk, T. K.**, Dechlorination of chloroorganics by a white-rot fungus, *Tappi J.,* 68, 98, 1985.
22. **Eaton, D. C.**, Mineralization of polychlorinated biphenyls by Phanerochaete chrysosporium, *Enzyme Microbiol. Technol.*, 7, 194, 1985.
23. **Bumpus, J. A., Tien, M., Wright, D., and Aust, S. D.**, Oxidation of persistent environmental pollutants by white rot fungus, *Science,* 228, 1134, 1985.
24. **Bumpus, J. A. and Aust, S. A.**, Biodegradation of DDT [1,1,1-trichloro-2,2-bis (4-chlorophenyl)ethane] by the white rot fungus Phanerochaete chrysosporium, *Appl. Environ. Microbiol.*, 3, 2001, 1987.

25. **Arjamand, M. and Sandermann, H., Jr.**, Mineralization of chloroaniline/lignin conjugates and of free chloroanilines by the white rot fungus Phanerochaete chrysosporium, *J. Agric. Food Chem.*, 33, 1055, 1985.
26. **Mileski, G. J., Bumpus, J. A., Jurek, M. A., and Aust, S. D.**, Biodegradation of pentachlorophenol by the white rot fungus Phanerochaete chrysosporium, *Appl. Environ. Microbiol.*, 54, 2885, 1988.
27. **Bumpus, J. A.**, Biodegradation of polycyclic aromatic hydrocarbons by Phanerochaete chrysosporium, *Appl. Environ. Microbiol.*, 55, 154, 1989.
28. **Bollag, J. M., Liu, S. Y., and Minard, R. D.**, Asymmetric diphenol formation by a fungal laccase, *Appl. Environ. Microbiol.*, 38, 90, 1979.
29. **Suflita, J. M. and Bollag, J. M.**, Oxidative coupling activity in soil extracts, *Soil Biol. Biochem.*, 12, 177, 1980.
30. **Suflita, J. M. and Bollag, J. M.**, Polymerization of phenolic compounds by a soil-enzyme complex, *Soil Sci. Soc. Am. J.*, 45, 297, 1981.
31. **Minard, R. D., Liu, S. Y., and Bollag, J. M.**, Oligomers and quinones from 2,4-dichlorophenol, *J. Agric. Food Chem.*, 29, 250, 1981.
32. **Liu, S. Y., Minard, R. D., and Bollag, J. M.**, Coupling reactions of 2,4-dichlorophenol with various anilines, *J. Agric. Food Chem.*, 29, 253, 1981.
33. **Liu, S. Y., Minard, R. D., and Bollag, J. M.**, Soil-catalyzed complexation of the pollutant 2,6-diethylaniline with syringic acid, *J. Environ. Qual.*, 16, 48, 1987.
34. **Katase, T. and Bollag, J. M.**, Transformation of trans-4-hydroxycinnamic acid by a laccase of the fungus Trametes versicolor and its significance in humification, *Soil Sci.*, 151, 291, 1991.
35. **Hoff, T., Liu, S. Y., and Bollag, J. M.**, Transformation of halogen-, alkyl-, and alkoxy-substituted anilines by a laccase of Trametes versicolor, *Appl. Environ. Microbiol.*, 49, 1040, 1985.
36. **Claus, H. and Filip, Z.**, Effects of different soil constituents on the activity of some phenoloxidases, in *Hazardous Waste: Detection, Control, Treatment*, Abbou, R., Ed., Elsevier, Amsterdam, 1988, 1651.
37. **Claus, H. and Filip, Z.**, Effects of some solid soil constituents and other environmental factors on the activity of phenoloxidases, *DECHEMA Biotechnol. Conf.*, 3, 899, 1989.
38. **Claus, H. and Filip, Z.**, Enzymatic oxidation of some substituted phenols and aromatic amines, and the behaviour of some phenoloxidases in the presence of soil related adsorbents, *Water Sci. Technol.*, 22, 67, 1990.
39. **Serban, A. and Nissenbaum, A.**, Humic acid association with peroxidase and catalase, *Soil Biol. Biochem.*, 18, 41, 1986.
40. **Khaziyev, F. K. and Gulko, A. Y.**, Some properties of the humus peroxidase complex, *Pochvovedeniye*, 2, 30, 1990.
41. **Claus, H. and Filip, Z.**, Behaviour of phenoloxidases in the presence of clays and other soil-related adsorbents, *Appl. Microbiol. Biotechnol.*, 28, 506, 1988.
42. **Klibanov, A. M., Tu, T. M., and Scott, K. P.**, Peroxidase-catalyzed removal of phenols from coal-conversion wastewaters, *Science*, 221, 259, 1983.
43. **Klibanov, A. M. and Morris, E. D.**, Horseradish peroxidase for the removal of carcinogenic aromatic amines from water, *Enzyme Microbiol. Technol.*, 3, 119, 1981.
44. **Duguet, J. B., Dussert, B., Bruchet, A., and Mallevialle, J.**, The potential use of ozone and peroxidase for removal of aromatic compounds from water by polymerization, *Ozone Sci. Eng.*, 8, 247, 1986.
45. **Maloney, S. W., Manem, J., Mallevialle, J., and Fiessinger, F.**, Transformation of trace organic compounds in drinking water by enzymatic oxidative coupling, *Environ. Sci. Technol.*, 20, 249, 1986.
46. **Atlow, S. C., Bonadonna, A. L., and Klibanov, A. M.**, Dephenolization of industrial waste water catalyzed by polyphenol oxidase, *Biotechnol. Bioeng.*, 26, 599, 1984.
47. **Letts, D. and Chase, J., Jr.**, Chemical modification of mushroom tyrosinase for stabilization to reaction inactivation, in *Advances in Experimental Medicine and Biology*, Vol. 42, Dunlap, B., Ed., Plenum Press, New York, 1974, 317.
48. **Vilanova, E., Manjon, A., and Iborra, J. L.**, Tyrosinase hydroxylase activity of immobilized tyrosinase on Enzacryl-AA and CPG-AA supports: stabilization and properties, *Biotechnol. Bioeng.*, 26, 1306, 1984.

49. **Davis, S. and Burns, R. G.,** Decolorization of phenolic effluents by soluble and immobilized phenoloxidase, *Appl. Microbiol. Biotechnol.,* 32, 721, 1990.
50. **May, S. W. and Li, N. N.,** Encapsulation of enzymes in liquid membrane emulsions, in *Enzyme Engineering,* Vol. 2, Pye, E. K. and Wingard, L. B., Eds., Plenum Press, New York, 1974, 77.
51. **Sarkar, J. M., Leonowicz, A., and Bollag, J. M.,** Immobilization of enzymes on clays and soils, *Soil Biol. Biochem.,* 21, 223, 1989.
52. **Berry, D. F. and Boyd, S. A.,** Decontamination of soil through enhanced formation of bound residues, *Environ. Sci. Technol.,* 19, 1132, 1985.
53. **Leonowicz, A. and Trojanowski, J.,** Induction of laccase by ferulic acid in basidiomycetes, *Acta Biochim. Pol.,* 22, 291, 1975.
54. **Lamar, R. T. and Dietrich, D. M.,** In situ depletion of pentachlorophenol from contaminated soil by Phanerochaete spp., *Appl. Environ. Microbiol.,* 56, 3093, 1990.
55. **Lyr, H.,** Detoxification of heartwood toxins and chlorophenols by higher fungi, *Nature (London),* 195, 289, 1962.
56. **Dupius, G., Johri, B., Bandoni, J., and Towers, G. H. N.,** Cinamylphenols as inhibitors of fungal growth, *Can. J. Microbiol.,* 18, 929, 1972.
57. **Dec, J. and Bollag, J. M.,** Microbial release and degradation of catechol and chlorophenols bound to synthetic humic acid, *Soil Sci. Soc. Am. J.,* 52, 1366, 1988.
58. **Aust, S. D.,** Degradation of environmental pollutants by Phanerochaete chrysosporium, *Microb. Ecol.,* 20, 197, 1990.
59. **Majchercyk, A., Zeddel, A., Kelchebach, M., Loske, D., and Hüttermann, A.,** Degradation of PAH by white rot fungi, *BioEngineering,* 2, 27, 1993.
60. **Mahro, B. and Kästner, M.,** Der mikrobielle Abbau polyzyklischer aromatischer Kohlenwasserstoffe (PAK) in Böden und Sedimenten: Mineralisierung, Metabolitenbildung und Entstehung gebundener Rückstände, *BioEngineering,* 1, 50, 1993.
61. **Nannipieri, P. and Bollag, J. M.,** Use of enzymes to detoxify pesticide-contaminated soils and waters, *J. Environ. Qual.,* 20, 510, 1991.

Chapter 31

Position Paper of Part II

F. Andreux, J.-M. Bollag, S. A. Boyd, Z. K. Filip, P. M. Huang, J. M. Oades, P. S. C. Rao, N. Senesi, W. F. Spencer, A. T. Stone, and B. K. G. Theng

Natural organics in the environment range from simple molecules to polymeric compounds of varying complexity and reactivity. The remains of living organisms are transformed through a series of biotic and abiotic reactions. In these transformations, surfaces which may be organic or inorganic, inanimate or living, often play a major role. By interacting with these surfaces, natural organics are modified, stabilized, or transformed, resulting in specific features which have varying stabilities and residence times, and are organized at molecular and even larger scales. Such features determine soil types and properties, directly control the composition of the soil solution, and indirectly affect the quality of water from catchments.

A wide range and variety of anthropogenic compounds enter the environment, either intentionally (e.g., plant protection substances) or inadvertently. Their incorporation into, and retention by, surfaces modify their mobility, persistence, and toxicity. Conversely, these compounds are expected to modify or to compete with biotic and abiotic processes of degradation and synthesis of natural organics. More generally, human activities such as farming modify basic soil properties in unpredictable ways.

Part II is the longest section of this book. The theme of this section is the behavior and fate of natural and man-made organic compounds in the environment, with special emphasis on soil systems. Apart from two to three chapters that have a field dimension, the results and discussions refer to fundamental research detailing key mechanisms.

I. NATURAL ORGANICS

The oral and poster contributions on natural organics were mostly concerned with organic constituents of soils, notably their nature and stability, role in soil processes, and biotic and abiotic transformations.

A. CHEMICAL NATURE AND STABILITY

Research into soil organic matter (SOM) has turned from extracting and characterizing more or less arbitrary fractions, such as humic (HA) and fulvic acids (FA), to identifying pools of SOM, and the fluxes between various pools (J. M. Oades). These pools need to possess well-defined chemical and biological functions, and to be scored along the time scale, that is to say, in terms of residence time. Emphasis was placed on at least three categories of components:

1. "Inert" or "passive" or "stable" organic matter. Inert organic matter (IOM) was associated with the clay fraction, and composed of substituted polymethylene chains (B. K. G. Theng and K. R. Tate). In a Spodosol from New Zealand, the IOM was partly intercalated into a mica-beidellite structure. The interlayer species, which was completely protected against microbial and enzymatic attack, and even resisted oxidative degradation by H_2O_2, was dated at 6800 years BP.
2. "Young" or "transient" organic matter. They are mostly made of compounds excreted by microorganisms and plants. Some of these, such as microbial extracellular polysaccharides, may bind to soil minerals (C. Chenu). Others are released to the soil solution,

0-87371-914-X/95/$0.00+$.50

as reported by W. J. Slattery and P. M. Mele in the case of aliphatic acids. In Andisols from Japan, the concentration of water-soluble carboxylic acids of low molecular weight was shown to increase with depth, and correlated with the content of amorphous Al and Fe hydroxides (M. Tani et al.).

3. Unevenly transformed and stabilized remains. Plant remains and soil biota have a short turnover time (≤1 year) unless they are stabilized by interactions with mineral constituents. The extent and strength of these interactions can be investigated by soil particle size fractionation. The organic material in the different size fractions can be characterized according to the relative abundance of certain structures (e.g., aromatic, aliphatic) and functional groups (e.g.. carboxylic, quinonoid, and O-alkyl) (J. M. Oades; B. K. G. Theng and R. K. Tate; C. M. Monreal et al.). Changes in plant cover and soil management were also reflected in the distribution of organic matter among the different particle size fractions (C. M. Monreal et al.).

Two questions were raised. (1) How do microbial cells and plant remains interact with clays and other minerals? (2) Are humic-like polymers the principal agents in the formation of organomineral complexes and microaggregates in soil? (J. M. Oades; M. Schnitzer). Structural aspects of humic substances (HS) were not discussed in detail. However, their formation by oxidation of phenolic monomers (P. M. Huang), and their metal-complexing ability were described (M. Schnitzer; J. M. Xu and K. N. Yuan). Moreover, it was pointed out that nonextractable HS, i.e., humins, which predominate in most soils, and their modes of interaction with mineral components and adhering microorganisms deserve closer attention.

B. MAJOR ROLES

The distribution of SOM among the different size fractions was considered an important soil characteristics, especially in the case of changes in plant cover and soil management (C. M. Monreal et al.). The influence of undecomposed organic residues on sulfur mineralization was described by L. Deleon et al. They established that addition of ground, dried, leaves to some Costa Rican soils increased mineralization of sulfur at rates that were significantly correlated with the content of native SOM. Results presented by J. M. Xu and K. N. Yuan showed that HS differing in metal-complexing ability were extracted from some Chinese soils, using various reagents. As might be expected, Fe and Al were more strongly bond to HS than Ca. However, neither the nutrient-carrying properties of HS nor their possible influence on the activity of microbial populations were considered.

Plant and microbial polysaccharides play a major role in the formation and stabilization of soil aggregates. The nature of the binding between mineral colloid surfaces and polysaccharides was questioned, and the diverse roles of the latter (e.g., as adhesive agents, energy sources, water-holding material, nutrient and enzyme carrier, and ionic barriers to toxic metals) were reassessed (C. Chenu). Other agents are also involved in aggregation processes: the role of polymethelene compounds in rendering some sands hydrophobic ("nonwetting"), and of lignin-derived aromatic compounds in stabilizing clay-size fractions, was described (J. M. Oades).

C. MAJOR TRANSFORMATIONS

Only two works dealt with the impact of agricultural practices, whereas all other studies referred to laboratory experiments. In general, short time scales were preferred. If functional organic pools can be identified, the key consideration would be the mechanism and speed with which SOM passes from one pool to another.

The dynamics of SOM were often considered in terms of a competition between biotic and abiotic (catalytic) degradation processes. The ability of short-range ordered minerals in catalyzing the transformations of organic compounds was stressed (P. M. Huang). Iron and Mn oxides were shown to catalyze the transformation of amino acids (M. C. Wang).

When the oxidation and degradation of phenolic and quinoid rings took place, they could prevent the catalytic deamination of free amino acids by pyrogallol, as observed for several amino acids, with the exception of DL-methionine.

Microbial activity in soil was assessed through counting (W. J. Slattery and P. M. Mele) and respiration (C. C. Ainsworth et al.). The former workers compared the effect of retaining or burning stubble (following cropping) on microbial population in northeastern Victoria, Australia. Over 14 years of minimum tillage, no increase in organic carbon was found in the stubble-retained, as compared with the burned plots. The difference between fungi, actinomycetes, and total aerobes in the stubble-standing treatments was higher than in the stubble-burned counterparts. The concentration of aliphatic acids was higher in the burned than in the standing treatments, whereas the reverse situation was obtained for aromatic acids. The same authors (C. C. Ainsworth et al.) found that sorption on Al_2O_3 reduced the availability of aromatic substrates to *P. cepacia* and *P. putida.* The extent of reduction depended on the microbial species, and on the type and rate of desorption of the acid.

Enzymatic transformations of organic compounds were investigated, using free and immobilized enzymes (IE). Acid phosphatase was immobilized on chitosane by G. Y. Shin et al., using glutaraldehyde as a cross-liking agent. Adding the IE to manure and soils did not increase the rate of hydrolysis of organic phosphate. However, when the IE was added to soil immediately after substrate incorporation (that is, before the native soil enzymes could mineralize the phosphate), the hydrolysis rate increased. Microbial aspartase immobilized on Ca-montmorillonite was more active than the free enzyme in deaminating aspartic acid (A. Naidja and P. M. Huang). In addition, these authors observed that the "dirty" and purified clays equally stimulated the deamination of the enzyme, and prevented it from denaturation in the pH range of most soils.

II. ANTHROPOGENIC ORGANICS

The behavior and fate of anthropogenic organics in soils were discussed in terms of their adsorption/retention or degradation/detoxification, although some papers dealt with both aspects. The choice of pollutants was varied: phenols and quinones were often used as representative organic compounds, while other experiments were carried out with herbicides and chlorinated aromatic molecules.

A. ADSORPTION/RETENTION PROCESSES

These processes were investigated at different energy levels and on different adsorbing surfaces, such as clay minerals and synthetic or naturally occurring metal oxides. The adsorption of uncharged organic pollutants was greatly enhanced by rendering the adsorbent surface hydrophobic. A simple and effective way of obtaining this is by exchanging the inorganic counterions at the surface for long-chain quaternary ammonium cations (S. A. Boyd).

A few papers described the sorption and reaction of anthropogenic organics on soils (A. E. McGrath and J. G. McColl). The role of SOM and its constituents in the adsorption and binding of hydrophobic organic species by clays and soils was emphasized (P. S. C. Rao et al.; F. Andreux et al.). The abiotic mechanisms of the interactions of organic pollutants with HS were reviewed by N. Senesi. The effects of surface adsorption/retention mechanisms on the mobility in the soil water phase (C. Marucchini et al.), and volatilization from soil (W. F. Spencer et al.) of introduced organic species, were also considered.

B. DEGRADATION/DETOXIFICATION MECHANISMS

Abiotic degradation often complements, rather than compete with, biotic processes. The abiotic decomposition of anthropogenic organics could be enhanced by their adsorption on clay mineral surfaces through various mechanisms (G. G. Ristori and P. Fusi). Surfaces of "hard-metal catalysts" such as Al, Ti, and Fe(III) oxides facilitated the hydrolysis of ester bonds in organic pollutants via mechanisms analogous to metal ion catalysis in homogeneous solutions (A. T. Stone and A. Torrents).

The question arose whether clay minerals and metal oxides were real catalysts in the conventional sense, or only powerful oxidants, and conclusions were drawn from new experiments with different analytical techniques (J.-M. Bollag et al.). Although catalysts are frequently defined as material which accelerate chemical reactions without themselves undergoing change, this is too optimistic a definition. The properties of all real catalysts do change with use. Actually, many substances classified as catalysts are destroyed either as a result of the process that gives them their catalytic activity or because of subsequent combination with the products. From a practical point of view, a catalyst is a substance that changes the rate of a desired reaction, regardless of the fate of the catalyst itself. When polyaromatic phenols and quinones were oxidized by MnO_2, the metal ion was reduced to Mn^{2+}, and insoluble humic-like polymers were formed (G. Whelan and R. C. Sims).

The degradation of polyphenolic compounds in the presence of birnessite increased with an increase in light intensity and temperature, indicating that photochemical and thermal effects were important in degradation processes (J. S. K. Lee and P. M. Huang). Quartz sands and kaolinite were shown to enhance the microbial synthesis of phenoloxidases. On the other hand, the activity of these enzymes was reduced to different extents by adsorption on clays and clay-humus complexes (Z. K. Filip and H. Claus).

Only one paper dealt with improving simultaneously the biotic degradation of pesticides in the field and agricultural practices, by comparing conventional vs. minimal tillage (D. Levanon). In a sandy loam from Quebec, and under laboratory conditions, E. Salvano et al. observed that the rate of mineralization of native SOM increased with increasing doses of formulated atrazine. The mobility of metolachlor and the new herbicide terbuthylazine in soil was investigated by C. Marucchini et al., using lysimeter techniques: both herbicides disappeared from soil following first-order kinetics.

III. EXPERIMENTAL APPROACHES

Organic constituents in soil are being investigated more in terms of their functions (e.g., stability, availability, and accessibility to microorganisms and enzymes) than of their structure. This change of emphasis also applies to HS. Because of their stability, however, HS are actively involved in the adsorption and retention of anthropogenic compounds. Both quantity and quality are important in this respect. Even the internal physical structure of SOM must be taken into account in constructing conceptual and predictive models of the behavior and fate of organic pollutants in soil (P. S. C. Rao et al.).

Whether HS are obtained from natural soils and waters or synthesized by chemical or biological processes, their complex structure is still not well understood. Supplementation of efficient standard separation techniques with high-resolution methods is therefore urgently needed (N. Senesi; J.-M. Bollag et al.). For both natural and anthropogenic compounds, extraction techniques should aim at improving not only the yield, but also the purity of the recovered materials. In this regard, supercritical fluid extraction (SFE) shows considerable promise (H. H. Cheng et al.). Thus, ferulic acid and its transformation products, which were bound to MnO_2 and resisted extraction with organic solvents, were released from the complex by SFE.

Particle size fractionation remains a useful method for differentiating natural organics and understanding detailed localization of pesticide residues that are bound to them in soil aggregates. The use of labeled (especially with ^{14}C, ^{13}C, and ^{15}N) compounds, and the application of new and improved spectroscopic techniques (e.g., solid state ^{13}C-NMR and pyrolysis field ionization mass spectrometry (Py-FIMS)) can yield information on the localization, transformation, and persistence of organic pollutants in soil (J. M. Oades; B. K. G. Theng and K. R. Tate; C. M. Monreal et al.; F. Andreux et al.).

IV. SUMMARY AND RECOMMENDATIONS

Natural organics in the environment represent a vast array of molecules, from transient simple molecules, which are present in the soil solution and in surface- and groundwaters, to diversely stabilized polymers, including insoluble, inert SOM. In this section, more emphasis was placed on studying the properties of these various components, than on elucidating their sometimes unclear chemical nature.

Concerning transient components, few studies comparing the composition and fluxes of solutes present in soil solutions of the more representative ecosystems are available. Development of such studies would improve our knowledge about the major organic classes (especially aliphatic and aromatic acids with high complexing ability) involved in the extraction of nutrients through rock weathering, their migration throughout the soil profile, and finally their transfer to the plant-root system.

The sequences of reactions which occur between natural organics and mineral surfaces, from the very start of decay processes to their insolubilization in soil microaggregates, follow general mechanisms that need detailed studies, for each class of organic component. Studies about the role of soil minerals at the different steps of these transformations are still fragmentary. There are numerous works about the complexing ability of organic acids and soluble HS in aqueous media, whereas relatively less is known about sorption complexes of biological and humic macromolecules on clay and oxide mineral surfaces. In the case of enzymes, the way the retention and activity of a given enzyme can be modified by different types of clay and oxide surfaces should be studied with minerals isolated by mild procedures from soils or waters, rather than with purified minerals. Similarly, the impact of a given surface type on interconnected enzymatic processes occurring successively, rather than on only one elementary process, should be investigated. The same applies for microbial transformations, as well as for abiotic reactions. In the case of humin, almost nothing is known about its chemical structure, and its participation in the processes of storage and supply of mineral and organic nutrients to soil organisms.

In general, in the different media, the influence of mineral surfaces on the biotic and abiotic reactions occurring in the environment should be ideally graded according to both the type of mineral and type of substrate. Such an approach would probably provide scientists with a more complete insight into the role of soil minerals on the transformation and turnover rates of the principal chemical forms of C, N, and P, and possibly S, in soils and other environmental compartments.

The amount, quality, and probably the physical structure of natural SOM primarily control the adsorption and retention of nonionic organic pollutants by soils. Organic species bound to soil will then be protected, transformed, or degraded by biotic and abiotic processes. The mechanisms of transformation/degradation need to be clarified, and the resultant products ("bound residues") characterized. In many instances, anthropogenic and synthetic organics can serve as probe molecules, and be used in reactions that mimic those occurring with natural compounds, in order to characterize reactive sites in soils and soil fractions.

The microbial (enzymatic) degradation of organic pollutants in the environment, especially in soils, is strongly influenced by the nature and properties of the mineral and organic components. Adsorption by and binding to mineral and organic surfaces can delay or even inhibit the degradation of organic pollutants. On the other hand, many short-range ordered minerals and metallic oxides act to assist microbes and endogenous soil enzymes in numerous oxidation and ring cleavage reactions. Such oxides have a prevalent role in the oxidative conversion of many organic pollutants under anoxic conditions. Techniques to determine the distribution of anoxic habitats, their seasonal variations, and microbial population need therefore to be developed and improved.

Separation and analytical methods require the same sophisticated technical tools for both natural and anthropogenic organics. In preparing humic-like substances from phenolic monomers, increased attention needs to be paid to reaction kinetics, as well as to the yield and chemical composition of the products. Particle size fractionation of soil without chemical pretreatment, coupled with modern spectroscopic techniques (NMR, Py-FIMS) still offer the best means of investigating the distribution and properties of natural and anthropogenic organics in soil. However, no detailed structural studies could probably be successful without prior solubilization and extraction. Calibrated supercritical fluid extraction, using different fluids and conditions, may prove to be a valuable tool in this respect.

INDEX

A

B

D

E

F

G

H

I

J

K

L

M

N

O

P

Q

R

T

U

V

W

X

Y

Z